An introduction to
GLOBAL ENVIRONMENTAL ISSUES

The global environment and our perceptions of it are changing very rapidly. Theories and empirical research are becoming increasingly sophisticated. Causes, effects, and solutions are pursued, often independently, across a variety of disciplines.

An Introduction to Global Environmental Issues provides a full description of our scientific knowledge of environmental systems and processes. The authors build on this factual base to analyse the world's major environmental concerns. The human dimension – cultural, economic and political factors and impacts – is explained and integrated into this discussion.

Contemporary case studies are drawn from every part of the world. State of the art research on specific problems and solutions is clearly presented and explained.

A superb range of figures and colour photographs illustrate the text. Each chapter contains worked case studies, highlighted key words, boxed technical information, a summary of essential points and an annotated guide to further reading. A full glossary (including synopses of seminal events and publications such as the Earth Summit and *Agenda 21*) and questions for essays/class discussion are also provided.

The book will prove an essential introduction to environmental issues for students beginning courses or options in Environmental Science/Studies and Geography. It will also prove invaluable for students in both the Natural and Social Sciences seeking a broader understanding of these issues.

Kevin T. Pickering is Reader in Geology at University College London. Lewis A. Owen is Lecturer in Geology and Geography at Royal Holloway, University of London.

An introduction to
GLOBAL ENVIRONMENTAL ISSUES

Kevin T. Pickering

and

Lewis A. Owen

LONDON AND NEW YORK

First published 1994
by Routledge
11 New Fetter Lane, London EC4P 4EE

Simultaneously published in the USA and Canada
by Routledge
29 West 35th Street, New York, NY 10001

© 1994 Kevin T. Pickering and Lewis A. Owen

Typeset by Solidus (Bristol) Limited

Printed and bound in Great Britain by
Butler & Tanner Ltd, Frome and London

British Library Cataloguing in Publication Data
A catalogue record for this book is available from the British Library

Library of Congress Cataloging in Publication Data
is also available

ISBN 0–415–10227–8 (hbk)
0–415–10228–6 (pbk)

Whoever you are! you are he or she for whom the
earth is solid and liquid,
You are he or she for whom the sun and
moon hang in
the sky.

'A Song of the Rolling Earth'
from Walt Whitman, 'Leaves of Grass'

Contents

Colour plates

Black and white plates

Tables

FIGURES

Preface

The past few decades have seen an increasing awareness of environmental issues. This started in the late 1950s and took off in the 1960s, being particularly inspired by such causes as Ban the Bomb, Anti-Apartheid, and Amnesty International, and with the growing vegetarian culture. The most famous anti-nuclear group to emerge at this time was the Campaign for Nuclear Disarmament (CND). The anti-war demonstrations stimulated mass media coverage as many people took to the streets to protest about the atrocities in Vietnam, Korea, and Cambodia, and many other parts of the world.

Many of these movements were, and still are, considered anti-establishment, and were not taken seriously by many in power. In general, the protests were given equivocal publicity, probably because many demonstrations ended in violence. It was during this period that seminal environmental works were published such as Rachel Carson's book *Silent Spring* (1962), which highlighted the effects of the uncontrolled use of highly toxic pesticides. The evolution and establishment of the 'green' movement is well outlined in Steven Yearley's book *The Green Case: A Sociology of Environmental Issues, Arguments and Politics* (1991).

Environmental awareness, however, is not a new phenomenon. For several centuries, scientists have been aware of anthropogenic effects on the natural world. In Europe, the popularization of environmental science, or natural history as it was then called, first caught the public imagination with the work of early explorer-naturalists in the late eighteenth and nineteenth centuries. This included the voyages of Charles Darwin, Alfred Wallace, Edward Forbes, and Louis Agassiz. They discovered the extent of biological diversity, the striking variations between communities of plants and animals throughout the world, and the particular ecological niches of many rare species. Their detailed observations, especially those of the Victorians, facilitated the laying of the foundations for present-day environmental science. These explorer-naturalists developed revolutionary concepts such as natural selection and adaptation to the environment.

These Victorian postulates, paradigms, hypotheses, and theories challenged and altered perceptions about the natural world. In the late nineteenth and early twentieth centuries, public awareness of human interference with biological systems began to emerge with the growth of conservation groups or societies. The Royal Society for the Protection of Birds (RSPB) was formed in 1889, the UK National Trust in 1899, and the Royal Society for the Conservation of Nature in 1912. These societies tended to focus on the protection of particular species or habitats. It was not until the 1930s that the ideas of communities and the complex interrelationships between the organic and inorganic systems began to emerge.

The term ecosystem was defined by the German zoologist Ernst Haekel in 1869 to incorporate the entire science of the relationships between organisms and the physical environment. Thus emphasizing the relationship between the organic and inorganic world, Sir Arthur Tansley in 1933 developed these concepts much further, and can rightly be considered as the father of the modern science of ecology. Ecosystems are all-encompassing environmental systems maintained in a state of dynamic equilibrium by both negative feedback and to a lesser extent positive feedback, which control the flow of energy and matter between the inorganic and organic realms.

After an initial flurry of research activity, progress on understanding environmental issues was curtailed during the First and Second World Wars, and it was not until the late 1950s that the scientific community began to study ecological systems. These studies were aided by new technologies and a more rigorous, quantitative approach to gathering data on such issues as pollution. Also, there were rapid advances in the field of photography, for example with the development of machines such as the scanning electron microscope. Improved, machine-based methods of accurate chemical analysis replaced the laborious wet-chemistry techniques that used to take days to weeks, rather than minutes to hours. The introduction of computers offered quick statistical analyses of data.

During the past few decades, many international scientific programmes concerned with ecological issues have been initiated. These included the Arid Zone Programme (Unesco, 1951–64), the International Biological Programme (ICSU, 1964–74), the Man and the Biosphere Programme (Unesco, from 1971), and the International Antarctic Glaciological Project (from 1969). The United Nations (UN) established the United Nations Environmental Programme (UNEP) to stimulate environmental work amongst organizations and agencies of the United Nations with other established bodies. Examples include links between UN groups such as the World Health Organization (WHO), the Food and Agriculture Organization (FAO), the International Labour Organization (ILO), the International Atomic Energy Agency (IAEA), and the World Meteorological Organization (WMO), with other organizations such as the Organization for Economic Co-operation and Development (OECD), the European Commission, and the Council of Europe. In 1983, the World Commission on Environment and Development (WCED) was established as an independent body to report to the UN General Assembly on the relationship between the environment and economic development. The WCED and UNEP collaborated through a committee chaired by Gro Harlem Brundtland, who was then the Norwegian Prime Minister, and which culminated in the WCED's report in 1987, *Our Common Future*, providing an agenda for sustainable economic growth.

With the proliferation of such organizations, awareness of environmental issues spread from the scientific and academic communities to mainstream political parties and the public. Pressure groups and 'green parties' are now taken more seriously. Greenpeace is probably the best-known international environmental pressure group which became well known in the 1970s and 1980s for its swashbuckling direct action. In 1992, for the first time in its 21-year history, Greenpeace International saw its annual funding by public contributions drop, with the result that it had to reduce its budget by US $9 million to US $27 million, sell the largest of its seven ships (the *Gondwana*), and make plans for the redundancy of up to 25 per cent of its 500 campaigners. The US contributions dropped considerably after the Gulf War because of Greenpeace's opposition to the fighting.

Daily, the shopping malls/supermarkets, garages/gas stations and convenience shops are marketing environmentally friendly products on a routine basis. Additive-free, 'natural' food products form a multi-billion dollar business. Politicians are using green issues in their manifestos. In the UK, for example, the number of paid-up members in environmental societies is higher than those in trade unions. Hopefully, political parties will become even more genuinely concerned about global problems, and will increasingly incorporate environmental issues into electoral pledges, manifestos and action. International green parties themselves have become particularly successful during recent years. In Sweden the 'Miljopartiet de Grona' is estimated to have the support of up to 10 per cent of the electorate, while in Switzerland the 'Grüne Partei' has nine of the 200 seats in the National Chamber, and in the UK the Green Party expects to achieve more than 5 per cent of the vote at the next General Election. The green parties still have a long way to go before they will compete with their well-established rivals. They still suffer from images of radical, disorganized, one-issue pressure groups. The electorate at large tends not to see such groups as having a well-thought-out portfolio of policies across the board which include those on sound economic, fiscal, educational, and health matters. However, there is little doubt that the environmental revolution is well under way and that over the next few decades people will witness great changes in society's perception of the environment. These changes may be as radical as many encountered during the agricultural and industrial revolutions of previous centuries, and they will probably come about much more rapidly.

Conferences on the environment are now a feature of international diplomacy. In June 1992 in Rio de Janeiro, Brazil, the Earth Summit (United Nations Conference on Environment and Development) took place and followed the first conference of its kind in Stockholm in 1972. The summit was not an unqualified success but virtually all the participating nations agreed to try and work more closely in tackling global environmental issues.

After the initial flurry of green books, there has been a period of intense public debate and a more considered, balanced perspective on many issues such as the enhanced greenhouse effect, or the pollution caused by the Gulf War in 1991. Following the Rio Earth Summit in 1992, governments are taking stock of their commitments to the Climate Change and Biodiversity conventions; also in 1992 the influential Intergovernmental Panel on Climate Change (IPCC) updated its predictions on global climate change due to human activities; the collapse of the Soviet Union has brought about an uncertain world order with the status of nuclear weapons and arms treaties becoming clouded, and discussions are under way for a new protocol on acid rain to replace the '30% Club'. These are just a few of the issues addressed in this book. Along with courses on environmental issues, many undergraduate and other student courses now incorporate something on these issues. The environmental arguments have become much more sophisticated over the past five years.

PREFACE

We have tried to produce a book that is readable and stimulating. It explores some of the basic science and structure behind the most important environmental issues and links these issues to contemporary socio-economic and political considerations. This is not, however, nor was it ever intended to be, a sociological perspective on the environment – something that is beyond the scope of this book. We are involved in teaching or researching aspects of the subject matter of this book as scientists, and so we have tried to produce a book that has a clearly scientific angle.

We conclude with a manifesto for living, because it is impossible to discuss these issues dispassionately when the consequences of human actions in the environment affect all of us now and in the future. But, we have tried to write in such a way as to present the important factual information and then comment. The book will be useful for teachers in colleges and higher education institutes as a source for discussion and reference material.

Organization of the book

Over the ten chapters of this book, we examine the major global environmental issues. Technical information, or case studies where appropriate, are boxed in order to permit a student to take a less detailed overview of a topic and return later to the more complex and/or detailed treatment of a topic. Throughout, we emphasize the concern for the survival of our species and other vulnerable species.

For anyone who is really interested in pursuing further the range of issues covered in this book, we would recommend *Planet Under Stress: The Challenge of Global Change*, edited by Mungall and McLaren (1990), *Global Environmental Change*, edited by Mannion (1991), Radford's *The Crisis of Life on Earth* (1990), Simpson's *The Times Guide to the Environment* (1990), *Global Environmental Issues*, edited by Smith and Warr (1991), Kemp's *Global Environmental Issues: A Climatological Approach* (1990), Goudie's *Environmental Change: Contemporary Problems in Geography* (1992), and *It's a Matter of Survival* by Gordon and Suzuki (1991). There are books which explore global environmental issues from a specifically sociological and political perspective. Amongst these, we would recommend *Green Development: Environment and Sustainability in the Third World* by Adams (1990), *The Green Case: A Sociology of Environmental Issues, Arguments and Politics* by Yearley (1991), *A New World Order: Grassroots Movements for Global Change* by Ekins (1992), *Refashioning Nature: Food, Ecology and Culture* by Goodman and Redclift (1991), and Anderson's *Alternative Economic Indicators* (1991). The emphasis is different in each book and, although there is overlap, each covers somewhat different issues. This book has been written with a slant towards the physical world and global environmental issues, but with short sections on the socio-economic, cultural, and political aspects. This has been a deliberate intention of ours, since there are a number of good, recently published, environmental books that take a more specifically social science perspective, referred to above.

Finally, we thank the many individuals who have in some way contributed to this book, either in conversations with us and through reviewing parts of, or entire, earlier drafts. In particular, we owe a large measure of gratitude to David Kemp, Alastair Dawson, David Evans, Louise Pickering, Vicky Myers,

Jim Best, Val Saunders, Dorrik Stow, Steve Temperley, Sarah Davies, Catrin Jones, and Cathy Hayward for reviewing this book and making many helpful comments, to Jill Keegan for help with the quotes, and to Justin Jacyno for producing much of the artwork and at a uniformly high standard. Kevin Pickering acknowledges the help and advice which was given by the UK Parliamentary Office of Science and Technology staff in Westminster, London, where this book was completed during tenure of a Westminster Fellowship whilst on sabbatical leave in 1993. We thank especially Judith Bates, in the Parliamentary Office of Science and Technology, for her critical and thoughtful comments which have much improved the text. We would like to thank Liv Gibbs for carrying out the exhausting task of copy-editing and, at Routledge, Tristan Palmer, Sue Bilton, Sarah Lloyd, Ann Southgate, and Susie Hilsdon, without whose support and encouragement this book would not have appeared.

The authors and publishers would like to thank the following for permission to reproduce copyrighted material:

Virago Press and Random House for Maya Angelou, 'On the Pulse of Morning', opposite p. 1; Joan Baez and Carlin Music Corporation for 'Warriors of the Sun' on p. 64; Sony Music Publishing for Bob Dylan, 'A Hard Rain's a Gonna Fall', © Special Rider Music/For the territory of UK & Eire, Sony Music Publishing, 17/19 Soho Square, London W1V 6HE; Faber & Faber Ltd and Harcourt Brace for lines from T. S. Eliot, 'Little Gidding' on p. 250; Faber & Faber Ltd for Louis Macneice, 'Prayer before Birth' on p. 280.

NASA/Lunar and Planetary Institute for Plates 1, 10, 8.4; Professor B. F. Windley for Plate 4; Dr M. Collison for Plate 5; NOAA/NESDIS/NCDC/ SDSD for Plate 6; Magnum for Plates 9, 17, 18, 32, 35, 38, 42, 43, 48; Comstock for Plates 11, 22, 23, 24, 30, 33, 34, 36, 40, 47; Greenpeace for Plates 12, 13, 14, 16, 39, 41, 44, 50; Rex Features for Plates 19, 20, 21, 6.2; Münchener Ruck for Plate 27; SABA Katz Pictures for Plate 28; US Geological Survey for Plate 29; Geotechnical Control Office, Hong Kong for Plate 31; Mike Eden for Plates 37 and 9.1; K. C. G. Owen for Plate 2.3; Royal Geographical Society for Plates 2.5B, C and D; Professor E. Derbyshire for Plate 2.6; Jeremy P. Richards, University of Leicester for Plate 4.2B; Oxfam for Plates 5.2, 5.3, 8.5, 10.1A–D, 10.2, 10.6, 10.8, 10.9; Rob Potter for Plate 10.3B; Gary Nichols, University of London for Plate 10.9A; John Underhill, Yvonne Cooper (photographer) and the Department of Geology and Geophysics, University of Edinburgh for Plate 2.5A; Vlaso Milankovitch for Plate 2.5E.

A Rock, A River, A Tree
Hosts to species long since departed,
Marked the mastodon,
The dinosaur, who left dried tokens
Of their sojourn here
On our planet floor,
Any broad alarm of their hastening doom
Is lost in the gloom of dust and ages.

But today, the Rock cries out to us, clearly,
forcefully,
Come, you may stand upon my
Back and face your distant destiny,
But seek no haven in my shadow,
I will give you no hiding place down here.

You, created only a little lower than
The angels, have crouched too long in
The bruising darkness
Have lain too long
Facedown in ignorance,
Your mouths spilling words

Armed for slaughter.
The Rock cries out to us today,
You may stand upon me;
But do not hide your face.

Maya Angelou, 'On the Pulse of Morning'
(Read by the poet at the inauguration of
William Jefferson Clinton, 20 January 1993)

CHAPTER 1
Introducing Earth

Who would survive?

If all the nuclear weapons in the world's arsenals were detonated (of which the USA and former Soviet Union possess more than 50,000), the Earth would continue and with it life in some form. Some species would survive and new species would come into existence. But, in this doomsday scenario, one thing is virtually certain: human beings and other vulnerable species would be obliterated. Humans would not survive. 'To be, or not to be: that is the question.' This indeed is the question over our survival as human beings, together with the survival of many endangered species, and fragile ecosystems. These simple and profound words from *Hamlet* by William Shakespeare echo through all human actions over the environment. This is very much a book about human survival and the continuation of the natural world as it now exists, not the survival of the planet. Volcanic eruptions cannot be controlled yet, but humankind can control the pollution of the atmosphere, oceans, and land.

Human activities have exposed many parts of the natural environment to considerable risks. Science and technology are used to understand and harness the world's resources but not always for the greater good of humankind. Short-term economic gain tends to outweigh most other considerations. Civilization is now so advanced that we are able to study in considerable detail our planet and universe, an exciting and stim-ulating endeavour. Scientists can examine the Earth at all scales, from the subatomic using high-energy physics, to cosmic scales using the most sophisticated telescopes and spacecraft. Images of Earth from Space are now common-place (Plate 1). Sophisticated global climate modelling and predictions about future climate change are becoming commonplace.

Too many want too much

Too many humans want too much, whether it is food, land, power or influence. Over-population and waste are the two biggest problems facing our generation. Other issues tend to stem directly or indirectly from these two problems. The way these global issues are tackled will determine the legacy that is bequeathed to future generations. These are the unavoidable issues of the 1990s, the twenty-first century and beyond, to be tackled now. There are those who would not agree, through religious and other beliefs, that over-population is a problem; they might claim that the real problem is the manage-ment of the resources on Earth, not the number of people. This issue is extremely contentious, but the Earth could be managed with much less risk if there was less demand for the limited, finite, global resources and the natural environ-ment was under less stress from planners, devel-opers, and colonizers. Indeed, in February 1992 the US National Academy of Sciences and the Royal Society of London published a joint document on global problems in which the world population growth is considered a central issue. The joint document, the first ever produced by these two academies, took two years to write and expresses 'deep concern' over the links between the estimated growth of the world's population of 100 million a year (based on the 1991 report of the UN Population Fund) and the way in which human activities are causing 'major changes in the global environment'. Without a change in this growth in population and the present pattern of human activity, according to the document, 'science and technology may not be able to prevent either irreversible degradation of the environment or continued poverty for much of the world'.

The global issues addressed in this book should cause us to ask how we can make our planet more habitable. There are no easy answers to these weighty questions. We encourage you to consider where your priorities lie in helping to shape the priority issues for the rest of the 1990s and into the twenty-first century. And we would hope that, having formed opinions on these issues, you will act in whatever capacity you see fit, however insignificant it may seem in the global scheme of things. As fellow travellers on Spaceship Earth, the issues cannot be ducked for long without forfeiting our right to criticize what industrialists and politicians do.

Having no opinions about global issues is tantamount to sticking our heads in the proverbial sands of time. And, as surely as we are here now, those sands of time will run out on us unless we can manage the planet better than we currently do.

To manage the Earth more efficiently, and husband the natural resources with less waste, there is a need to understand the planet. This book is concerned with understanding the inextricable links between the living and the inanimate world, the way in which the forces of nature shape our daily lives, and the actions which can be taken for humanity to become more in harmony with the pulse of the Earth.

As Earth scientists, we wear these labels in our professional careers as university lecturers. As human beings concerned with environmental issues, we have used our scientific training and expertise to express opinions from a perspective which combines scientific explanations with our emotional involvement with the world in which we live. The information presented in this book does not lead to only one conclusion and a unique course of action. This book has been written to provoke debate and action. It is not a cosy cornucopia of facts to be digested and regurgitated at leisure. It is aimed at bringing the major global environmental issues into focus in order for you to decide whether human beings can influence and change the course of our history.

Are **acid rain** and an **anthropogenically** enhanced greenhouse effect inevitable, are they natural processes, and can scientists and policy makers ameliorate their effects? What were past climates like and how does such knowledge help predict future climate changes? Are human activities polluting vulnerable **ecosystems** beyond recovery? Is society wasting energy resources? Are there economically viable alternative energy resources to the traditional **fossil fuels**? Does society want nuclear power? Are nuclear weapons acceptable in a civilized world? Is it possible to predict natural hazards and so to reduce their often devastating effects? How does human activity affect the landscape? Can the world's growing population be adequately fed? These are issues considered in this book. The final chapter examines ways in which the Earth is managed, including a look at such diverse topics as population growth, the destruction of the rainforests and agriculture, and it is there that we suggest that there are things which can be done to make the planet more habitable – to increase the chances of human beings and other vulnerable species surviving longer. The reader may well disagree with our shopping list of action. If so, then one of the main aims of this book will have been achieved – to provide a critical and provocative look at global environmental issues.

Studying Earth

Through studies and observations scientists have become increasingly aware of the relationships and interactions between the Earth and the Solar System, or the universe, the inorganic and the organic world. No matter how detailed the studies, there are always new principles and phenomena to be discovered. Some relationships are so complex that scientists can hardly begin to understand them, yet others seem very simple. The laws of physics and chemistry allow the description of many natural phenomena. Physicists concern themselves with the ultimate origin of matter and time, and they endeavour to quantify this absolutely fundamental problem of science. Yet, the complexity of the living, organic world still defies such elegant mathematics. This point is well made by Richard Dawkins, a zoologist from Oxford University, in his book *The Blind Watchmaker*, in which he describes the 'sheer hugeness of biological complexity and the beauty and elegance of biological design'.

In order to understand the world better, the various intellectual attempts at understanding the natural world need to be simplified. No single person, however intelligent, could hope to comprehend the range and depth of all knowledge, let alone interpret it. So, society has developed with different areas of knowledge, such as mathematics, philosophy, music, history, medicine, and literature, each with its own so-called experts or practitioners. Many of these subjects are extremely specialized, but nevertheless they are not independent of each other.

These specialities include subjects that form the focus for much of this book, such as geology, geography, **meteorology**, **hydrology**, oceanography, botany, zoology, **geodesy**, **pedology**, and many others.

Earth in Space

The Earth is one of nine planets that orbit the Sun. These heavenly bodies, together with their moons and a belt of fragmented planets – the asteroids – constitute the Solar System. The Solar System containing planet Earth is just one of about 10^{11} (100,000,000,000) which form our galaxy, the Milky Way. This, in turn, is one of 10^{11} galaxies in the Universe, all with similar numbers of planets and stars to our own galaxy. Our planet, therefore, is estimated as just one of at least 10^{22} planets travelling in space, held in their orbits by the gravitational forces that exist between the planets and stars.

Cosmic distances are phenomenally large, for example, the distance from the Earth to the Sun is 150 million km. These distances are so large that they are measured in **light years**, which is the distance light travels in one year. In just one second, light travels 300,000 km. It takes around eight minutes for light to reach the Earth from the Sun, while it would take 100,000 years for light to travel across the diameter of our galaxy. Humankind is travelling on a spacecraft planet as it revolves around the Sun at speeds of about 107,000 km per hour. The Sun travels around the galaxy at about 300 km per second, which itself is travelling at enormous speed outwards from the centre of the Universe which is still expanding after its creation in the **Big Bang**. The creation of the Universe probably occurred some 10,000 million to 20,000 million years ago. This contrasts with the date of 4004 BC for God's creation of the world proposed by the Irish archbishop of Armagh, Ussher (1581–1656). The noise from the Big Bang is still reverberating around Space as a constant background level of radiation. So, humans are cosmic passengers embarking on a journey at enormous speeds within the vastness of Space. The story of the creation of the Universe, and with it the Earth, is eloquently told in Stephen Hawking's best-selling book, *A Brief History of Time*.

Humankind is currently living through the growing pains of the Space Age which really began in the 1960s. On 12th April 1961, Yuri Gagarin's historic Space flight aboard his Vostok capsule began the era of extra-terrestrial human travel. The dream of countless earlier generations was fulfilled in this Soviet mission. The race to land a man on the Moon was on. The USA, with strong Presidential backing, especially from John F. Kennedy, and massive public investment, were the first to land astronauts on the Moon in the Apollo 11 mission in 1969, flown by Michael Collins, Edwin 'Buzz' Aldrin, and Neil Armstrong. Upon landing, Armstrong spoke to Mission Control: 'Houston. Tranquility Base. The Eagle has landed'; on 21st July 1969 at 3.56 a.m. BST, Armstrong was the first man to walk on the Moon, when he stated 'That's one small step for a man, one giant leap for Mankind'.

On 12th April 1981, exactly 20 years to the day after Yuri Gagarin's flight, the space shuttle *Columbia* was launched by the USA. But, tragedy was to strike the space shuttle programme, when in 1986, just 73 seconds after lift-off, the space shuttle *Challenger* exploded, and with it the cosy myth of Space travel becoming routine into the 1980s. The accident also caused people to ask about the cost of Space travel, not only in terms of astronauts' lives, but also in relation to broader human costs versus benefits. Is it morally defensible to spend billions of US dollars on a space programme when so much of the world's population has pitifully inadequate food and shelter?

Humankind has the technical ability to explore Space, yet human suffering, starvation and disease seem as prevalent as they ever were. Over thousands of years, humans have developed a rich and diverse human culture through many civilizations. Despite this technological age, with its enormous advances and achievements, the human species remains as aggressive as ever.

The twentieth century has witnessed two world wars and many regional conflicts. There were more than 10 million deaths in the First World War and more than 55 million in the Second World War. Artificially created **radio-activity** has been harnessed for peaceful use as an energy resource, but also used to kill tens of thousands of people in Hiroshima and Nagasaki. Nuclear weapons could have been used in 1961 over the blockade of Berlin, or in 1963 during the Cuban missile crisis, or in Vietnam in the late 1960s. They could have been used on other occasions but were not; we cannot be sure that they will not be used at some future date. International diplomacy, while undoubtedly more sophisticated than in previous centuries, remains incapable of stopping wars in many parts of the world. Many of the issues addressed in this book can only be tackled in a climate of international diplomacy, confidence, and good will. In the decaying Soviet Union, the momentous events of the third week in August 1991, with the abortive military *coup* to overthrow President Mikhail Gorbachev, followed by his resignation on Christmas Day 1991, with the handing over of power to Boris Yeltsin as President of Russia, and of the new commonwealth, symbolized the formal breakup of the Union. The events of the next few years in the dismembered Soviet Union may lead to greater superpower co-operation over the environment. It must be hoped that the danger of a nuclear war has receded somewhat after the death of Soviet communism, but this is by no means certain.

The outer layers of the Earth

Earth scientists divide the outer layers of the Earth into four main spheres or realms (Figure 1.1), which are the: **lithosphere**, comprising the outer layers of the solid Earth, as rocks, sediments and soils; **atmosphere**, the layers of gases that extend from the Earth's surface up to about 100 km to the outer boundary of our planet; **hydrosphere**, the layer of water that covers our planet, from a maximum depth of more than 11 km in the oceans to shallower and less extensive bodies of water such as lakes and rivers (this group also includes snow and ice that form glaciers and ice sheets, and the water within the

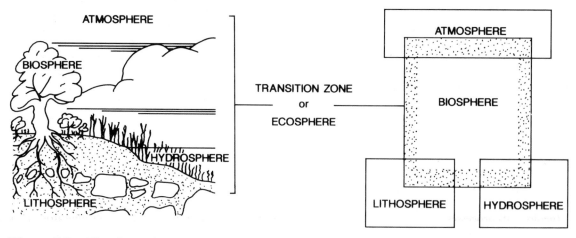

Figure 1.1 *The relationship between lithosphere, hydrosphere, atmosphere, biosphere, and ecosphere. Adapted from White et al. (1986).*

soils and rocks beneath us); **biosphere**, a term first made famous by the Swiss geologist Suess for the thinnest layer, comprising organic matter, which is generally only up to a few metres thick covering much of the land surface (this layer, at its thickest, reaching several tens of metres in rainforests, also extends into the atmosphere as creatures fly and plant spores are blown by the wind, and deep into the lakes, seas, and oceans into the deep-sea trenches). Human beings are part of the biosphere, and exist by interacting with the other three 'spheres'.

Even though the world can be split into these four sections, they are closely interrelated: all are part of the **ecosphere**. **Ecology** is the study of the ecosphere, what many people loosely refer to as the study of the environment. It is undoubtedly one of the most difficult subjects to study because it utilizes information and ideas from just about every other subject, from science to politics, from economics to culture.

Life on Earth

The Sun's rays provide energy to drive the ocean currents and atmospheric processes (the weather), helping to distribute gases, water, and the heat around the Earth. The Earth's rocks furnish the vital nutrients and water essential for life, and its surface provides the substrate for life. Humans, the most complex and sophisticated of all organisms, live on Earth as an integral part of it; people are short-stay passengers, and are affected by its internal, surface, and atmospheric processes. More importantly, human activity may be profoundly altering the Earth's atmospheric systems and climate. Although each person only rides a short way with Earth, the pollution left behind can have a long-term influence.

On cosmic scales, planet Earth seems insignificant, but it may be unique. It is the only planet that scientists are aware of which is capable of supporting human life. Various chemical and physical arguments suggest that our Earth is about 4.6 billion years old. Life on Earth is incredibly diverse. It has been estimated that there are as many as 1.4 million formally described species of animals and plants on Earth (Wilson 1989). There are many more that await detailed study and the conferment of formal species names. Conservative estimates put the actual number of species on Earth closer to 4 million. Biological groups receiving relatively little attention from scientists who study the diversity of biological systems include many

---

The critical role played by biological activities in releasing O_2 to the atmosphere is **oxygenic photosynthesis**, in which the water molecule is split to release pure oxygen:

$$CO_2 + H_2O \rightarrow CH_2O + O_2$$

It is the free oxygen released in this reaction that, over geological time, increased the levels of atmospheric oxygen to those now present. The burial of organic matter (shown here in its simplest chemical formula, CH_2O in the above equation) in sediments allows the release of free oxygen into the atmosphere. A corollary of the burial of organic carbon and biogenic sulphide is that electrically charged sulphate ion (SO_4^{2-}) and ferric iron ion (Fe^{3+}) levels increased at the Earth's surface and atmosphere. Oxidation of the Earth's atmosphere and surface environments was facilitated both by biological and sedimentary (geological) processes.

Although the Earth's atmosphere has changed to one in which free oxygen is present, the earliest life evolved in a very different atmosphere. There is a 3,800 million-year isotopic record of life on Earth, something that Earth scientists have discovered by examining the carbon contained in lithified sediments, or sedimentary rocks, at Isua in west Greenland. Carbon is the key element for life, and it occurs as a mixture of two stable isotopes, carbon-12 (^{12}C) and the heavier carbon-13 (^{13}C), along with a relatively short-lived radioactive nuclide of carbon-14 (^{14}C). Scientists have demonstrated that all the common photosynthetic pathways discriminate against ^{13}C, therefore living organisms show a very strong preference for the lighter carbon isotope, ^{12}C. Consequently, the heavier isotope, ^{13}C, tends to remain in the Earth's surface reservoir of oxidized carbon, mainly as dissolved bicarbonate in seawater. An increased ratio of ^{12}C to ^{13}C is found in sedimentary organic matter dating back 3.5 to 3.8 Ga (thousand million years), an indicator of the principal carbon-fixing reactions involved in photosynthesis (see review by Schidlowski 1988).

Box 1.1

Oxygenic photosynthesis and an atmosphere rich in free oxygen

insects such as mites, fungi, and organisms inhabiting the deep ocean. More than half of the total number of species of flora and fauna inhabit the rainforests with their moist tropical climate.

The rainforests account for approximately 6 per cent of the land surface where rainfall is in excess of 200 cm per year. The number of individuals of any species inhabiting the rainforests are also amazing. In just 1 g of soil, there may be as many as 100,000 algae, 16 million moulds and fungi, and several billion bacteria. Up to 5,000 species of organisms can inhabit just one rainforest tree. In a single acre of rainforest in Panama, it is estimated that there are as many as 40 million animals, not counting the bacteria, fungi, and moulds. Perhaps the other major ecosystems with extremely large **biodiversity** are the coral reefs (Plate 2). Here, myriads of organisms occupy the **ecological niches** where there is a plentiful supply of food, and well-oxygenated warm waters.

The Earth provides the life support system for this diverse and abundant array of organisms. The atmosphere filters out potentially lethal **radiation** from the Sun, yet at the same time allows some of the radiation to penetrate through and provide the energy for plants to construct tissues of carbohydrates from **carbon dioxide** (CO_2) and water (H_2O) in the process of **photosynthesis** (see Box 1.1). These plants, in turn, are a food for the animal kingdom. In addition, the atmosphere provides the carbon dioxide, oxygen (O_2), and much of the water vapour needed for the basic functions of animal life.

The oxidation of the Earth's crust early in Earth history and the associated increase in atmospheric oxygen have been linked to the accumulation of reduced carbon in sedimentary rocks. By studying the carbon **isotope** composition of sedimentary organic carbon and carbonate, Des Morais *et al.* (1992) have shown that during the **Proterozoic** period, 2.5–0.54 Ga (thousand million years ago), the organic carbon reservoir grew in size relative to the carbonate reservoir. They further showed that this increase and the transition to an oxidizing atmosphere took place mainly during intervals of enhanced global **seafloor spreading**, continental breakup and rifting, and **orogeny** in what is broadly referred to as **tectonic processes**. Around 3.0–2.4 Ga, relatively small **continental plates** or cratons welded together to form the first relatively large and stable continental plates. These processes provided the templates for the accumulation of large amounts of sediments, and set the stage for the growth of carbonate platforms 2.6–2.3 Ga. Although there is evidence to suggest oxygenic photosynthesis in the algal mats called **stromatolites**, the net accumulation of atmospheric O_2 was virtually zero, because there was very little burial of organic carbon. Approximately 2.2–2.1 Ga, the large continental plates began to disintegrate by rifting and seafloor spreading, and the breakup allowed the development of **anoxic** basins in which organic matter could accumulate and be buried. Also, at this time, there appears to have been significantly enhanced erosion and continental run-off, inferred from the rise in seawater strontium isotope values (Box 1.2).

Evidence from **palaeosols** suggests that prior to about 2 Ga atmospheric O_2 levels were low, but such a situation would not have inhibited the efficient remineralization of organic matter in microbial mats, as is the case today. A large part

Box 1.2

Strontium isotopes through geological time

Strontium (Sr) occurs as a trace element in rocks. The composition of seawater is directly influenced by the interaction of Sr with rocks close to or at the Earth's surface. Carbonate sediments precipitated from seawater preserve the original ratio of the radiogenic isotope of ^{87}Sr to its non-radiogenic isotope ^{86}Sr in seawater. Analysis of marine carbonates reveals a temporal variation in the $^{87}Sr/^{86}Sr$ throughout geological time. As Sr supplied to the oceans displays characteristics of those sources, measurements of $^{87}Sr/^{86}Sr$, taking into account such factors that the proportion of these isotopes has changed throughout the evolution of the oceans and continental-crust, provide an indication of the changing relative importance of sediments supplied from various sources. Present seawater has a Sr concentration of about 8 ppm, a residence time of 10^6-10^7 years (mixing time in the oceans is 1.6×10^3 years), and a $^{87}Sr/^{86}Sr$ value of 0.7092. Since the oceanic mixing time for Sr is much less than its residence time, the Sr isotopic composition is homogeneous at any given geological time.

of the early free O_2 was probably consumed in reactions associated with the more voluminous **hydrothermal systems** on the seafloor, associated with extensive **magmatic** and volcanic activity. Seawater sulphate ion (SO_4^{2-}) levels, therefore, would have been much less than in the modern oceans.

The views of Des Morais *et al.* (1992) challenge the widely held belief that major rapid changes in biological evolution controlled the long-term increase in oxygen levels in the atmosphere. The development of oxygenic photosynthesis took place at least 600 million years ago, prior to the accumulation of significant amounts of O_2 in the atmosphere. **Eukaryotic organisms**, which require O_2 to **biosynthesize** the essential **lipids**, appear about 2.1 Ga.

Oxygenic photosynthesis must have provided a mechanism capable of sustaining a dramatic increase in atmospheric O_2 levels, but the timing and magnitude of the O_2 accumulation was regulated by tectonic processes controlling erosion and sedimentation (ibid.).

Evolution and extinction

Palaeontology has provided evidence that throughout the past 570 million years of Earth history, there have been catastrophic extinction events when large numbers of species are known to have become extinct. The causes of such fatalities remain uncertain and are often keenly debated. Climatic shifts from global warming to global cooling periods, or meteorite impacts, are common amongst the favoured causal processes.

Five major mass extinction events are known to have occurred in geological time at approximately: 439 Ma (million years ago), at about the boundary of the Ordovician and Silurian periods, when there was a major glaciation; 375 Ma, late in the Devonian period; 240 Ma, at the boundary between the **Permian** and **Triassic** periods; 210 Ma, in the Triassic period; and at 65 Ma, the boundary between the **Cretaceous** and **Tertiary** periods, best known because it included the demise of the dinosaurs. There are other extinction events known throughout these time periods, but compared to the five mass extinction events, they were relatively small. The greatest mass extinction event known occurred

240 Ma, at the end of the Permian period, when an estimated 80–95 per cent of all known marine animal species were wiped out. The geological record tells palaeontologists that evolution is slow, at least by the yardstick of human longevity, and that environmental changes can cause a dramatic reduction in biodiversity, which only recovers in time spans measured in millions of years. Human activity is threatening this diversity, probably in a way that has not happened since the more recent mass extinction events.

The greatest threats to species have come in the relatively isolated environments such as islands and lakes. In these areas, perturbations to the environment can cause rapid extinctions. The South Atlantic island of St Helena was robbed of its unique plant flora in the nineteenth century because of deforestation. The destruction of the tropical rainforests is probably our greatest crime against the diversity of species and fauna on Earth. It is estimated that about a half of the bird species have been obliterated from Polynesia because of hunting and the destruction of the rainforests. After the wholesale removal of large areas of rainforests, people are beginning to realize what damage has been done, but concerned individuals and organizations are still a long way from persuading the exploiters of the rainforests to desist from destroying these ecosystems.

Against this panoply of environmental issues, arguably the main problem is over-population of the planet. In the short term, the ingenuity of human endeavour is required to create a sustainable planet for the present and near-future predicted population levels. In the longer term, the world population must be reduced. This is not easy. Cultural, ethical, religious, and socio-economic factors are inextricably interwoven in this issue.

Interdependence

To begin to appreciate the complexity and inter-relationships between the lithosphere, atmosphere, hydrosphere, and biosphere, a look at an ostensibly simple system provides an insight into the interwoven links. Consider a poppy in a field. A plant such as a poppy will anchor itself into the soil layers, the lithosphere, using its root system, and it will obtain many important nutrients from minerals in the soil that have been derived from the **weathering** of rocks. At the same time, the poppy obtains carbon dioxide from the atmosphere to build up carbohydrates to form tissue. It obtains the water necessary for life from the groundwater in the soil, part of the hydrosphere. When the poppy reproduces, atmospheric processes such as the wind help disperse its seeds and so facilitate fertilization and propagation. In life, the poppy is an integral part of the organic layer, the biosphere, yet at the same time it is in all four spheres, as part of the ecosphere. The poppy will also contribute to the atmosphere by producing oxygen during the process of photosynthesis. If, during life, one of the spheres is severely altered – for example, the hydrosphere becomes depleted in water or the soil (part of the lithosphere) becomes depleted in vital nutrients – then the plant wilts and dies. Upon death, the poppy becomes part of the lithosphere to add new nutrients to the soil, and it may even become fossilized in a rock to form a fossil fuel, such as **lignite** or coal. Animals need oxygen during metabolism when they produce carbon dioxide which plants use to produce carbohydrates. But, if there is no free oxygen being given up into the atmosphere because it is no longer being produced, then oxygen becomes depleted and animals die.

From such extreme scenarios, the so-called domino effect of one deleterious action fuelling another can be appreciated. Many scientists refer to our planet as being in a state of delicate balance. If this 'balance of nature' is upset by altering the inputs to the natural systems, the consequences or outputs may be detrimental to our well-being. The Earth can be visualized as a system with inputs and outputs. The systems concept was originally developed by the biologist Ludwig von Bertalanffy in the 1920s, and was later adopted in 1949 by the new science of **cybernetics**. Some of the outputs may become inputs again, that is they feed back into the system. These inputs may further enlarge the output which, in turn, may feed back into the system again, and so on. Where the original effect is magnified, such loops are called **positive feedbacks**. When a feedback results in a decrease in the output, the feedback is said to have a negative effect as a **negative feedback**. Negative feedback and positive feedback mechanisms are very important in understanding how the Earth's natural systems work. All environmental systems are both open and in **dynamic equilibrium**. That is, there is an input of energy and matter and a corresponding output of energy and matter which are in some way balanced. This balance is controlled by negative feedback mechanisms. Environmental systems are resistant to positive feedback, which is evident by the time delay between the input and output or response. When positive feedbacks take effect, the response is usually in the form of major environmental change. Over geological time, the major stimulus causing positive feedbacks in the ocean–atmosphere system has been climatic change. Many examples will be considered throughout this book.

Thresholds

Some processes or events will not occur until the amount of input has reached a certain level. These levels are referred to as thresholds. A pain threshold, for example, is familiar to everyone. People with higher pain thresholds will stay in a hot shower longer than those with lower thresholds. There comes a point, however, when even people with high thresholds have to jump out of the shower as the water becomes too hot. There are many similar thresholds in nature, when the consequences of some input process are a sudden sharp change of output. Though very important, thresholds for many events are not known or are poorly understood.

As an example, the discovery in 1985 of a significant depletion or 'hole' in the **ozone layer** in the **stratosphere** over Antarctica provides an illustration of the fact that the depletion of ozone has now crossed some sort of threshold set of conditions that had previously maintained a continuous ozone layer over this region. This hole was not anticipated by the scientific community, and it should make us curious to discover any other thresholds in the ocean–atmosphere system that human activities might have caused us to cross or approach.

Earth – a self-regulating organism

In the late 1960s, James Lovelock, a distinguished scientist and Fellow of the Royal Society of London, developed a hypothesis which he called the **Gaia hypothesis**, named after the Greek word for the Earth Goddess. Lovelock and his colleagues suggested that the Earth is a self-regulating system, one able to maintain the climate, atmosphere, soil, and ocean composition in a stable balance favourable to life. The inputs and outputs are delicately balanced and controlled by feedback mechanisms. The Gaia hypothesis proposes that life itself produced and kept the very gases essential for its own existence at a constant level, and that life itself kept the natural systems of the Earth in a state fit for life. The Gaia hypothesis also explained the extinction of species as a result of those species failing to contribute to maintaining the natural systems and therefore the balance of Gaia. The hypothesis also helped to explain why the surface temperature of the Earth has remained relatively constant over the last 4 billion years, since life first emerged from the primordial organic soups and gases of our planet, despite the fact that the Sun's heat has increased by 25 per cent. Over the same period, the overall carbon dioxide level has dropped, reducing the heat-holding potential of the Earth. These were all consequences of botanical activity, the fixation of carbon dioxide from the atmosphere by photosynthesizing organic matter. The amount of oxygen has remained constant over the past 200 million years, the result of the balance of the complex interactions of organisms and the inorganic components of Earth. Lovelock and his colleagues believe that if human activity continues to disturb the **geosphere**, by disturbing the natural balance of Gaia, and if human activities are not harmonized with the natural processes of Gaia, then this life-support machine will no longer sustain us and we will become extinct, to be replaced by new species.

To help illustrate the Gaia hypothesis, Watson and Lovelock (1983) developed the 'daisy world model'. They imagined a world inhabited only by black and white daisies. In their model the Sun began to warm up, something that actually happened during the early history of the Earth. A lifeless world also warmed up because of the greater heat energy being emitted from the Sun. In the daisy world, however, the black daisies absorbed more of the incoming radiation, and were thus favoured because of their more suit-

able survival strategy, at least during the early days of the faint Sun. As the Sun continued to heat up, however, the black daisies became unsuited to the warmer world, and then the white daisies began to compete more successfully since they reflected more sunlight, which also provided a negative feedback by helping to cool the planet's surface. In such a changed world, the white daisies could become more abundant than the black daisies. Eventually the Sun would become so bright that all the daisies would die, unable to reflect the large amounts of solar radiation reaching the surface of the planet.

This simple model shows how evolving life on Earth could modify climate through both negative and positive feedbacks. Lovelock and his fellow workers suggest that similar processes took place on Earth throughout geological time, and that the Earth will continue to self-regulate itself, if human activity does not cause changes in global climate, by affecting the biosphere and atmosphere–ocean system, which are faster than any negative feedbacks that might otherwise moderate the system.

Not all scientists agree with the Gaia hypothesis. A contrary view is that the Earth's atmosphere has evolved by chance chemical reactions and degassing from the **mantle**. Lovelock argues, however, that this and the traditional evolutionary theories (both Darwinian and **punctuated evolution**) are inadequate as theories because they invoke a passive role for biota throughout Earth history. Lovelock believes that the **biota** played, and continues to play, an active role in controlling their environment. This perspective led Lovelock to propose that life itself is the major control on the atmosphere–ocean system on Earth, maintaining a habitable planet through complex feedbacks.

Today, the consensus of scientific opinion lies somewhere between a Gaian perspective and an appreciation of non-biological, often random processes that contribute to creating this world. Many scientists criticize the Gaian hypothesis because it is untestable. Though parts of the hypothesis can be tested by observation, it is generally impossible to prove that an organism can produce its own environment rather than be conditioned by it. To test the hypothesis fully, scientists would need to travel back in time to the origin of life on Earth and then have a fast-forward replay of evolution tied to other natural events on Earth. The Gaia hypothesis cannot be used to predict specific future changes, for example in the environment. These criticisms mean that the Gaia hypothesis cannot become accepted as a theory, but remains a series of interesting speculations. The hypothesis, however, provides an interesting perspective on life on Earth, and a set of ideas for active debate amongst those concerned with the environment.

Time and rates of change

The rates at which processes take place must also be considered. This is one of the hardest factors to appreciate because our concept of time is influenced by the experience of a single human lifetime, for most a mere 70 or so years. Time in terms of the Earth history is measured in thousands, tens of thousands, millions, or even billions of years.

Geologists believe the age of the Earth is about 4,600 million years. The first bipedal **hominid** (*Australopithecus afarensis*) evolved about 3.75 million years before present (BP), while true modern humans (*Homo sapiens sapiens*) have only been in existence for about 200,000 years. For this reason geologists divide time into a number of geological periods, defined where possible by global events (Figure 1.2). The present period, for example, is called the **Quaternary** (see Chapter 2), with a beginning defined by evidence to suggest that it marks the start of the last major, abrupt, global cooling at about 2 Ma.

Time can also be split up as if it were 1 year old and the important events are shown as calendar dates to help us appreciate time on our scale of experience. To quote a phrase used by Stephen Gould in his book, *Wonderful Life*, 'By the turn of the last century, we knew that the earth had endured for 4 billion years, and that human existence occupied but the last geological millimicrosecond of this history – the last inch of the cosmic mile, or the last second of the geological year, in our standard pedagogical metaphors.'

Chaos theory, the unpredictability of events

In recent years the mathematics of **chaos theory** have been applied to the Earth's natural systems. Many scientists believe that systems, such as weather patterns, are in a state of chaos. Chaos dictable nature of the physics of motion, for example, means that in any new circumstance, however closely it appears to mirror the original set of circumstances, it will always produce a unique outcome. The weather systems were first modelled quantitatively by the meteorologist Edward Lorenz, using simple computers. Using a weather model, he found that he always obtained different results from almost identical inputs. From mathematically plotting the outputs from his model he produced a curve which defied mathematical definition. This showed that no event could be predicted and that the smallest change of inputs could produce very different results. This very unpredictability, however, means that chaos is in a sense predictable; that is, you know in advance that predictions are impossible! The shape of the graph produced by the experimental results of Lorenz looked similar to a butterfly and so his principle became known as the **butterfly effect**. Analogies were drawn with natural processes. Using this principle, for example, the consequences could be such that if a butterfly flapped its wings in Beijing, it would cause turbulence in the atmosphere, which could ultimately lead to the formation of a hurricane in Florida. Such small occurrences could have profound, large, and even catastrophic effects. Many scientists are now beginning to explore the interrelationships within natural systems using the mathematical principles of chaos theory. Any turbulent system, such as the motion of air masses or ocean currents in which turbulent eddies form and decay, lends itself well to chaos theory.

All these natural processes and responses need to be understood if the consequences of human actions are going to be appreciated. It is imperative, therefore, that there is a good understanding of just how the Earth works, the intimate interrelationships between the organic and inorganic elements of this planet, the mutual relationship between organisms and their environments, and the reciprocal relationship between human beings and the Earth. Not only is this interesting and satisfying to natural, instinctive curiosity, but also if the human species is going to survive, then it is vital that the life-support machine planet Earth is maintained and functions efficiently.

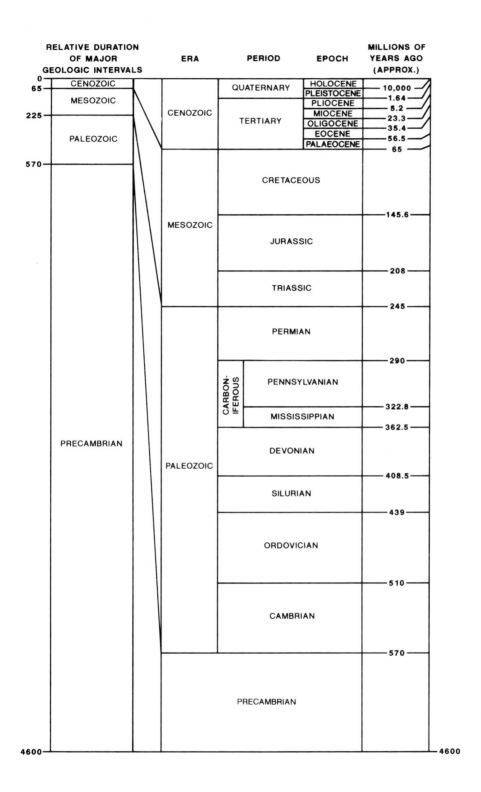

Figure 1.2 *The geological time scale. Time is divided up in a way that reflects the major events in the evolution and/or extinction of species of animals and plants. After Harland et al. (1989).*

Earth scientists need to be able to examine and appreciate the world at different scales, quite simply because we exist and make use of the natural world at a variety of scales. They need to be able to explain the chemical reactions that take place on an atomic scale and to relate these to the effects on larger, mesoscopic to macroscopic scales. Scientists also need to understand just how well scale models of processes accurately reflect and mimic larger-scale phenomena. Ultimately, there is the need to appreciate the consequences of the sum of small-scale processes on a global scale. The relatively new study of **fractal geometry** is fast becoming a potential means to do this.

The destruction of the ozone layer, a protective gaseous layer in the atmosphere which shields the Earth's surface from the harmful effects of the Sun's radiation, is a good example of the range of scales at which scientists can see processes operating that are interrelated. The chemical reactions that lead to the destruction of this protective ozone layer take place on the atomic scale, as compounds such as **chloro-fluorocarbons** (**CFCs**) combine with ozone to break it down. This leads to regional effects such as the depletion of ozone over the Antarctic during the spring and late summer which, in turn, allows more radiation to reach the surface of the Earth. Radiation at the short **wavelength** part of the **electromagnetic spectrum** can be harmful to organisms, and may destroy animals and plants, especially important bacteria, and can cause mutations and cancers.

Anthropogenic emissions of **greenhouse gases**, such as carbon dioxide, may contribute to global warming, which in turn may lead to increased melting of the Antarctic ice sheet, and

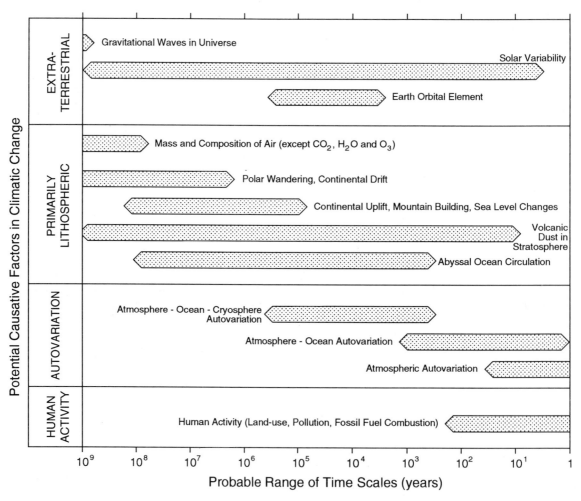

Figure 1.3 *Processes involved in environmental change and their time scales. After Williams et al. (1993).*

a **eustatic** rise in sea level. A rise in sea level is enhanced by the thermal expansion of the ocean waters because of the global warming. The sequence of events is much more complex than this simplified version of a potential run-away **greenhouse effect**, with feedback mechanisms operating either to enhance (positive feedbacks) or to diminish (negative feedbacks) the global warming. Many of these processes and principles are examined in detail throughout this book.

Figure 1.3 summarizes the main processes involved in environmental change and their scales (Williams *et al.* 1993), providing a useful overview and reference point for much of what is discussed throughout this book. Whilst understanding the causes and effects of global environmental change, it is not as easy to predict the actual timing of any abrupt, and possibly catastrophic changes – due, at least in part, to the inherent chaos in any system.

Key points

CHAPTER 1: KEY POINTS

1 There are many questions that need to be addressed for the effective management of the Earth. Amongst these, the principal ones include: can human activities and population growth be sustained without extreme environmental damage? How will human activities change global climate? What lessons can be learnt from studying past climates and then applied to help predict future global climate change? Are human activities polluting the various ecosystems beyond recovery, what are the threshold levels, and how reversible is the human damage to the natural environment? Is society squandering energy resources? Are there economically viable energy resources other than traditional fossil fuels, and how can these be exploited best? Does society understand the long-term implications of utilizing nuclear power, and if this energy resource is to be used then just how safe is it? Are nuclear weapons inevitable and acceptable in a civilized world, and is their proliferation unstoppable or perhaps even desirable? Is it possible to predict accurately and mitigate the effects of natural hazards? How does human activity affect and alter the natural landscape?

2 The Earth is one of 10^{22} planets in the universe, and may be unique in supporting life as we know it. Biological diversity is enormous and is sustained by energy from the Sun and the Earth's internal energy systems. Biological activity plays a critical role in releasing free O_2 to the atmosphere by oxygenic photosynthesis. Evidence from chemical isotopes in sedimentary rocks, from Isua in west Greenland, suggests that life existed on Earth *c.* 3,800 million years ago. A study of carbon and strontium isotopes in sedimentary rocks suggests that free O_2 started to accumulate in substantial quantities in the Earth's atmosphere about 2,000 million years ago as anoxic basins began to form by plate tectonic processes which allowed organic

carbon to be buried. Prior to this, oxygen was held in carbonate rocks as the so-called 'carbonate reservoir'. Oxygenic photosynthesis took place at least 600 million years ago, and provided a mechanism capable of sustaining atmospheric-free O_2 levels. The Sun provides the energy to drive photosynthetic, atmospheric, and hydrological systems.

3 The outer layers of the Earth comprise the atmosphere, biosphere, hydrosphere, and lithosphere. These are interrelated as the ecosphere. A 'systems approach' allows their interdependence to be studied, from which it is possible to appreciate both 'negative' and 'positive' feedbacks. The Gaia hypothesis describes the Earth as a self-regulating organism, able to sustain itself in equilibrium without any long-term major changes, and maintaining climate, and the composition of the atmosphere, soils, and oceans. Processes and events within these 'spheres' or systems may change from one level or condition to another when an input has reached a 'critical threshold'. Chaos theory proposes that natural systems are fundamentally unpredictable. Time scales and rates of change vary from hundreds of million of years to fractions of a second, and such rates and magnitudes must be appreciated when considering environmental systems. The evolution of life is a continuous process punctuated by periods of rapid change marked by the appearance of new species and the disappearance or extinction of other species. It is these relatively rapid changes in life on Earth that allow a sensible division of geological time.

4 Environmental awareness has evolved since the interest in natural history during the nineteenth century, to the study of ecology and the development of international scientific programmes and organizations in the 1950s, and recently with the concern for human-induced global climate change.

Chapter 1: Further reading

Bradshaw, M. and Weaver, R. 1993. *Physical Geography: An Introduction to Earth Environments*. London: Mosby.
Comprehensive and well-illustrated textbook outlining the principles of Earth systems at an introductory level suitable for high school and first year undergraduates. Atmosphere–ocean systems are described in terms of their dynamics; plate tectonics is introduced; processes of geomorphology are described; aspects of human interaction with the natural environment are discussed, and ecological systems are outlined in which there is a useful emphasis on soil dynamics and the characteristics of biomes.

Broecker, W.S. 1987. *How to Build a Habitable Planet*, New York: Eldigio Press.
An extremely readable introduction to the origin and evolution of the Earth. Broecker manages to make seemingly complex scientific arguments simple and interesting. This book is highly recommended as an introductory book for both students and teachers wishing to understand some basic geochemical arguments about the Earth.

Dawkins, R. 1986. *The Blind Watchmaker*. Harlow, Essex: Longman Scientific & Technical.
An examination of the evolution of life, which inspires the reader with a vision of existence and the elegance of biological design and complexity. Dawkins argues for the truism of Darwinian theory and shows for example how modern views such as punctuated evolution are part of neo-Darwinian theory. An excellent supplementary book for many courses in the natural sciences and environmental studies.

Gleick, J. 1987. *Chaos*. London: Heinemann.
A readable account of the historical development and the elementary principles of the science of chaos.

Lovelock, J.E. 1988. *The Ages of Gaia: A Biography of Our Living Earth*. Oxford: Oxford University Press.
The follow-up book to *Gaia: A New Look at Life on Earth* (1982), which elaborates on the Gaia view of Earth. This book examines the interaction between the atmosphere, oceans, the Earth's crust, and the organisms that evolve and live on Earth. Lovelock discusses recent scientific developments, including those on global warming, ozone depletion, acid rain, and nuclear power. This book provides a thought-provoking look at interdependence, and the role of negative and positive feedbacks in controlling the evolution and adaptability of life.

Manahan, S.E. 1993. *Fundamentals of Environmental Chemistry*. Michigan: Lewis Publishers.
A comprehensive and well-written textbook aimed at students having little or no background in chemistry. This book gives the fundamentals of chemistry and environmental chemistry needed for a trade, profession, or curriculum of study requiring a basic knowledge of these topics. It also serves as a general reference source. This book will appeal to those involved in college and university studies where the environmental course has a relatively strong science base, and is unlikely to appeal to those in the social sciences and geography.

Nebel, B.J. and Wright, R.T. 1993. *Environmental Science: The Way the World Works*, 4th edn. Englewood Cliffs, New Jersey: Prentice-Hall.
An undergraduate environmental textbook with a central theme of sustainability. There are four sections in this book: Part I, What ecosystems are and how they work; Part II, Finding a balance between population, soil, water, and agriculture; Part III, Pollution; Part IV, Resources: biota, refuse, energy, and land. The text has various elements that provide teaching aids, e.g. learning objectives, review questions etc. While we found this book useful, it has the somewhat irritating presentation style of very well-drawn and sophisticated diagrams alongside over-simplistic, naive, artwork. The book is aimed at college students taking environmental courses.

Yearley, S. 1992. *The Green Case: A Sociology of Environmental Issues, Arguments and Politics*. London: Routledge.
A comprehensive account of the basis for 'green' arguments and of their social and political implications. Yearley examines the reasons for the success of leading campaign groups (such as Greenpeace), and analyses developments in green politics and green consumerism. He also emphasizes the serious ecological problems in the developing world, and argues that these problems are inextricably linked with debt and their need for development. A well-written sociological perspective, and a recommended supplementary book for those interested in the broader aspects of global environmental issues.

Blow, winds, and crack your cheeks! rage! blow!
You cataracts and hurricanoes, spout
Till you have drench'd our steeples, drown'd the cocks!
You sulphurous and thought-executing fires,
Vaunt couriers to oak-cleaving thunderbolts,
Singe my white head! And thou, all-shaking thunder,
Strike flat the thick rotundity o' the world!
Crack nature's moulds, all germens spill at once
That make ingrateful man!

Spoken by Lear.
William Shakespeare, *King Lear*, Act III, Scene ii.

The night has been unruly: where we lay,
Our chimneys were blown down; and, as they say,
Lamentings heard i' the air; strange screams of death,
And prophesying with accents terrible
Of dire combustion and confus'd events
New hatch'd to the woeful time. The obscure bird
Clamour'd the livelong night: some say the earth
Was feverous and did shake.

Spoken by Lennox, Nobleman of Scotland
on the night Macbeth dies.
William Shakespeare, *Macbeth*, Act II, Scene iii.

The Earth's climate has not always been as it is today. There have been times in the geological past when the global climate was warmer or considerably colder than at present. The geographic and temporal distribution of organisms, preserved as fossils, and the particular chemical signatures and sediment types available for study, show that the Earth's climate has fluctuated over geological time. As an example, 4.5–3.5 million years ago (Ma), parts of eastern Antarctica were a lot warmer. During the Tertiary period of Earth history, from about 65 Ma but prior to 1.64 Ma, boreal forests were growing in the Canadian High Arctic as far north as 78°N, now preserved as fossil forests (Plate 2.1). Although it is now known that there have been substantially different climates in the past, the exact causes of such variations remain unclear.

Beside the intellectual curiosity that drives humankind in search of knowledge about past climates on Earth, about how major climatic change may come about, and the rates at which such changes could occur, it is possible to begin to make sensible predictions and models about negative and positive feedback processes in controlling global climate change. Put more simply, the geological record provides an unprecedented insight into the circumstances in which the greenhouse effect and **icehouse effect** occur, and the opportunity to assess the potential impact of human activities in controlling climate change.

Climates, both past and present, are studied by many people: meteorologists trying to improve weather prediction and construct climatic models for the future, archaeologists wishing to understand the climatic conditions that prevailed during the early development of human life around the globe, and Earth scientists endeavouring to unravel the history of our planet and the dynamics of Earth surface processes. Public interest in global warming, acid rain, the potential effects of a **nuclear winter**, and how other forms of chemical pollution in the atmosphere or oceans affect climate have all

Plate 2.1 *Fossil tree stump preserved at 79°N on Axel Heiberg Island in the Canadian High Arctic. This provides evidence for the existence of high latitude boreal forests in polar regions during Tertiary times.*

17

contributed to a resurgence of interest in past climates, primarily as a key to predicting future climatic change.

Earth scientists have suddenly found themselves at the centre of media attention. Large sums of money are now available for research into past climates. Computer-based climatic models, commonly referred to as **general circulation models (GCMs)**, are in vogue. The past few years have witnessed a concerted effort to understand causal factors which contribute to global climate change.

Scientists studying past climates over hundreds of thousands of years used to consider themselves involved in Tertiary or Quaternary studies (see Chapter 1), or Environmental studies, but suddenly newer labels are in fashion which emphasize the oceanic or climatic aspects of their work, such as **palaeoclimatology** or **palaeo-oceanography**. It has been more than a change of label. A quiet revolution in the Earth sciences has crept up on all of us and, arguably, there is a real need for a whole new area of study to be recognized formally and budgeted for, at least at the level of higher education. Palaeoclimatology as a scientific subject is truly interdisciplinary, regularly and necessarily involving many different Earth scientists, chemists, biologists, physicists, astronomers, and mathematicians. It is, perhaps, more than any other current scientific pursuit, the youngest science looking for universal recognition.

Earth scientists have now established that global climatic changes occur on time scales up to hundreds of millions of years, but they have not yet developed well-constrained cause-and-effect models for global changes in climate. One of the main ways to understand past climates and the nature of climate change over the past few hundred thousand years is through the study of ice cores (Box 2.1), therefore increasing attention is being focused on the climatic signatures preserved in such cores.

This chapter considers some of the major, dramatic changes in the Earth's climate at a few selected time intervals. In terms of climate change, humanity is currently in a particularly interesting period of geological time, the Quaternary, often referred to as the present **Ice Age**. During the period, which extends back for over 1.64 million years (Harland *et al.* 1989), the Earth's climate has cooled down and undergone a series of rapid fluctuations between warm and cold phases. It is important to understand the nature of these changes if Earth scientists are to resolve the effects of human activities and natural variation in the climatic system. Particular attention, therefore, is given to the nature and study of the Quaternary period in this chapter. Whatever the exact cause, or causes, of the past sudden shifts in Earth climate, the one thing that Earth scientists are certain of is the catastrophic consequences for life on Earth at such times. Clearly, just as current political thought and, hopefully action, is built upon the lessons that history teaches, so humankind should attempt to gain a better understanding of Earth history in order to appreciate the potential that exists, either natural or human-made, for destroying various animal and plant life on this planet. Human activities

Box 2.1

Ice cores

Ice cores provide a unique archive of past climatic conditions, including atmospheric chemistry. Complete ice cores record annual and seasonal changes in atmospheric gases, chemicals such as acids, trace metals, and wind-borne dust which were sealed into the falling snow and buried to form ice. The stable isotopic composition of the ice (see Box 2.3) depends on the air temperature at the time the snow formed and accumulated, thereby providing a means of calculating past atmospheric temperatures.

Increasingly, scientists wishing to document past climatic conditions, and understand the causes and effects of climate change, are analysing the chemical and physical nature of ice cores. Ideally, ice cores are drilled in parts of the world where there is likely to be an undisturbed and continuous signature of past climates, for example in the Greenland ice sheet and in Antarctica. Examples of ice cores include:

1 the American 'Thule' core, drilled to the bottom of the Greenland ice sheet between 1963 and 1966 near Thule in north-western Greenland, retrieving a 120,000-year record, and a 100,000-year record, also reaching the underlying bedrock, drilled near a radar station in south-eastern Greenland between 1979 and 1981;

2 the American Byrd core from West Antarctica, drilled in 1968, and giving a record of the past 70,000 years;

3 the 2,083 m long Vostok ice core from East

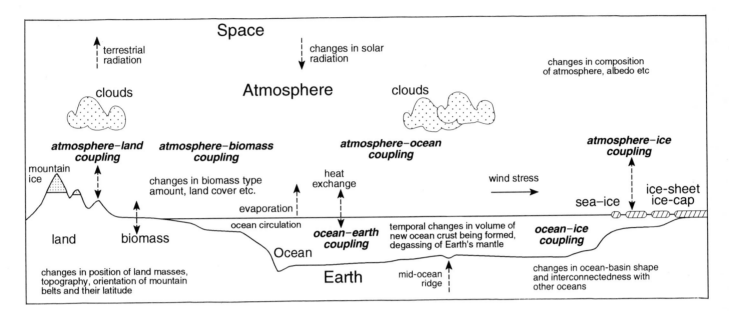

Figure 2.1 *Principal components of a climate system.*

may be exerting a forcing effect on world climate (see Chapter 3).

Causes of global climate change

In order to consider present and past climate, it is important to have at least a rudimentary understanding of the principal components of any climate system (Figure 2.1), the structure of the Earth's atmosphere (Figure 2.2), an idealized global wind circulation (Figure 2.3), and the ocean conveyor belt (Figure 2.4) which

distributes heat around the Earth's surface (Figure 2.5). Up to about 100 km above sea level, the Earth's atmosphere comprises an essentially uniform mixture of gases but with substantially varying proportions of water vapour, concentrated mainly in the troposphere (Table 2.1).

The Sun has a direct and important influence on the atmosphere–ocean system; indeed they are literally solar-powered. Short-term changes in global climate, on a scale from tens of thousands to hundreds of thousands of years, appear to be a result of slight changes in the distribution and amount of solar radiation, or **solar flux**, reaching the surface of the Earth. Such changes

Box 2.1
Ice cores

Antarctica, drilled by the Soviets in the early 1980s and analysed jointly with French scientists, recording the past 160,000 years;

4 a 3,029 m long core drilled from 1990 to 1992 on the summit of the Greenland ice sheet, to its base, by the Greenland Ice Core Project (GRIP), under the aegis of the European Science Foundation (with researchers from Belgium, Britain, Denmark, France, Germany, Iceland, Italy, and Switzerland), going back 250,000 years – the first to contain information from two ice ages and the intervening three warm interglacials.

After the burial and compaction of snow and its transformation to ice, the layers of ice may be subject to disturbances because of ice flow, tensional stresses in the ice, and exhumation by

the stripping away of younger layers to form an ice surface. Thus, the dating of ice cores requires considerable care. Ice cores are dated using various techniques. The latest GRIP core was dated to 14,500 years ago by counting the annual layers. The counting was made possible by the acid and dust content of the ice core. Summer snow contains peak amounts of acid, whereas dust content peaks during the winter and spring seasons. For the GRIP core, calculations using two well-dated 'fixed points' were employed to calibrate the rest of the ice core record, that is, the cold period about 11,500 years ago (called the Younger Dryas) which followed the last glaciation, and the very cold interval 113,000 years ago, after the Eemian interglacial.

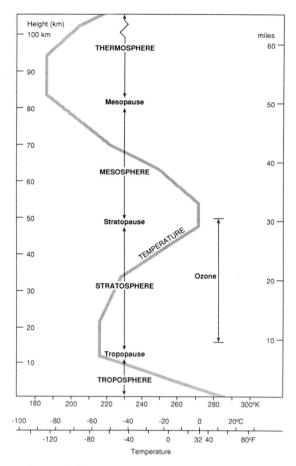

Figure 2.2 *Thermal structure of the atmosphere to a height of approximately 110 km (68 miles).*

in solar flux result from variations in the orientation and proximity of the Earth to the Sun. These factors can be thought of as external controls on climate.

Broecker and Denton (1990) advocate a mutual interaction between global climate and ocean current circulation. They suggest that warming in the northern hemisphere prompts biological activity, and the consequent production or release of carbon dioxide from the oceans to the atmosphere. In turn this changes the ocean circulation, together with the way in which heat energy is transferred through the oceans. Such changes in the thermal structure of the oceans induce the formation of the **North Atlantic Deep Water (NADW)**, a deep ocean current that is currently active, but did not flow as strongly during glacial times. The formation of the NADW involves the upwelling of north-flowing waters of high salinity from depths of about 500 m, and as these cold waters rise to the

surface they replace the warmer surface waters that flow southwards, aided by the strong winter winds (Figure 2.5). As the surface water travels northwards, it loses heat energy and cools, which together with its high salinity leads to an increase in water density, and it therefore begins to sink to abyssal depths (in the vicinity of Iceland) and then flow south, across the equator, towards Antarctica and into the Pacific Ocean. This 'Atlantic Conveyor', as it has become known, releases vast amounts of heat energy during this process, approximately equivalent to about a quarter to one third of the direct input of solar energy to the surface of the North Atlantic. The volume of flowing water is immense, roughly equivalent to 20 times the combined flow of all the world's rivers. Scientists now believe that towards the end of a glacial period, when the NADW begins to form, it fashions a different pattern of global oceanic circulation, and redistributes the heat energy in a manner different from that of the present day. Such changes in ocean circulation and heat exchange between the oceans and atmosphere may have had a profound effect on global climate and help drive the rapid climatic changes (see also the summary in Street-Perrott and Perrott 1990).

Beside **Milankovitch cyclicity** – the way in which the orbital parameters of the Earth moving around the Sun influence global climate change (see Box 2.2) – it has also been suggested that climatic fluctuations on a time scale up to thousands of years may result from variations in solar activity, generally measured as a function of **Sun-spot** activity. Records of Sun-spot activity since about 1700 show a cyclicity of roughly 11 and 100 years. By dating samples of wood using the radioactive isotope of carbon, [14]C (produced in the atmosphere by the interaction of cosmic rays and atoms of the nitrogen isotope ^{14}N), a 9,000-year record of solar activity has become available to us. During periods of increased solar activity, more particles are emitted from the Sun as a solar wind which effectively holds back more of the cosmic rays and, therefore, less ^{14}C is produced in the Earth's atmosphere. Data gathered during the last 200 years show that variations in Sun-spot activity closely correlate with the ^{14}C record. Correlating Sun-spot cycles with historical data has led to uncertainties and conflicting views about the cause of short-term fluctuations in global climate. Tropical temperature records for the period 1930–50, for example, show a positive correlation with Sun-spot activity, but a negative correlation between 1875 and 1920. It has been suggested that there could have been a correlation for this latter time interval, but that it is masked by variations in stratospheric ozone concentrations. Ozone appears to be more abundant about two years before a Sun-spot minima resulting in strato-spheric warming which, in turn, weakens the subtropical anticyclones and **Westerlies**. Cool and dry weather then follow slightly out of phase with the Sun-spot cycle.

The cold winters of the **Little Ice Age** have been correlated with 100-year Sun-spot cycles, corresponding with a so-called 'quiet Sun' or 'Sun-spot minima'. Similar low winter temperatures occurred during the nineteenth century. Sun-spot maxima correlate with high annual temperatures. It is predicted that the twenty-first century will be a Sun-spot minima, whereas the twentieth century is currently in a Sun-spot maxima. The Earth may, therefore, return to Little Ice Age conditions during the next century, if this is not offset by human-induced global warming (Thompson 1992).

At the first ever joint meeting between the (British) Royal Society and the French Academie des Sciences in London in February 1989, Charles Sonett (University of Arizona, USA) suggested that the ^{14}C record shows a dominant 200-year cycle, modulated by shorter 80–90-year cycles and longer 1,000-year and 2,300-year cycles. The 200-year cycle in ^{14}C may well account for the Little Ice Age recorded throughout Europe in the seventeenth century, which is linked to a quiet period of Sun-spot activity. Other element isotopes produced by the bombardment of cosmic rays with particles in the Earth's atmosphere show cyclic variations in abundance. Beryllium in the isotope beryllium-10 ([10]Be) forms in this way and settles to the ground unabsorbed by living organisms, and in cores from the Antarctic it shows a cyclic varia-tion in abundance of about 194 years – close to the 200-year cycle interpreted from the ^{14}C record.

Measurements of atmospheric turbidity, for most practical purposes considered as an indi-cation of atmospheric dustiness or dirtiness, have shown enormous increases (30–50 per cent) since the beginning of this century. Bryson

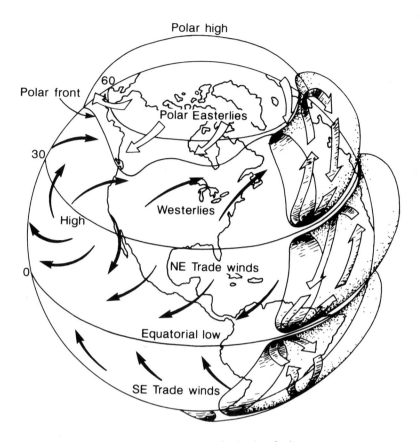

Figure 2.3 *Idealized global atmospheric circulation.*

Table 2.1 *Composition of the atmosphere.*

Constituent	Chemical formula	Abundance by volume*
Nitrogen	N_2	78.08%
Oxygen	O_2	20.95%
Argon	Ar	0.93%
Water vapour	H_2O	variable (%–ppmv)
Carbon dioxide	CO_2	340 ppmv
Neon	Ne	18 ppmv
Helium	He	5 ppmv
Krypton	Kr	1 ppmv
Xenon	Xe	0.08 ppmv
Methane	CH_4	2 ppmv
Hydrogen	H_2	0.5 ppmv
Nitrous oxide	N_2O	0.3 ppmv
Carbon monoxide	CO	0.05–0.2 ppmv
Ozone	O_3	variable (0.02–10 ppmv)
Ammonia	NH_3	4 ppbv
Nitrogen dioxide	NO_2	1 ppbv
Sulphur dioxide	SO_2	1 ppbv
Hydrogen sulphide	H_2S	0.05 ppbv

*ppmv = parts per million by volume; ppbv = parts per billion by volume
Source: After Henderson-Sellers and Robinson (1986).

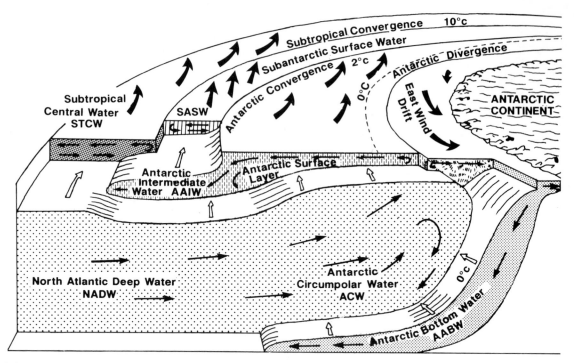

Figure 2.4 *Schematic illustration to show the principal water masses in the Southern Ocean in proximity to Antarctica. The water masses have different temperatures and densities, and move as discrete currents. There is upwelling of cold, nutrient-rich water where the surface currents diverge, whereas 'downwelling' takes place where currents converge. The Antarctic Bottom Water (AABW) flows into the Atlantic Ocean. After Williams* et al. *(1993).*

(1968) estimates that these increases may lead to a 3.5–6.5°C lowering of the Earth's surface temperature. During the 1960s, it was from evidence such as this that the view that the Earth may return to glacial conditions within the near future was fostered. Ironically, any cooling this might initiate will probably be offset by warming induced by anthropogenically produced greenhouse gases. Today, however, there is more concern about global warming than cooling.

Large igneous provinces and climate change

Over time-intervals measured in tens of millions of years, global climate is strongly influenced by the amount of new oceanic crust being produced at oceanic spreading centres (such as the Mid-Atlantic Ridge or the East Pacific Rise, linear, mainly submarine, mountain chains, and associated central depressions or graben formed by the extrusion of new and warm basaltic magmas

and lavas), and also from so-called mantle plumes. Mantle plumes rise diapirically through the Earth's mantle, and are caused by the detachment of mantle melts or magmas from depths in the Earth of between 670–650 km (the transition between the lower and upper mantle), and possibly even from sources as deep as the core-mantle boundary in the Earth to produce so-called 'super-plumes'. At the Earth's surface, the expression of such mantle plumes is the eruption of large volumes of basaltic igneous rocks to produce so-called 'large igneous provinces' with diameters of up to about 1,400 km. Mantle plumes are about 200°C hotter than the surrounding mantle through which they rise, and therefore are commonly associated with large-scale uplift or doming of the Earth's crust.

An ancient example of a large igneous province produced by a mantle plume acting like a blow torch on the base of the Earth's crust is the 'Tertiary North Atlantic Igneous Province', represented above sea level by parts of Iceland and north-west Scotland, and which was erupted over a very short geological time-interval

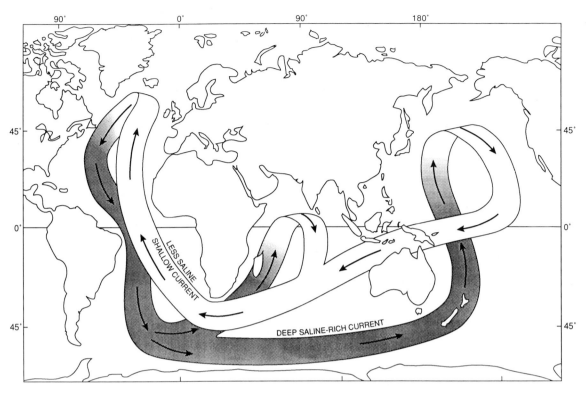

Figure 2.5 *The thermohaline (heat-salt) conveyor belt in the oceans. Dark bands show flow of deep, cold, and salty water; light bands show return surface flow. The deep currents begin in the North Atlantic, in the East Greenland Sea, then move southwards from the Atlantic into the Pacific Ocean. The upper, warmer current may begin in the tropical seas around Indonesia, and includes the strong flow out of the Gulf of Mexico.*

approximately 55 million years ago. Other examples include the Ontong-Java Plateau in the western Central Pacific Ocean where an estimated 12–15 km³ of igneous rock were erupted annually, or the approximately 65 Ma Deccan Plateau basalts, India, where an estimated 2–8 km³ of igneous rock was erupted annually (Coffin and Eldholm 1993). Given that the estimated global network of mid-ocean ridges has produced 16–26 km³ of new oceanic crust each year over the past 150 million years, these large igneous provinces have created new crust at rates comparable to, or greater than, that of seafloor spreading. It has been estimated that a single flood basalt event which generates 1,000 km³ of lava, typical of the 16 Ma Columbia River igneous province in the western USA, is associated with the emission of 16×10^{12} (trillion) kg CO_2, 3×10^{12} (trillion) kg of sulphur, and 30×10^9 (billion) kg of halogens (e.g., F, Cl, Br) (Coffin and Eldholm 1993). Since large volumes of gases such as CO_2 and SO_2 are emitted from the Earth's mantle, any dramatic increase in the rate of generation of oceanic crust and associated

mantle degassing (and/or accelerated global igneous activity) over short time-intervals will have a profound forcing effect on global climate. A good example of this effect occurred during the Cretaceous period of Earth history, where igneous activity peaked around 120 million years ago, with very large-volume volcanic activity centred in the Pacific Ocean basin. This Cretaceous igneous activity appears to have been associated with a greenhouse period of Earth history, when global mean annual temperatures were much warmer than today (in the order of 10°C warmer), global sea level was higher (by more than 100 m), and organic-rich black muds accumulated in many parts of the world's oceans in oxygen-poor waters – created by the decreased rate of ocean-current circulation in the warmer climate and, therefore, its reduced ability to dissolve oxygen and ventilate the world's oceans.

Many of the large igneous provinces appear to be associated with large-scale or mass extinction events in Earth history. For example, the biggest extinction event known throughout Earth history occurred 248 Ma when about 95 per cent of all marine species were wiped out in a mass extinction event which coincided with the eruption of the voluminous Siberian Traps, a major igneous province. While a large meteorite impact may have been the principal cause, the eruption of the Deccan Traps, about 65 Ma, may have contributed to the major extinction event that witnessed the demise of the dinosaurs.

Modelling global climate and climate change

In attempts to understand better the nature of global climate change, scientists are developing computer models to replicate present climatic conditions, and to predict future changes in climate. Climate models are used by many research groups to evaluate the effects of the various positive and negative feedbacks that can influence climate change (Figure 2.7). In effect, such models are less sophisticated versions of the weather forecasting models which appear on the

world television networks. The various computer-based general circulation models represent the atmosphere as a finite number of stations both in geographic locations around the world, and three-dimensionally as vertically stacked points in the atmosphere. In many GCMs, the oceans tend to be represented as stations with a defined sea-surface temperature, although more sophisticated models are beginning to divide the ocean into vertical slices. From all these atmosphere–ocean stations, a three-dimensional grid of points is fed into a computer program, whose physical states are mathematically linked to neighbouring points. The computer program then runs, and the numerical relationships are allowed to evolve in discrete temporal steps until predetermined conditions are satisfied – for example, a certain time period has evolved. Because the more sophisticated computer programs require very large amounts of memory and relatively lengthy running times, super-computers are well suited to GCMs.

An important aspect of GCMs is that they are only models, and the output can only be as good as the data which are input – they are approxi-

Box 2.2

Milankovitch cyclicity

The Yugoslavian astronomer Milutin Milankovitch (Plate 2.4E) calculated how summer radiation at latitudes 55°, 60°, and 65° North varied during the past 650,000 years, then mailed his graphical results to the great German climatologist, Wladimir Koppen. Koppen immediately wrote back to Milankovitch to say that the data could reasonably be matched to the periodicity of the Alpine glaciations that were reconstructed by Penck and Bruckner some 15 years earlier. In 1924, the 'Milankovitch curves' were published in Koppen and Alfred Wegener's book *Climates of the Geological Past*, which allowed Milankovitch's work to reach a wide scientific audience. Milankovitch then began work on calculating radiation curves for eight latitudes ranging from 5°N to 75°N, and published this work in 1930 in a volume entitled *Mathematical Climatology and the Astronomical Theory of Climate Change*. Milankovitch was not to rest there. He set about the task of calculating just how much the ice sheets would respond to a defined change in solar flux or solar radiation, and the results were published in 1938 in a volume called *Astronomical Methods for Investigating Earth's Historical Climate*. In 1941, Milankovitch published the comprehensive results of his life's

work as a unifying theory linking the astronomical control on variations in the amount of solar radiation reaching the surface of the Earth and climatic change, in his book *Canon of Insolation and the Ice Age Problem*.

The work of Milankovitch and others has provided a major and fundamental contribution to the Earth sciences, where in an attempt to understand better the forcing mechanisms for global climate change, Earth scientists are utilizing various astronomical studies that reveal three scales of global climate change caused by temporal variations in the nature of the Earth's orbit around the Sun (Figure 2.6). These external controls or orbital factors are: (i) changes on a scale of about 19,000–23,000 years caused by variations in the Earth precession (the tilt of the Earth relative to the plane or ecliptic on which the Earth moves around the Sun); (ii) changes on a time scale of about 41,000 years caused by changes in variations in the obliquity of the Earth (tilt of the Earth's axis of rotation); and (iii) changes on a scale of 100,000 and 400,000 years caused by changes in the Earth's eccentricity (the shape of the Earth's orbit, cyclically changing from more circular to more elliptical and back again).

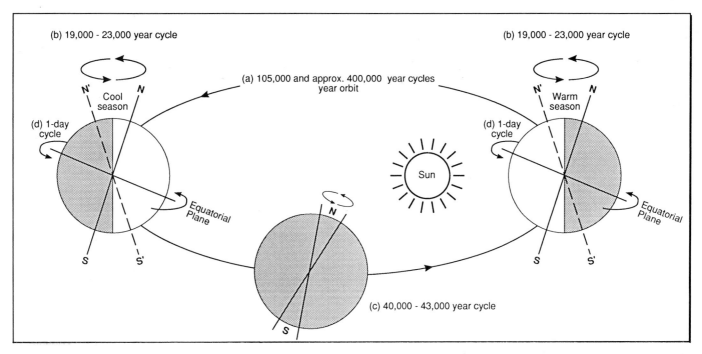

Figure 2.6 *The variability in the Earth's orbit around the Sun at various time scales measured in tens of thousands of years, and referred to as 'orbital parameters'. The temporal variation in orbital parameters causes long-term changes in the amount of solar energy reaching the surface of the Earth, which in turn can result in significant changes in global climate, referred to as Milankovitch cyclicity, so-named after one of the first people to propose a link between changes in the Earth's orbit and global climate. After Peltier (1990).*

mations to what may actually happen. For example, the atmosphere and oceans are continuous fluids, but they are represented as finite points in the model. In most GCMs, gridpoints typically involve horizontal separations of 500 km (100 km in more refined models), and with time steps of say 30 minutes. Cloud cover and cloud type, for example, are parameterized so that their evolution is described by substantial approximations to the physical and chemical processes affecting them, ideally in a manner that preserves the important spatially averaged properties of the variable. Ocean circulation and the way in which heat is transferred within the ocean–atmosphere system is a current area of research, and therefore not included in most GCMs. To improve GCMs, much more research is required such as **sensitivity analyses** of GCMs to many poorly understood variables, for example cloud types and cloud-forming processes, and heat transfer in the oceans.

As GCMs are developing, mainly for predicting future potential climate, Earth scientists are beginning to make use of such models to try and understand past climates. Currently, there are

three main GCMs: the Canadian Climate Centre model, the US Geophysical Fluids Dynamics Laboratory model, and the UK Meteorological Office model. Figure 2.8 summarizes some of the results from these GCMs. Much of the variation between the various GCM results is due to the different weightings given to various assumptions.

Volcanic activity and global climate change

Volcanic activity influences both long- and short-term global climate (Plate 2.2). On a scale of many millions to tens of millions of years, increased volcanic activity can emit enormous volumes of greenhouse gases and increase the rate at which new oceanic crust is generated at spreading centres, such as the present-day Mid-Atlantic Ridge or East Pacific Rise. Increased emissions of greenhouse gases can lead to substantial global warming. The enhanced production of thermally warm and buoyant oceanic

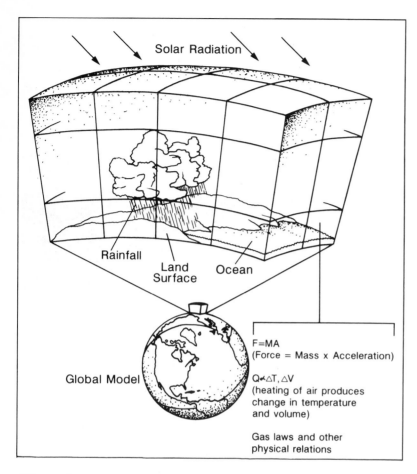

Figure 2.7 *Schematic diagram of global climate system, to illustrate the way in which the Earth's atmosphere–ocean system, and land surface area, is divided into thousands of boxes with sides typically extending several hundred kilometres in latitude and longitude, and with heights of a few kilometres in altitude. In a general circulation model (GCM), the computer treats each box as a single element as it calculates the evolving global climate. The GCM imposes seasonal and latitudinal changes of incoming solar radiation, the height and shape of the continents, and other external conditions which affect the behaviour of the atmosphere. In GCMs, for example, the equations may be solved in hourly increments over at least 20 years of simulated time to generate an output which is statistically 'accurate'. Such large and time-consuming calculations require the use of super-computers. After Ruddiman and Kutzbach (1991).*

crust causes a shallowing in the mean water depth in the oceans which, in turn, leads to a flooding of the land surface, seen as a rise in sea level. Both these effects occurred together during the Cretaceous period of Earth history, with the result that during that greenhouse phase global or eustatic sea level was up to a couple of hundred metres higher than at present.

In the shorter term, down to periods of a year, volcanic eruptions can eject large volumes of gases and ash which have relatively short-term effects on climate. Large eruptions can pump out enough ash into the higher levels in the atmosphere to cause a reduction in the solar flux to the Earth's surface.

Figure 2.9, based on data from the North Atlantic sediment core V23-82 and on oxygen isotope stages, summarizes the major volcanic eruptions during the Late Quaternary in relation to summer sea-surface temperatures. There is also an expanded part for the past 100 years, showing the relationship between major historic volcanic eruptions and the Northern hemisphere mean annual temperature anomaly. From this figure, it is possible to infer that immediately following some major volcanic eruptions, there is a drop in mean annual temperature, for example associated with Krakatoa and Mount Pelée. Lamb (1972) noticed that the wettest and coldest summers over the past three centuries coincided with time intervals of enhanced volcanic activity and, also at such times, Arctic sea-ice appears to have been more extensive and persistent.

The increased volcanic activity in the late 1940s and mid-1960s could account for the cold winters during this period. The eruptions of Mount St Helens in the USA (June 1980) and El Chichon in Mexico (April 1982) appear to have exerted only a short-term effect on reducing solar radiation, thus not all volcanic activity causes significant climatic change. The increased use of motor vehicles from the late 1940s onwards may also have increased atmospheric dust by **combustion-dust loading**. This may have been more important in cooling the winters during the middle of this century. Today, motor vehicle emissions have been greatly reduced in developed countries by improved legislation, but as anyone who has travelled in the developing world knows, there is little or no pollution control in big cities such as Delhi, Bangkok, and Beijing.

During the Late Quaternary, the Toba eruption in northern Sumatra, dated at *c.* 73,500 years BP, was probably the largest volcanic eruption, by order of magnitude (Chesner *et al.* 1991). The eruption has been correlated with the oxygen isotope stratigraphy (Ninkovich *et al.* 1978). Ash from the Toba eruption was transported up to 2,500 km west of Sumatra and deposited on land as far away as India (Stauffer *et al.* 1980, Ninkovich *et al.* 1978). A total of 10^{15} g each of fine ash and sulphuric acid were believed to have been emitted (Rampino and Self 1992). It is argued that the eruption of such large amounts of ash led to an increase in atmospheric turbidity and global cooling in the order of 3–5°C over a period of several years. This may have initiated rapid ice growth and corresponding lowered global sea levels which, in turn, could have enhanced global cooling and greater sea level falls attributed to the transition from oxygen isotope Stage 5a to Stage 4. Rampino and Self (1992) emphasize, however, that the Toba eruption occurred after the start of global sea-level fall in the transition of Stage 5a to 4, suggesting that other factors were important in initiating the global climatic shift to cooler conditions. The Toba eruption, however, at least appears to have provided a contributory causal factor which probably helped drive global cooling.

Ninkovich *et al.* (1978) and Fisher and Schmincke (1984) have postulated that the column height of the **tephra** from the Toba eruption may have reached an altitude of 50–80 km, although others have suggested more modest heights of 27–37 km (Rampino and Self 1992). Wood and Wohletz (1991) have argued that the eruption may have produced an **ignimbrite** eruption column which could have reached heights of 32–23 km. These lower eruption height estimates support the arguments that the mass of sulphuric acid injected into the atmosphere may be more important in influencing global climate change than the actual physical power of the eruption. In order to quantify the role of sulphuric acid **aerosols** in influencing global climate change, further research is necessary.

Continental positions, mountain-building events and global climate change

The very long-term changes in global climate, over hundreds of millions of years, appear to be strongly controlled by the position of the continents. As the plates which make up the outer surface of the Earth relentlessly move around, at speeds typically measured in millimetres to centimetres per year, so the size and position of

Plate 2.2 *Mount Fuji, Honshu, Japan. Volcanic eruptions may cause global climate change, both in the short and long term.*

continents or land area changes. The theory which explains the movement of these plates is known as **plate tectonics**. At times in Earth history, there have been super-continents (e.g., with names such as **Pangea** and **Gondwana**) when many continental plates were locked together. At other times, the distribution of continents has been more like it is today, with many large continents separated by oceans. The size and distribution of these continents, for example centred over polar or equatorial latitudes, profoundly affects global climate, as does the rate at which ocean basins floored by oceanic crust are created – on a time scale measured in tens of millions of years. At times when there was fast production of new oceanic crust at **mid-ocean ridges** (or spreading centres), greater amounts of heat energy were released from within the Earth together with more greenhouse gases. The result of this enhanced heat exchange between the solid Earth and hydrosphere–atmosphere–biosphere is that it could have caused past greenhouse periods in the Earth's history. These factors can be thought of as internal controls, that are entirely a consequence of processes within the Earth's heat engine.

Some scientists believe that mountain-building episodes can give rise to ice ages. Ruddiman and Kutzbach (1991), and more recently Raymo and Ruddiman (1992), for example, have proposed that the uplift of Tibet,

(a) DJF $2 \times CO_2 - 1 \times CO_2$ surface air temperature: CCC

(b) DJF $2 \times CO_2 - 1 \times CO_2$ surface air temperature: GFHI

(c) DJF $2 \times CO_2 - 1 \times CO_2$ surface air temperature: UKHI

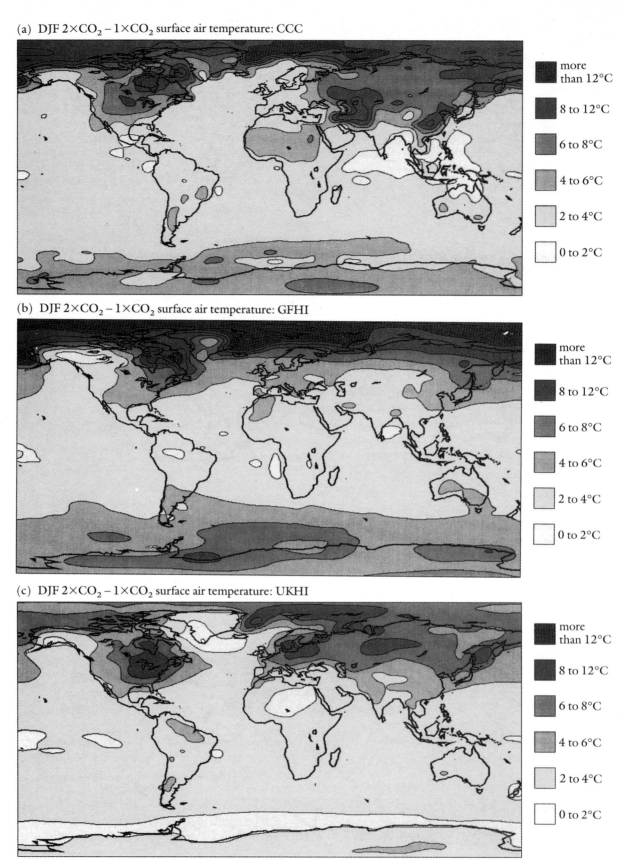

more than 12°C

8 to 12°C

6 to 8°C

4 to 6°C

2 to 4°C

0 to 2°C

Figure 2.8 *GCM output. The change in surface air temperature (10-year means) due to doubling carbon dioxide for (left side) December–February, and (right side) June–August, as simulated by three*

(d) JJA 2×CO$_2$ – 1×CO$_2$ surface air temperature: CCC

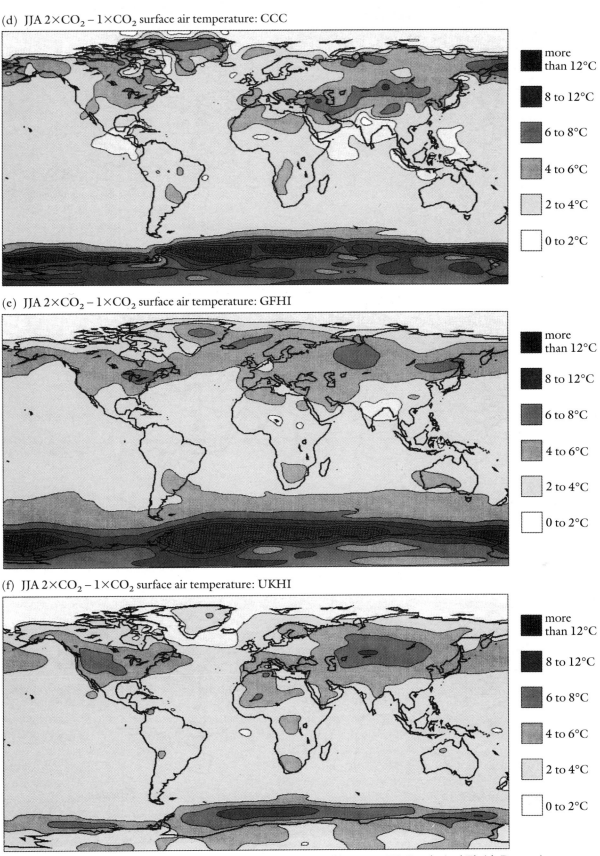

■	more than 12°C
■	8 to 12°C
■	6 to 8°C
■	4 to 6°C
■	2 to 4°C
□	0 to 2°C

(e) JJA 2×CO$_2$ – 1×CO$_2$ surface air temperature: GFHI

■	more than 12°C
■	8 to 12°C
■	6 to 8°C
■	4 to 6°C
■	2 to 4°C
□	0 to 2°C

(f) JJA 2×CO$_2$ – 1×CO$_2$ surface air temperature: UKHI

■	more than 12°C
■	8 to 12°C
■	6 to 8°C
■	4 to 6°C
■	2 to 4°C
□	0 to 2°C

high-resolution models: (a) CCC: Canadian Climate Centre, (b) GFHI: US Geophysical Fluids Dynamics Laboratory, and (c) UKHI: United Kingdom Meteorological Office. After IPCC (1992).

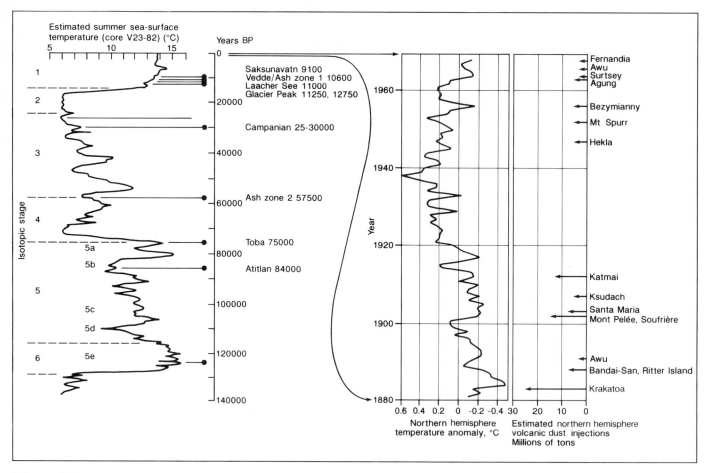

Figure 2.9 *Timing of major volcanic eruptions in the Late Quaternary in relation to summer sea-surface temperatures from North Atlantic core V23-82 and oxygen isotope stages, and in relation to historical records of the Northern hemisphere temperature anomalies during the last 100 years. Estimates of the amounts of ejected volcanic dust are given for eruptions during that period. Notice the close coincidence between major volcanic eruptions, rapid changes of sea-surface temperatures and negative temperature anomalies. After Dawson (1992), and Decker and Decker (1989).*

the Himalayas, and the American south-west, caused large areas of land in low latitudes to reach a height that altered global atmospheric circulation patterns, which helped induce global atmospheric cooling (Plate 3). In addition, they argue that increased uplift exposed more rock which then underwent accelerated rates of chemical and physical weathering. During many weathering reactions, CO_2 is extracted from the atmosphere to react with the decomposing minerals and form bicarbonates. These bicarbonate compounds are soluble in water and were carried in solution finally to be deposited as sediments in the oceans. Also, the uplift increased river gradients causing the rivers to erode more deeply and carry sediment to the sea at greater rates, and the uplift could have increased storminess along the mountain front leading to more rainfall, and

faster-flowing rivers. In essence, there is a net removal of CO_2 from the atmosphere during the chemical reactions associated with the breakdown of rock-forming minerals, a process that can therefore reduce any potential greenhouse warming, and hence encourage a global cooling. Such tectonic processes of mountain-building, or orogeny, could provide a negative feedback to the ocean–atmosphere system.

Control on global climate by micro-organisms in the world oceans and seas

There are scientists who believe that as global warming commences, marine **plankton** may show a multiplying effect. As a counterpoint, there are also equally eminent scientists who believe that as atmospheric CO_2 levels begin to rise, the rate at which the marine plankton absorb the greenhouse gas CO_2 may actually decrease with the result that the rate of warming increases. This latter scenario is an example of a positive feedback mechanism. Since the oceans contain about 20 per cent more carbon than the total land plants, animals, and soil, the oceans with their biota probably represent the principal factor in controlling global atmospheric CO_2 levels. At present CO_2 released by human activities adds about 7 ± 1.2 gigatonnes of carbon per year (GtC a^{-1}) to the atmosphere, of which about 2 GtC a^{-1} is believed to be sequestered in the oceans, and in a steady-state, phytoplankton fix about 35–50 GtC a^{-1}, representing a significant part of the natural carbon cycle (Falkowski and Wilson 1992). A current source of considerable scientific debate is focused on the potential ability of changing ocean productivity to sequester, or 'draw down', any increased (anthropogenically created) CO_2 in the surface waters and, therefore, act as a buffer on global climate change. Records of mainly coastal water data, spanning the period 1900 to 1981 for the North Pacific, indicate that although very minor changes in phytoplankton **biomass** have occurred over the 70-year time interval, they are too small to have had a significant effect on the rise in atmospheric CO_2 concentrations (ibid.).

Unfortunately, this 'multiplier' effect is poorly researched and, in past GCMs, has tended not to be an important part of most computer models. Indeed, in 1989, the five principal computer-based models for predicting global climates did not take account of the positive feedback mechanism due to plankton – that is, the four programs in the USA and one in the UK at the Meteorological Office, Bracknell. Current models assume that 50 per cent of the CO_2 injected into the atmosphere as a consequence of the burning of fossil fuels is 'drawn down' into the oceans by marine plankton where it is stored. Clearly, the significance of plankton in controlling climate may well invalidate this assumption and lead to underestimates of global warming rates.

It is now believed that the past glaciations during the **Pleistocene** period ended with slight changes in the solar flux to the Earth's surface caused by variations in the Earth's orbit, known as Milankovitch cyclicity. Such small changes in the amount of solar energy reaching the Earth's surface were multiplied by the decreased ability of the marine plankton to absorb CO_2.

The glacial events, which lasted as long as 100,000 years, therefore switched off rather

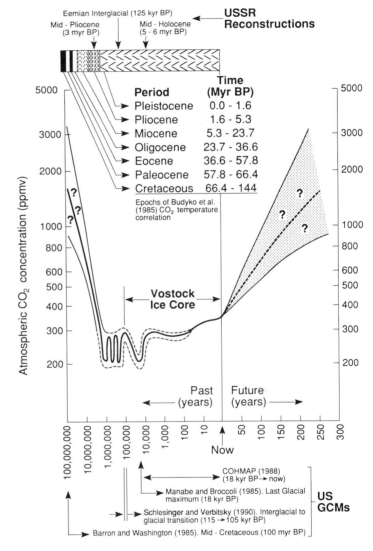

Figure 2.10 *Variations of atmospheric carbon dioxide concentration for the past 100 million years and the projected future 300 years. The upper scale (top left) shows only palaeoclimates as reconstructed by Russian researchers, keyed against a table showing the name and duration of each period. The lower scale shows periods simulated by general circulation models, both for the past (to the left of 'Now') and for the future (shown to the right of 'Now'). Note that the future scale is linear in contrast to the scale for the past which is exponential. Palaeoclimatic changes were at least partly due to the greenhouse effect led by fluctuating carbon dioxide levels. Human activities could create global greenhouse conditions similar to those which occurred naturally in the past. After Hoffert (1992).*

rapidly – for example ice cores from south Greenland revealed a 7°C rise in just 50 years following the last major glaciation (Dansgaard *et al.* 1993). Indeed, the idea that past increases in atmospheric CO_2 levels might be responsible for global warming was suggested by the research results of Shackleton (1977), where he showed that in cores from the Pacific Ocean, CO_2 gas levels increased after slight changes in the Earth's orbit but prior to the start of an increase in global temperatures. A major source of this CO_2 appears to be the marine plankton.

In central Antarctica, samples of ice taken from the Vostok ice core have provided one of the longest palaeoclimatic records, for example, including variations in atmospheric CO_2 concentrations which are set within a much longer geological time framework, and projected 300 years into the future, in Figure 2.10. It is also from the Vostok ice core that the first historical record of **biogenic** sulphur emissions from the southern hemisphere oceans has been gleaned. Legrand *et al.* 1991 have demonstrated that at the end of the last Ice Age, levels of methyl sulphonic acid, produced by marine plankton, decreased significantly at the same time as atmospheric CO_2 levels increased. These findings strongly support the role of plankton as a major factor in controlling atmospheric CO_2 levels and, therefore, climate. It has also been shown that the concentrations of methanesulphonate and non-sea-salt sulphate, products of the atmospheric oxidation of dimethylsulphide from plankton in the oceans, vary systematically over a complete 160,000-year **glacial–interglacial** cycle. During the later stages of the glacial period, there were increased oceanic emissions of dimethylsulphide compared to the present day. At around 13–14 ky BP, the end of the last glaciation, mean methanesulphonate levels changed from about 31 to 5 parts per billion by volume (pbbv), and non-sea-salt sulphate dropped from 222 to 102 ppbv. The enhanced producivity from the biota in the oceans, and correspondingly increased emissions of dimethylsulphide from the plankton, appear to have taken place between 18,000 and 70,000 years ago. So, the ocean–atmosphere sulphur cycle, linked to marine plankton, is extremely sensitive to global climate change. The biogenic aerosols play an important part in forcing global climate change by altering the cloud **albedo** and distribution, or because of their direct effects on absorbing and re-radiating solar radiation.

Techniques for studying past climates

Palaeoclimatologists looking back in time on a scale of hundreds of years have historical records as well as an enormous range of sophisticated scientific techniques to probe past climates. Many techniques are available, and their applicability depends upon the age of the sediments and fossils, and each is associated with varying degrees of confidence or error bars.

To interpret the record over hundreds of thousands of years, scientists have to rely on various subtle techniques, and obviously without recourse to human records. Looking even further back throughout geological time, on a scale of millions to hundreds of millions of years, the available data for confident climatic reconstructions become more uncertain, the techniques utilized become more subtle, and the assumptions made become critical. Despite the apparently impossible odds, Earth scientists are able to use a vast range of different data and techniques to interpret ancient climates (palaeoclimates).

Just what are the tools of the trade for deciphering past climates? Careful study of ancient sediments, which are now lithified sediments or rock, can show the type of environment that they accumulated in, for example desert, glacial, river, lake, coastal, shallow, or deep marine settings. Particularly diagnostic sediments include coals, **evaporites** such as **halite** and **gypsum**, **till**, carbonate reefs and **sedimentary ironstones**. If the sediments have a good magnetic record locked into the microscopic metallic mineral grains, then it may be possible to unravel their latitudinal position on the surface of the Earth when they accumulated, for example whether they were deposited in the equatorial, temperate, or polar regions.

Fossils

The remains of dead organisms (fossils) are extremely important in understanding ancient environments and past climates. Large colonies of reef corals, for example, suggest low-latitude/equatorial, warm, clear waters as off the Bahamas or Great Barrier Reef today. Fossils are also vitally important in helping to date ancient sediments accurately, something that is essential in any discussion on what the Earth's climate was like at various times in the geological past.

The analysis of pollen as an aid in the interpretation of **palaeoenvironmental** change is one of the most widespread methods adopted by palaeoclimatologists. Pollen grains extracted from ancient organic deposits such as peat provide information regarding changes in vegetation through time. Pollen grains are easily preserved because they are protected by a coat called **sporopollenin**. Pollen grains are identified under optical and scanning electron microscopes and the species determined by examining their shape and surface textures. The percentages of different pollen grains are estimated under the optical microscope and plotted graphically on pollen diagrams. This allows **palynologists** to determine the changes of vegetation down through a section of a sedimentary deposit, that is, through geological time.

Care must be taken when interpreting pollen data, as pollen may have travelled large distances, and such data do not necessarily represent the climatic conditions at the location they occur in. Also, pollen may be derived from older sediments, eroded out, and then redeposited with younger sediments and younger fossils. Nevertheless, with all these caveats, a careful pollen analysis provides an excellent tool for interpreting palaeoenvironments, especially when many pollen sites are compared with other kinds of palaeontological information which include fossil mosses, **diatoms**, and insects.

Over the last 20 years, fossil insects have provided an exciting new means of studying environmental change throughout the Quaternary. These include bugs, flies, bees, dragon flies, and beetles. The **Coleoptera** provide the best value because they have very robust **chitinous exoskeletons** that tend to survive with their original chemical signature. They are well preserved in a wide variety of deposits, and they can often be identified from isolated fragments of the body including head, thorax, wing covers, and genitalia.

Unlike pollen, fossil beetles are commonly preserved at or in very close proximity to where they lived. They are the most studied and collected group of insects, colonizing almost every terrestrial, freshwater, and intertidal environment. Many species show a marked preference for a particular environment, where humidity, temperature, vegetation, water conditions, and substrate satisfy a rather limited range. They are, therefore, good indicators of palaeoclimate and particularly **palaeotemperature**. It has been shown that in north-west Europe subtle variations in temperature over the last 50,000 years, because of the **stadials** and **interstadials**, can be picked out by the dominance and presence of various beetle species (Coope 1986).

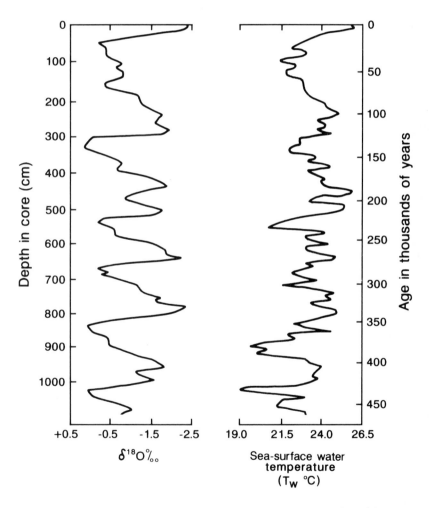

Figure 2.11 *Variation of sea-surface water temperature inferred from the oxygen isotopes recovered from the shells or tests of fossil floating, or free-swimming (planktonic), marine micro-organisms, expressed in parts per thousand and, by convention, expressed as $\delta^{18}O$ values based on the reference Caribbean core V12-122. After Imbrie* et al. *(1973).*

Most studies of past climates have focused on rock types that contain abundant fossils. There has been a tendency to neglect the ancient dry (arid) regions simply because they yield less data.

Chemical methods

The chemistry, including isotope studies, of sediments and fossils as a tool for trying to understand past climates, and estimating palaeotemperatures, oceanographic, and atmospheric conditions, is coming of age. Many chemical techniques are now available, and their use and interpretation is the subject of considerable current research. Box 2.3 summarizes the underlying chemical rationale behind using some of the most commonly employed isotopic techniques.

The calcareous skeletons of planktonic **foraminifera** are commonly chosen for isotopic analysis because these organisms live in surface waters and, therefore, they provide one of the best measures of surface-water temperature: in turn, sea-surface temperatures can be linked to global temperatures. Determination of past ocean temperatures, using isotopes, also involves an estimate of the volume of water stored in ice sheets and the oceans, and the change in volume per millionth part in isotopic composition of a shell which is precipitating out of seawater, and in isotopic equilibrium with the seawater. The calculations of ice volumes, however, are prone to many errors. Calculations of the water stored in ice sheets during the **Last Glacial Maximum**, for example, range from 47×10^6 km^3 (Dansgaard and Tauber 1969) to 100×10^6 km^3 (Craig 1965). Oxygen isotope curves predominantly represent fluctuations in the global ice and ocean volumes. Work on **benthic** foraminifera may be more truly representative of oceanic volume changes, since waters near the ocean floors remain relatively unaffected by local temperature changes, so that such temperature-dependent variations in oxygen isotopes can be disregarded (Dansgaard 1984, Shackleton 1987).

Detailed studies of ^{18}O values (see Box 2.3) in

Box 2.3

Chemical isotope methods in palaeoclimatology

Oxygen isotopes

The ratio of the heavier ^{18}O to lighter ^{16}O isotope in the remains of planktonic micro-organisms and benthic organisms, such as foraminifera, reflects the isotopic composition of seawater at the time their tests formed, assuming that the shells have not undergone any chemical alteration after reaching the seafloor and being buried. With some caveats, changes in isotopic composition reflect changes in the relative proportions of the lighter to the heavier isotope of oxygen locked up in ice sheets and glaciers giving a measure of global ice volume, which can be used to infer seawater (generally surface water) temperatures, and past global climate. Thus, the secular change in oxygen isotopes in fossils and sedimentary rocks can be used to infer past water temperature. During glacial periods, large volumes of seawater are locked up in polar ice caps. The lighter ^{16}O isotope is preferentially incorporated into the ice crystals, because water vapour formed by evaporation of liquid water is enriched in ^{16}O so that the global seawater becomes relatively enriched in ^{18}O. The marine organisms that secrete calcium carbonate ($CaCO_3$) shells using oxygen atoms from seawater will have varying ratios of ^{16}O to ^{18}O, which reflect changing polar ice volume or climate.

The isotopic composition of oxygen is expressed in terms of differences in ^{18}O/^{16}O relative to a standard called SMOW (Standard Mean Ocean Water), with reference to a large volume of distilled water distributed by the US National Bureau of Standards (NBS), such that: ^{18}O/^{16}O (SMOW) = 1.008 ^{18}O/^{16}O (NBS-1). The isotopic composition of oxygen in a sample is expressed as per mil (‰) differences relative to SMOW such that: $\delta^{18}O = [(^{18}O/^{16}O)_{sample} - (^{18}O/^{16}O)_{SMOW} / (^{18}O/^{16}O)_{SMOW}] \times 10^3$. Positive values of δ^{18}O indicate enrichment of a sample in ^{18}O, whereas negative values indicate depletion. The SMOW standard tends to be used for δ^{18}O values in waters and silicates, whereas for carbonate oxygen the PDB (Upper Cretaceous Peedee Formation belemnite fossil, South Carolina) standard is commonly used. $\delta^{18}O_{SMOW} = 1.03086 \, \delta^{18}O_{PDB} + 30.86$. The δ^{18}O in polar snow and ice depends principally upon the temperature of formation of the precipitation. The isotopic composition of oxygen in a carbonate sample is determined from the CO_2 gas obtained by reaction with 100% phosphoric acid, normally at 25°C. Using the oxygen isotopes ^{16}O and ^{18}O for palaeotemperature studies is also a well-tried and

marine **microfossils** over the past 350,000 years, in the Quaternary period, have revealed fluctuations in climate over time scales of tens to hundreds of thousands of years (Shackleton and Opdyke 1973, Chappell and Shackleton 1986). Figure 2.11 shows the variations in the sea-surface temperature calculated from the $\delta^{18}O$ values measured from a core collected from the Caribbean. The last Ice Age can be seen as higher ^{18}O values from just over 110,000 to 20,000 years ago. This isotopic signal thus provides a record of glacial and interglacial stages. By convention, odd numbered stages represent interglacials and even numbers glacials. The record shows that there have been more cycles than so far identified from other lines of evidence in the continents. It also shows that glacial stages are about five times longer than interglacials, and their terminations are rapid. Furthermore, the record shows small perturbations in the average climate during glacials and interglacials, that is, stadials and interstadials, respectively. For the last 15,000 years, there have been dramatic climatic changes on a scale from a few hundred to a few thousand years, spanning the deglaciation from the last glacial phase into the present interglacial.

Amongst the more recent ways of gathering a high-resolution record of past climate change through the Quaternary Period is the recovery of continuous ice cores from ice caps (see Box 2.1). Earth scientists can measure the chemical properties of trapped air bubbles, oxygen and deuterium isotopes (a heavy isotope of hydrogen), and dissolved and particulate material in the ice. Perhaps the best-known of these cores is the Vostok ice core, which was drilled in east Antarctica and recovered over several years from the Soviet station, Vostok. The ice core totalled 2,083 m in length and extended back in time 160,000 years (Barnola *et al.* 1987, Genthon *et al.* 1987). A study of the CO_2 in air bubbles trapped within the ice core has shown that during the last interglacial period, about 125,000 years ago, average atmospheric temperatures were probably around 2°C warmer than at any period since the ice sheets started melting approximately 18,000 years BP (Figure 2.12). During the last interglacial, it seems that

---

Box 2.3

Chemical isotope methods in palaeoclimatology

tested technique. The $\delta^{18}O$ values from marine shelly material made of calcium carbonate ($CaCO_3$) are routinely used to infer palaeotemperatures and palaeoclimates. Oxygen isotope composition of pre-Carboniferous (> 360 Ma) normal marine carbonates, cherts and phosphates (including fossil brachiopod shells), for example, suggests that early Devonian (ca. 390 Ma) low-latitude seawater was at 25 ± 7°C (Gao 1993), somewhat similar to modern oceans, at least for some of this time period. Similar high $\delta^{18}O$ values have also been obtained for older Ordovician and Silurian samples (Wadleigh and Velzer 1992).

In palaeoclimatology, past temperatures are calculated from isotopic data in carbonates (eg., Anderson and Arthur 1983) as follows: $T°C_{water}$ = 16.9 − 4.2 ($\delta^{18}O_{calcite\ PDB\ scale}$ − $\delta^{18}O_{water\ SMOW\ scale}$) + 0.13 ($\delta^{18}O_{calcite\ PDB\ scale}$ − $\delta^{18}O_{water\ SMOW\ scale}$)^2. There are other equations for aragonite and phosphate oxygen.

Carbon isotopes and changes in biomass productivity

The use of ^{12}C and ^{13}C isotopes for interpreting the photosynthetic strategy that fixes fossil organic matter is a relatively well-understood tool. The carbon delta ($\delta^{13}C$) value can even be used to study herbivore diets since the isotopic ratio is passed on to the grazing animal and is deposited in the animal's bone collagen, which has a greater preservation potential than the softer organic matter. Small organisms with shelly matter, such as snail shells, contain sufficient organic material to enable analysis of their palaeoclimatic signature. Indeed, snail shells have been used to extract carbon isotope signatures for understanding the climate in the Negev Desert, Israel, 3,000–4,000 years ago. Ancient, well-preserved bone material in fossil vertebrates makes it possible to interpret the climatic conditions under which that animal lived.

Measurements of $\delta^{13}C$ values from the CO_2 trapped in air bubbles in an ice core from Byrd Station, Antarctica, have shown that during the Last Glacial Maximum atmospheric concentrations of CO_2 were 180–200 parts per million by volume (ppmv), much lower than the pre-industrial values of about 280 ppmv.

Nitrogen isotopes

The isotopic ratios of nitrogen are just beginning to be utilized. Nitrogen which is fixed, for example, by symbiotic bacteria in leguminous

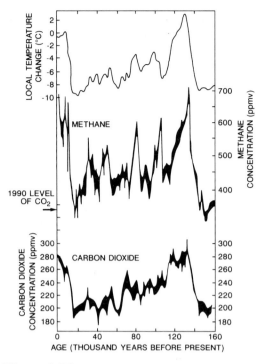

Figure 2.12 *Analysis of trapped air in the Vostok ice core to show the methane and carbon dioxide concentrations during the past 160,000 years. Notice the close correlation between methane and carbon dioxide with local temperatures over this period. After Lorius et al. (1988).*

the peak mean global temperatures could have been similar to those of the projected anthropogenically created greenhouse period.

Similar techniques have been applied by Earth scientists to the shells of microfossils going much further back in time to produce palaeotemperature curves stretching back 140 million years. The curves are derived from the data obtained from the shells of planktonic, near-surface organisms, and deeper-water species. From such graphs, it can readily be seen that seawater, and therefore mean Earth surface temperatures, were somewhat warmer 80–140 Ma, during the Cretaceous period of Earth history.

There are other ways of studying past climates beside using the isotopes of various elements. Earth scientists have used the concentration of toxic metals such as copper, zinc, and cadmium (see Box 2.3) in cores from the Antarctic ice to assess the past, natural, atmospheric conditions in the **troposphere** during the last 40,000 years (Batifol *et al.* 1989). The highest values of these toxic metals occurred during the Last Glacial Maximum of the last Ice Age some 25,000–16,000 years ago.

The source of these higher levels of copper, zinc, and cadmium, during the Last Glacial Maximum, are believed to be wind-borne dust

Box 2.3

Chemical isotope methods in palaeoclimatology

plants contains about the same $^{15}N/^{14}N$ ratio as the ambient atmosphere. Most non-symbiotic plants, however, possess up to five or more parts per thousand (ppt) ^{15}N. Thus, the nitrogen isotopic signature in fossil organic matter allows some insight into the contribution by nitrogen-fixing organisms to its decay or preservation. Nitrogen ratios may prove useful to palaeo-climatologists because the biological fixation of nitrogen described here tends to decrease as soils become drier.

An example of the use of nitrogen isotopes in studying past climatic–oceanographic conditions is in determining the causal factors for the formation of deep-marine layers (from cores collected in *c.* 1,375 m water depth) of organic-rich sediments (sapropels, with up to 4.5 per cent by weight organic carbon) in the eastern Mediterranean, from off the mouth of the Nile, during the Holocene and Upper Pleistocene to about 450,000 years BP (Calvert *et al.* 1992). The $\delta^{15}N$ record, which closely follows those of the organic carbon trends but as an inverse relationship, displays large and systematic variations, with an amplitude up to 9‰, and with

the systematically lighter values in the sapropels reaching 0.3‰ and the heaviest values being confined to the glacial stages. Amongst the possible explanations for the accumulation of the sapropels, the more plausible include: (1) enhanced preservation of organic carbon in anoxic bottom waters with reduced rates of renewal of the deep water, possibly due to lower sea level associated with the Last Glacial Maximum, and/or because of reduced salinity in surface waters linked to increased run-off of surface waters; (2) a greater flux of organic matter to the seafloor associated with increased primary production related to increased surface-water run-off. Calvert *et al.* (1992) found that the sapropels contain significantly lighter nitrogen isotope ratios ($^{15}N/^{14}N$) than the intercalated marls (calcareous muds). They concluded that the large differences could not be due to either variable mixtures of marine and terrestrial organic matter with different isotopic compositions, or to differences in the type and extent of post-depositional alteration. A terrestrial contribution to the sapropels is minor, since the $\delta^{13}C_{organic}$ values (mean 21.0 ± 0.82‰) are identical to those which would be preferentially concentrated in the troposphere during drier climatic conditions associated with glacial phases. There have been increased concentrations of these metals since about 13,500 years ago, however, which may be, at least in part, due to volcanic and/or biogenic activity, and increased metal contents in sediments over the last few centuries due to increased industrialization and pollution.

Tree rings and recent changes in climate

Studies of tree rings can be used to infer past climates. An example of this approach is the work undertaken by Earth scientists examining west European oaks and their tree ring characteristics back to 1851 (Kelly *et al.* 1989). Temperature, barometric pressure and precipitation (rainfall) data are available for the last 150 years or so from the study area.

The width of tree rings is related to the rate of growth, which in turn tells us something about the overall climatic conditions in any particular year. By studying many trees across a wide area, it is possible to see if there are years in which a significant proportion of trees show similar changes in growth-ring width. Using these techniques on west European oaks, it has been shown that the years in which there was greater growth of tree rings tended to be associated with enhanced cyclonic activity over the middle latitudes of western Europe, accompanied by an increase in precipitation. Temperature variations appear not to have played a significant role in the growth of the tree rings.

Changes in the growth rates of tree rings can be related to past climate. By studying the chemical isotopes of the **cellulose** in the tree rings, it is possible to interpret the past composition of the atmosphere and the hydrosphere. As a reliable and absolute time scale is developed, so this technique is becoming a very powerful means for understanding the changes in global climate brought about by the change from the last major glaciation (Pleistocene) to our present warmer (**Holocene**) period.

Tree-ring time scales are now being established which go back nearly 10,000 years. By using tree remains from the oak (*Quercus robur, Quercus petraea*) and pine (*Pinus sylvestris*) that have accumulated in the river terraces of south-central Europe, a team of German scientists have compiled a 'dendrochronological' (tree ring)

Box 2.3

Chemical isotope methods in palaeoclimatology

in plankton from the present Mediterranean, and there is no gradient in the isotope values in cores recovered at varying distances from the Nile river, the main source of any terrestrial sediment input. The variation, however, is consistent with a greater utilization of dissolved nitrogen during the accumulation of the sapropels, that is, the formation of the sapropels was associated with high productivity of plankton in surface waters causing a higher flux of organic matter to the seafloor.

Cadmium/calcium ratios and seawater temperatures

Studies of deep-ocean benthic foraminifera have demonstrated that there is a relationship between the amount of dissolved cadmium (Cd) in seawater and the Cd/Ca ratio in biogenic calcium carbonate (Boyle 1988), something that has also been shown for scleractinian corals from the Galapagos Islands (Shen *et al.* 1987). Other studies have confirmed that Cd/Ca ratios in fossil shell material can provide insights into past oceanic circulation and, therefore, palaeoclimates.

Upwelling of nutrient-rich waters in the oceans is driven by temperature differences between air masses over the land and oceans. These relationships have been used by van Geen *et al.* (1992) in a study of the Cd/Ca ratio in the shell of the benthic foraminifera *Elphidiella hannai* (from sediment cores in the mouth of San Francisco Bay) which is proportional to the Cd concentration in coastal waters, in order to calculate the past changes in mean upwelling intensity along the west coast of North America. The benthic foraminifera *Elphidiella hannai* inhabits coastal waters shallower than about 50 m along the west coast of North America. This study has revealed that the foraminiferal Cd/Ca ratio decreased by about 30 per cent from 4,000 years ago to the present day, probably because of a reduction in coastal upwelling. These changes were interpreted as reflecting the weakening of the northwesterly winds that drive upwelling, associated with the decreased summer insolation of the northern hemisphere by about 8 per cent over the past 9,000 years as a consequence of systematic changes in the Earth's orbit around the Sun.

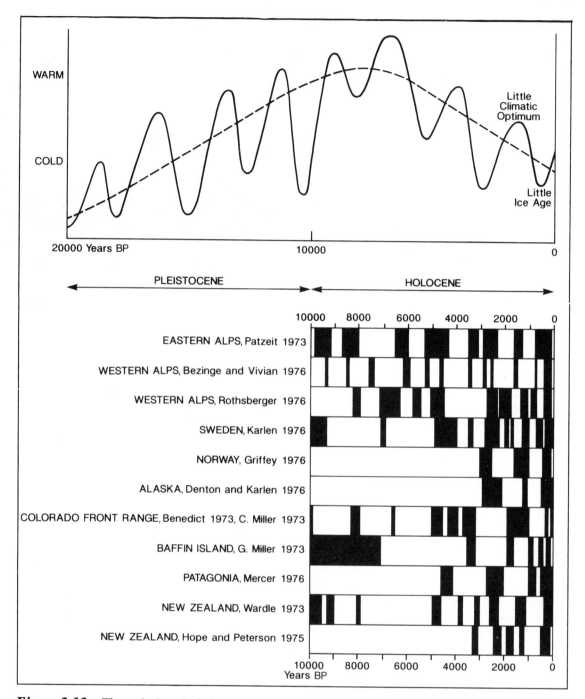

Figure 2.13 *The variation of relative temperature during the past 20,000 years and the advance of glaciers from selected regions of the globe (shaded). The dashed curve shows how solar radiation varied as a precession-related Milankovitch cycle for a latitude of 65°N. After Pielou (1991), and Grove (1979). (References cited in Grove 1979).*

record of the past 9,928 years and past 1,604 years, using the oak and pine, respectively (Becker *et al.* 1991). By calibrating these **dendrochronologies**, an absolute time scale can be established. Such correlations have led Becker

and his colleagues to suggest that the last significant cold phase (commonly referred to as the Late Glacial or **Younger Dryas**) must have ended at a minimum of 10,970 years BP.

The effects of volcanic eruptions on global

climate is recorded in tree ring signatures. Detailed explanations of the methodologies and examples of chronologies can be found in Fritts (1976) and Schweingruber (1989). LaMarche and Hirschboech (1984) were able to correlate frost rings in bristlecone pines in the western USA with major volcanic eruptions on a global scale. Baillie and Munroe (1988) correlated exceptionally narrow tree rings in Northern Ireland and California dating to 1627/8 BC with the eruption of Santorini in the Aegean Sea. This eruption was originally dated using Late Minoan Stage 1a pottery at about 1500 BC, although **radiocarbon dating** suggested a slightly earlier date (Bell and Walker 1992). The tree ring date is further confirmed by an acidity peak in the Dye 3 Greenland ice core (Hammer *et al.* 1987). It has been argued that the massive collapse of the Minoan civilization on Crete, 120 km away from Santorini, was related to this eruption (Watkins *et al.* 1978).

Using ice cores from Greenland, Hammer *et al.* (1980, 1981) have produced evidence for volcanic activity over the past 1,500 years. Their studies were based on the acidity levels in annual ice layers as established by electrical conductivity measurements, which reflect the amount of sulphuric acid washed out of the atmosphere in any one year – a function of the amount of volcanic aerosol present in the atmosphere at that time. By comparing their data with tree rings and isotope data, they were able to correlate the acidity with records of temperature variations in the northern hemisphere. The close correlation between ice core acidity and Late Holocene glacier variations led Porter (1986) to suggest that sulphur-rich aerosols emitted by volcanic eruptions are one of the main driving forces of global cooling.

Extent of glaciers, ice caps, landforms, and sediments

Particularly important in the study of palaeo-environmental change is the reconstruction of the former extent of ice bodies such as valley glaciers and ice sheets. During glaciations, when the Earth's climate was much colder, precipitation was dominated by snowfall. Over the years, the compacted and buried snow became thick

Plate 2.3 *Yosemite National Park, illustrating the evidence for former glaciations. The deep U-shaped and hanging valleys were once full with glacial ice which helped to erode them to their present form.*
Courtesy of K.C.G. Owen.

enough to change its structure and form glacier ice. As a result, valley glaciers and ice sheets formed, increased in size and flowed across the continents. These glaciers eroded the landscape and deposited glacial debris to form a rich variety of landforms. In response to the changing global climate, there have been many advances and retreats of glaciers, some of which may be globally synchronous, but others appear to have been more localized. Figure 2.13 shows such data for the past 20,000 years (adapted from Grove 1979, Pielou 1991), together with the relative temperature changes.

In some areas, ice sheets were very extensive. During the last glaciation, for example, the Laurentide ice sheet stretched from Banks Island southwards, flowing from three main ice domes which were located over the south-east of Hudson Bay, the north of Hudson Bay, and Keewatin. The glaciers in the Arctic were constrained in their high latitude by severe aridity and actually only advanced about 20–30 km southwards. During the last major glaciation, ice covered most of northern Europe, extending south to the north German Plain from the Fennoscandian ice sheet, and south to the English Midlands for the British ice sheet (Figure 2.14). It was from evidence such as this for the former extent of continental ice during past glaciations, partic-

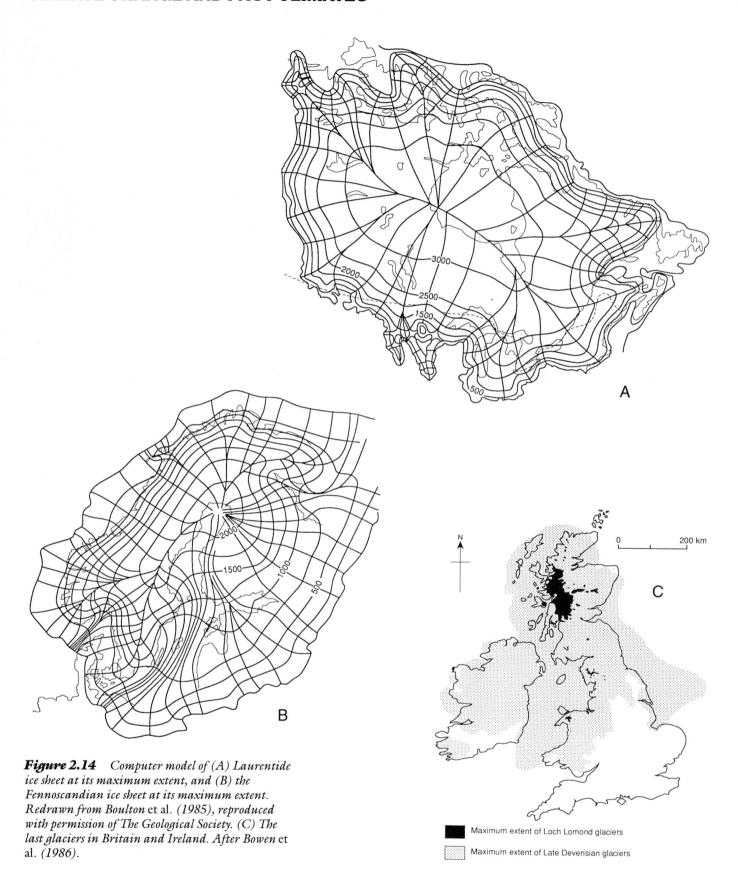

Figure 2.14 *Computer model of (A) Laurentide ice sheet at its maximum extent, and (B) the Fennoscandian ice sheet at its maximum extent. Redrawn from Boulton* et al. *(1985), reproduced with permission of The Geological Society. (C) The last glaciers in Britain and Ireland. After Bowen* et al. *(1986).*

■ Maximum extent of Loch Lomond glaciers

Maximum extent of Late Devensian glaciers

ularly on the continents of South America, Africa, Australia, and India, that led Alfred Wegener, in 1915, to propose that the continents had drifted around the surface of the Earth. Wegener used such information to reconstruct the supercontinent of 'Gondwanaland'. These ideas were embodied in his theory of **continental drift**, which provided many of the early ideas that were incorporated into the present theory of plate tectonics.

Mapping and geochronological dating of glacial landforms provides information on the former extent and temporal variation relating to past climate. Plate 2.2 illustrates the type of glacial erosional landforms which enable geomorphologists accurately to reconstruct former ice thicknesses and extents. Research has shown that several periods of ice advance can be identified for most high- and mid-latitude regions of the world. Many of these occurred at the same time, suggesting global changes in climate. Figure 2.14 shows the expansion of glaciers from selected parts of the world, and shows the degree to which glacial advance can be correlated. Of particular interest are the fluctuations during the past few centuries, especially during the seventeenth century which was a cold period known as the Little Ice Age (see Grove 1988). Christmas cards that use paintings from this time show a great deal of snow and ice – the picturesque white Christmas. Even the River Thames in London froze over.

From this and other types of data discussed earlier, it is possible to reconstruct, with a relatively high degree of accuracy, estimates of temperature changes over the past several hundred thousand years, from which it is possible to begin to understand the nature of changes in global climate.

Sea level change

During times when there are substantial ice caps on Earth, sea-level changes appear to occur at frequencies of 10^4–10^5 years, and with amplitudes from 10 to more than 100 m, resulting from the expansion and contraction of continental ice sheets, apparently at Milankovitch frequencies. A puzzle, however, has been to explain such fluctuations in global (eustatic) sea

Plate 2.4 *Rapid sea level change can result in the development and preservation of coastal features such as spits, barrier islands and lagoons, for example, as seen here at Chesil Beach in the UK.*

level even at times during Earth history when there appears to have been no significant continental ice (e.g., the Triassic, **Jurassic**, and Cretaceous Periods). Jacobs and Sahagian (1993) argue that these latter sea level fluctuations, producing smaller rises and falls in sea level (up to about 10 m), result from periodic (Milankovitch frequency) climate-induced changes in lake and groundwater storage.

Raised beaches and coral reefs provide important information regarding sea level changes throughout the Quaternary, and reflect the amount of water stored as glaciers during a glaciation, and the volume of water released into the oceans when ice sheets melted (Figure 2.15). If the entire Greenland ice sheet (with an estimated 2.82×10^6 km^3 of ice) melted, global sea level would rise by about 6 m. If the entire Antarctic ice cap melted, global sea level would rise by approximately 60 m. Additionally, raised shorelines may allow reconstructions of former ice thickness. This is because the growth of ice sheets and glaciers depresses the Earth's crust due to their extra weight. When the ice melts, the Earth's crust responds to the released stress by rebounding upwards, in a process known as **glacio-isostatic rebound**. In coastal areas, as the crust uplifts, coastal regions and **raised beaches** are uplifted to form raised shorelines. These raised shorelines can be dated by **radiometric** ages on fossil shells and other organic matter. In addition, rising sea levels may lead to the development of landforms such as spits, barrier islands, and lagoons. An example of one of the most spectacular constructional landforms in England is shown in Plate 2.4. This has formed since the last glacial as rising sea levels pushed sediment progressively inland to form an impressive coastal barrier.

It transpires that the amount of uplift is directly proportional to the thickness of ice. The uplift history, however, is complex because as the ice sheets melt, sea level also rises. To determine the absolute amounts of uplift, curves for global sea level changes have been constructed using shorelines and coral reefs in geologically stable areas that were not glaciated (Jelgersma 1966, Fairbanks 1989).

Further complications have to be taken into account in constructing global sea level curves, such as secular variations in the global **geoid** due to subtle changes in the Earth's gravitational field induced by plate tectonic processes, and the growth or decay of ice sheets.

One of the most common perceptions held by many scientists and non-scientists is that global warming will lead to the melting of polar ice sheets, with a concomitant rise in global sea level. Moderate temperature rises, however, could cause increased precipitation in high latitudes, resulting in greater amounts of water being locked up as snow on the polar ice caps. In the latter scenario, there would be a global fall in sea level. Snow accumulation rates in Antarctica are known to be dependent upon the mean annual air temperature above the surface **inversion layer** (Robin 1977), something that is consistent with the lower accumulation rates during the Last Glacial Maximum (Lorius *et al.* 1985). The total annual water budget for Antarctica is several times greater than that of Greenland, with the snow that falls on the grounded ice being equivalent to approximately 5 mm per annum of global sea level change (Jacobs 1992). Over the past century, the observed rise in global sea level has been in the range 1.0–2.4 mm per year, with a 'best guess' estimate of about 1.5 mm per year (IPCC 1990). At the lower limit, most observed sea level rises could be explained by the thermal expansion of the oceans, together with the melting of temperate and Greenland margin glaciers (ibid.).

Two lines of evidence may suggest the growth of polar ice sheets, that is, satellite altimeter measurements over Greenland (Zwally *et al.* 1989), and positive correlations between net snow accumulation and increased air temperature (Morgan *et al.* 1991). Satellite altimetry measurements are limited in duration, and can be compromised by a number of factors, including the changing distance from moisture sources. Jacobs (1992) concluded that it is too early to say whether the Antarctica ice sheet is shrinking or growing (Plate 4).

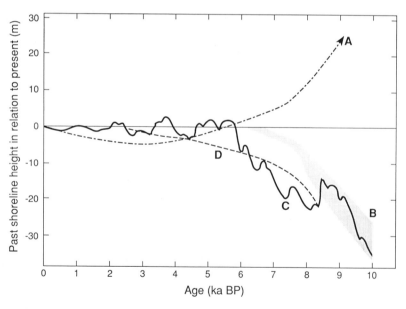

Figure 2.15 *Holocene sea level records for (A) Baffin Island, (B) eastern Australia, (C) a synthesis of several areas, and (D) Netherlands. These data are associated with error bars that are not shown, but emphasize local rather than global (eustatic) changes in sea level, although many areas show a rise in sea level due to post-glacial melting of ice and thermal expansion of the ocean waters. High-latitude areas, such as Scandinavia and Arctic Canada, show a relative fall in sea level (e.g., Baffin Island) associated with the isostatic rebound of the continents after removal of considerable thicknesses of ice by melting. After Williams* et al. *(1993).*

Quaternary climates

Historical perspective

Towards the end of the eighteenth century, Earth scientists such as the Scottish geologist, James Hutton (often referred to as the father of the science of geology: Plate 2.5A), and John Playfair were among the first to develop a theory of glaciation to explain many of the geological phenomena that were then explained by the **diluvial theory**. The glaciation theory had already been presented to the Swiss Society of Natural Sciences in 1837 by its young president, Louis Agassiz (Plate 2.5B), but was not destined to become widely accepted until the 1860s. There were a number of competing theories beside the diluvial explanation. In 1833, the English geologist Charles Lyell (Plate 2.5C) explained the features now known as being of glacial origin, such as erratics and drift deposits, as the products of floating icebergs. In 1840, however, the Reverend William Buckland, Professor of Mineralogy and Geology at Oxford University, and Charles Lyell eventually accepted the arguments by Agassiz for glaciation. Until then, Buckland had been committed to the **catastrophe theory**. Indeed, in 1863 Archibald Geike (Plate 2.5D) proposed a multiple glaciation hypothesis to explain the surficial glacial deposits in Scotland, a view which is generally accepted to this day.

As far back as 1909, the Alpine glaciations and interglacial periods were interpreted as alternating warm and cold stages by the German geographers Albrecht Penck and Eduard Bruckner. In 1842, the French mathematician Adhemar invoked changes in the orbit of the Earth around the Sun as being the main reason for such climatic change, while in 1864 the Scottish geologist James Croll hypothesized that changes in the Earth's orbital eccentricity could be the cause of the Ice Ages, a theme he elaborated upon in his book *Climate and Time*, published in 1875. Without a very precise means of dating the climatic changes and linking them to orbital parameters, these ideas lay dormant. It was not until well into the twentieth century, between 1920 and 1940, that these astronomical interpretations for climatic changes on Earth found support and widespread acceptance throughout the scientific community in the work of Milutin Milankovitch (Plate 2.5E) (see Box 2.2).

A brief history of the Quaternary

The Quaternary is defined by Earth scientists as the relatively recent period of geological time spanning the last 1.64 million years of Earth

Plate 2.5 *Portraits of selected scientists who have made fundamental contributions to the development of the science of palaeoclimatology:*
(A) James Hutton;
(B) Louis Agassiz;
(C) Charles Lyell;
(D) Archibald Geike;
(E) Milutin Milankovitch.
B, C and D courtesy of the Royal Geographical Society; A courtesy of the Department of Geology and Geophysics, University of Edinburgh; E courtesy of Vlaso Milankovitch.

history: many scientists studying the Quaternary would argue that this period should extend back to 2.6 Ma (discussed in detail by Shackleton *et al.* 1990). Floral and faunal evidence suggests that there was an abrupt change from warm to cold climatic conditions anywhere between 2.6 Ma and 1.64 Ma, depending upon which data are used. In 1985, at Vricia in Calabria, Italy, the International Commission on Stratigraphy formally defined the base of the Quaternary period as where a claystone horizon containing the first appearance of **thermophobic** foraminifera directly overlies a **sapropel**. The identification of ice-rafted debris in cores from the Antarctic deep sea, however, has placed the onset of glaciation as far back as 3.5 Ma (Opdyke *et al.* 1966), together with other lines of evidence, although dating of diatom-bearing **glaciomarine** strata in east Antarctica suggests that there was an extensive deglaciation of Antarctica at approximately 3 Ma (Barrett *et al.* 1992). In the North Atlantic region, a study of foraminifera linked to oxygen isotope data recovered from a deep-sea drilling site revealed evidence for the onset of glaciation associated with progressively deteriorating climatic cycles, and ice sheet initiation, at about 2.5 Ma (Shackleton *et al.* 1984). The last 10,000 years of this time interval, defined as the time period following the last glaciation, is referred to as the Holocene, and from 1.64 million to 10,000 years ago as the Pleistocene. Further back in geological time, at least five other major global Ice Ages are known, two in the late Precambrian (4600 Ma to 570 Ma) and three in the Phanerozoic (570 Ma to the present).

The last glaciation was a period of extreme cold on Earth when the polar ice caps were more extensive than today, much of the continents were covered by continental glaciers and ice caps, and sea level was much lower than at present. Following the Last Glacial Maximum, deglaciation in Antarctica was well advanced by about 10,000 years BP, and by 6,000 years BP was complete (Colhoun *et al.* 1992).

During past glaciations, global sea level was lower because large volumes of sea water were frozen as ice. It has been suggested that the role of Antarctic ice in contributing to global sea level fall at the Last Glacial Maximum was dependent on the thickness and extent of peripheral ice (ibid.). Models suggest that there was probably a thickening of 500–1,000 m which induced a sea level drop of about 25 m. Evidence from raised beaches in the Ross embayment and east Antarctica shows that sea level dropped by only 0.5–2.5 m. This discrepancy suggests that either sea level fell less than present estimates suggest, or that ice volumes in the northern hemisphere must have been considerably larger, to account for a global sea level lowering during the Last Glacial Maximum. The last major glaciation, when ice cover was at its maximum extent about 18,000 years ago, ended fairly abruptly as the mean temperature of the Earth's surface increased. Large-scale melting of polar and con-

tinental ice ensued and the return of this water mass to the oceans and seas, together with its thermal expansion, led to a rise in sea level of up to approximately 120 m in some parts of the world. Ice core studies at Vostok and Dome C, Antarctica, suggest that during the Last Glacial Maximum, the surface of central Antarctica was 200–300 m lower than at present (ibid.).

The last major glaciation is known by different names throughout the world, for example the **Devensian** in Britain, the **Wisconsin** in North America, and the **Weichselian** in mainland Europe. The Ice Age was not a single event, but a number of closely spaced cold-period glaciations (glacials) with durations in the order of 100,000 years separated by intervening warmer periods referred to as interglacials (as opposed to brief warm intervals within glacial stages, known as interstadials), lasting in the order of 10,000–20,000 years. During the Quaternary, this pattern of glacial and interglacial periods seems to have repeated itself at least 10 times. Indeed, a chronology of glaciations in the USA (Figure 2.16) reveals many more glaciations, even extending back into the Pliocene period at about 3 Ma.

Table 2.2 shows the correlations of synonymous names for the various Pleistocene phases in the northern hemisphere. Between about 11,000 and 10,000 years ago, there was a brief return to near-glacial conditions in the Younger Dryas. This event interrupted the change from the Pleistocene glacial to Holocene warmer climates. Indeed, a study of high-resolution ^{18}O isotope records from benthic and planktonic foraminifera in two radiocarbon dated cores from the Sulu Sea, western Pacific, has shown that the Younger Dryas was a global event which occurred synchronously and as far afield as in the surface and deep waters of the northern Atlantic Ocean and the Sulu Sea in the western Pacific, and was associated with low atmospheric CO_2 concentrations (Kudrass *et al.* 1991). The Younger Dryas was probably the result of the sudden increased rate of melting of the Laurentian ice sheet with large volumes of cool meltwater entering the oceans and affecting atmospheric temperatures. Recent studies on oxygen isotopes, in an ice core from Camp Century in Greenland, suggest that the Younger Dryas terminated very abruptly, possibly even within a few decades (Johnsen *et al.* 1992). In

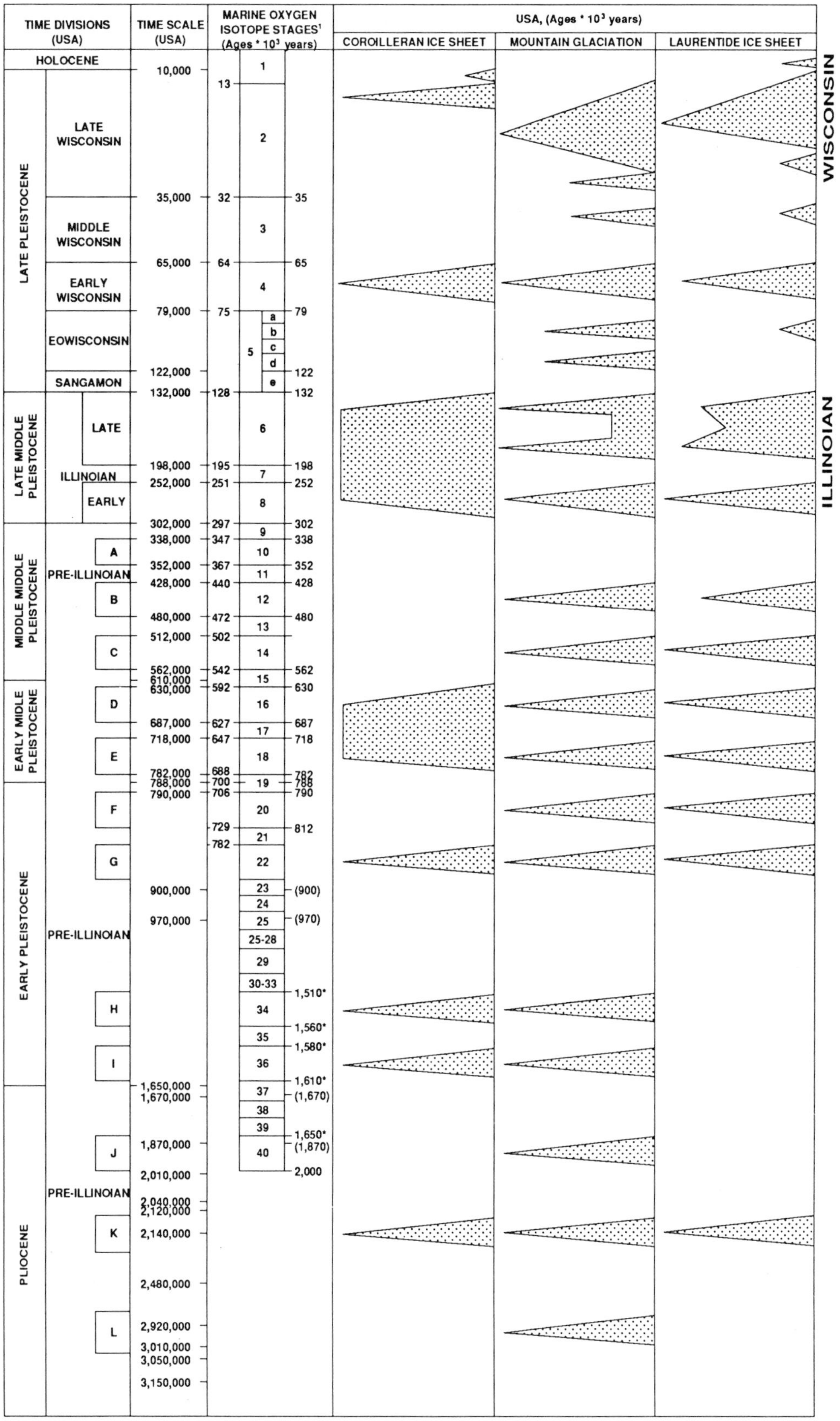
TIME DIVISIONS (USA)
TIME SCALE (USA)
MARINE OXYGEN ISOTOPE STAGES[1] (Ages * 10^3 years)
USA, (Ages * 10^3 years)
COROILLERAN ICE SHEET
MOUNTAIN GLACIATION
LAURENTIDE ICE SHEET
HOLOCENE
LATE PLEISTOCENE
LATE WISCONSIN
MIDDLE WISCONSIN
EARLY WISCONSIN
EOWISCONSIN
SANGAMON
LATE MIDDLE PLEISTOCENE
LATE
ILLINOIAN
EARLY
MIDDLE MIDDLE PLEISTOCENE
PRE-ILLINOIAN
EARLY MIDLE PLEISTOCENE
EARLY PLEISTOCENE
PRE-ILLINOIAN
PLIOCENE
PRE-ILLINOIAN
A
B
C
D
E
F
G
H
I
J
K
L
10,000
35,000
65,000
79,000
122,000
132,000
198,000
252,000
302,000
338,000
352,000
428,000
480,000
512,000
562,000
610,000
630,000
687,000
718,000
782,000
788,000
790,000
900,000
970,000
1,650,000
1,670,000
1,870,000
2,010,000
2,040,000
2,120,000
2,140,000
2,480,000
2,920,000
3,010,000
3,050,000
3,150,000
13
32
64
75
128
195
251
297
347
367
440
472
502
542
592
627
647
688
700
706
729
782
1
2
3
4
5
a
b
c
d
e
6
7
8
9
10
11
12
13
14
15
16
17
18
19
20
21
22
23
24
25
25-28
29
30-33
34
35
36
37
38
39
40
35
65
79
122
132
198
252
302
338
352
428
480
562
630
687
718
782
788
790
812
(900)
(970)
1,510*
1,560*
1,580*
1,610*
(1,670)
1,650*
(1,870)
2,000
WISCONSIN
ILLINOIAN

Figure 2.16
The chronology of glaciations in the USA. The main glacial advances are shaded. Notice that the glaciations occurred as early as the Pliocene Period. After Goudie (1992).

Table 2.2   *Sequence of Pleistocene phases in the northern hemisphere.*

		Rhine estuary	Britain	Alpine foreland	European Russia	North America
GLACIAL PLEISTOCENE		WEICHSELIAN	DEVENSIAN	WÜRM	VALDAI	WISCONSIN
		Eemian	Ipswichian	Riss-Würm	Mikulino	Sangamon
		SAALIAN	WOLSTONIAN	RISS	MIDDLE RUSSIAN	ILLINOIAN
		Holsteinian	Hoxnian	Great Interglacial	Likhvin	Yarmouth
		ELSTERIAN	ANGLIAN	MINDEL	WHITE RUSSIAN	KANSAN
		Cromerian	Cromerian	Günz-Mindel	Morozov	Aftonian
		MENAPIAN	BEESTONIAN	GÜNZ	ODESSA	NEBRASKAN
PLIOCENE		Waalian	Pastonian	Donau-Günz	Kryshanov	
		EBURONIAN	BAVENTIAN	DONAU		
		Tiglian	Antian			
		PRETIGLIAN	THURNIAN			
			Ludhamian			
			WALTONIAN			
		'Pre-Glacial'				

Glacials in upper case; Interglacials in lower case
Source: After Goudie (1992).

Box 2.4

Climatic instability in the Quaternary

The rates at which global climate changes occur, together with their abruptness, are beginning to be appreciated. Rapid fluctuations of $\delta^{18}O$ values have been recognized for some time as typical of at least parts of the Quaternary, for example from the Dye 3 and Camp Century Greenland ice cores for between 80,000 and 30,000 years ago (Figure 2.17), indicative of rapid changes in ice volume and, possibly, temperature. Isotopic and chemical analyses from the GRIP (Greenland Ice-core Project) ice core from Summit, central Greenland, suggest that in Greenland, between approximately 135,000 and 115,000 years ago, during the last interglacial (known as the Eemian interglacial in Europe and correlated with the Sangamon in North America, which was warmer than the present case), there were intervals of severe cold conditions which began extremely rapidly, and lasted from decades to centuries (Dansgaard *et al.* 1993, GRIP Members 1993). The past 10,000 years have witnessed a relatively stable, interglacial climate, but prior to this during the last Ice Age which lasted about 100,000 years, and in the transitional period, global climate change was abrupt and erratic. The GRIP team have shown that changes of up to 10°C occurred within a couple of decades – possibly even less than a decade.

Other examples of rapid climate change come from uranium/thorium dating of carbonate lacustrine (lake) sediments in the dry valleys along the western margin of the west Antarctic ice sheet, and show that there were rapid and marked retreats of grounded ice between 130,000 and 98,000 years ago (Denton *et al.* 1989). The apparently sudden and sporadic, possibly chaotic collapse of the west Antarctic ice sheet over the past million years led MacAyeal (1992) to develop a computer 'finite-element' model of ice sheet flow and mass balance that reproduces the present-day flow regime of the ice sheet. He pointed out that the distribution of basal till, which helps lubricate ice sheet movement, possesses inherently irregular behaviour. Examples such as these emphasize the complexity and problems associated with trying to model the former extent of ice, and demonstrate the need for proxy data, such as those obtained from sediments, their chemistry, and landforms, in order to test various postulates and models.

effect, the Younger Dryas was a brief cool interval in a warmer period, and is referred to as a stadial. Evidence is now emerging for abrupt changes in global climate during the Quaternary, during periods of climatic instability (see Box 2.4). These climatic changes are related to glacial–interglacial cycles which, in turn, are related to changes in the global carbon cycle. This supports the view that there are strong links between climate, biogeochemical cycles, and metabolic processes in organisms.

CO_2 and CH_4 concentrations in the atmosphere have also changed considerably during past glacials and interglacials. During interglacials, there is approximately 25 per cent more CO_2 and 100 per cent more CH_4. These changes in the concentrations of atmospheric gases have important implications for understanding the global carbon cycle. It suggests, for example, that organic productivity was greatest during glacial periods, thereby providing a sink for carbon, for example, in the oceans. Such changes in CO_2 and CH_4 concentrations from glacial to interglacial periods appear to have taken place suddenly, that is within a few hundred years (Jouzel *et al.* 1987, Stauffer *et al.* 1988). The precise causes of these changes in atmospheric gas concentrations, and the threshold conditions that precipitated a switch from glacial to interglacial period, remain poorly understood. The release of CH_4 stored as methane gas hydrates in **permafrost** may have provided a significant contribution to the rapid rise in atmospheric CH_4, and CO_2, leading to the global temperature rise at the end of the last major glaciation about 13,500 years ago. The release of CH_4 would have led to a strong positive feedback which could have had the net effect of amplifying the emission of greenhouse gases. This warming, driven by methane release from various reservoirs, may have induced the release of CO_2 from the oceans to the biosphere, thereby stabilizing the interglacial carbon cycle at a different level of productivity. The study and understanding of these changes are important, because a small anthropogenically induced warming could further thaw permafrost and release CH_4 from methane gas hydrates.

Data for the past 160,000 years from the Vostok ice core (Petit-Maire *et al.* 1991) suggest that tropical wetlands are a leading influence on variations in atmospheric CH_4 levels. During

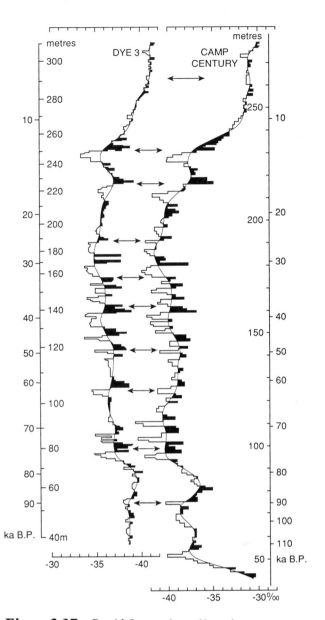

Figure 2.17 *Rapid fluctuations of ice volume (? temperature) during the last glacial period, from about 80,000 to 30,000 years ago, derived from $\delta^{18}O$ profiles obtained from two Greenland ice cores, Camp Century and Dye 3. Note rapid bimodal fluctuations. Carbon dioxide measurements also reveal variations between two states, in general agreement with the $\delta^{18}O$ data. After Oeschger and Mintzer (1992), originally published by Dansgaard et al. (1982).*

glacial maxima, CH_4 levels have naturally fluctuated around 350 parts per billion by volume (ppbv), compared to 650 ppbv during the warm interglacial periods. The CH_4 record from the

Vostok ice core shows four significant temporal periodicities at 110, 38, 24, and 19 kyr, in agreement with the orbital parameters of the Earth, that is, eccentricity (100 kyr), obliquity or tilt (41 kyr) and precession (23 and 19 kyr). This correlation led to the suggestion that orbitally driven changes in monsoon rainfall exert a crucial role in controlling CH_4 emissions from low-latitude, tropical wetlands (ibid.). The precession and eccentricity of the Earth are the principal controls on long-term variations in insolation in the tropics, whereas obliquity or tilt become increasingly important with higher latitudes.

Three deep ice cores recovered from the Greenland ice cap show $\delta^{18}O$ profiles which reveal irregular but well-defined episodes of relatively mild climatic conditions, or interstadials, that occurred during the middle and

Plate 2.6 *Approximately 2.4 million years of sedimentation of loess exposed at Gansu Province, central China. This provides the most continuous continental record of Quaternary climatic change.*

Courtesy of Prof. E. Derbyshire.

later parts of the last glaciation (Johnsen *et al.* 1992). The oxygen isotope record from these cores suggests that the interstadials lasted from 500 to 2,000 years, and their irregular development has been interpreted in the context of complex behaviour of the North Atlantic Ocean circulation (ibid.).

During the last 200,000 years up until the Last Glacial Maximum, there was increased global aridity which led to the most extensive spread of deserts and sand dunes in low latitudes (Sarnthein 1978). Regions such as the western Sahara and Sahel were, therefore, once much more extensive. This conclusion is supported by the work of Hovan *et al.* (1989), who examined the influx of **aeolian** (wind-blown) sediments in a deep-sea core from the north-west Pacific Ocean, at a site about 3,500 km down wind from central China. They were able to show increased quantities of wind-blown sediment in the core, and relate this to enhanced wind action during a more arid climatic period, which was linked to glacial stages as determined by the oxygen isotope curve. In addition, they were able to relate this influx of aeolian sediments to a sequence of wind-blown silts in Xifeng, China. These wind-blown silts, known as **loess**, contain fossil soils (palaeosols) (Plate 2.5). Palaeosols are thought to develop mainly during interglacials, and correspond to times of decreased aridity (wetter intervals), and faster rates of accumulation of aeolian sediments. This evidence supports the view of increased aridity during glacial periods.

There are many ways in which Earth scientists can read the history of the dramatic and cyclic changes in the Earth's climate over the Quaternary. One method is to study the type and relative abundance of plant spores or pollen in ancient sediments. Recently, French scientists have used the pollen record from sediment cores in eastern France (La Grande Pile and Les Echets) to reconstruct a 140,000-year continental climate sequence (Guiot *et al.* 1989). Mook and Woillard's (1982) work on a core of pollen-rich laminated sediments at La Grand Pile is particularly important as it provides a continuous continental pollen record reflecting climatic change over the past 140,000 years. Sixteen radiocarbon dates help provide a detailed chronology. The onset of a cold period at 70,000 BP was identified, marked by the disappearance of deciduous forests, which was correlated with

Plate 1 *Earth rising above the surface of the Moon. The 1969 landing on the Moon provided a new perspective of the planet.*
Courtesy of NASA/Lunar and Planetary Institute.

Plate 5 *Cretaceous/ Tertiary boundary interval exposed at Ricks Place, Montana, showing the meteorite impact layer which contains shocked quartz and high concentrations of iridium, together with other chemical anomalies.*
Courtesy of Dr M. Collison.

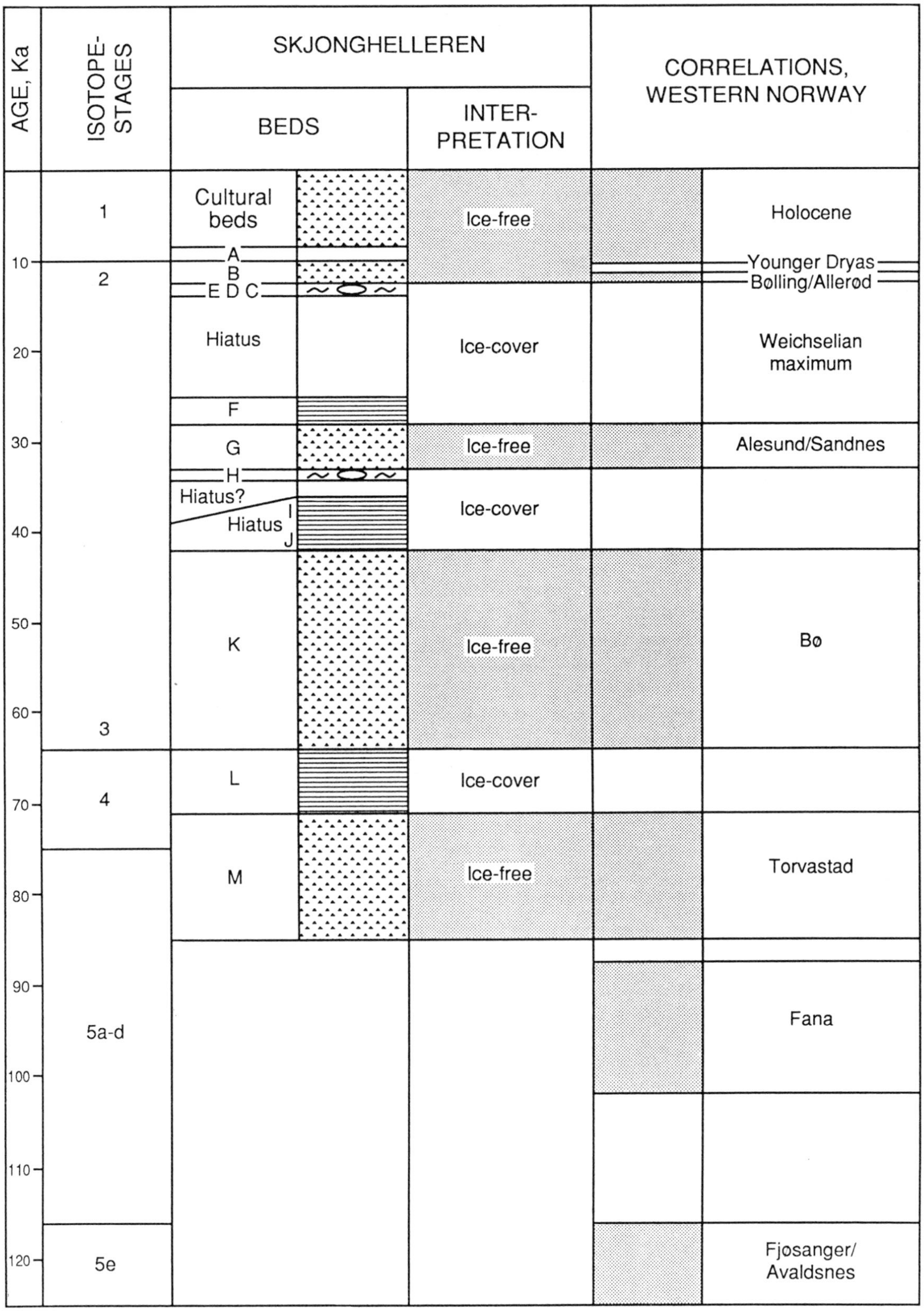

AGE, Ka	ISOTOPE-STAGES	SKJONGHELLEREN			CORRELATIONS, WESTERN NORWAY
		BEDS		INTER-PRETATION	
	1	Cultural beds		Ice-free	Holocene
10	2	A			Younger Dryas
		B			Bølling/Allerød
		E D C			
20		Hiatus		Ice-cover	Weichselian maximum
		F			
30		G		Ice-free	Alesund/Sandnes
		H			
		Hiatus?		Ice-cover	
40		Hiatus I J			
50		K		Ice-free	Bø
60	3				
70	4	L		Ice-cover	
80		M		Ice-free	Torvastad
90	5a-d				Fana
100					
110					
120	5e				Fjøsanger/Avaldsnes

Figure 2.18 Summary of the stratigraphy in Skjonghelleren and its correlation with western Norway for the past 120,000 years. After Larsen et al. (1987).

the transition from oxygen isotope Stage 5a to Stage 4, that is, the Early Weichselian/Middle Weichselian transition (ibid.).

The fundamental assumption behind this, and similar palaeoclimatic studies, is that corresponding vegetation and pollen types existed in similar ecological niches to their counterparts today. The validity of such assumptions needs much more research before scientists can feel confident about the interpretations, but they represent reasonable criteria from which to begin palaeoclimatic studies. Present-day plants may have different climatic requirements from those ancient plants, the variability of past climates may not exist today, and human activities have undoubtedly made a unique impact on modern plant life. With these provisos in mind, it is possible to develop some palaeoclimatic models based on fossil plant material in the geological record.

Cave sediments at Skjonghelleren, western Norway, provide good evidence for multiple glaciation during the Weichselian. Larsen *et al.* (1987) have identified evidence for three glaciations over the past 70,000 years. The caves were formed during high sea level stands by wave action. These comprise three units of beds of glaciolacustrine sediments, which formed subglacially, and that are interbedded with structureless blocky deposits (called diamictons), formed by the collapse of the roof during ice-free periods. The entire sequence was deposited during the Weichselian stage (Figure 2.18). Some of the diamictons contain bones and teeth of birds, mammals, and fish which have been dated using radiocarbon methods, along with dates on **spe-**

leothems using uranium isotope dating techniques. These radiometric dates cluster around 30,000 BP, the end of the Alesund interstadial, and between 12,000 and 10,000 BP.

The pollen records for the last 140,000 years, based on data from eastern France, suggest that the Holocene and the last interglacial (known as the Eemian) were the warmest and most humid climates of that period (Guiot *et al.* 1989). The main time interval of global ice growth commenced before 110,000 years, which is defined as the end of the Eemian. If the growth of continental ice sheets between latitudes 50° and 60°N implies a cold and humid climate, as suggested by climatic models, then the pollen data indicate three major periods of ice development in Europe during this time interval. The oldest occurred as a very humid and markedly cold climate towards the end of the Eemian (*c.* 110–115 ka) which immediately predated the even colder and drier Melisey I stadial (*c.* 103–110 ka).

The next period of major ice development in Europe occurred towards the end of the St-Germain I interstadial, which was very humid and moderately cold, and that was succeeded by the cold, dry Melisey II stadial (*c.* 83–92 ka). The third major ice growth occurred at the end of the St-Germain II interstadial and into the start of the substantially colder (and moderately humid) Lower Pleniglacial, prior to the second very cold, dry part of this major stadial (*c.* 45–72 ka).

Temperate conditions, not unlike those of the present day, especially in terms of temperature, appear to have existed during the St-Germain I

Box 2.5

Mineralogy and climate change

Temporal variations in the type of minerals being fed through rivers into large deltas can be used to determine climatic changes. By looking at the changing ratios of two silicate minerals named pyroxene and amphibole, Foucault and Stanley (1989) have elucidated palaeoclimatic changes in East Africa during the last 40,000 years or so in the time period referred to as the Late Quaternary. The Nile river system, formed mainly by the drainage from three large rivers – the White Nile, Blue Nile, and Atbara – flows across nearly 35° of latitude from south of the Equator to the Mediterranean. The river drains a vast area of mixed climates from tropical, humid, to warm, arid conditions. During the Quaternary, changes

in global climate caused the climatic belts to migrate large distances with the effect that there were changes in the sediment yield of the river, as well as the mineralogy and grain size of sediments reaching the Nile delta. Detailed studies of sediment mineralogy in age-dated cores from the three tributary rivers and the main Nile were made in the context of the drainage basins with their different geology and climates/vegetation cover.

Decreased amounts of pyroxene relative to amphibole in the sediments of the Nile delta and main Nile and eroded from volcanic rocks in the Ethiopian plateau, suggest increased vegetation cover with a more humid climate in the Blue Nile

and II interstadials, *c.* 72–84 ka and 93–104 ka, respectively. Perhaps the most surprising finding from these pollen data is that these temperate climatic phases during the St-Germain I and II interstadials have not been recognized in sediment cores from the Antarctic ice cores, Pacific Ocean records, Atlantic Ocean deep-water temperature estimates, or the northern European record. It has been suggested that this apparent discrepancy could be due to steeper thermal gradients than occur today between the poles and equator (ibid.).

There is growing evidence to suggest rapid advances of the **Laurentide ice sheet** in North America, and that the 5,000–10,000-year intervals between the events are inconsistent with Milankovitch orbital frequencies. This evidence comes from layers of ice-rafted sediments, known as **Heinrich layers**, in the North Atlantic ocean. The most recent six of these layers, which accumulated 70,000–14,000 years ago, indicate marked decreases in sea-surface temperature and salinity, reduced fluxes of foraminifera to the seafloor, and enormous discharge of icebergs from eastern Canada, which was produced as glaciers entered the sea and began to **calve** over short time intervals (Bond *et al.* 1992). Melting icebergs drifting across the North Atlantic ocean must have been a major factor in reducing the salinity in the surface waters, implied by the $\delta^{18}O$ values, and the salinity drop would probably have been sufficient to shut down the **thermohaline circulation** of the North Atlantic. The ice-rafted sediments on the seafloor, including detrital carbonate (limestone and dolomite with a provenance in eastern Canada), delineate the

path of the icebergs and show that they must have travelled more than 3,000 km, a distance that in itself suggests extreme cooling of the surface waters, and substantial volumes of drifting ice. Indeed, the Heinrich layers all show a dominance of the left-coiled planktonic foraminifera species *Neogloboquadrina pachyderma*, which indicates a deep southward penetration of polar water. The actual cause of the ice sheet surging remains unclear, but it has been suggested that shortly after sea-surface temperatures and foraminifera fluxes to the seafloor began to decline, ice streams in eastern Canada and possibly in north-western Greenland advanced rapidly, leading to massive calving as ice fronts reached maximum seaward positions (ibid.). The colder sea-surface water temperatures, created by the release of the large volumes of ice, would have slowed melting rates and facilitated the long-distance transport of ice-rafted sediments.

Other lines of evidence also suggest sub-Milankovitch, short-term climatic shifts, for example sediment cores from the eastern equatorial Pacific have revealed vast 1.5–4.4 Ma laminated diatom mats which rapidly accumulated at rates exceeding 10 cm per year over distances stretching for more than 2,000 km (Kemp and Baldauf 1993).

In East Africa, there is a good correlation between lake sediment stratigraphy, geochemistry, and lake water levels, all of which can be linked with Late Quaternary global climatic fluctuations. Indeed, there are good case studies of links between mineralogy and climate change (Box 2.5). Street-Perrott and Perrott (1990) showed that periods of low lake levels generally

---

and Atbara drainage basins. More humid conditions would probably have led to a longer rainy season and a greater cover of vegetation which, in turn, would have reduced erosion of sediments. Now, even if the wetter conditions led to an increase in river discharge, the sediment load carried by the Blue Nile and Atbara would have decreased. The decreased sediment load would result in a reduced supply of pyroxene to the main Nile. On the other hand, increased proportions of pyroxene supplied to the main Nile and delta probably indicate reduced vegetation cover, accelerated rates of erosion of the Ethiopian plateau and a more arid climate.

Measurements of African lake levels in the

Ethiopian Rift and Plateau, tied to this mineralogical data, suggest that high lake levels, lower pyroxene values and a more humid, wetter climate prevailed in north-east Africa *c.* 40,000–17,000 and 7,000–4,000 years ago, and from 1,500 years ago to the present day. Low lake levels, increased abundance of pyroxene and a more arid climate existed *c.* 17,000–7,000 and 4,000–1,500 years ago. The significance of changing lake levels in response to fluctuating global climate has been well discussed by Street-Perrott and Perrott (1990).

Box 2.5
Mineralogy and climate change

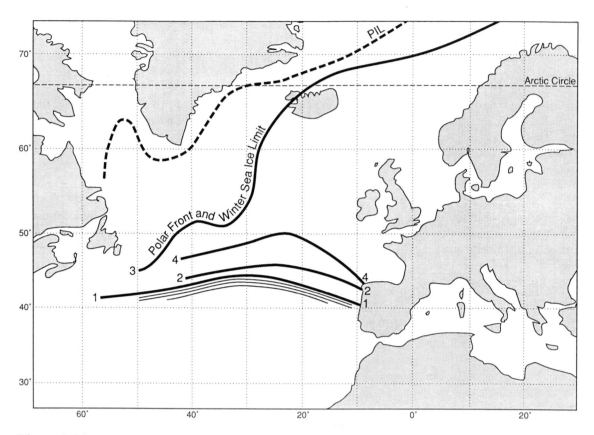

Figure 2.19 *Position of the Polar Front, and limit of winter sea ice, during the period c. 20,000–10,000 years BP. 1 = 20,000–16,000 years BP; 2 = 16,000–13,000 years BP; 3 = 13,000–11,000 years BP; 4 = 11,000–10,000 years BP. Thin lines represent the pronounced thermal gradient to the south of the Polar Front. PIL = approximate southern limit of pack ice at present day. After Ruddiman and McIntyre (1981).*

occurred between 20,000 and 13,000 BP, 11,000 and 10,000 BP, and 8,000 and 7,000 BP. They attribute the last two low stands in lake level to prolonged periods of aridity produced during times of anomalously low sea-surface temperatures in the North Atlantic. These low temperatures may have been caused by large volumes of glacial meltwater entering the North Atlantic during deglaciation and increasing the ocean salinity stratification. Such changes could then suppress the formation of North Atlantic Deep Water (NADW), and further lower the sea-surface temperature, leading to decreased rainfall and, therefore, lower lake levels. During periods when Laurentide meltwaters flowed into the Gulf of Mexico, the production of NADW would return and lake levels would show a corresponding rise. During the Last Glacial Maximum, much of the North Atlantic would have had a cover of ice, the production of NADW would have been impeded, and arid

conditions would have prevailed over much of Africa and America.

Ruddiman and McIntyre (1981) have discussed the changes in the position of the Polar Front, and the limit of sea ice, as a response to global climate change during the Late Pleistocene to early Holocene (Figure 2.19). From such data, it appears that there was a northward migration (retreat) of the Polar Front *c.* 20,000–11,000 BP, followed by a re-advance at approximately 11,000 BP, attributed to the cooler period referred to as the Younger Dryas.

Human evolution in the Quaternary Period

A study of the Quaternary Period is important in understanding human evolution and colonization, including the human impact on the natural environment. It was a little time prior to the onset of the Ice Age that the first bipedal hominids evolved (3.75 million years BP). These were known as *Australopithecus afarensis*. The most famous fossil find was unearthed by Louis Leakey in Ethiopia in the mid-1970s and became known as 'Lucy'. It is believed that the genus *Homo* evolved from *A. afarensis* about 2 million years ago. The first species was *H. habilis*, but within another 500,000 years *H. erectus* evolved. *H. erectus* probably organized themselves into groups for hunting and gathering food, as well as making tools and utilizing fire. These were the forerunners of our modern, organized society. Many scientists believe that *H. erectus* was the ancestor of modern humans (*Homo sapiens sapiens*) and evolved about 200,000 years ago. Neanderthals (*H. sapiens neanderthalensis*) are also believed to have evolved from *H. erectus*, but became extinct about 30,000 years BP.

By 50,000 years BP, *H. sapiens sapiens* had spread to Australia. They arrived in the Americas between 14,000 and 12,000 years BP and by the start of the Holocene had colonized every continent, except Antarctica. Their migration was undoubtedly influenced by climatic change, often aided by the extensive coastal regions that were created as a result of the fall in sea level caused by water being locked up in the ice sheets during the last glacial.

Towards the close of the last glacial, about 15,000 years BP, hunter-gatherer communities began to develop, and these groups began to clear land for farming and settlements. They were the first humans to initiate the process of deforestation. This practice of forming organized settlements and land clearing began in many regions, particularly in the Near East, Central Asia, and South America. In the Near East, by 10,000 years BP, the domestication of plants and animals was well established. By about 9,000 years BP Jericho, one of the earliest permanent settlements, was established, associated with cultivating cereals, wheat and barley. The domestication of animals became more sophisticated. Approximately 8,000 years BP, chickens, for example, were domesticated from the red jungle fowl of South-East Asia, and horses were first domesticated in the Ukraine about 4,000 years BP.

Such changes led to the modification of the landscape, vegetation, soil, and water courses, as modern humans cleared more forest for farming and the establishment of permanent settlements. The need for tools also had a profound effect on the environment, as more trees were required for charcoal to aid in the smelting of metal ores. In the Near East, by 7,000 years BP, copper was being smelted, which then gave way to arsenic bronze by 5,000 years BP, and eventually iron about 4,000 years BP.

This societal organization provided a more secure environment for human survival, and even produced surplus food. Improved living conditions meant that humans could spend time in pursuits other than basic survival, for example in making jewellery and ornaments. Barter thus became possible. Religious activities also developed. Between 5,000 and 4,000 years BP, large monuments were being constructed. These included the pyramids in Egypt (*c.* 4,500 years BP), the Minoan palaces of Crete (4,000 years BP), and the construction of Stonehenge in England (*c.* 4,000 years BP).

Wood (1992) reviewed the evolution of *Homo* in the light of recent advances in techniques for absolute dating (e.g., Grun and Stringer 1991), and also reassessed some of the evidence from fossils. He argues that a simple unilineal model for the evolution of humans, where *H. habilis* succeeded the australopithecines and then evolved via *H. erectus* into *H. sapiens* is untenable. Despite such arguments amongst the experts, no clear consensus on human evolution has emerged, so the actual pattern of human evolution and colonization of the Earth remains unclear.

The important point through this deviation into human history, is that throughout the Quaternary there is evidence for studying the human impact on the natural environment – as far back as the Late Pleistocene and Early Holocene. The human impact has to be considered when studying climatic and palaeoenvironmental change during this time period. At present, there is great debate regarding the extinction of many species of animals, as well as major changes in natural vegetation, that occurred near the end, and after, the last glacial stage. The fossil record for the last interglacial shows a decline in diversity of species. In Europe, during the last interglacial, abundant elephants, rhinos, bison, and giant deer were present. In Australia, a more diverse marsupial fauna existed, including giant wombats, giant kangaroos, a diprotodont (a marsupial somewhat like a hippopotamus), and in New Zealand there were giant birds. In each continent outside Africa, these faunas disappeared as complex human societies evolved. In Australia the marsupials were greatly reduced by 30,000 years BP, while in North America three-quarters of the genera disappeared by about 11,000 years BP. The most recently colonized regions of the world, such as Madagascar (*c.* 1,500 years BP) and New Zealand (1,000 years BP), saw the extinction of large flightless birds like the rocs and moa, respectively.

A detailed study of the Quaternary period of Earth history allows us to assess the possible relationship between the growth of human society and the extinctions of various species, together with any environmental changes, towards the end of the last glacial stage. It may be that the extinctions and changes in the natural environment occurred entirely independently of human activities, because of natural processes that exerted a more profound influence, for example, the changes in the ocean–atmosphere system brought about by the end of the last glaciation.

Meteorite impacts on Earth and global climate change

The Cretaceous–Tertiary (K–T) boundary event 65 Ma

Another example of abrupt global climate change occurred about 65 million years ago when a giant **meteorite** impacted on the Earth's surface. This is particularly interesting because it provides Earth scientists with information on how external, cosmic processes may lead to major climate change, and the extinction of faunas and floras.

Approximately 65 million years ago (64.5 ± 0.1 Ma) as dated using an argon laser probe technique on Haitian **tektites** at the US Geological Survey in Denver, Colorado, a phenomenal catastrophe hit the Earth, the consequences of which were fatal for many organisms. An estimated 70 per cent of the flora and fauna on Earth became extinct. Such is the significance for the evolution of life on Earth that Earth scientists define the time era after 65 Ma as the Tertiary, and the immediately preceding time interval as the end of the Cretaceous Period, hence the K–T boundary event. The 'K' is from the German spelling 'Kretaceous'.

The most widely known event at the K–T boundary was the extinction of the dinosaurs. Their demise allowed the humble mammals to inherit the role of dominance from the dinosaurs, and paved the way for human beings. There have been many theories to explain the extinction of the dinosaurs, but here only the most plausible event is presented, an explanation subscribed to by most Earth scientists, which is the impact of a massive meteorite.

Towards the end of the 1970s, an Earth scientist named Walter (L.W.) Alvarez was researching the rates at which ancient marine clay-rich sediments were laid down around the

Box 2.6

K–T boundary meteorite impact site

Geochemical analyses of the K–T boundary clays suggest that the site of meteorite impact was in the deep oceans, penetrating 3–5 km into the oceanic crust. The shocked quartz, however, indicates at least a thin cover of land-derived, continental material overlying the oceanic crust. Few people have suggested a precise site for the enormous meteorite impact, but potential sites, based on age, dimensions, and shape, that have been proposed include the Amirante Basin, west Indian Ocean, the Nicaragua Rise in the Caribbean Sea, and the Manson impact crater in Iowa with its diameter of 35 km and date of 65 Ma.

In March 1991, new evidence was presented to the Lunar and Planetary Science Conference in Houston, Texas, in favour of an impact site in the Yucatan peninsula in the southern Gulf of Mexico (Ravven 1991). A particularly thick layer rich in spherules was interpreted as resulting from the ejection of vaporized and melted material from the meteorite impact which was spread over a very large area. Similar spherule layers, albeit much thinner, have been identified at many K–T boundary sites. In northern Yucatan, Mexico, the prime candidate for the site of the meteorite impact is the 200 km diameter Chicxulub impact structure which contains deformed or 'shocked' rock or lithic fragments that are similar to those found worldwide at the K–T boundary, an observation that may favour a single meteorite impact rather than a comet shower (Sharpton *et al.* 1992). The impact structure is associated with

K–T boundary near a place called Gubbio in Italy. Chemical analyses of these clays revealed an unexpected abundance in a chemical element called iridium, now known as an **iridium anomaly**. Alvarez and his co-workers interpreted this anomaly as a result of an enormous meteorite impacting on the Earth at the end of the Cretaceous Period. This hypothesis was published by Alverez in 1980 in the American journal *Science*. The size of this meteorite may have been about 10 km across, and upon impact had an estimated explosive energy equivalent to 100 million megatons of TNT, or roughly 10,000 times the world's total nuclear arsenal (Ravven 1991).

Plate 5 illustrates a typical example of the types of sediments present at the K–T boundary which are exposed in Montana, USA. The iridium anomaly was discovered in other rocks of the same age from around the world, but always in marine sediments. A popular interpretation, therefore, was that it was caused by chemical reactions in seawater which preferentially extracted iridium into the sediments. This notion was shattered in 1981 with the discovery of the same iridium anomaly in terrestrial sedimentary strata dated at 65 Ma in New Mexico. However, the geochemical iridium **spikes** at extinction horizons, commonly associated with spikes in the other platinum group elements (Ru, Rh, Pd, Re, Os, Pt, Au), can be the result of post-depositional redistribution in the sediments because of changes in **redox** conditions at or near the seafloor (Colodner *et al.* 1992). Such geochemical spikes and ratios may, therefore, be characteristic but not diagnostic of a cosmic source.

Other elements, beside the platinum group elements, were found to be enriched in sediments occurring at the K–T boundary, for example nickel, chromium, cobalt, and iron, all of which have been interpreted as the result of a large meteorite impact. Although these **siderophile elements** occur in varying quantities on Earth, their relative abundances and concentrations at the K–T boundary are quite unlike those of typical terrestrial rocks, but similar to those encountered in certain types of meteorites.

Another line of evidence in favour of an impact event at the K–T boundary is the presence of highly deformed or 'shocked' quartz in which the crystal structure is believed to have suffered very rapid strain during a meteorite impact. Such varieties of the mineral quartz (which only form at extremely high pressures, such as caused by a meteorite impact) are found in the sediments of the K–T boundary at Raton Pass, Mexico. These quartz minerals are coesite and stishovite, which require respectively 20 kilobars and 110 kilobars of pressure to form (1 bar is equivalent to 1 atmosphere, which equals $1\,kg\,cm^{-2}$).

Chondritic meteorites also contain abundant diamonds of 3–5 microns in size, something that prompted the search for similar small diamonds in the sediments at the K–T boundary. The boundary clay from Red Deer Valley, Alberta (known as the 'Knudson Farm' locality) has yielded a white fraction containing 97 per cent more carbon, which is absent from the

---

igneous rocks (andesites) produced by the impact, which have been radiometrically dated by $^{40}Ar/^{39}Ar$ techniques to 65.2 ± 0.4 Ma (ibid.), in good agreement with the recently reported date of 64.98 ± 0.05 Ma (Swisher *et al.* 1992).

It may be that there was not one but several meteorite impact sites at the K–T boundary, something proposed in 1988 by Eugene Shoemaker of the US Geological Survey at Flagstaff, Arizona. A comet passing close to the Sun could have fragmented and caused several meteorites to impact on Earth at several locations. The multiple impact hypothesis might explain the fact that some Earth scientists now recognize a number of impact sites, for example the most plausible candidate sites include the Caribbean,

the 35 km diameter Manson crater (Iowa), and the 105 km diameter Popigai crater (Siberia), all dated to approximately 65 Ma.

In the Brazos river, Texas, and the New Mexico sites, there is also evidence of tsunami (tidal wave) activity, possibly caused by a meteorite impact, but this event is about 230,000–330,000 years after the principal K–T boundary extinctions (Montgomery *et al.* 1992). If, as some scientists suspect, the K–T boundary meteorite impact was not actually a single event, but perhaps many smaller impacts associated with a very large, main impact, then the Brazos river section, although slightly younger in age, may represent a part of the K–T meteorite shower events.

Box 2.6
K–T boundary meteorite impact site

surrounding layers. Two Canadian scientists have demonstrated that this carbon-rich material is almost certainly very small diamonds (Carlisle and Braman 1991) and have therefore provided additional supportive evidence for the meteorite hypothesis.

There are a number of other lines of evidence pointing towards an extra-terrestrial, meteorite impact cause for the event at the K–T boundary, including the presence of **spheroids** in the sediments. These sand-size minerals are believed to result from the crystallization at high temperatures of material melted by a meteorite impact and rapidly ejected into the air and water. Sites where these spheroids, mainly of the mineral feldspar, occur include the K–T boundary clays at Caravaca in southern Spain, Petriccio in Italy, El Kef in Tunisia, and the central Pacific Ocean (see Box 2.6).

An intriguing aspect of the K–T boundary event is the evidence which is emerging for global fires. The percentage of carbon in sediments at the K–T boundary is much greater than expected, with the carbon occurring as fluffy aggregates of 0.1–0.5 micrometer graphite. Fluffy graphitic carbon is similar to charcoal that is produced from forest fires today. Analysis of clay samples from the K–T boundary at Woodside Creek, New Zealand, Stevn's Klint, Denmark, and Caravaca, Spain, led Woolbach *et al.* (1985) to suggest a worldwide flux of carbon of about 10,000 times greater than the present day and 1,000 times greater than in the underlying Cretaceous and overlying Tertiary sediments. The source of this graphitic carbon is unlikely to have been the meteorite, but a massive impact event could have caused devastating fires that raged throughout enormous areas of land. These fires could have ignited various shallowly buried deposits of fossil fuels such as coal which would have released even more carbon into the land and atmosphere at that time. The meteorite impact need not have been on land to cause such catastrophe. An oceanic impact could still have led to enormous fireballs and expanding clouds of rock vapour.

Many suggestions exist as to just how the dinosaurs became extinct, something that is now known to have occurred over a few million years. Perhaps the most reasonable interpretation is that the meteorite impact ejected huge volumes of very fine material into the upper atmosphere, together with the soot and other materials contributed by global fires. Such clouds would have been very effective in absorbing sunlight (solar energy) to stop it reaching the surface of the Earth. The atmosphere would also have become extremely polluted by the emission of very large amounts of gases from the wildfires to produce **pyrotoxins**.

Evidence from the remains of plants that were living at the time of the impact event can even give us a clue as to the season and month when the devastating meteorite hit the Earth. A study at the US Geological Survey in Denver, Colorado, of aquatic leaves in the K–T boundary section near Teapot Dome, Wyoming, shows the preservation of detail that can only be achieved experimentally in aquatic leaves by freezing. The impact of a huge meteorite would be expected to produce large amounts of light-attenuating debris in the atmosphere and, therefore, an **impact winter**. Reproductive stages reached by the fossil aquatic plants at the time of death suggest that the freezing took place in approximately early June, that is in the early summer in the northern hemisphere (Wolfe 1991).

The other line of evidence for a protracted time interval of meteorite falls around the K–T boundary comes from the geochemistry. There are amino acids which are extremely rare on Earth, but known to occur in meteorites. Research by two American scientists, Kevin Zahnie and David Grinspoon (1990), into the K–T boundary site at Stevn's Klint, Denmark, has demonstrated that the concentration of these organic compounds shows an increase over about 50,000 years prior to the K–T boundary, followed by a fall-off afterwards, again over approximately 50,000 years. They suggested that if the amino acids only came with the big meteorite impact, then they would have been incinerated at the K–T boundary. Interestingly, the amino acids only occur in the few centimetres above and below but not in the boundary clay itself. In order to explain this anomaly, it was suggested that the amino acids were deposited with the dust from a giant comet trapped in the inner Solar System, a fragment of which formed the K–T impactor. The amino acids would have been swept up by the Earth prior to and following the meteorite impact and therefore preserved in sedimentary layers, but those deposited at the K–T event would have been destroyed.

At the K–T boundary, there is also a change in the strontium isotope composition of seawater recorded from foraminifera in an exceptionally thick, palaeontologically well-characterized, K–T section exposed at Bidart in south-west France. Here, there is a rapid increase in $^{87}Sr/^{86}Sr$ (strontium isotopes) of ocean water about 1.5–2.3 million years before the boundary (Nelson *et al.* 1991). **Bioturbation** cannot explain these changes before the K–T boundary because vertical mixing by such processes typically involves up to about 10 cm of uncompacted sediment, equivalent to about 5 cm of compacted sediment. The studied section at Bidart is two to three orders of magnitude thicker, with the chemical anomaly appearing approximately 90 m, reaching a maximum 40–50 m, below the K–T boundary. The actual change in strontium isotopic signature of the ocean water is explained by a 10 per cent increase in strontium supply to the oceans from the continents over about 1 million years. Such a change could be due to increased erosion of sediments from the land, induced by a major change in global climate (ibid.) that began prior to the K–T event.

In effect, the Earth could have suffered many of the effects of a nuclear winter at the K–T boundary. A few years of darkness and freezing temperatures at the Earth's surface would have ensued. Plants would have been unable to photosynthesize **chlorophyll** and would have died. Animals, particularly the 'higher' species, would have died both as a direct consequence of the meteorite impact and through starvation. Much of the complex food chains may have broken down as plants died. Using the scenario of a huge meteorite impact, it has been calculated that light levels would certainly have been too low for photosynthesis to occur for about 200 days, and that global temperatures at the surface of the Earth would have been below freezing because of the lack of sunlight penetrating the dense and poisonous atmosphere.

Even life in the oceans appears to have been killed by the meteorite impact. Microscopic organisms that secreted shells or plates of calcium carbonate (calcite) suffered extinction. A group of such organisms to become extinct at the K–T boundary were various species of calcareous plankton. The shells of dead organisms on the seafloor appear to have been subject to dissolution in shallow marine waters where such

processes would not be expected. Geochemical evidence suggests that at the K–T boundary, the oceans suffered an unprecedented depletion in the element calcium, so essential for life. The actual cause of this decalcification of the ocean waters remains puzzling.

A possible cause may have been a dramatic shallowing of the depth at which material made of calcite dissolves in the world's oceans, known as the **carbonate compensation depth (CCD)**. At present, particles of calcite such as the tests or shells of dead microscopic organisms that are settling slowly through the water column begin to dissolve in the open oceans at depths of about 5.4 km in the Atlantic Ocean and 4.3 km in the Pacific Ocean. It has been suggested that at the K–T boundary, the position of the CCD rose to within the **photic zone** in less than a couple of hundred metres of water depth with the result that organisms such as the calcareous plankton were unable to secrete their calcite shells. The result would indeed have been catastrophic with mass extinctions.

One possible reason for this decalcification is that as the huge meteorite travelled through the atmosphere and became very hot, high-temperature shock waves were generated and nitrogen oxide was formed just as occurs today when lightning causes shock heating. Nitric acid (HNO_3), along with other acids, would form and cause acid rain to fall with an estimated **pH** of 0–1 (strongly acidic). Such acid rain would rapidly cause a critical decalcification of the upper ocean waters and the CCD would rise substantially.

The acid rain hypothesis finds additional support in the high levels of nitrogen found in many of the sediment samples analysed from the K–T boundary. Furthermore, the ratios of the strontium isotopes $^{87}Sr/^{86}Sr$ show a sharp increase at the K–T boundary, something that is predicted by very acidic rainwater dissolving large quantities of continental granites and releasing the abundant ^{87}Sr isotopes into the water cycle from these rocks. Of course, the very acidic rainwater could have been a direct cause of enormous fatalities and mass extinctions of species.

The possibility that the impact of a huge meteorite at the K–T boundary generated nitrogen oxide (NO) is supported by data from a much smaller meteorite shower in 1908 called the Tunguska meteor fall. This meteorite fall is estimated by Turco (1981) to have caused a substantial depletion in ozone from the ozone layer. It has been calculated that as much as 30 million tonnes of NO could have been produced, and that approximately 45 per cent of the **ozonosphere** in the northern hemisphere was destroyed. Clearly, the much greater magnitude of a meteorite impact at the K–T boundary would have had more devastating consequences than the 1908 Tunguska meteor fall.

If the ozone layer was destroyed by the K–T boundary event, then the immediate result of the meteorite impact would have been that lethal doses of **ultraviolet radiation** and heat from the Sun would have struck the Earth's surface. Such radiation levels could have caused mass extinctions.

Not all Earth scientists believe in the impact theory. Archibald (1993) argues that the commonly quoted mass extinctions at the K–T boundary are misleading, and that the actual numbers of species that died out were between 52 and 72 per cent rather than the commonly quoted 75 per cent. He suggests that many species did not actually become extinct in the true sense that their entire gene pool was wiped out, but rather that species disappeared locally. Anderson (1993) argues that many dinosaurs could survive the darkened skies and global cooling associated with a nuclear winter brought about through an asteroid impact, based on current work being undertaken on dinosaur fossils from Dinosaur Cove, in the Otway Range 220 km west of Melbourne, Australia. During the Cretaceous Period, when the dinosaurs of Dinosaur Cove lived, the **palaeoaltitude** lay between 70 and 80°S, and the region would have experienced between six weeks to four and a half months of continuous darkness. Anderson argues, therefore, that dinosaurs may have been more able to adapt to environmental stress than has previously been thought.

Officer (1993) has suggested that there is evidence to show that dinosaurs actually died out before the iridium anomaly, and that the impact therefore could not have been the principal cause of their extinction. He proposes instead that volcanic eruptions and global sea level changes may have been more important in causing the extinction of the dinosaurs. Volcanic eruptions can cause significant climatic changes, although some of the largest known eruptions, such as

Toba in 75,000 BP, did not cause any species extinctions. Such eruptions may also produce large amounts of iridium, although detailed studies of the amounts which can be produced by volcanic activity remain poorly quantified. Swinburne (1993) also argues that other fossil groups that are lumped into the total number of species which became extinct at the end of the Cretaceous Period, such as inoceramid and rudist **bivalves**, actually died out 2 and 10 million years prior to the K–T boundary event, respectively. Thus, their extinction cannot be attributed to a single impact event at the K–T boundary. The problem with arguments such as these against a meteorite impact is that they ignore the cumulative evidence for an impact, but merely show that any one aspect of the evidence could be interpreted in other ways. Furthermore, these arguments commonly involve exploiting the uncertainties in precise dating of events.

The K–T boundary event was indeed a catastrophe for life on Earth, and something that only the most hardy and fortunate species survived. The Earth's ecosystems were stressed almost to the limit. The meteorite impact event brought immediate devastation with acid rain and an 'impact winter' of prolonged freezing temperatures because of the dust particles blocking out much of the sunlight, together with a possible depletion of stratospheric ozone to contribute further to global problems. In short, at the K–T boundary, for the survivors of the actual impact, life on Earth experienced global acid rain, an impact winter, and an ensuing period with perhaps little suitable vegetation as part of any diets. Not surprisingly, this chain of events was more than could be borne by more than 70 per cent of the species of flora and fauna.

Other geological mass extinction events

The K–T boundary is the most studied and publicised mass extinction event, primarily because of the extinction of the dinosaurs, but other comparable events have occurred in the Earth's history. Indeed, the biggest known extinction event in Earth history occurred at the end of the Permian Period of geological time –

about 250 Ma. Unlike at the Cretaceous–Tertiary boundary, there is no iridium anomaly or any other evidence for a meteorite impact in sediments at the Permian–Triassic boundary, so a different explanation is required.

At the boundary between the Permian and Triassic Periods, it appears that there was a super-continent formed by much of the continental land masses being welded together in equatorial to low latitudes. This super-continent, called Pangea, was dominated by arid, desert conditions with intense evaporation which produced extensive areas where thick accumulations of salt or evaporite minerals formed, because of the frequent desiccation of standing bodies of water. The amalgamation of many of the continental land masses into a single super-continent meant that the global amount of shelf-sea area was greatly reduced. Consequently, the competition for suitable marine ecological niches by many organisms was intense, and the demand on available nutrients far exceeded that which was available. A crisis for life on Earth occurred, and resulted in mass mortalities and extinctions of more than 95 per cent of the species then living. The Permian–Triassic extinction event is an important example of how plate tectonics can provide an explanation for mass extinction events, linked to regional climatic conditions and not necessarily global climate alone.

Iridium anomalies, however, have been identified as being associated with other mass extinction events at the Precambrian–Cambrian (570 Ma) and Ordovician–Silurian (435 Ma) transitions, within the Devonian, Frasnian–Famennian Stage (365 Ma), and within the Carboniferous, Mississippian–Pennsylvanian Stages (325 Ma) of Earth history.

Throughout the geological column, other examples of meteorite impact events are being identified. Beneath Chesapeake Bay and the adjacent Middle Atlantic Coastal Plain on the US east coast, there is a 60 m thick boulder bed interval containing a mixture of sediments of different ages, distributed over an area of > 15,000 km^2, which is matched to a layer of equivalent age impact material recovered from a deep-sea drilling site on the New Jersey continental slope (Deep Sea Drilling Project or DSDP Site 612) and is interpreted as the result of a meteorite impact in the late Eocene (Poag *et al.* 1992). The tektite glass (part of the impact ejecta, including shocked quartz) from the DSDP Site 612 has been radiometrically dated by ^{40}Ar/^{39}Ar methods to be 35 ± 0.3 Ma (Obradovich *et al.* 1989). The candidate impact site for this Eocene event has also been identified by **seismic** reflection profiling across the continental shelf, and is represented by a 15–25 km wide impact crater with a central 2–3 km wide zone of disturbed sediments about 40 km north-north-east of DSDP Site 612, also extending several kilometres down (Poag *et al.* 1992).

Younger rocks with iridium anomalies, possibly caused by meteorite impact, include an 11 Ma event in the Miocene Period where iridium levels are 15 times greater than the background values. More work needs to be undertaken on this latter event to see what caused it. Not all iridium anomalies, or mass extinctions of species, have to be caused by meteorite impacts. Whatever the trigger for these mass extinction events, they would have been associated with changes in global climate so severe as to make sustainable existence impossible for the species of fauna and flora that became extinct.

A hot desert at the bottom of the Mediterranean Sea 5–6 Ma

The discovery that the Mediterranean ocean basin dried up to become a desert came as a dramatic and fascinating discovery to a team of Earth scientists drilling and recovering cores in the Mediterranean Sea in 1970. The scientists were part of an international team on board the scientific drill ship *Glomar Challenger* taking part in the Deep Sea Drilling Project, aimed at understanding more about the world's ocean basins. This is an example of how plate tectonic processes and climatic conditions conspired to exert a dramatic effect on the climate of a very large region, the Mediterranean, and in this important respect it differs from the previous case studies of global climate change.

How did the Mediterranean ever become an enclosed ocean basin that could dry up? Some 20 Ma, the plate containing Arabia (the Arabian Plate) impinged against the Eurasian Plate to the north, to cut off the Mediterranean from a closing ocean to the east named Tethys. Once the Mediterranean became landlocked (enclosed), the only connection to the large oceans was to the west through the narrowing seaway that separates North Africa from Europe at the Straits of Gibraltar. The climate became drier, and without an open, wide, marine seaway connecting it to other oceans over a period of about 1 million years the Mediterranean virtually dried up.

Today, the evidence of this desert lies up to 3,000 m below the surface of the sea. This incredible discovery is in gypsum layers, which were formed by evaporation of the Mediterranean seawater under desert conditions about 5–6 Ma. The extreme evaporation of such a large volume of saline water led to the accumulation of more than 1,000 m of evaporite salt deposits. Seismic surveys of the sedimentary layers below the Mediterranean reveal a bright reflecting surface known as the 'M' reflector which is this layer of salt.

The sea floor of the Mediterranean some 5–6 Ma lay some 2,000 m below the global sea level, e.g. west of Gibraltar, in the Atlantic Ocean. Such an enormous drop in sea level led to the rivers draining into the Mediterranean excavating deep, steep-sided valleys or ravines into the underlying sediments and rocks. Using sophisticated geophysical techniques to look at the subsurface rock strata, Earth scientists have identified buried river gorges up to about 1 km below the present land surface containing ancient river gravels and sands and associated with rivers such as the Nile and Rhône.

As the Mediterranean evaporated, the waters became stagnant and **hypersaline**. Most organisms simply could not cope with the hostile environment and died. The Mediterranean basin became, in effect, a 'Death Valley' with a series of salt lakes that periodically dried up completely. Calculations of the volume of evaporite minerals compared to the typical 35 g of dissolved salts in every litre of seawater suggest that something like 30–35 times the volume of water in the present Mediterranean would have been necessary to form the 1 km thick salt deposits. The

only way this could have happened was through periodic flooding of the Mediterranean by incursions of salty seawater which then evaporated to leave yet more evaporite minerals. So, the Mediterranean cannot have been completely isolated from the world's oceans.

About 5 Ma, the dam which separated the Mediterranean from the Atlantic Ocean was finally breached. Seawater cascaded down the world's most impressive waterfall at a rate of approximately 40,000 km^3 per year, taking about 100 years to fill the Mediterranean. The waterfall at Gibraltar was 100 times larger than the Victoria Falls on the Zambesi river. The salinity crisis (called the **Messinian salinity crisis** after the geological time period when it occurred) thus came to an end. Very deep water again covered the seabed which had been dry land. The hot desert climate at the bottom of the Mediterranean was reclaimed by the sea.

Conclusion

Throughout this chapter, the evidence for past changes in global climate has been considered. The evidence is multi-faceted and extensive, varying in the amount of information, type of data, and the confidence with which the interpretations are made. Furthermore, whilst the causal factors and rates of global climate change still require much more research, it is clear that evidence from the geological time scale reveals climatic conditions that were much more extreme than those experienced by humans. However, a concern for the natural environment that exists currently, together with attempts to make better predictions for future climate change, can only be made with continued research, both into past climates and by gathering detailed observations of present atmospheric, ocean, and land physio-chemical conditions.

CHAPTER 2: KEY POINTS

1 The Earth's climate has changed throughout geological time and is still undergoing change. Palaeoclimatology is the study of past climate. There have been at least six major cold periods, or Ice Ages, throughout geological time. Since about 2.4 million years ago, global climate has cooled considerably, and the Earth has entered the present Ice Age, referred to as the Quaternary Period, during which global climate has fluctuated between cold (glacial) stages and warm (interglacial) stages, with less intense warm (interstadial) and cold (stadial) periods.

2 Natural causes of global climate change result from: (i) internal Earth processes such as plate tectonic processes, which lead to a redistribution of land masses and altitude which, in turn, influence global atmospheric, hydrologic, and biologic systems, together with volcanic activity which may cause changes in atmospheric aerosols and gases; (ii) processes external to the Earth, such as Milankovitch cyclicity resulting from variations in the Earth's orbital parameters around the Sun, and solar variations such as Sun-spot activity, all of which lead to variations in the amount of solar insolation to the Earth's surface, thereby causing changes in the atmosphere–ocean system, for example, changes in biomass production and burial; (iii) catastrophic events such as large meteorite impacts, which may cause large-scale extinction events and thereby open up ecological niches for existing or new species to inhabit and evolve within.

3 Palaeoclimatology is studied using many different methods and techniques. Petrological techniques use characteristic sediment and rock types to interpret past climates (evaporite minerals, glacial deposits, etc.). Palaeontological techniques, such as the use of pollen spores, provide proxy data on global climate. Chemical methods include the study of stable oxygen isotopes, for example, from foraminifera in deep-sea sediment cores, and air bubbles trapped in glacial ice, which provide an indication of seawater temperature and, indirectly, an estimate of the relative amounts of seawater stored as glacial ice. Stable carbon isotopes in fossil organic matter can be used to evaluate changes in biomass production, which is a function of both regional and global climate. Stable nitrogen isotopes in fossils may be used as a proxy indicator of the contribution of nitrogen fixation by leguminous plants, again strongly influenced by global climatic conditions. Concentrations of various trace metals such as cadmium (Cd) in fossils (commonly expressed as a cadmium/calcium ratio) provide an insight into seawater temperatures, and by extrapolation oceanic circulation patterns and global climate. The distribution of wind-blown fine sediment, or loess, is an indicator of global aridity. Variations in the thickness of tree rings provide important information on past changes in climate, at least on an annual basis for the past 9,928 years. Glacial erosional and depositional landforms provide evidence for the extent of former ice sheets, a proxy for global climate. Raised beaches indicate the extent and position of former sea levels which are a function of both global climate and tectonics.

4 The Quaternary Period is most often used in the prediction of future global climate change because most data remain available from all the geological periods for study. When the Quaternary Period and the onset of the last Ice Age began is debated, but probably occurred about 2.4 million years ago. Glacials were periods of extensive ice cover lasting between 100,000 and 200,000 years, whereas interglacials, lasting 10,000 to 20,000 years, were much warmer periods, some being warmer than the present interglacial. These fluctuations in global climate are probably controlled mainly by Milankovitch cyclicity – the orbital characteristics of the Earth around the Sun. The glacial–interglacial cycles were complex with rapid transitions and perturbations in climate. The most studied transition is the last glacial (Devensian–Wisconsin–Weichselian) to the present interglacial (Holocene). During this transition, there was a brief return to near-glacial conditions (Younger Dryas stadial). Studies of the isotopes in ice cores (e.g., Vostok ice core), palaeontology, sedimentology, and geomorphology provide important information on the rates of change of global atmospheric conditions and their resultant effects, including increased biological productivity, lower global sea levels, and increased aridity during the last glacial. Humans evolved during the Quaternary Period, the first biped hominid (*Australopithecus aferensis*) 3.75 million years ago, the first *Homo* (*H. habilis*) 2 million years ago, and modern humans (*H. sapiens sapiens*) about 200,000 years ago. The development of human culture has affected the global biota and climate.

5 Global climate has been influenced by meteorite impacts. A major extinction event, which included the dinosaurs, took place about 65 million years ago at the Cretaceous–Tertiary (K–T) boundary. This major event is believed to be the result of one or more meteorites colliding with the Earth. Evidence for one or more meteorite impacts is provided by the iridium anomaly which is present in rocks at the K–T boundary, along with other platinum group elements showing concentrations characteristic of extra-terrestrial bodies or meteorites, the common occurrence of spherules of molten glass and very high-pressure minerals (coesite and stishovite) atypical of conditions at or close to the Earth's surface, high concentrations of combusted organic carbon (charcoal or fluffy carbon), and concentrations of rare amino acids which are more common in meteorites. Several locations have been suggested for the impact crater(s), with the most favoured site having been near the Yucatan peninsula in the Gulf of Mexico. The meteorite impact(s) caused global fires, enhanced levels of atmospheric aerosols, and reduced sunlight, which, in turn, led to global cooling, and pyrotoxins having extremely serious effects on the most evolved life forms such as the dinosaurs. Other effects of the meteorite impact appear to have included very acidic rain, a depletion of the stratospheric ozone layer, and the decalcification of the oceans.

6   Other mass extinctions have occurred throughout geological time, some of which may also have been due to meteorite impacts, but at least some of which were caused by other processes leading to global climate change. The greatest extinction event known in Earth history, which occurred 250 million years ago at the close of the Permian Period

and the start of the Triassic Period and involved the extinction of about 95 per cent of all living species, does not appear to have been associated with a meteorite impact but, rather, the growth of a super-continent in low/equatorial latitudes which caused a dramatic reduction in the area of favourable ecological niches, an unquenchable demand for nutrients and the exhaustion of nutrients sufficient to sustain the biomass. These circumstances conspired to lead to a crisis for life on Earth and mass extinctions.

7   There are examples of spectacular regional changes in climate caused by plate tectonic processes. About 5–6 million years ago, the Mediterranean became landlocked as a result of plate tectonic processes, with the result that the Atlantic Ocean waters were sealed off from those of the Mediterranean in the region of the Straits of Gibraltar. The Mediterranean evaporated and changed the regional climate to desert conditions; the evaporation of the seawater, probably periodically replenished by catastrophic flooding from the Atlantic Ocean, caused the accumulation locally of up to about 1 km in thickness of salts or evaporite minerals. This event is referred to as the Messinian salinity crisis. About 5 million years ago, the Straits of Gibraltar were breached by the Atlantic Ocean waters which then flooded back into the Mediterranean.

8   An understanding of past global and regional climate change, the causes, processes, and effects, is important to humankind in order to distinguish natural from human-induced climate change.

Chapter 2: Further reading

Bradley, R.S. 1985. *Quaternary Paleoclimatology – Methods of Paleoclimate Reconstruction.* London: Unwin Hyman.
A comprehensive textbook suitable for undergraduates and researchers wishing to appreciate the various methods used in the reconstruction of past climates. Topics covered include: the nature of global climate change; dating methods; ice core studies; the study of marine sediments; non-marine geological evidence; non-marine biological evidence; pollen analysis; dendroclimatology; and historical data.

Dawson, A.G. 1992. *Ice Age Earth: Late Quaternary Geology and Climate.* London: Routledge.
A detailed review of the fluctuations in the Earth's climate during Late Quaternary time. Suitable for undergraduate students and researchers interested in the complex and dynamic changes that affected the Earth's surface and atmosphere during this period. Topics considered in depth include: ocean sediments and ice cores; general circulation models for the Late Quaternary; glaciation and deglaciation during Late Quaternary time; Late Quaternary environments; Ice Age aeolian activity; Late Quaternary volcanic activity; crustal and subcrustal effects; Late Quaternary sea level changes; and Milankovitch cyclicity in exerting a control on global climate.

Gates, D.M. 1993. *Climate Change and Its Biological Consequences.* Sunderland, Mass.: Sinauer Associates, Inc.
An extremely readable textbook on climate change and its biological consequences, with clear diagrams.

The book is aimed at college/undergraduate students, and is in eight chapters: climate change: cause and evidence; past climates; plant physiognomy and physiology; past vegetational change; forest models and the future; ecoystems; agriculture, droughts, and El Nino; what to do?

Hsü, K.J. 1983. *The Mediterranean was a Desert: A Voyage of the Glomar Challenger.* New Jersey: Princeton University Press.
Written by one of the co-chief scientists on the deep-sea drilling vessel *Glomar Challenger*'s voyage to the Mediterranean in 1970 which first showed the Messinian salinity crisis when the ocean basin dried up. This very readable book describes the evidence that led to the proposal that the Mediterranean had evaporated. It introduces geological concepts with a minimum of terminology to explain the significance of the discovery and describes the technical problems encountered in undertaking such work. An interesting introduction to the excitement associated with discoveries made by Earth scientists who are involved with drilling into the sediments and rocks in the deep oceans.

Imbrie, J. and Imbrie, K.P. 1979. *Ice Ages: Solving the Mystery.* Harvard: Harvard University Press.
A very readable account of the causes and effects of Ice Ages. Strongly recommended to any student and teacher who wants a good historical background in global climate change.

Mannion, A.M. 1991. *Global Environment Change: A Natural and Cultural Environmental History.* Harlow, Essex: Longman Scientific & Technical.
A synopsis of how natural and cultural agents have transformed the Earth's surface during the past 3 million years. It concentrates on aspects of environmental change induced by human activities as humans evolved through the Late Quaternary, as societies emerged, animals and plants began to be domesticated, agricultural systems spread, and as industrial development has increased during recent times.

Williams, M.A.J., Dunkerley, D.L., Deckker, P. de, Kershaw, A.P. and Stokes, T. 1993. *Quaternary Environments.* London: Edward Arnold.
Comprehensive and well-illustrated text which examines the environmental changes that have taken place throughout Quaternary time. Useful for undergraduate students as well as a reference source for teachers and researchers. Emphasis is placed on the interactions between geological, biological, and hydrological processes that have caused environmental change throughout this period and have resulted in the present environments.

We are the Warriors of the Sun
We are the Warriors of the Sun

If it's true 'bout no more water but the fire next time
Will the Children of the 80s be ashes or live to their prime
If we don't heed reasonable people and their warnings
of days to come
We'll all be incinerated Warriors of the Sun

We'll be there to feed the hungry and to tend the sick
We'll be there when the night gets black and the
going gets thick
We'll be there to carry your feeble, your hopeless and
your weary ones
We are the Warriors of the Sun

The black angel of Memphis is by our side
He walked and he talked in truth until the day
that he died
He said, 'It ain't what you can do for me, ah but what
can I do for thee?'
And he took us to the mountaintop and he set us free
           (We are the Warriors of the Sun)

We may be crazy, and it may be our final run, yeah,
We are the Warriors of the Sun

Everybody knows that the whales are smarter than we
Probably that's why we call them the 'King of the Sea'
We killin' everything on dry land, why don't we just let
the fishes be

Some of us are Greenpeace Warriors of the Sea
           (We are the Warriors of the Sun)
           (We are the Warriors of the Sun)
           (We are the Warriors of the Sun)

Joan Baez, 'Warriors of the Sun'

This chapter examines the two main issues relating to global atmospheric change, of atmospheric ozone depletion and emissions of greenhouse gases and, therefore, provides a contrast with the generally more local atmospheric pollution caused by acid deposition, or acid rain (see Chapter 4). Although the impact on global atmospheric change caused by human activities is emphasized, natural processes are also discussed.

Stratospheric ozone depletion

Ozone in the upper atmosphere – the stratosphere – is part of an important naturally occurring shield around the Earth. The ozone layer or shield acts as an absorbant for a large amount of the ultraviolet radiation reaching the Earth from the Sun, and is involved in controlling the thermal structure of the stratosphere by absorbing incoming ultraviolet solar radiation and the outgoing longer wavelength radiation from the Earth's surface.

In polar regions, stratospheric ozone depletion during the winter months occurs mainly through the catalytic action of chlorine which is freed by chemical reactions which take place on polar stratospheric cloud (PSC) particles. In contrast, at middle to low latitudes where the solar illumination is more intense, and because PSCs are absent, the rate of ozone destruction is influenced by a combination of different catalytic

reactions. The relative importance of the possible chemical reactions which lead to stratospheric ozone depletion, and the precise controls on influencing such depletion, remain controversial. For example, gas phase models of the atmosphere suggest that nitrogen oxides, rather than chlorine and associated chemical species, are more important in destroying stratospheric ozone (Fahey *et al.* 1993). *In situ* measurements of stratospheric sulphate aerosol, reactive nitrogen, and chlorine concentrations at middle latitudes confirm the importance of aerosol surface reactions that convert active nitrogen to a less reactive, reservoir form, resulting in mid-latitude stratospheric ozone being less vulnerable to active nitrogen but more vulnerable to chlorine species. The effect of aerosol reactions on active nitrogen depend on the rates of gas phase reactions, therefore following volcanic eruptions aerosol concentrations will only have a limited effect on ozone depletion at these latitudes (ibid.).

Ozone is an effective greenhouse gas, particularly in the upper and middle troposphere of the atmosphere. Ozone (O_3) is formed in the atmosphere, where a series of complex chemical reactions are catalysed by the action of sunlight on carbon monoxide (CO), methane (CH_4), nitrogen oxide radicals (NO_x), and non-methane hydrocarbons (Figure 3.1).

A reduction in the amount of ozone in the upper atmosphere means that more solar radiation reaches the troposphere and Earth's

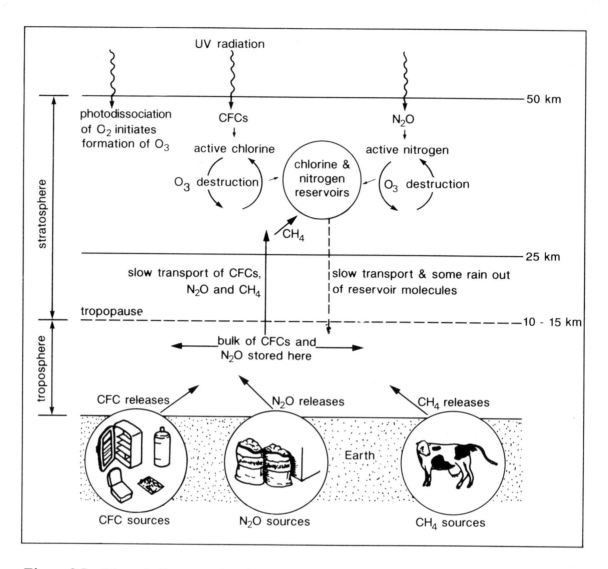

Figure 3.1 *Schematic diagram to show the principal sources of atmospheric ozone, and the main reactions that cause ozone depletion in the stratosphere. After Smith and Warr (1991).*

Box 3.1

UV-B and the ozone shield

The importance of stratospheric ozone is its role in controlling the ultraviolet-B (UV-B) reaching the Earth's surface. The ozone layer absorbs part of the outgoing longwave radiation and re-radiates it back to the troposphere below, and to the Earth's surface. UV-B is a normal component of sunlight, with up to about 0.5 per cent of the energy reaching the Earth's surface under a clear sky at noon comprising biologically active UV-B radiation, but any significant increase in UV-B radiation above natural levels is potentially harmful to human health and the environment.

Naturally, the absolute intensity of UV-B radiation reaching the Earth's surface is influenced by many factors, including the angle of the incident sunlight, principally controlled by the seasons and the time of day. Estimating UV-B intensity at the Earth's surface cannot be done from measuring stratospheric ozone levels alone, therefore it is important to obtain accurate UV-B data to establish long-term trends and causal factors. UV-B which passes through the stratosphere may be absorbed and scattered by air pollution, including ozone, in the lower atmosphere.

surface which, in turn, leads to greater surface warming. Reduced O_3 levels in the stratosphere, however, also mean that this part of the atmosphere becomes cooler since it now absorbs less long wavelength and solar radiation, and emits less to the troposphere – the result is that the Earth's surface will tend to cool. It so happens that the warming due to incoming solar radiation, related to the ozone column in the atmosphere, and the cooling because of the long wavelength radiation, related to the actual vertical distribution of the ozone, are similar in magnitude. So, the juggling act between the magnitude of the ozone-related cooling or warming of the atmosphere and Earth's surface is critically affected by the magnitude of any change in the ozone concentration and distribution – obviously strongly influenced by latitude, altitude, and the seasons. Furthermore, the creation and destruction of O_3 in the stratosphere is affected by the reactive chemical elements of oxygen, hydrogen, nitrogen, and the halides (e.g., chlorine and bromine). UV-B radiation (see Box 3.1) probably influences the recycling of organic matter in the surface layers of the oceans, because the processes are mediated by bacterioplankton, which are affected by solar radiation. Elevated levels of incoming solar ultraviolet-B (UV-B) radiation, due to the destruction of the stratospheric ozone layer, could lead to reduced bacterial activity in the surface layers of the world's oceans, with an accompanying increase in the concentrations of labile dissolved organic matter because bacterial uptake of this is suppressed (Herndl *et al.* 1993).

On 11th January 1992, high levels of O_3-destroying chlorine chemicals were recorded over Moscow, Amsterdam, and London. Early in 1992, a combination of anthropogenically created pollutants and a cocktail of chemicals from volcanic eruptions caused an unprecedented problem in the upper atmosphere. The news was released on 3rd February 1992 by both the European Ozone Research Co-ordinating Unit and US government scientists. Measurements revealed that chlorine-based chemicals were present in concentrations great enough to create a complete depletion of O_3 in the ozone layer – the layer that protects people from being sunburnt by the ultraviolet radiation. Skin cancer and eye cataracts can be caused by excessive exposure to UV radiation.

A hole in the Earth's protective ozone layer

In 1977, the British Antarctic Survey observed and recorded a zone of stratospheric O_3 depletion, which is commonly referred to as a 'hole', in the naturally occurring ozone layer at 20–30 km above the Earth's surface. It was not until a decade later, however, that concern arose about the possible implications of this observation. Farman *et al.* (1985) were the first to show that the springtime values for total stratospheric ozone concentrations at the British Antarctic Survey stations (Argentine Islands at 65°S, 64°W, and Halley Bay at 76°S, 27°W) had fallen significantly since 1957. They emphasized that lower stratospheric circulation had not changed and, therefore, the decreased stratospheric ozone levels were attributable to a chemical cause resulting in ozone depletion. They further suggested that the very low temperatures which prevail in midwinter, until after the spring equinox, make the stratosphere over the Antarctic region uniquely sensitive to the destruction of O_3 in chemical reactions involving chlorine molecules. Antarctic ozone depletion generally occurs between altitudes of 12 and 22 km, the main region of stratospheric cloud formation. The size of this zone or hole of depleted O_3, which has fluctuated over the years, appears to be getting bigger. The hole exists because the ozone-producing reactions have been inhibited or reduced in activity, possibly as a result of the excessive generation of certain ozone-destroying anthropogenic emissions of chlorofluorocarbons (CFCs) and other chemical species.

In the stratosphere, anthropogenic chlorine is converted to chemically reactive forms that lead to a depletion of the ozone, with particularly large O_3 losses during the springtime in Antarctica. Heterogeneous chemistry in stratospheric clouds, followed by the action of sunlight, converts the stratospheric chlorine from relatively inert forms to the much more reactive forms, of which ClO is dominant. Enhanced ClO is now known to precede the Antarctic and Arctic O_3 depletion (Waters *et al.* 1993). It has been suggested that the O_3 loss in the south, long before the development of the Antarctic O_3 hole, can be masked by the influx of O_3-rich air (ibid.). Although there is a decline in the anthropogenic emissions of gases that put chlorine into the stratosphere, the effects of these anthropogenic emissions will continue to increase over the next decade, and will remain for about a century at levels higher than those which were initially responsible for the Antarctic O_3 depletion.

The British research base on Antarctica, Halley, has monitored the meteorological conditions in this region since 1957, and up until 1977 there appeared to be no cause for alarm – climatic conditions appeared stable and the O_3 seemed intact. In 1979, however, a thinning of the O_3 layer was noted but its significance went

unappreciated, probably because the British base was the only meteorological station in the world to record these changes and, at the time, the results were considered anomalous, probably because of outdated instrumentation giving erroneous results! The stage was set for the dramatic discoveries in the 1980s.

On 7th October 1987, the American Nimbus 7 satellite which was monitoring the O_3 layer over Antarctica recorded a substantial depletion of the O_3 layer at a height of 16.5 km, with a 97.5 per cent destruction of the amount of ozone measured on 15th August 1987 (Farman 1987). This depletion, equivalent to more than the area of the USA, had developed from the Antarctic spring with more than 50 per cent of the O_3 over Antarctica being destroyed within 30 days.

In the Antarctic spring of 1991, balloon-borne observations showed local ozone reductions approaching 50 per cent in magnitude which were observed at altitudes of 11–13 km (lower stratosphere) and 25–30 km (upper stratosphere) above the South Pole and McMurdo Station – these reductions being in addition to the normal springtime reductions at altitudes of 12–20 km (Hofmann et al. 1992) (Plate 6). Until then, ozone depletion had not been observed at these altitudes, and by September 1991 the net result was an ozone column 10–15 per cent less than recorded in previous years. It was also observed that this depletion coincided with penetrations into the lower stratosphere polar vortex of increased concentrations of sulphate aerosol particles from the volcanic eruptions that took place in 1991 (e.g., the eruption of Mount Hudson, Chile, at 46°S between 12th and 15th August, and Mount Pinatubo, Philippines, at 10°N on 15th and 16th June). The most plausible explanation for this ozone depletion in the 11–13 km altitude layer is that it occurred because of 'heterogeneous reactions' in the poleward-drifting volcanic cloud (ibid.).

Attention naturally has turned from the Antarctic to include the Arctic. Are there signs of a hole in the O_3 layer there? It was not until 1989 that a clear affirmative came (ibid.) (Plate 6). Not only were scientists able to detect the type of stratospheric clouds that allow the O_3-destroying reactions to occur, but they were also able to measure the beginning of ozone deple-

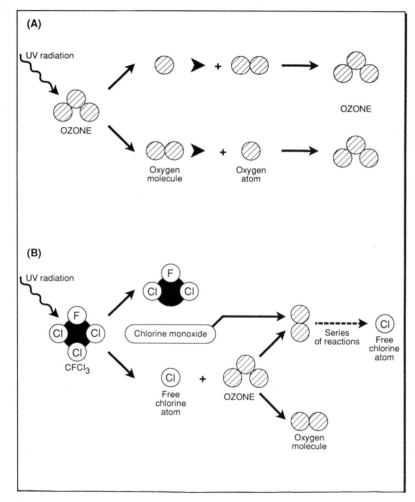

Figure 3.2 (A) The naturally occurring chemical processes leading to the formation and decomposition of ozone in the atmosphere in the presence of ultraviolet radiation. (B) The decomposition of ozone initiated by chlorine atoms released during the breakdown of a commonly occurring, anthropogenically generated CFC believed to be harmful to the atmosphere (CFCl₃). Not all the two-atom (diatomic) molecules of oxygen combine to form ozone, and the free chlorine atoms that are liberated are potentially capable of initiating further reactions that lead to the breakdown of ozone.

tion at a height of 22–26 km. This followed the coldest January in the North Pole stratosphere for at least 25 years.

A major problem to be solved by further research is to establish exactly where in the 20–40 km zone above ground level the thinning of the ozone layer is most dramatic. Heterogeneous reactions in the lower stratosphere present the greatest risks to the ozone layer since it is here that most of the protective ozone is concentrated.

Ozone that is present in the stratosphere is know as high-level or stratospheric ozone, while O_3 that is present in the troposphere is known as low-level or tropospheric ozone. Tropospheric O_3, which controls the chemical cycling of atmospheric trace gases and exerts an important effect on global climate, is supplied naturally by downward transport from the stratosphere and, depending upon the local levels of NO_x, is produced by **photochemical** reactions. Stratospheric O_3 is decreasing, whereas above polluted regions in the northern hemisphere, tropospheric O_3 is increasing and often rises above the natural background levels. Over Europe, tropospheric O_3 concentrations may have more than doubled in the last 100 years (Volz and Kley 1988). At sufficiently high concentrations, tropospheric O_3 is damaging to life and is probably partially responsible for forest dieback near industrialized centres. Tropospheric O_3 is produced as a waste product from automobile exhaust and many industrial processes. A consequence of this is that the concentration of tropospheric O_3 is at its greatest around large industrial cities, and poses a major threat to public health. Tropospheric O_3 is also a greenhouse gas and may be an important contributor to global warming if it is produced in large quantities.

Ozone loss is most pronounced during the northern hemisphere winter months or the Antarctic spring. Over the Antarctic, the ozone layer is destroyed by so-called heterogeneous reactions. These are reactions of chemicals in different states, for example as between gas and liquid, gas and solid, or solid and liquid. Such heterogeneous reactions take place on the surface of crystals in freezing clouds in the stratosphere. The catalysts for these reactions are CFCs produced by human activities. Since these reactions were not predicted, their discovery came as a surprise in the 1980s. The reactions are so rapid that 95 per cent of the destruction of the ozone layer in any year occurs in the first few weeks of the beginning of each Antarctic spring. A simplified series of reactions that lead to the breakdown of atmospheric ozone are illustrated in Figure 3.2.

The particular clouds in the stratosphere where the ozone is destroyed over Antarctica apparently only form at temperatures below about –80°C, although there is another cloud type responsible for ozone depletion which forms at –72°C. This latter cloud type is confined to polar air and is nine times more abundant than the colder clouds. Furthermore, the warmer clouds are widespread over the Arctic whereas the cold types only occur over Antarctica. The warmer clouds are believed to contain fewer reactive chemicals than the cold types. For example, the warmer clouds do not contain hydrochloric or sulphuric acids, but they do contain nitric acid that can trigger the heterogeneous reactions so harmful to the ozone layer.

Conventional wisdom puts the blame for the depletion of the ozone layer over Antarctica as due almost entirely to the accelerated anthropogenic emissions of certain reactive gases, such as the chlorine compounds CFCs. But in the USA, at Boulder, Colorado, a group of scientists from the National Oceanic and Atmospheric Administration (NOAA) suspect that natural fluctuations in the sea-surface temperature in the eastern equatorial Pacific may be a major control on the concentrations of O_3 in the atmosphere. Their research in the eastern Pacific over the past 25 years has shown that between 1962 and 1975 when the eastern equatorial Pacific cooled, the global atmospheric O_3 budget increased (Joyce 1991). Then, between 1976 and 1988, when the eastern equatorial Pacific waters warmed, the global atmospheric O_3 budget decreased. So, there may be a good correlation between sea-surface temperature and atmospheric ozone levels, but the mechanism by which they are linked remains unclear. Of course, finding natural, non-anthropogenic, cause-and-effect relationships between the levels of ozone and sea-surface temperatures is not a recipe for complacency in controlling human emissions of various gases.

In the stratosphere over Europe, the concentration of O_3 is decreasing at a rate approximately twice as fast as previously thought

(Brown 1991). In a UK report published in July 1991 by the Stratospheric Ozone Review Group, it was stated that the concentration of O_3 in a wide band from the latitude of southern England to about latitude 30° North decreased by 8 per cent between 1979 and 1990. The potential problem of O_3 depletion is not confined just to the Antarctic and Arctic – the effects may be greatest at the poles, but the knock-on effects of O_3 depletion over other parts of the globe, such as Europe, are now being appreciated. While the additional CO_2 would warm the lower atmosphere, it could cool the lower stratosphere and increase the formation of clouds that convert the potential O_3-depleting species to their active forms (Austin *et al.* 1992), that is, enhance the stratospheric cloud chemistry that leads to the destruction of O_3 by chlorine from anthropogenically produced CFCs. In a numerical 3D simulation of the northern hemisphere winter stratosphere, it has been shown that a doubling of the atmospheric CO_2 concentration – something that is likely to happen in the next century if steps are not taken to avert global warming – could lead to the formation of an O_3 hole in the Arctic and over northern Europe comparable to that observed over Antarctica, with almost 100 per cent local depletion of the O_3 in the lower stratosphere (ibid.). The upper stratosphere would be affected to a lesser degree, and the Arctic would still have greater protection each spring compared to the Antarctic. But, since there are many more people living in northern Europe and Canada at high latitudes, the risk of skin cancer, cataracts, and other hazards would be enhanced. It was estimated that only about 20 per cent of the winters might produce an ozone hole over the Arctic, and persist into April or even May of such years (ibid.).

CFCS, HCFCs, HFCs, halons

Human activities involve the use of aerosols in agriculture, industry, and domestic situations which release CFCs. Chlorine in CFCs has been linked to stratospheric ozone depletion. The chemical stability of CFCs gives them long atmospheric lifetimes, and because they provide a long-term source of chlorine in the strato-

sphere CFCs are seen as posing a serious source of atmospheric pollution which could contribute to global climate change. Peak chlorine loading on the atmosphere will be reached over the next five years and, depending upon the exact date for phasing out CFCs, the loading should return to present-day levels some time between the years 2000 and 2010. Scientists participating in the United Nations Environment Program (UNEP)/World Meteorological Organization to assess the role of CFCs in contributing to stratospheric ozone depletion have given a high priority to minimizing the future risks of ozone depletion by phasing out such harmful anthropogenically created chemicals.

CFCs are widely used in the electronics industry where, for example, CFC-113 is a solvent used in more than 100 specialized applications. Pre-Industrial Revolution levels of CFCs were zero, so the emission of these molecules into the atmosphere is entirely due to human activities. While large parts of industry have attempted to develop alternative substances, there are many who believe that the electronics industry has been particularly slow in responding to the need for considerable research and development into replacement chemicals. Human activities, however, still result in the current annual production of 10^6 tonnes of CFCs, but the world consumption of CFCs 11, 12, and 113 is now 40 per cent less than in 1986, which is considerably less than the quantities permitted under the Montreal Protocol: the 1990 London amendments to the Montreal Protocol require further reductions.

Other chemical compounds that are believed to be destroying the stratospheric ozone layer include the oxides of bromine which are much more potent than the equivalent quantities of chlorine compounds. Reactions of bromine monoxide (BrO) and chlorine monoxide (ClO) can destroy ozone even in the absence of sunlight which generally initiates such destructive reactions. Another set of reactions with ClO and BrO produces OClO, believed to be one of the gases responsible for the destruction of the ozone layer over Antarctica and the Arctic in the spring. The fumigant methyl bromide is a major ozone depleter in the upper atmosphere, and worries over its adverse effects on health and safety (toxic by inhalation, and it can cause pulmonary oedema, and disorders of the central nervous system) led the Netherlands to cut back drastically on its use between 1981 and 1989. Of the total annual global production of about 67,000 tonnes of methyl bromide, the USA uses about 43 per cent (26,000 tonnes), of which 22,300 tonnes are used as a soil fumigant. In November 1992, an international agreement was reached at a meeting of the Montreal Protocol held in Copenhagen to freeze the production and consumption of methyl bromide at 1991 levels to take effect from 1st January 1995. Besides the natural emissions of methyl bromide, anthropogenic emissions may account for 0.05–0.01 per cent of the observed annual global ozone depletion of 4–6 per cent and could increase to about one-sixth of the predicted ozone loss by the year 2000 if annual methyl bromide production increases at the current rate of 5–6 per cent (Buffin 1992).

Two of the **halons** that are particularly responsible for destroying the ozone layer are $CBrClF_2$ and $CBrF_3$. $CBrClF_2$ is now 2 parts per trillion by volume of the atmosphere, and since the early 1980s it has increased at a rate of 12 per cent per annum. The halon $CBrF_3$ is now 1.3 parts per trillion and increasing at 5 per cent per annum (Singh et $al.$ 1988). $CBrF_3$, amongst all the chemicals that are destroying the protective ozone layer, is perhaps the most effective and efficient of all the CFCs known at present. The British Antarctic Survey scientists now believe that the principal chemical culprits that are destroying the ozone layer are two particularly widely used compounds of bromine (1211 and 1301) which have a long residence time in the stratosphere (Thompson 1992).

Alternatives to CFCs are being developed and marketed. HCFCs and HFCs, for example, contain hydrogen in the structure and, unlike CFCs, have short atmospheric lifetimes and tend to be destroyed in the lower atmosphere by natural processes. HFCs contain no chlorine and, therefore, do not contribute to stratospheric ozone depletion, whereas HCFCs contain relatively small amounts of chlorine which provides some contribution to stratospheric ozone depletion. As examples of potential substitutes for various CFCs, HFC-134a could replace CFC-12 in refrigeration, air-conditioning, certain foams, and medical aerosols, HCFC-123 could replace CFC-11 in refrigeration and air conditioning, HCFC-141b for CFC-11 in energy-efficient insulating foams and solvent cleaning, HCFC-

124 for CFC-114 and HFC-125 for CFC-115 in certain refrigeration uses, and HCFC-225ca/cb for CFC-113 in solvent cleaning (in precision engineering and electronic industries).

The relative ozone depletion potentials (ODPs) of various CFCs, HCFCs, and HFCs, calculated over their full lifetimes in the atmosphere, are compared in Table 3.1. Although HCFCs and HFCs appear to break down relatively easily in the lower atmosphere, the ultimate breakdown products are acidic compounds which will contribute to acid rain at minimal levels, but will not contribute to the formation of photochemical smogs in urban areas. The hydrogen, chlorine, and fluorine released by the breakdown products of HCFCs and HFCs should be removed from the atmosphere by dissolution in cloud water followed by precipitation as rain within an average of around two weeks. Other trace amounts of the potentially harmful breakdown products such as carbonyl and trifluoroacetyl halides are expected to remain in the atmosphere for a few months, where they should be incorporated into cloud water, the oceans, and land surface, and hydrolysed to CO_2 and trifluoracetic acid, respectively, and form the corresponding hydrochloric and hydrofluoric acids (AFEAS and PAFT Member Companies 1992). Further independent research is needed to evaluate any potentially harmful environmental impacts from these breakdown products of the alternative fluorocarbons.

Table 3.1 *Ozone depletion potentials (ODPs) of the principal CFCs, HCFCs, and HFCs.*

	Chemical	*ODP*
CFCs	11	1.0
	12	1.0
	113	0.8
	114	1.0
	115	0.6
HCFCs	22	0.055
	123	0.02
	124	0.022
	141b	0.11
	142b	0.065
	225ca	0.025
	225cb	0.033
HFCs	32	0
	125	0
	134a	0
	152a	0

Source: AFEAS and PAFT Member Companies (1992).

GLOBAL ATMOSPHERIC CHANGE

The greenhouse effect – global warming

Most of this chapter examines global warming, or the so-called 'greenhouse effect', a phenomenon that has become widely reported over the last few years. It was first observed in 1896 independently by the Swedish chemist Arrhenius, and the American geologist Thomas C. Chamberlain. It is interesting to note that Arrhenius suggested that by doubling the natural atmospheric levels of CO_2, average temperatures would rise by about 5–6°C. This phenomenon has been termed the greenhouse effect because it was originally thought that greenhouses are heated in a similar manner. The Sun's rays passing through the glass of a closed greenhouse include shorter wavelength (ultraviolet) radiation which is absorbed by objects inside, which in turn re-radiate the heat but at longer wavelengths (**infrared radiation**) to which the glass is nearly opaque. The heat is therefore trapped in the greenhouse with the net result that there is a sharp rise in temperature, together with more condensation. The condensation of water particles on the glass then leads to some cooling, but without ventilation and in bright sunlight the greenhouse can reach intolerable temperatures. The commonly cited analogy is not perfect, because the warming of air in a greenhouse is mainly due to the trapped air inside which is unable to mix with the cooler air outside, but it represents a crude way of looking at the global greenhouse effect.

To investigate the extent to which human activities have begun to affect global climate and warm the planet, in 1990 a major review of the scientific evidence was conducted by the Intergovernmental Panel on Climate Change (IPCC), in preparation for the World Climate Conference which took place in November of that year. This report was followed by an update in 1992 in which, although some of the earlier predictions were downwardly revised, the findings remained essentially the same – that anthropogenic emissions of greenhouse gases are contributing to global warming. Perhaps the most significant shift in perspective by the IPCC between their 1990 and 1992 reports concerns the rate at which concentrations of greenhouse gases are increasing, which is the principal control on how fast the world might be warming. Under the IPCC 1990 'business-as-usual' scenario, they estimated that the CO_2 doubling milepost could be reached as early as the year 2025, but the more recent forecasts now predict that this doubling will be delayed until the year 2050 or beyond. Initial estimates of global warming and the rise in sea level (caused both by an expansion of the world's oceans because warmer water occupies a greater volume, and through melting of polar ice) suggested a rise of 10–30 cm by the year 2030, and 33–75 cm by the year 2070, compared to present sea level. The most recent estimates, however, have downwardly revised these figures, and suggest a global sea level rise of 2–4 cm per decade due to thermal expansion of ocean waters alone, and an additional 1.5 cm per decade contributed by melting ice caps and glaciers.

Recently, Manabe and Stouffer (1993) have examined the century-scale effects of increased atmospheric CO_2 on the ocean–atmosphere system. Using computer models to investigate climate sensitivity (see Box 3.2), they conclude that by doubling and quadrupling the concentration of atmospheric CO_2, global mean surface air temperature may increase over 500 years by 3.5°C and 7°C respectively, with corresponding global sea level rises of 1 m and 2 m due to thermal expansion of ocean waters alone. Any melting of ice sheets could make these values

Box 3.2

Climate sensitivity

The concept of climate sensitivity has been developed to provide an indication of the amount of global warming that could result from the increased concentration of greenhouse gases. Climate sensitivity is commonly expressed as the global temperature rise that would result from a doubling of CO_2 concentrations relative to pre-industrial levels. Climate sensitivity is estimated from general circulation models (GCMs), which are computer models that simulate the actual or real behaviour of global climate systems. In order to deal with the complexity of real climate systems over land and the oceans, GCMs need to be run on the fastest available super-computers. Current IPCC estimates of climatic sensitivity is that it is between 1.5 and 4.5°C, with a best estimate of approximately 2.5°C.

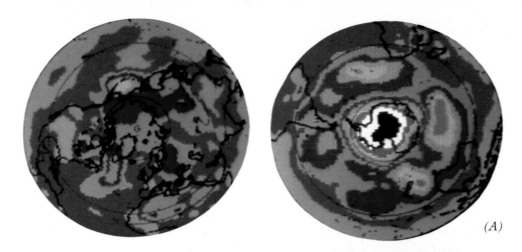

(A)

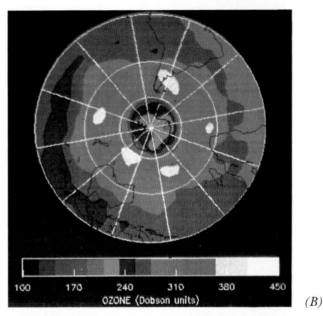

(B)

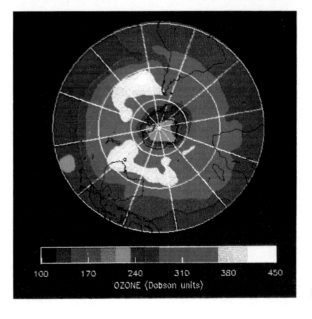

(C)

Plate 6 Maps of the ozone column based on the data from the Total Ozone Mapping Spectrometer (TOMS) on board NASA's Nimbus 7 satellite. (A) Maps for the Antarctic and Arctic for 1989. These show greater depletion of the ozone column in the Antarctic. (B & C) The variation in ozone depletion in the Antarctic summers for 1990 and 1991. Note that the amount of ozone present is measured in Dobson units. If the atmosphere was compressed under a pressure of 1,000 millibars, it would be about 8 km thick, the thickness of oxygen would be about 1.5 km and ozone about 3 mm. One Dobson unit is a hundredth of a millimetre of ozone in such a compressed atmosphere at standard pressure and temperature. *Courtesy of NOAA/NESDIS/NCDC/SDSD.*

Plate 7 Boracay Island in the Philippines is one of many small islands which are threatened by the possibility of global sea level rise. Most people on these islands live only a few metres above the present sea level because it is difficult to build on the higher adjacent steep slopes. Fresh water is obtained from the ground. Rising sea levels would not only cause flooding of settlements and fields, but also lead to the intrusion of saline groundwater under the island, thereby reducing and contaminating the present water resources.

Plate 8 Icebergs frozen into sea ice in Otto Fjord, northern Ellesmere Island, Canadian High Arctic. These icebergs form as glaciers and calve when they enter the sea. This process constrained the extent of glaciation in high latitudes during the Last Glacial. Global warming may lead to widespread melting of sea ice which in turn will lead to a decrease in aridity in the Arctic. Increased precipitation, primarily as snowfall, will lead to the growth of glaciers, rather than their melting.

Plate 9 Burning oil wells and US troops in action in Rumallah, Kuwait, during the Gulf War in 1991. *Courtesy of Abbas/Magnum.*

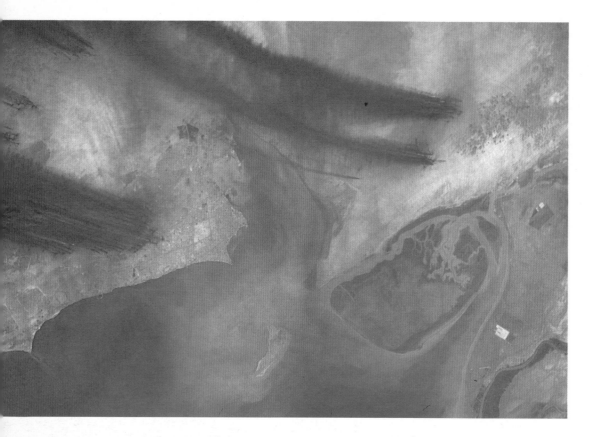

Plate 10 *Satellite image of the Kuwait oil fires created during the Gulf War in 1991, and highlighting the extent and density of the local atmospheric pollution.* Courtesy of NASA/Lunar and Planetary Institute.

Plate 11 *Photochemical smog over Los Angeles.* Courtesy of Comstock.

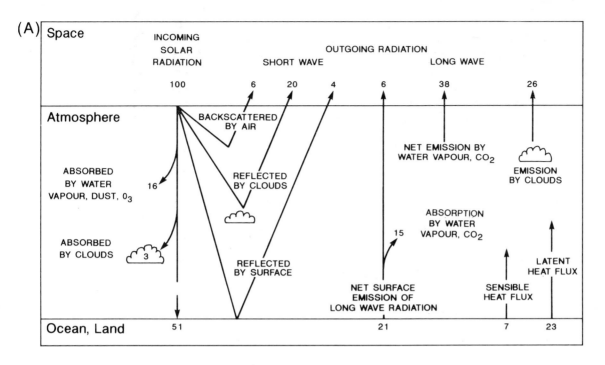

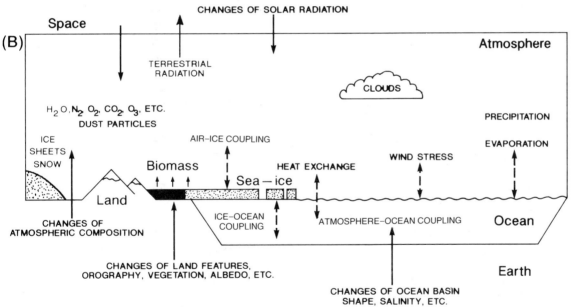

Figure 3.3 (A) Mean annual radiation and heat balance of the atmosphere relative to 100 units of incoming solar radiation. (B) The major components of the climate system, illustrating feedback mechanisms which play an important role in climatic variations. After US Committee for GARP 1975, in IPCC (1990).

much larger. Also, under such a scenario, a quadrupling of atmospheric CO_2 levels could force the oceans into a new stable state in which thermohaline circulation would have ceased entirely, and the **thermocline** deepened substantially. Such changes in the structure of the oceans would prevent the **ventilation** of the deep ocean, with a likely profound effect on the carbon cycle and biogeochemistry of the coupled system. Deep-ocean anoxia would prevail, and considerable amounts of the carbon budget could be locked up in organic-rich muds,

GLOBAL ATMOSPHERIC CHANGE

in turn severely stressing biomass production in surface waters, and thereby leading to mass mortalities and, possibly, extinction events in marine organisms. According to the IPCC (1990), a quadrupling of greenhouse gases could occur under their business-as-usual scenario if greenhouse gas emissions continued unchecked until the end of the twenty-first century, so these possibilities are potentially realizable.

The natural balancing act

The air around the planet provides a balanced mix of gases and water vapour to sustain life at a comfortable level. Amongst the planets in the Solar System, the Earth is unusual because it contains substantial amounts of methane (CH_4) and ammonia (NH_3) that should not coexist with oxygen. CH_4 burns in oxygen, and this therefore means that some mechanism must be present that maintains the CH_4 in the atmosphere; this mechanism is the metabolic processes of life itself. Scientists have looked for the presence of methane in distant planets as the indicator of extra-terrestrial life. The atmosphere is largely responsible for keeping the Earth's surface temperature range at levels which generally keep water, so essential for life, in a liquid

state and provide a climatic range conducive to existence as at present.

A comparison of the Earth and the Moon shows that the Moon, which lacks an atmosphere and is at roughly the same distance from the Sun as the Earth, has a much greater range of surface temperatures. The side of the Moon facing the Sun can reach 100°C, while on the opposite side night temperatures drop to a staggering –150°C. In fact, the average lunar temperature is –18°C, at which level the incoming heat energy or solar flux roughly balances the energy radiated from the Moon's surface. In contrast, the average temperature at the Earth's surface is 15°C.

The much warmer surface temperature of the Earth compared to that of the Moon is a product of the atmosphere acting as a blanket to keep a large amount of the solar energy trapped on Earth. Without the atmosphere, the Earth's surface temperatures would be similar to that of the Moon.

The Earth's atmosphere provides the all-important insulation to sustain the rich diversity of life on Earth. It is easy, however, to overlook the fact that organisms have evolved to survive best in the Earth's climate as it exists without the intervention of humans to change this equilibrium. If human activities exert a forcing effect on the global climate to cause an irreversible and rapid climatic change, either through a warming or cooling, then many organisms, including

Box 3.3
Sulphate ions, negative feedback, and climatic cooling

The climatic significance of sulphate aerosols derived from anthropogenic SO_2 emissions results in three main effects: (1) possible increase in longevity of individual clouds; (2) a reduction in incoming short wavelength solar radiation under clear-sky conditions; and (3) a possible increase in cloud reflectivity because the sulphate aerosols act as cloud condensation nuclei (Wigley and Raper 1992). Sulphate aerosols are very short-lived, therefore increases or reductions in anthropogenic emissions have a rapid effect on global warming. There are large uncertainties associated with estimating the radiative forcing caused by sulphate aerosols, for example, the quantitative links between changes in the mass of sulphate aerosol and changes in the number of cloud condensation nuclei are poorly understood. Such uncertainties mean that arguments about the exact consequences of atmospheric sulphate aerosols derived from anthropogenic SO_2 emissions should

be treated speculatively, with more research being necessary in order to improve the understanding of cause and effect.

Sulphate ions (SO_4^{2-}) represent the dominant aerosol species in the Antarctic atmosphere, as well as being a significant part of the ice and snow. The sulphate ion is the end member of the oxidation processes in the atmospheric sulphur cycle. On a global scale, anthropogenic emissions of sulphur dioxide (SO_2) make the largest contribution to the amount of atmospheric sulphur (see Chapter 4). When anthropogenic emissions are excluded, the greatest concentration of tropospheric sulphate is in tropical regions where the emission of biogenic dimethylsulphide is highest. The influence of volcanic eruptions in contributing to increased levels of tropospheric sulphate is greatest over regions such as Indonesia and the western part of the North Pacific. Increased levels of tropospheric

74

humans, may not survive. This is the essence of human interest and concern over the greenhouse effect.

The surface of the Earth emits radiation, mostly in the longer wavelength or infrared part of the spectrum, and at a power of about 390 Watts per square metre, which represents about one-half of the total incoming solar energy. The weather in the troposphere is the consequence of this energy exchange, or balancing act, between incoming and outgoing radiation. Figure 3.3 illustrates the radiation and heat balance of the atmosphere and the major components which contribute to the climatic system.

Certain gases – the greenhouse gases – such as water vapour (H_2O), carbon dioxide (CO_2), methane (CH_4), ozone (O_3), nitrous oxide (N_2O), CFCs, and halons absorb and emit infrared radiation at wavelengths which are longer than for visible light, that is at wavelengths greater than 12 microns. CFCs tend to absorb infrared radiation on the shorter wavelength side of 12 microns where, without their existence, there would be a crucial window for infrared radiation to escape from the Earth back into Space.

The temperature of the troposphere decreases with height above the ground at a rate of about 6.5°C per km, known as the environmental lapse rate. The overall pattern is for the lower, warmer layers of the atmosphere to absorb radiated

energy from the ground and the rising air, and to prevent a lot of this energy from being re-radiated upwards with the net result that there is a reduction in the amount of infrared radiation radiated back into Space from the Earth. Without this layer of greenhouse gases, the Earth would be much like Mars, cold with an average surface temperature of about –55°C and also lifeless. Another interesting comparison is between the Earth and Venus. Venus is approximately the same size as Earth, and probably has a very similar composition, but the atmosphere is different. Venus' is nearly all CO_2 with an atmospheric pressure at its surface of about 90 Earth atmospheres and temperatures of around 470°C; it is also devoid of water. In the past, Venus had water on its surface, but early in its history, as the Sun began to brighten, the surface of Venus became hot and water vapour rose to the upper layers of the atmosphere where it broke down to liberate hydrogen from the water which was then lost into Space. CO_2 increased in the atmosphere of Venus and produced a runaway greenhouse effect, warming the surface to its present-day temperature of 470°C. In contrast, on Earth, the runaway greenhouse effect was averted, probably through the evolution of life. There are, however, similar amounts of CO_2 on Earth and Venus, but on Earth much is stored in limestones and other carbonate sediments, oil and coal. The comparison is important because

sulphate may lead to global climate change because of changes in cloud albedo, their backscattering of incoming sunlight. As the mass concentration of aerosol sulphate increases, the number of cloud condensation nuclei also increase, but not in a simple way. The SO_4^{2-} produced through aqueous-phase oxidation becomes associated with pre-existing particles and does not appear to produce more particles directly. The size distribution of pre-existing aerosol particles, however, will be altered, which may lead to the generation of additional particles. The part of the emitted SO_2 available for creating new particles is that which is oxidized by hydroxyl radicals in the gas phase. Increasing tropospheric sulphate, therefore, should lead to a negative feedback (or negative forcing) of global climate, that is, a cooling effect. A large amount of current research with climatic models tied to real measurements is aimed at attempting to

understand the relative importance of positive climate forcing due to increases in CO_2 and other greenhouse gases versus negative feedback caused by SO_2 emissions.

Over the past 100 years, human activities have roughly tripled the global emissions of sulphur gases, with the result that there are enhanced sulphate aerosol concentrations of up to a factor of 100 in the northern hemisphere, but very small ones over the southern hemisphere oceans (Langer et al. 1992). Computer modelling of the distribution of tropospheric sulphate concentrations with actual mean monthly meteorological data suggests that at most 6 per cent of the anthropogenic emissions of sulphur is available for the formation of new aerosol particles; because about one-half of the sulphur dioxide is deposited on the Earth's surface, while most of the remainder is oxidized in cloud droplets with the result that the rest of the SO_4^{2-} becomes

Box 3.3
Sulphate ions, negative feedback, and climatic cooling

human activities have been releasing the CO_2 stored in the rock reservoirs over the last few decades at ever-increasing rates. In a worse-case scenario, Venus may be a celestial warning of things to come.

To maintain equilibrium temperatures at the surface of the Earth, the incoming thermal energy from the Sun must balance that emitted from the Earth – in this case, there must be some balance between the absorbed and emitted total infrared radiation. There is a natural greenhouse effect, resulting from non-anthropogenic causes, and one of the biggest challenges for those working on this problem is to appreciate the balance between the natural and anthropogenic controls on global climate change. Amongst the chemical processes conspiring to maintain equilibrium temperatures, sulphate ions in the atmosphere play a role in creating a negative feedback to any warning (see Box 3.3).

Seeing the greenhouse effect

Since detailed records started in 1973, ice and snow cover have decreased by about 8 per cent in the northern hemisphere, with a 2 per cent drop between 1983 and 1992 in the amount of Arctic sea ice cover. Scientists from the British Antarctic Survey (BAS), based in Cambridge, have been documenting somewhat disturbing indications of possible global warming. On the Antarctic peninsula below the King George VI ice shelf, field parties drilling into what should be solid ice have discovered many pockets of seawater at almost 0°C; also, the ice in this area has retreated about 48 km in the last 40 years. Clearly, if this situation is repeated around Antarctica, then it could have a devastating impact on the polar ice cap with rapid retreat and melting taking place. Furthermore, US and UK scientists estimate that the Antarctica ice sheet is shrinking annually by about 470,000 tonnes, and in Antarctica, since 1966 the Wordie ice shelf has disintegrated and reduced in area by about 1,300 km². Other early warnings of a change in global climate caused by the greenhouse effect include measurements by American satellites of an increase in surface temperature of the North Atlantic Ocean by 1°C. Widespread bleaching of tropical corals has been documented, and is regarded as arising from exceptionally warm ocean waters.

If the entire ice cap melted, then sea level around the globe could be expected to rise by 50–55 m. This scenario is extremely unlikely. If, however, as seems more probable, CO_2 levels double in the atmosphere into the next century, then a rise in sea level of decimetres can be expected (IPCC 1992). Whole areas of low-lying coastal and estuarine environments will be flooded and many islands will become totally

Box 3.3

Sulphate ions, negative feedback, and climatic cooling

associated with pre-existing particles. It has been calculated that the net effect is that the rate of formation of new sulphate particles may have doubled since pre-industrial times (ibid.).

The second major source is part of the natural rhythm of life on Earth – sulphur released into the atmosphere from the oceans through biological activity. This organic (biogenic) sulphur is produced by the activity of the countless millions of sulphate-reducing plankton and algae. Another important source of sulphur is volcanic emissions, when it is injected high into the stratosphere. The biogenic sulphur source produces a chemical known as dimethylsulphide, which is then oxidized to methane sulphonate and the non-sea-salt sulphate ion (SO_4^{2-}). The importance of the biogenic sulphate source in controlling global climate has been mooted for several years, but it is only recently that strong scientific evidence for this has been demonstrated. The evidence comes from both present-day atmospheric sampling over Antarctica and from ice cores that reveal past climatic events (see Chapter 2). A study of aerosol samples which were continuously collected at Mawson, Antarctica, between February 1987 and October 1989 has shown that the concentrations of both methane sulphonate and non-sea-salt sulphate have a strong seasonality peaking in the austral summer, which parallels the cycle of oceanic biogenic sulphur production (Prospero *et al.* 1991). So, there appears to be sound scientific evidence to link the Antarctic atmospheric sulphur cycle with biological processes in the southern hemisphere oceans.

The latest IPCC report (1992) suggests that the cooling effect of sulphur emissions may have offset a significant part of the greenhouse warming in the northern hemisphere during the past few decades.

submerged, for example some of the Maldive islands and many of the Pacific islands (Plate 7). It could be that the magnitude of the sea level changes will not be constant throughout the world because the Earth's rotation causes greater effects in the low latitudes/tropics.

Many uncertainties remain as to whether global warming has actually started, with some evidence, for example from Arctic geotherms, suggesting that it is already under way (see Box 3.4). Indeed, the IPCC emphasize uncertainties in predicting the timing, magnitude, and regional patterns of climate change, which are the result of important areas of ignorance concerning the sources and sinks of greenhouse gases, together with their control on interactions in the ocean–atmosphere system, such as in affecting cloud type and cloud cover, polar ice sheet development, ocean circulation, etc. Such large areas of ignorance and uncertainty, and the potentially very serious implications for the natural environment which might result from the anthropogenic emissions of greenhouse gases, have resulted in universal interest both in detecting any actual global warming and the causal factors. With these factors in mind, the following section considers the principal chemical culprits that contribute to global warming.

The chemical culprits

Human activities release gases that have a warming effect such as carbon dioxide (CO_2), methane (CH_4), nitrous oxide (N_2O), and tropospheric ozone (O_3), together with other potentially harmful chemicals including carbon tetrachloride, which was used extensively in dry

Table 3.2 *Characteristics of greenhouse gases.*

Gas	Major contributor?	Long lifetime?	Sources known?
Carbon dioxide	yes	yes	yes
Methane	yes	no	semi-quantitatively
Nitrous oxide	not at present	yes	qualitatively
CFCs	yes	yes	yes
HCFCs, etc.	not at present	mainly no	yes
Ozone	possibly	no	qualitatively

Source: IPCC (1990).

cleaners in North America and Europe. The IPCC (1990) listed the main greenhouse gases (Table 3.2), those influenced by human activities (Table 3.3), and their radiative effects, together with the indirect trace gas chemical–climate interactions (Table 3.4).

These gases absorb infrared radiation in the range 7–13 microns which is part of the 'window' through which more than 70 per cent of the radiation emitted from the surface of the Earth escapes into Space. The 1990 IPCC report showed the dramatic increase in the principal greenhouse gases that have resulted from human activities (Figure 3.4). Using the 1992 IPCC update, the greenhouse gases show the following annual increases: 0.5 per cent for CO_2, 0.5 per cent for CH_4, 4 per cent for CFCs, and 0.25 per cent for nitrous oxide. Their molecule-for-molecule global warming potential, normalized to CO_2 (= 1), which remains unchanged, is (with the IPCC 1990 estimates in parentheses): CH_4 = 11 (21); CFCs = 0 (> 1,000); N_2O = 260 (290). CO_2 emissions are commonly measured according to the carbon content, in millions of tonnes of carbon (MtC), where 1 tonne of carbon is equivalent to 3.67 (or 44/12) of carbon dioxide.

Measurements of a series of vertical temperature profiles in boreholes drilled into the permafrost in northern Alaska have revealed anomalously high temperature gradients in the upper 100 m of the permafrost (Lachenbruch and Marshall 1986). Since heat transfer in permafrost is almost exclusively by conduction, because there is no flow of groundwater to transfer heat by other means, it has been suggested that these anomalous temperature gradients indicate that global warming is already under way. Analysis of the data, using heat-conduction theory, suggests that warming in the order of 2–4°C has occurred this century, probably over the past few decades. This evidence is particularly convincing because it provides a long-term signal of major climatic change and measures the direct thermal consequence of any change in atmospheric temperature conditions. Data of this type, gathered by drilling and coring ice, are extremely useful since the information can be recovered when and where desired, but the process is costly.

Box 3.4
Arctic geotherms

GLOBAL ATMOSPHERIC CHANGE

Table 3.3 *Atmospheric concentrations of key greenhouse gases influenced by human activities.*[1]

Parameter	CO_2	CH_4	CFC-11	CFC-12	N_2O
Preindustrial atmospheric concentration (1750–1800)	280 ppmv[2]	0.8 ppmv	0	0	288 ppbv[2]
Current atmospheric concentration (1990)[3]	353 ppmv	1.72 ppmv	280 pptv[2]	484 pptv	310 ppbv
Current rate of annual atmospheric accumulation	1.8 ppmv (0.5%)	0.015 ppmv (0.9%)	9.5 pptv (4%)	1.7 pptv (4%)	0.8 ppbv (0.25%)
Atmospheric lifetime[4] (years)	(50–200)	10	65	130	150

1 Ozone has not been included in the table because of lack of precise data
2 ppmv = parts per million by volume; ppbv = parts per billion by volume; pptv = parts per trillion by volume
3 The current (1990) atmospheric concentrations have been estimated based upon an extrapolation of measurements reported for earlier years, assuming that the recent trends remained approximately constant
4 For each gas in the table, except CO_2, the 'lifetime' is defined here as the ratio of the atmospheric content to the total rate of removal. This time scale also characterizes the rate of adjustment of the atmospheric concentrations if the emission rates are changed abruptly. CO_2 is a special case since it has no real sinks, but is merely circulated between various reservoirs (atmosphere, ocean, biota). The 'lifetime' of CO_2 given in the table is a rough indication of the time it would take for the CO_2 concentration to adjust to changes in the emissions
Source: IPCC (1990).

Table 3.4 *Direct radiative effects and indirect trace gas chemical–climate interactions.*

Gas	Greenhouse gas	Is its tropospheric concentration affected by chemistry?	Effects on tropospheric chemistry?*	Effects on stratospheric chemistry?*
CO_2	Yes	No	No	Yes, affects O_3
CH_4	Yes	Yes, reacts with OH^-	Yes, affects OH, O_3 and CO_2	Yes, affects O_3 and H_2O
CO	Yes, but weak	Yes, reacts with OH^-	Yes, affects OH^-, O_3 and CO_2	Not significantly
N_2O	Yes	No	No	Yes, affects O_3
NO_x	Yes	Yes, reacts with OH^-	Yes, affects OH^- and O_3	Yes, affects O_3
CFC-11	Yes	No	No	Yes, affects O_3
CFC-12	Yes	No	No	Yes, affects O_3
CFC-113	Yes	No	No	Yes, affects O_3
HCFC-22	Yes	Yes, reacts with OH^-	No	Yes, affects O_3
CH_3CCl_3	Yes	Yes, reacts with OH^-	No	Yes, affects O_3
CF_2ClBr	Yes	Yes, photolysis	No	Yes, affects O_3
CF_3Br	Yes	No	No	Yes, affects O_3
SO_2	Yes, but weak	Yes, reacts with OH^-	Yes, increases aerosols	Yes, increases aerosols
CH_3SCH_3	Yes, but weak	Yes, reacts with OH^-	Source of SO_2	Not significantly
CS_2	Yes, but weak	Yes, reacts with OH^-	Source of COS	Yes, increases aerosols
O_3	Yes	Yes	Yes	Yes

*Effects on atmospheric chemistry are limited to effects on constituents having a significant influence on climate
Source: IPCC (1990).

These greenhouse gases cause radiative forcing, a measure of their ability to perturb the heat balance in a simplified model of the Earth–atmosphere system. The concept of **global warming potentials (GWPs)** has been conceived to provide a simple way of describing the potential of greenhouse gas emissions to influence future global climate by radiative forcing, which is controlled by various parameters such as the amount of gas emitted, its infrared energy absorption properties, and the amount of time (residence time) of each gas in the atmosphere. The 1992 IPCC report, using the revised GWPs, estimated that the contribution made by the main greenhouse gases to global warming breaks down as follows: 72 per cent for carbon dioxide, 18 per cent from methane, and 10 per cent due to nitrous oxide. The following sections review the principal greenhouse gases.

CFCs, HCFCs, and HFCs

Global warming potentials (GWPs) relative to a CO_2 molecule have been calculated for the principal CFCs, HCFCs, and HFCs, and are presented in Table 3.5. Initial research suggested that CFCs, because they are greenhouse gases, are important contributors to global warming, but their ability to destroy stratospheric ozone and thereby contribute to global cooling suggests that CFCs do not provide a net contribution to global warming, that is, their global warming and cooling potentials more or less cancel out.

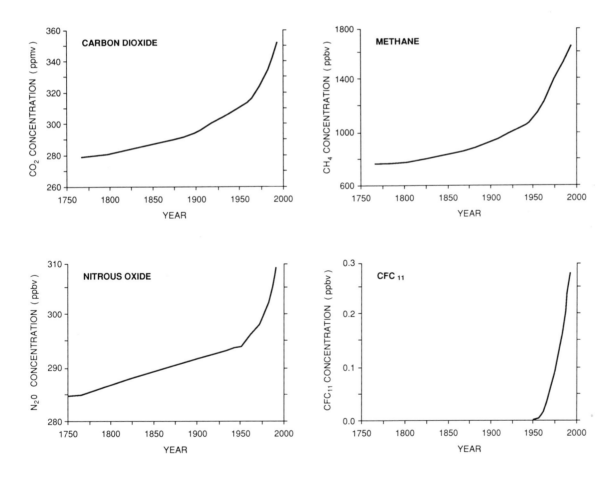

Figure 3.4 *Changes since the middle of the eighteenth century in the atmospheric concentration of carbon dioxide, methane, nitrous oxide, and the commonly occurring CFC, CFC-11. Over the past few decades there has been a very large increase in the atmospheric concentrations of CFCs, which were absent before the 1930s. After IPCC (1990).*

GLOBAL ATMOSPHERIC CHANGE

Table 3.5 *Global warming potential of principal CFCs, HCFCs and HFCs, compared with CO_2 and CH_4.*

Compound	Estimated atmospheric lifetime	GWPs for various integration time horizons*		
		20 yrs	100 yrs	500 yrs
CO_2	**	1	1	1
CH_4***	10.5	35	11	4
CFC-11	55	4,500	3,400	1,400
CFC-12	116	7,100	7,100	4,100
CFC-115	550	5,500	7,000	8,500
HCFC-22	15.8	4,200	1,600	540
HCFC-123	1.7	330	90	30
HFC-125	40.5	5,200	3,400	1,200
HFC-141b	10.8	1,800	580	200
HFC-225ca	2.7	610	170	60
HFC-225cb	7.9	2,400	690	240
HFC-134a	15.6	3,100	1,200	400
HFC-152a	1.8	530	150	49

*'Integration time horizon' is the timespan over which GWPs are calculated for this study from the cumulative radiative forcing over a given integration time horizon
**The decay of carbon dioxide concentrations cannot be reproduced using a single exponential decay lifetime, thus there is no meaningful single value for the lifetime that can be compared directly with other values in this table
***GWP values include the direct radiative effect and the effect due to carbon dioxide formation, but exclude any effects resulting from tropospheric ozone or stratospheric water formed as methane decomposes in the atmosphere
Source: AFEAS and PAFT Member Companies (1992).

Water vapour

Water vapour is the principal greenhouse gas. It absorbs light waves strongly in the range 4–7 micrometres, whereas CO_2 absorbs in the band 13–19 micrometres. The concentration of water vapour in the troposphere is determined internally within the global climate system, and on a global scale it is unaffected by anthropogenic sources and sinks. The coldest and hottest places on Earth are also the driest, such as the deserts, with central Asia being the coldest and driest and central Australia the hottest and driest. At night, energy escapes into space to make these places cold, while during the day the lack of cloud cover allows more solar radiation to reach the ground and thereby make it hot. These places have least water and are therefore least able to maintain an equable climate throughout the day.

Carbon dioxide

Carbon dioxide (CO_2) is one of the main greenhouse gases, and is of greatest concern as a controllable gas emission caused by human activities. Following the Industrial Revolution, the combustion of fossil fuels, together with deforestation, has caused an increase in the concentration of atmospheric CO_2 by 26 per cent. Between 1950 and 1980, CO_2 emission increased by an estimated 586 per cent in the developing countries, 337 per cent in the former Soviet Union and eastern Europe, 91 per cent in North America, and 125 per cent in Western Europe, the rest being made up by other developed countries. Figure 3.5 shows the countries responsible for the main emissions of CO_2. It is interesting to note that the principal responsibility for producing these emissions rests with the developed and industrialized countries. The rapid industrialization of developing countries such as China, which has poor controls on environmental pollution, pose a major threat to the environment (Plate 3.1). The USA is the largest emitter of CO_2, accounting for 24 per cent of global emissions. The UK accounts for 3 per cent, of which 96 per cent comes from the burning of fossil fuels for energy use, mostly from electricity generation (Department of the Environment 1992a). The peak emission of 190 million tonnes in 1979 was followed by a drop, but from 1984 to 1987 there was an 18 million tonne increase to 171 million tonnes. In 1987, power stations accounted for 37 per cent of the total emissions of CO_2, with 20 per cent from industry, 16 per cent from transport, 14 per cent from domestic combustion, 13 per cent from offices and other sources. Just over 50 per cent of the 1990 CO_2 emissions were accounted for by the use of private cars, and more than 25 per cent by the industrial, commercial, and public sectors' use of road transport.

The European Community's (EC's) total CO_2 emissions represent approximately 13 per cent of global CO_2 emissions compared with 5 per cent for Japan and 25 per cent for Eastern Europe and the former Soviet Union (CONCAWE 1992).

Transport uses nearly one-third of the total global energy consumption and contributes around 25 per cent of the world CO_2 output, as well as CFCs, methane, and nitrous oxide

(Greenpeace 1990). North America and Europe each possess more than one-third of the world's vehicles, which was 400 million in 1985. In the EC, 26 per cent of the total anthropogenic CO_2 emissions come from transport. Despite these figures, there is still relatively little concerted effort at energy efficiency and reducing the harmful emissions of greenhouse gases from car fumes.

The generation of CO_2 as a result of the combustion of fossil fuels suggests that if present trends continue, then its concentration will double every 50 years. In 1989 and 1990, the combustion of fossil fuels added an estimated annual 6.0 ± 0.5 gigatonnes of carbon (GtC) to the atmosphere (of which the main contributory nations are shown in Figure 3.6), compared to 5.7 ± 0.5 GtC in 1987; the estimated total release of carbon in the form of CO_2 from the oilwells of Kuwait in 1991 was 0.065 GtC, equivalent to about 1 per cent of the total annual anthropogenic emissions (IPCC 1990, 1992).

Plate 3.1 *Heavy industrialization along the Yellow River in central China, emitting large quantities of greenhouse gases. The growth of China's industry and its poor environmental legislation poses one of the greatest threats in terms of reducing greenhouse gas emissions.*

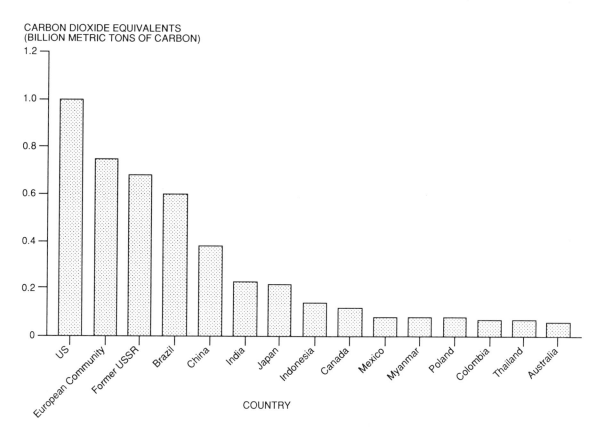

Figure 3.5 *The countries with the highest greenhouse gas net emissions for 1987. After World Resources Institute (1990).*

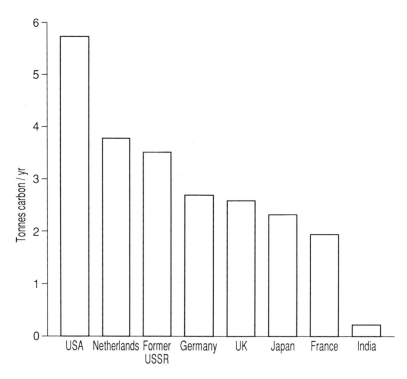

Figure 3.6   *Annual fossil fuel carbon emissions in 1989, expressed as tonnes of carbon, for selected countries. India, as a developing country, is shown for comparison alongside the richest developed countries.*

Monitoring of the increase in CO_2, however, only shows an increase that is 50 per cent of the predicted level and this difference is ascribed to the ameliorating or buffering effect of reactions in the world's oceans and atmosphere. Some researchers have estimated that a doubling of the present CO_2 level to 600 ppmv is estimated to cause an average rise in global temperature of about 8°C (Maddox 1989). These figures do not take account of any feedback mechanisms which might serve to reduce the predicted temperature rise, for example an increased abundance of clouds with their cooling effects.

To understand the details of these feedback systems, it is necessary to identify the major global storage and transfer of carbon. The carbon cycle is illustrated in Figure 3.7 and Appendix 4.

Dying forests may contribute to the greenhouse effect. This is because decaying vegetation releases CO_2 and H_2O, and also as part of the respiration process trees convert CO_2 to O_2. Deforestation could be sending an annual 4 billion tonnes of CO_2 into the atmosphere which would otherwise be taken up by plants during their metabolic processes, double the most com-

monly quoted estimates (Pearce 1989a). The IPCC reports (1990, 1992) quote an annual average net flux to the atmosphere of 1.6 ± 1.0 GtC from land use during the 1980s.

Methane

Methane (CH_4), an atmospheric trace gas involved in many chemical reactions in the troposphere and stratosphere, initially received relatively little attention as a greenhouse gas, but in the past few years this has been rectified. The 1992 IPCC report stated that a methane molecule has 11 times more global warming potential than a molecule of carbon dioxide, and the total annual anthropogenic and natural emissions of CH_4 are 500 teragrammes (Tg; 1 Tg = 10^{12} g). Since 1950, concentration levels of methane have been rising at 1 per cent per annum (10–16 parts per 10^9 by volume – ppbv) which is four times the rate of increase of carbon dioxide and could lead to methane becoming the principal greenhouse gas within 50 years (Pearce

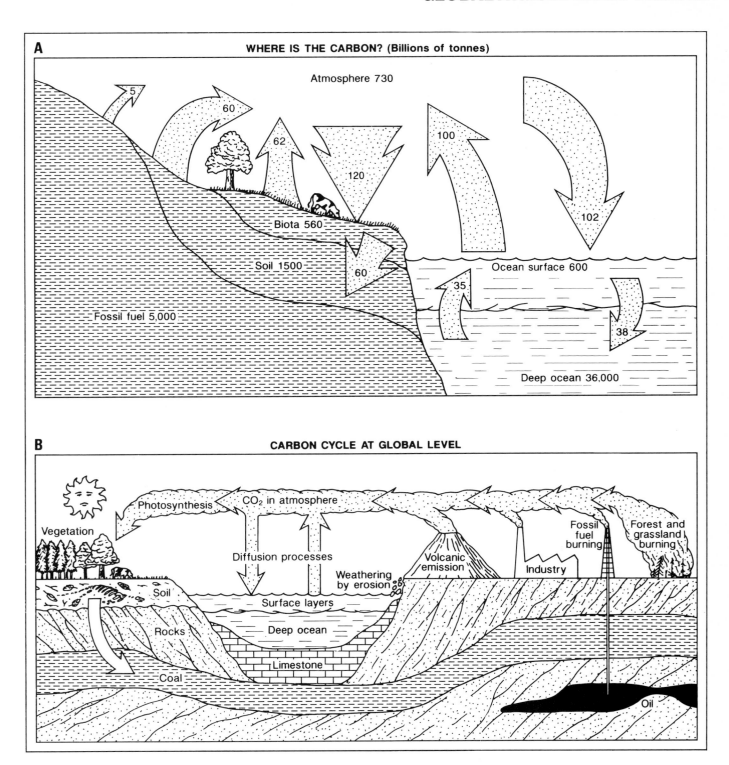

Figure 3.7 *Carbon cycle showing (A) the storage of carbon, and (B) the variety of natural and human-induced processes that move carbon around the cycle. Units shown are in billions of tonnes per annum. After Scurlock and Hall (1991).*

1989a). The IPCC report (1990) puts the annual increase of CH_4 at 14–17 ppbv, giving a present atmospheric concentration of 1,700 ppbv. Atmospheric CH_4 concentrations have more than doubled since the Industrial Revolution as a result of increased rice production, cattle rearing, biomass burning, coal mining, and the ventilation of natural gas (ibid.). Prior to the Industrial Revolution, CH_4 atmospheric concentration was almost constant at about 0.65 ppmv, whereas in 1990 it reached 1.72 ppmv (Badr *et al.* 1992a, b).

The main sources and sinks of CH_4 input into the atmosphere, expressed in global average Tg per annum, are given in Table 3.6 (IPCC 1992). The net annual average input of CH_4 to the atmosphere is estimated at about 32 Tg. A further source of CH_4 emissions are from natural gas leakages in distribution systems, and from livestock – including cows' farts! It has been estimated that the annual emission of methane from the world's cattle is close to 100 million tonnes (Pearce 1989b). If only we could harness this! Perhaps one of the most surprising aspects of the source of CH_4 production is the role

played by termites. There are an estimated 250,000 billion termites in the world which inhabit approximately two-thirds of the land area and consume something in the region of one-third of the global vegetation. In 1982, scientists from West Germany, USA, and Kenya discussed the importance of termites as contributors of CH_4 to the atmosphere and concluded that termites could account for about one-third of the annual global emission of methane or 150 million tonnes, although more recent evaluations of the contribution from termites have considerably revised this figure downwards to an annual 5 million tonnes.

Estimates of CH_4 emissions from various sources, however, remain uncertain, particularly compared to those for CO_2. Indeed, these IPCC figures exclude perhaps the largest source of CH_4 which may come from degassing of the mantle at mid-ocean ridges and from the bacterial breakdown of organic matter buried in sediments at continental margins and within lakes. Studies of carbon isotopes in CH_4 molecules suggest that approximately 100 Tg or 20 per cent of atmospheric methane was produced long ago and is currently escaping from melting permafrost, present as gas hydrates, coal seams, oil reservoirs, and rocks beneath the oceans and natural gas sources, that is, it is of fossil origin.

Table 3.6 *Estimated sources and sinks of methane (Tg per annum).*

	Annual release	*Range*
Sources		
Natural		
Wetlands (bogs, swamps, tundra, etc.)	115	(100–200)
Termites	20	(10–50)
Ocean	10	(5–20)
Freshwater	5	(1–25)
CH_4 hydrate	5	(0–5)
Anthropogenic		
Coal mining, natural gas & petroleum industry	100	(70–120)
Rice paddies	60	(20–150)
Enteric fermentation	80	(65–100)
Animal wastes	25	(20–30)
Domestic sewage treatment	25	?
Landfills	30	(20–70)
Biomass burning	40	(20–80)
Sinks		
Atmospheric (tropospheric + stratospheric) removal	470	(420–520)
Removal by soils	30	(15–45)
Atmospheric increase	32	(28–37)

Source: IPCC (1992).

Nitrous oxide

Nitrous oxide (N_2O) is an important trace gas in the atmosphere. The main anthropogenic source of nitrous oxide comes from fertilizers, fossil fuel combustion, and various synthetic chemical manufacturing processes such as nylon production. The global N_2O concentration has been rising at a rate of 0.2–0.3 per cent per annum, reaching about 310 ppbv in 1990 (Badr and Probert 1992). The increased atmospheric concentrations of N_2O are of concern because of its role in destroying the ozone layer as a result of producing nitric oxide in the stratosphere, and because N_2O contributes to the atmospheric greenhouse effect. Currently, estimates of individual N_2O sources and their emission rates are poorly constrained, with the IPCC (1992) estimated sources and sinks given in Table 3.7. Data from Antarctic ice cores show that atmospheric

Table 3.7 *Estimated sources and sinks of nitrous oxide (Tg per annum).*

	Range
Sources	
Natural	
Oceans	1.4–2.6
Tropical soils	
Wet forests	2.2–3.7
Dry savannas	0.5–2.0
Temperate soils	
Forests	0.05–2.0
Grasslands	?
Anthropogenic	
Cultivated soils	0.03–3.0
Biomass burning	0.2–1.0
Stationary combustion	0.1–0.3
Mobile sources	0.2–0.6
Adipic acid production	0.4–0.6
Nitric acid production	0.1–0.3
Sinks	
Removal by soils	?
Photolysis in the stratosphere	7–13
Atmospheric increase	3–4.5

Source: IPCC (1992).

N_2O concentrations were about 30 per cent lower during the Last Glacial Maximum, compared to the Holocene epoch (see Chapter 2), and with present-day N_2O concentrations unprecedented in the past 45 kyr, it has been suggested that the recent increases in atmospheric N_2O are due to human activities (Leuenberger and Siegenthaler 1992).

Tropospheric ozone

Tropospheric or low-level ozone is a greenhouse gas which is toxic to plants, humans, and other organisms. In the northern hemisphere, the growth in surface emissions of nitrogen dioxide and hydrocarbons leads to increased concentrations of ozone in the troposphere. A recent study by a group of British atmospheric modellers (Johnson *et al.* 1992) has shown that the radiative forcing of surface temperatures is most sensitive to changes in tropospheric ozone at a height of about 12 km, where aircraft emissions of nitrogen oxides are at a maximum, and where the model sensitivity of ozone to NO_x emissions is enhanced. The model also showed that the radiative forcing of surface temperatures is approximately 30 times more sensitive to the emissions of NO_x from aircraft than to surface emissions, and that the impact on global warming of increases in tropospheric ozone due to rises in the surface emissions of NO_x have been over-estimated by a factor of up to 5 (including in the 1990 IPCC report), because of errors in the calculations of the ozone budget. Compared to the northern hemisphere, the southern hemisphere is 60 per cent more sensitive to changes in the emissions of NO_x, since it receives only 18 per cent of the total emissions. In the atmosphere, hydroxyl ions are capable of ameliorating much of the harmful effects of gases such as NO_x by oxidizing them to less harmful substances (see Box 3.5).

A study by the American scientists Oltmans and Levy (1992) suggests that it is the natural processes, not the anthropogenic sources of pollution, which control the seasonal cycle of tropospheric ozone over the western North Atlantic, even though springtime daily average O_3 concentrations at Bermuda exceed 70 parts per billion (10^9) by volume (ppbv), and in 1989

The hydroxyl radical is the main cleansing agent in the atmosphere. It removes chemical compounds which are considered as pollutants by oxidizing them to less harmful substances. Amongst the gases that hydroxyl radicals deal with by oxidation processes are CH_4, CO, and formaldehyde (HCHO), the latter being converted to CO_2. Nitrogen oxides (NO_x) are oxidized to nitric acid, and SO_2 to SO_3^{2-} which dissolves in clouds to form sulphuric acid (H_2SO_4). After most of these reactions, hydroxyl radicals are returned to the atmosphere, and therefore are able to react again to cleanse atmospheric pollutants further.

Hydroxyl radicals are produced by the action of sunlight with ozone in the troposphere, with the greatest production in the equatorial regions. It is also produced by some reactions related to urban pollution. Levels of hydroxyl radicals in the atmosphere, however, are not well known and there is a fear that anthropogenically produced CH_4, CO, and NO_x will greatly reduce the levels and effectiveness of hydroxyl radicals as efficient atmospheric cleansing agents. Table 3.8 shows the estimated sources and sinks of carbon monoxide (IPCC 1992).

Box 3.5
Atmospheric cleansers: hydroxyl radicals

Table 3.8 *Estimated sources and sinks of carbon monoxide (Tg per annum).*

	WMO (1985)	Seiler & Conrad (1987)	Khalil & Rasmussen (1990)	Crutzen & Zimmerman (1991)
Primary sources				
Fossil fuel	440	640±200	400–1,000	500
Biomass burning	640	1,000±600	335–1,400	600
Plants	–	75±25	50–200	–
Oceans	20	100±90	20–80	–
Secondary sources				
NMHC oxidation	660	900±500	300–1,400	600
Methane oxidation	600	600±300	400–1,000	630
Sinks				
OH reaction	900±700	2,000±600	2,200	2,050
Soil uptake	256	390±140	250	280
Stratospheric Oxidation	–	110±30	100	–

NMHC = non-methane hydrocarbons
Source: IPCC (1992), references cited therein.

the hourly readings surpassed the Canadian Air Quality limit of 80 ppbv. Continuous measurements of tropospheric ozone from Bermuda (32°N, 65°W) and Barbados (13°N, 60°W) indicate that the high levels of O_3 are transported from the unpolluted upper troposphere at altitudes greater than 5 km above the northern USA and Canada. In support of their conclusions, Oltmans and Levy point out that in Barbados the seasonal and diurnal variations in surface O_3 are virtually identical to those measured at Samoa in the tropical South Pacific, far removed from anthropogenic sources of pollution, and where the low levels of NO_x ensure that natural processes control surface ozone levels. They also note that during the summer, when surface O_3 concentrations over the eastern USA can exceed 70 ppbv due to pollution, in Bermuda typical measurements are 15–25 ppbv.

Table 3.9 *Estimated contributions to sea level rise (cm) over the past 100 years.*

	Low	Best estimate	High
Thermal expansion	2.0	4.0	6.0
Glaciers/small ice caps	1.5	4.0	7.0
Greenland ice sheet	1.0	2.5	4.0
Antarctic ice sheet	−5.0	0.0	5.0
Total	−0.5	10.5	22.0
Observed	10.0	15.0	20.0

Source: IPCC (1990).

Climate and the greenhouse effect: a bleak future?

In 1861, John Tyndall of Manchester was certainly amongst the first people to suggest that the large amount of carbon dioxide produced by combustion could affect the radiation balance of the Earth. As the greenhouse effect takes a strong hold on the planet, perhaps by the middle of the next century, then the world climates and climatic belts will look very different from today. Temperatures near the poles have been estimated by some studies as getting up to 12°C warmer. Of course, not only will the temperature patterns look very different from the present, but also rainfall or precipitation patterns will change so that parts of the Earth will become drier and others wetter. The altered temperature and rainfall patterns will cause a dramatic shift in the position and extent of some vegetation belts while others will show little or no change.

Worldwide, climate change will bring about a shift in the more local climatic vegetation belts, with some narrowing and others widening. Predictions of a Mediterranean-type climate for Britain by the middle of the next century mean that the types of crops grown at present and the natural vegetation would change. Many of the grain crops, such as wheat, could be replaced by olives and grapes. This vision of a more equable climate for Britain might, at first, appear rather pleasant, but would these predicted changes

Table 3.10 *Estimates of future global sea level rise (cm).*

	Thermal expansion	Alpine	Greenland	Antarctica	Best estimate	Range[f]	To (year)
Gornitz (1982)	20	20 (combined)			40		2050
Revelle (1983)	30	12	13		71[b]		2080
Hoffman et al. (1983)	28–115	28–230 (combined)				56–345	2100
						26–39	2025
PRB (1985)	[c]	10–30	10–30	−10–100		10–160	2100
Hoffman et al. (1986)	28–83	12–37	6–27	12–220		58–367	2100
						10–21	2025
Robin (1986)[d]	30–60[d]	20±10[d]	to +10[d]	to −10[d]	80[i]	25–1,659	2080
Thomas (1986)	28–83	14–35	9–45	13–80	100	60–230	2100
Villach (1987)							
Jaeger (1988)[d]					30	−2–51	2025
Raper et al. (1990)	4–18	2–19	1–4	−2–3	21[g]	5–44[g]	2030
Oerlemans (1989)					20	0–40	2025
Van der Veen (1988)[h]	8–16	10–25	0–10	−5–0		28–66	2085

[a] From the 1980s
[b] Total includes additional 17 cm for trend extrapolation
[c] Not considered
[d] For global warming of 3.5°C
[f] Extreme ranges, not always directly comparable
[g] Internally consistent synthesis of components
[h] For a global warming of 2–4°C
[i] Estimated from global sea level and temperature change 1880–1980 and global warming of 3.5±2.0°C for 1980–2080
Source: IPCC (1990), references cited therein.

have deleterious knock-on effects on the food chains and animal life that rely upon the present balance? Again, what price must be paid to slow down this global warming?

There may be other important implications resulting from the global warming. Large volumes of water now locked up in the Antarctic and Arctic as ice sheets and glaciers may be released into the hydrosphere. Sea level may rise by an amount that would be significant, although experts differ in their estimates of this figure from over 10 cm to nearly 1.5 m. The lower estimates may seem insignificant to many people, but in fact even these relatively small sea level rises would cause the flooding of extensive areas of dry land. Table 3.9 gives the IPCC (1990) estimated contributions over the past 100 years to global sea level rise from the thermal expansion of the oceans, and the melting of glaciers, small ice caps, the Greenland ice cap, and the Antarctica ice cap to global sea level rise. The result of any significant rise in sea level would be the marginalization and destruction of large areas of coastal lowlands as agricultural land

and habitats for various flora and fauna diminish. IPCC (1990) estimates of future global sea level rise are given in Table 3.10.

There is debate on the feedback mechanisms associated with global temperature changes and atmospheric moisture content. Most current GCMs assume that global warming will be associated with an increase in atmospheric water vapour content or moisture. This assumption has been challenged by some scientists, who contend that global warming would increase air convection leading to a drying of the middle atmosphere, thereby providing a negative feedback to counteract any greenhouse effect. The consensus of scientific opinion, however, is that most current GCMs make appropriate allowances for the amplifying effect of water vapour – a view which appears to be supported by recent satellite observations.

A likely scenario is that rises in global temperature would lead to increased precipitation in the currently arid polar regions as the Arctic Ocean becomes more free of sea ice. The effect would be a reduction in aridity, which would

Plate 3.2 *View looking north from Axel Heiberg Island at 78° N across the frozen seas of Greely Fjord towards the Arctic Ocean. This sea only partially melts in summer as large 'leads' open (centre of the frame) and the surface of the ice begins to melt and form ponds. This ice plays an important role in the hydrosphere–atmosphere interactions at high latitudes. An understanding of its dynamics is critical for accurate modelling of global climatic change.*

lead to the growth of glaciers and ice sheets rather than their gross melting (Miller and de Vernal 1992). The geological data over the past 130,000 years support the idea that greenhouse warming, which is expected to be the most pronounced in the Arctic, coupled with decreasing summer insolation, may lead to more snow deposition than melting at high northern latitudes and thus to ice sheet growth (ibid.) (Plates 8 and 3.2).

A greenhouse conspiracy? Myth or reality?

Will global warming lead to the melting of the polar ice and the release of large quantities of seawater so that sea level rises and countries such as the Maldives are drowned? This section looks at some of the arguments and debate surrounding any global warming. There are four main pillars which are most frequently used to support the view that the Earth is currently experiencing an anthropogenically created greenhouse effect or global warming: first, that the Earth's climate record shows that global temperature has increased and sea levels have risen; second, that carbon dioxide has been the primary cause of these changes; third, based on predictions of climate models, that a doubling of atmospheric carbon dioxide will result in increased mean global temperature of 2–5°C; and finally, that the underlying physics is widely assumed to prove that carbon dioxide is a greenhouse gas and that further increases will result in rises in global temperature. Expert opinion remains divided on these issues.

One of the central issues focuses on the validity of the actual temperature measurements made over the last 100 years or so and the way in which they can be interpreted. The thermometer record, so it would seem, cannot be taken at face value. Worldwide, there are more than 60,000 measurements taken every day, amounting to a staggering 22 million annual measurements! But weather stations are not evenly distributed around the world; most are in the heavily populated, developed regions of the northern hemisphere, with far fewer in the southern hemisphere. Also, the oceans are almost unrepresented in this data set, yet they cover more than three-quarters of the globe. Even more significantly, most weather stations are sited in urban areas where temperatures are invariably warmer than the surrounding countryside. For example, Phoenix, Arizona, is frequently 10°C warmer than its suburbs. This temperature difference, known as the **heat-island effect**, is not due to global warming, but because urban areas release additional heat into the atmosphere. As urban areas have grown, so too have urban average temperatures risen through human activities. Some studies have even suggested that villages with as few as 300 inhabitants can cause urban warming of up to 0.3°C per decade, the amount proffered for global warming this century!

Even allowing for the heat-island effect, many critics argue that this effect is underestimated in the climate models that are used to support global warming. Historical measurements of sea-surface temperatures have also been unreliable. In the past, most measurements were made on water samples collected in canvas buckets lifted out of the sea on to a boat. During this process, some of the seawater evaporated and cooled the seawater in the uninsulated bucket. The result was an underestimate of the actual temperature. More recently, measurements were gathered from seawater in the intake for ship's engines. These results were roughly 0.5°C higher than the measurements using the earlier technique. Today, satellite measurements of sea-surface temperature are routinely taken, and provide both rapid and consistent data. Unfortunately, due to the poor sampling techniques in the past, there is not a reliable historical record of the long-term changes in sea-surface temperatures. Graphs of global temperature change over the past decade have been produced from satellite data, with a precision of about one-hundredth of a degree per month, and they do not appear to support the global warming hypothesis. Spencer, a physicist at the NASA Marshall Space Flight Center, University of Alabama, USA, has concluded from his analysis of this satellite data that 'over the entire ten-year period there was no net warming or cooling'. So, while the thermometer record shows an underlying upward trend in temperature for the last ten years, the satellite data appear to show that the Earth was warmer in the first half of the 1980s and cooler in the second part. A judicious choice of time frame in the last century can be used to suggest global warming or cooling. The temperature data are ambiguous at best, so the critics claim.

Another area of fierce debate centres around the predicted rise in sea level due to global warming. The popular press has carried figures of up to 20 m of sea level rise during the next century, but respectable scientific estimates are closer to a 0.5 m rise. Evidence for changes in sea level comes from tidal gauges, generally located in harbours and estuaries. Though thousands of measurements are available annually, controversy surrounds the interpretation of the data. An underlying cause for concern is whether the measurements chart the vertical movement of land relative to a fixed sea level, or the converse. Obviously, individual cases can be interpreted with varying degrees of certainty. After large

earthquakes, scientists are generally able to estimate the vertical movement of land. In the British Isles, for purely geological reasons, sea level is falling in the north of Scotland and rising in the south-east of England. The fall in the north-east of Scotland is due to the vertical rebound, or **isostasy**, after the weight of ice was removed from this region, along with Scandinavia and other northern land masses, following the last glaciation.

And what of the reports that the extent of sea ice is diminishing because of global warming and the melting of the polar ice? Submarines passing under the polar ice have reported that at specific locations the ice is now thinner than it was a decade ago. But satellite data gathered daily over the past 15 years do not appear to corroborate the notion of melting ice since they suggest no change. So, the sceptics of global warming argue that there is no evidence of an imminent greenhouse world with higher sea level. In the early 1970s, the media even talked of global cooling and the dawn of a new Ice Age!

Debate also surrounds the reliability of models which are used to make predictions about the future global climate. Sceptics argue that the uncertainties in these computer models, together with their lack of sophistication for simulating actual climatic conditions, render them at best inaccurate and at worst misleading. These arguments are not, in themselves, a case against global warming but rather an attempt to exploit the uncertainties that arise from modelling global climatic change.

The sceptics of global warming stress that the climate models tend to underemphasize the importance of negative feedback mechanisms which may stabilize any potential runaway greenhouse effect. Also, the term 'greenhouse gas' has misleading connotations when associated only with CO_2 and CH_4, because water vapour is actually the most common greenhouse gas, yet it is ignored in most articles. All these gases absorb and radiate heat energy in varying ways that depend upon many complex, interlinked factors such as their position in the atmosphere and the relative concentration of the cocktail of gases in the atmosphere. For example, convection currents complicate the heat budget of the atmosphere.

There are even experts who claim that an increase in atmospheric CO_2 could have beneficial effects on plant growth. It is interesting to note that plants evolved at a time when atmospheric CO_2 levels were probably 5–10 times greater than present levels. But what is good for plants may not be good for the human species and the continuation of civilization.

Perhaps the critical argument centres around the link between atmospheric CO_2 levels (and other greenhouse gases) and global climate. Detailed studies of atmospheric CO_2 levels and palaeotemperatures following the most recent deglaciation show that the rise in CO_2 levels significantly preceded the rise in local sea-surface temperatures (Shackleton 1990). These data were gathered from the ice core record and deep-sea sediment cores by techniques such as: (1) the UK-37 method, in which temperature is estimated from the ratio of various organic molecules (di-unsaturated to tri-unsaturated C_{37} alkadienones), which are specifically associated with a type of algae known as prymnestophyte algae or coccoliths, and (2) the identification of the influx of warmer water marine planktonic organisms such as the foraminifera *Globorotalia menardii*, which is a marine microfossil. So, changes in atmospheric CO_2 levels appear to drive changes in sea-surface and linked atmospheric temperatures as suggested by the proponents of global warming and not, as the sceptics would have it, the other way around. The link between atmospheric CO_2 levels and global temperature change appears robust.

In September 1990, the report of Working Group 1 of the Intergovernmental Panel on Climate Change (IPCC), set up jointly by the World Meteorological Organization and the United Nations Environment Programme, was published. After examining the scientific evidence, this very weighty document by leading international experts firmly came down in favour of global warming driven by a greenhouse effect:

We are certain of the following:
• there is a natural greenhouse effect which already keeps the Earth warmer than it would otherwise be;
• emissions resulting from human activities are substantially increasing the atmospheric concentrations of the greenhouse gases: carbon dioxide, methane, chlorofluorocarbons (CFCs) and nitrous oxide. These increases will enhance the greenhouse effect, resulting on average in an additional warming of the Earth's surface. The main greenhouse gas, water vapour, will increase in response to global warming and further enhance it.

(IPCC 1990)

Of course, the exact consequences of global warming remain uncertain, but one thing is certain – dismissing global warming or inaction can only serve to put an unacceptable risk on the survival of life on Earth, certainly for human civilization.

The IPCC 1990 document models a number of scenarios for predicted levels of change in the atmospheric concentrations of greenhouse gases, and the resulting changes in climate that might reasonably be expected to occur under the various 'options'. One of these predictions has been termed the 'business-as-usual' scenario, under which the emissions of greenhouse gases continue at the current rates. In this case, the IPCC estimates that during the next century: (a) global mean temperature will increase by 0.3°C per decade (with an uncertainty range of 0.2–0.5°C per decade), which is greater than that seen over the last 10,000 years, and (b) global mean sea level will rise by about 6 cm per decade (with an uncertainty range of 3–10 cm per decade), mainly because of the thermal expansion of the oceans and the melting of some land ice. These predictions suggest that global mean temperatures will be about 1°C above the present value by 2025, and global mean sea level will have risen by about 20 cm by 2030 (Figure 3.8).

More recently, revised projections of future global greenhouse gas warming have suggested that by the year 2100, with a rise of about 0.5°C by 2010, the increase relative to 1990 will vary (0.62–2.31°C, and 1.61–5.15°C respectively), depending upon whether CO_2 levels are 2 or 5.5 times the pre-industrial CO_2 concentrations (Schlesinger and Jiang 1991). Evidence for global warming is coming from places as remote as north-west Tasmania, at 1,040 m above sea level on the slopes of Mount Read and around Lake Johnston, where the width of growth rings from Huon pine trees (*Lagarostrobes franklinii*) well above their normal altitude range suggests that the temperature rise during the last 25 years has been much greater than at any time since AD

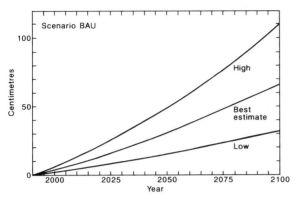

Figure 3.8 *Predictions for the amount by which global sea level will rise between 1990 and 2100 under three scenarios: a low and high rate of increased greenhouse gas emissions, and a business-as-usual scenario (BAU). After IPCC (1990).*

900 (Cook *et al.* 1991). The tree ring index, obtained by subtracting the growth from natural maturation from the thickness of the ring, can be used to interpret past climatic conditions. The tree ring index from the Tasmanian pines suggests a mean temperature rise of just over 1°C since 1965.

In 1992, the IPCC produced revised figures for the effects of greenhouse gas emissions (see Wigley and Raper 1992 and Box 3.6). These new figures result from taking into account new policies already implemented or proposed for controlling CO_2 emissions and **halocarbon** production, and allow for recent political changes. In addition, they are based on a wide range of socio-economic factors that influence the development of emissions in the absence of unilateral or multilateral efforts to reduce greenhouse gas emissions. The various scenarios that are presented differ from each other because they make different assumptions about, for example, population growth, economic growth, technological developments, resource limitations, fuel mixes, and agricultural development. The new climate models also include results for global mean

thermal expansion of the oceans, a principal component of future rises in sea level. From the range of possibilities, the IPCC shows low, middle, and high estimates of global mean temperature change, global mean sea level change and radiative forcing. The results are less severe than previous estimates but remain greater than the limits of natural variability. For example, middle estimates suggest that by the year 2100, global mean temperature will have risen by about 3.5°C, and global mean sea level will have risen by around 50–60 cm (ibid., fig. 4). The range of solutions for low, middle, and high temperature and sea level projections, based solely on the anthropogenic component of future change, show that over the period 1990–2100, warming will be between 1.7°C and 3.8°C, with corresponding sea level rises of between 22 cm and 115 cm (ibid.). These revised, reduced rates of projected future change are still four to five times those that occurred over the past century. The IPCC 1992 update, however, has downwardly revised the global warming potential of CFCs from being thousands of times more potent, molecule-for-molecule compared to carbon dioxide, to zero: this is because CFCs produce two opposing effects because of their role in destroying stratospheric ozone as well as being greenhouse gases.

Natural phenomena and global climate change

El Nino events and global warming

El Nino events are simply relatively large perturbations of a process that occurs annually in the Pacific Ocean. In a 'normal year', the variations in the atmosphere–ocean system produce a fairly predictable pattern of ocean currents in the southern Pacific Ocean, and in which the surface

---

Gas emissions are calculated by multiplying fuel consumption by a 'carbon emission factor'. The carbon emission factor is the amount of CO_2 released through the combustion of a specified quantity of fuel, e.g., 1 litre or 1 tonne. Since fuels contain varying amounts of carbon, they are associated with different carbon emission factors.

Predictions about future emissions are calculated by multiplying projected fuel consumption values (taking account of likely economic indicators, such as trends in fuel prices etc.) with the appropriate carbon emission factors. Clearly, errors are associated with such calculations, probably in the range 5–10 per cent.

Box 3.6
Calculating greenhouse gas emissions

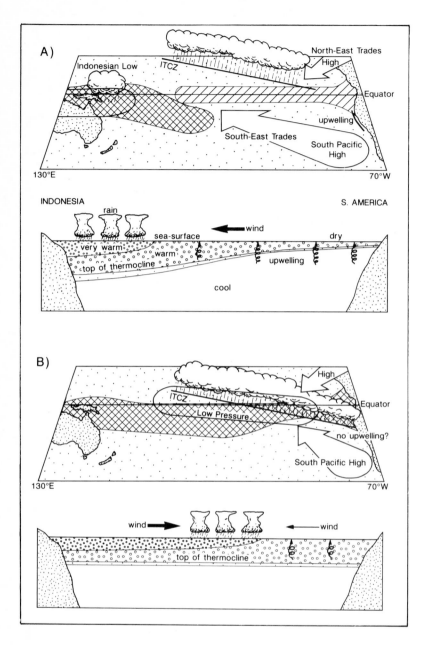

Figure 3.9 *Variations in the atmospheric systems, ocean temperature profiles and ocean currents in the southern Pacific Ocean during (A) a normal year, and (B) an El Nino Southern Ocean Oscillation (ENSO) event. In a normal year (A), the surface sea temperature is highest in the west (> 28°C) helping to induce the movement of strong, warm, maritime trade winds into Indonesia which creates heavy rainfall. On the western coast of South America, cold bottom waters upwell to provide fresh nutrients to surface waters. During an ENSO event, the surface water temperatures > 28°C develop much further eastwards, allowing the Inter Tropical Convergence Zone (ITCZ) to migrate southwards and suppress the Southeast Trade Winds or even reverse them. As a result, rainfall is heaviest in the central east Pacific; upwelling of cold, nutrient-rich, bottom waters is weakened and marine productivity is reduced. After Open University (1991).*

sea temperature is highest in the west (>28°C) which helps to induce the movement of strong, warm maritime South-East Trade Winds into Indonesia and with them heavy rainfall. A corollary of this is that cold, nutrient-rich bottom waters upwell to replenish surface waters off the western coast of South America. In contrast to such normal years, during an El Nino, also referred to as an El Nino Southern Oscillation event, surface water temperatures greater than 28°C develop much further eastwards and allow the Inter Tropical Convergence Zone (ITCZ) to migrate southwards and suppress the South-East Trades, or even reverse them. The result of these changes is that rainfall is heaviest in the central–east Pacific and upwelling of cold, nutrient-rich, bottom waters is weakened. The decreased upwelling leads to a reduction in marine productivity. With less bioproductivity, less CO_2 is sequestered from the atmosphere–ocean system by organisms and, therefore, can lead to greater concentrations of CO_2, a greenhouse gas, in the atmosphere. The El Nino effect is illustrated in Figure 3.9.

An El Nino event occurred in 1987 and 1988, and was associated with a change in the wind patterns and ocean currents in the Pacific Ocean, leading to severe droughts. This El Nino event ended in June 1989, with a decrease in the observed surge of CO_2 levels. The El Nino of 1987 caused the equatorial Pacific Ocean to warm by as much as 3°C, believed by some scientists (e.g., Ramanathan and Collins 1991) to be sufficient to cause a potential atmospheric warming. Such predictions are, of course, only as good as the computer models themselves and the data which go into them. Until the models are adequately tested, we have to be cautious in assuming that the ocean–atmosphere system will operate like a giant thermostat to regulate the mean global temperatures and global climate within very narrow limits.

El Nino events result in the release of large amounts of CO_2 into the atmosphere. At the meteorological observatory on the Hawaiian island of Mauna Loa, Keeling *et al.* (1989) have documented an increase in the rate of release of CO_2 into the atmosphere and showed that it rose by more than two-thirds in the course of two years: the result has been an increase in atmospheric CO_2 from pre-Industrial Revolution levels of 270 ppm to 350 ppm.

Recent observations during the 1987 El Nino showed that for the upper range of sea-surface temperatures, the greenhouse effect increases with surface temperature at a rate exceeding the rate at which radiation is emitted from the surface. In computer models, the atmospheric response to the so-called super greenhouse effect is the formation of highly reflective cirrus clouds which shield the ocean from the solar radiation (Ramanathan and Collins 1991). In effect, they act like a thermostatic umbrella around the Earth to regulate the temperature of the sea surface to less than 305° Kelvin. This model involves a negative feedback to regulate the surface temperature.

Volcanoes blow hot and cold

It is not only human activities that contribute to the gases that may cause either global warming or cooling by depleting the ozone layer. Natural causes may also be very important, for example volcanic eruptions. Volcanoes can emit huge quantities of greenhouse gases, including CO_2. Mount Etna in Sicily, for example, is amongst the world's most actively degassing volcanoes. Data from the eruptions of Etna, between 1975 and 1987, led Allard *et al.* (1991) to a conservative estimate of approximately 25 million tonnes of CO_2 per year, equivalent to the output from four 1,000 megawatt conventional coal-fired power stations, but still insignificant compared to the annual global emission of 5 GtC from the combustion of fossil fuels. Its SO_2 emission rate is also very high, at about 10 per cent of the global total for volcanic degassing. This emission from Etna is roughly an order of magnitude greater than that of Kilauea in the Hawaiian islands, another well-studied volcano. While this figure is indeed large, the CO_2 emissions from Etna are only 0.07 per cent of the annual anthropogenic CO_2 contribution to the atmosphere (ibid.). The global rate of CO_2 emissions from all the subaerial and submarine volcanoes is not precisely known, but it is probably somewhere in the region of 130–175 million tonnes per year (Gerlach 1991).

Volcanic eruptions, particularly the more explosive types, release chlorine (Cl) and fluorine (F) compounds into the stratosphere to produce 'halogen pollution'. Hydrogen chloride (HCl) and hydrogen fluoride (HF) are the main halogen compounds released during volcanic eruptions, with estimated annual yields of 0.4×10^6 to 11×10^6 tonnes of HCl, and 0.06×10^6 to 6×10^6 tonnes of HF. Of these gases 10 per cent are produced in the explosive types of eruptions alone where the exhalative gases are injected into the stratosphere. Of course, volcanic eruptions do not occur at regular time intervals, nor are they equally spaced around the Earth. Thus, if the atmospheric levels of anthropogenically created CFCs are at sub-critical concentrations, then it might conceivably only take one or two particularly large explosive volcanic eruptions to cause stratospheric ozone-destroying chemicals to exceed a critical threshold level, and cause an accelerated depletion of ozone. Such a scenario could lead to global cooling.

Earth scientists need to understand more about the role of volcanic eruptions in contributing to the overall levels of greenhouse gases and/or destroyers of the stratospheric ozone layer. This is because of the notion of a critical threshold level beyond which the consequences may be very grave for life on Earth. A very small additional amount of CFCs released by human activities could cause a very large change in climatic conditions. On a less alarmist level, scientists need to increase their understanding because further research allows them to gauge the 'natural' concentrations of chlorine and fluorine compounds in the stratosphere that result from volcanic eruptions and to use these figures as a benchmark or yardstick against which to calibrate the effects of human activities in destabilizing the atmosphere.

Volcanoes also eject sulphate particles into the lower stratosphere which could form surfaces on which heterogeneous reactions occur. Such crystal surfaces are therefore catalysts, just like the ice crystals in very cold clouds, and require further research to assess their role and potency in the reactions which deplete the ozone shield.

Not only are volcanic eruptions capable of emitting gases that can lead to global warming or cooling, but they may also precipitate a 'volcanic winter'. The eruption of Toba in Sumatra, 73,500 years ago, created the largest known volcanic event in the Quaternary: the eruption is estimated to have lofted approximately 10^{15} g each of fine ash and sulphur gases to heights of

27–37 km (Rampino and Self 1992). The injection of all this volcanic material into the atmosphere may have caused a decrease in the amount of solar radiation reaching the Earth's surface and, therefore, led to a global cooling, estimated as a decrease of 3–5°C lasting up to a few years (ibid.). Stable oxygen isotope data suggest that the eruption of Toba occurred during a period of rapid ice growth and falling global sea level, and it has been proposed that the eruption could have accelerated the deterioration in global climate (ibid.). The cool weather in 1992–3 may be a consequence of the eruption of Mount Pinatubo in the Philippines on 11th June 1991, which ejected very large amounts of volcanic dust into the upper atmosphere, thus reducing the solar flux to the Earth's surface.

The ash falls from Mount Pinatubo caused the deaths of several hundred people, and the evacuation of tens of thousands. The aerosol cloud was the largest since Krakatoa in 1883 which was estimated to be 25–30 megatons. Substantially greater than El Chichón (1982, 12 megatons) and Mount St Helens (1980, 0.5 megatons), the Pinatubo eruption injected the aerosol, comprising liquid droplets of approximately 25 per cent water and 75 per cent sulphuric acid, to heights of 15–25 km, i.e., that is, to the same level as the ozone layer. The aerosol absorbed most strongly at the infrared end of the spectrum and scattered solar radiation back into space. The stratosphere therefore warmed while the troposphere cooled. Observations by Vogelmann *et al.* (1992) on the biologically effective ultraviolet light (UV-BE) at the Earth's surface showed increased surface sunlight intensity immediately following the eruption, from which they calculated that the effect of ozone depletion outweighed that of increased scattering. Also, satellite imaging of stratospheric ozone revealed a depletion of as much as 15–25 per cent at high latitudes, and in November 1991 a depletion of 20 per cent was observed over Boulder, Colorado (Westrich and Gerlach 1992, Brasseur 1992). This marked ozone depletion was originally thought to have been caused by volcanogenic chlorine. This cause, however, was probably minor because of the rapid down-flushing of chlorine as hydrochloric acid. The depletion is now considered to be the result of frozen sulphuric acid droplets which acted in a similar way to the stratospheric ice particles in polar regions, providing surfaces for chlorine-releasing chemical reactions involving anthropogenic CFCs which caused ozone loss.

Volcanic eruptions appear capable of creating both positive and negative feedback in the global climate system, depending upon their timing in relation to overall global cooling or warming due to other causes (see Chapter 2). There is a useful review of climate change and volcanic activity over the past 500 years by Bradley and Jones (1992), and the subject is discussed in detail in a UGU Special Report by the American Geophysical Union (1992a). In summary, the average surface cooling resulting from major volcanic eruptions is in the order of 0.2–0.5°C for a short duration, up to perhaps about five years after a major eruption (Self *et al.* 1981, Rampino and Self 1992), although local temperature reductions over similar time periods may be as high as 1.5°C (Porter 1986). Proxy data, such as acidity levels in ice cores, by measuring electrical conduction provide a temporal comparison for possible volcanic events, and their possible relationship with any global cooling (Hammer *et al.* 1980, Rampino and Self 1992).

Long-term climatic stability?

Throughout much of the Earth's history the planet's climate has shown a long-term stability. At least as far back as 3.8 billion years ago, liquid water has existed on the Earth's surface, as inferred from the metamorphosed sedimentary rocks in Isua, west Greenland (Kasting 1989).

The concept of climatic stability merely carries the connotation of the continued presence of liquid water on the Earth's surface and the continued presence of life. Periods of substantial global cooling and major Ice Ages, and other times when the mean surface temperature was much warmer than today, can be traced back through the geological history of the Earth. In both extremes, however, liquid water existed as oceans, rivers, and lakes, and life was sustained.

The remarkable feature of the Earth's long-term climatic stability since 3.5 billion years ago is that throughout this time there has been an overall increase in the amount of solar energy or flux reaching the surface of the Earth. Figure 3.10 shows the changes in the brightness of the

Sun over geological time, and it also illustrates the associated temperature changes and evolution of life. Note the relative stability in atmospheric oxygen and carbon dioxide levels throughout the past 570 million years. Armed with this knowledge, we might predict that the surface of the Earth should be getting warmer irrespective of human intervention. Yet, apart from the last century or so with human pollution of the atmosphere, the Earth does not appear to have warmed up steadily through geological time. It appears that this increase has been offset by a decrease in the concentration of atmospheric carbon dioxide caused by a negative feedback mechanism, the reduction in CO_2 levels in the atmosphere.

Studies of theoretical changes in the amount of solar energy reaching the Earth, using reasonable ranges of values, suggest that such solar variability could not compete with the anthropogenic greenhouse gases as causes of global warming (Hansen and Lacis 1990). In other words, if the Sun were to radiate slightly less heat energy to the Earth, within the range of expected natural fluctuations, then the cooling that could result would not be enough to offset the overall warming effect caused by anthropogenically generated emissions of greenhouse gases.

Simple energy-balance calculations (using the climatic models referred to as the Budyko/Sellers type, published in 1969), predict that a 2–5 per cent decrease in solar output could result in a runaway glaciation on the Earth, yet solar fluxes 25–30 per cent lower early in the Earth's history (Gough 1981) apparently did not produce such an effect (Caldeira and Kasting 1992). A favoured explanation to circumvent this paradox is that the partial pressures of CO_2, as a result of higher rates of volcanic degassing, possibly associated with slower rates of silicate weathering in rocks, generated a large enough greenhouse effect to keep the Earth warm. It has been argued, however, that the oceans can freeze to form sea ice much more rapidly (< 1 year) than the rate at which CO_2 can accumulate in the atmosphere (> 10^5 years), therefore if such a transient global glaciation occurred in the past when solar luminosity was low, it may have been irreversible, because of the formation of highly reflective CO_2 clouds. Thus the Earth might not be habitable today had it not been warm during the first part of its history (ibid.). In fact the

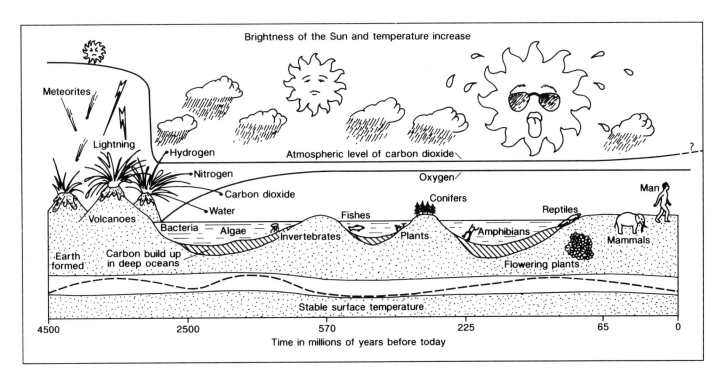

Figure 3.10   *Schematic development of the Earth's atmosphere and life, and the variability in the solar brightness and relative temperature throughout the Earth's history. After Watson (1991).*

presence of sedimentary rocks in the early Archean rocks from Isua, west Greenland, demonstrates that liquid water was present on the Earth's surface as early as 3.8 Ga, when solar luminosity was as much as 25 per cent less than at present. High amounts of atmospheric ammonia (NH_3) and CO_2 can account for the warm climate.

Self-regulating mechanisms for global climatic stability

To understand the way in which the Earth manages to maintain an overall stability of global climates, it is necessary to consider the self-regulating or negative feedback mechanisms. To do this, a bit of elementary chemistry which governs or regulates the long-term concentrations of CO_2 in the atmosphere is needed. By long term, we mean time periods greater than about 100,000 years which is the residence time of carbon in the oceans or the time taken for an 'average' carbon atom introduced into the oceans to be removed, for example by being locked into a rock such as limestone. The long-term control of atmospheric CO_2 involves the 'carbonate–silicate geochemical cycle' which is a measure of the way in which gaseous CO_2 exchanges with carbon dioxide contained in carbonate rocks. The last mechanism involves the chemical weathering of silicate minerals and the accumulation of carbonates. Gaseous CO_2 is returned to the atmosphere as silicate minerals are formed or as carbonates are **metamorphosed** (subjected to intense heat and pressure) to release CO_2.

Silicate weathering depends on temperature because warmer conditions encourage the chemical reactions that break down the silicates. The rate of chemical weathering is increased with greater rainfall, also strongly influenced by surface temperature. Thus, a decrease of temperature at the Earth's surface should be accompanied by a reduced rate of silicate weathering which, in turn, should induce an increased concentration of CO_2 in the atmosphere and an accompanying increase in surface temperature due to the greenhouse effect. In other words, these feedback mechanisms should mean that

the atmospheric CO_2 levels and climate form a self-regulating system.

Applying this logic to the converse scenario of an increase in the surface temperature of the Earth, then the rate of silicate weathering would increase and remove CO_2 from the atmosphere and thereby reduce the Earth's surface temperature. This is the essence of a negative feedback mechanism which, in this case, involves the greenhouse gas carbon dioxide and the greenhouse effect.

An area of disagreement concerns the possible implications for global climate change precipitated by a thawing of the entire Arctic permafrost. In such a scenario, some scientists postulate that the thawing could release CH_4 and CO_2 in sufficient quantities to make a substantial contribution to global warming. A countervailing argument, invoking a negative feedback, is that under a warmer CO_2-rich global climate, there would be enhanced tree and other vegetational growth which would slow or limit global warming. More research is needed in order to understand which sequence of events is likely.

Recent discoveries by Earth scientists have shown that wetter clouds can dampen the greenhouse warming (Slingo 1989). By altering the variables that go into computer-simulated climatic models, research suggests that near their freezing level, liquid water clouds replace ice crystal clouds. Liquid water clouds dissipate more slowly than ice crystal clouds, and they reduce the amount of solar energy absorbed, thereby reflecting more of this energy back into space. The net result is a smaller degree of warming than might otherwise occur if there was no change in cloud type. In fact, some computer runs with cloud type taken into account lead to estimated equilibrium surface warming, due to doubling CO_2 emissions, being reduced from 5.2 Kelvin to 2.7 Kelvin – a decrease of almost 50 per cent. Computer models that simulate greenhouse warming will need to be improved in order to take account of this sort of data so that better predictions are possible.

Clouds are probably the most important self-regulating mechanism in controlling feedback mechanisms in the ocean–atmosphere system, and prevent a runaway greenhouse or icehouse world. Their warming effect may be many times greater than doubling the CO_2 levels of a region.

Also, their overall cooling effects makes the Earth 10–15°C cooler than it would otherwise be if the planet were cloudless. There are two major sets of clouds. First, there are tropical clouds which reflect sunlight back into the atmosphere to cool the system, but they also exert a greenhouse warming effect. Second, there are middle- to high-latitude clouds of the northern and southern hemispheres which have a net cooling effect. If these two sets of clouds were induced to behave differently, then this could have dramatic consequences on global heat budgets, regional weather systems, and cause changes in precipitation patterns. Clouds also provide extra hydroxyl molecules which are capable of oxidizing CH_4 and NO_x (not a greenhouse gas), thereby removing the greenhouse gases from the atmosphere and reducing their greenhouse effect.

The rainforests also play an important role in regional water balance and the distribution of clouds. Trees may even control the rate and timing of cloud nucleation, by emitting a variety of cloud condensation nucleii to help produce local convective systems that may be as much as 5 km in diameter. The destruction of the rainforests could cause major perturbations in the global weather systems by disrupting the water balance, and by increasing atmospheric levels of CO_2. Estimates of deforestation, therefore, must be made to model global warming effectively, and understand the consequences of human actions. Indeed, many environmentalists advocate increased afforestation as a way of sequestering the increased anthropogenic emissions of CO_2.

Biological processes also serve as important self-regulating mechanisms for climatic stability. Indeed, some scientists maintain that the biological control is actually far more important than the inorganic chemical reactions in developing and maintaining an equable cocktail of atmospheric gases. One of the most ardent proponents of such a viewpoint is Lovelock (1988).

Finally, global warming could trigger a cascade of natural hazard effects, both directly through the meteorological processes associated with climate change, and indirectly because of rising sea level. Figure 3.11 summarizes the potential natural hazards.

International action on global atmospheric pollution

Atmospheric ozone depletion

In September 1987, the leading industrial nations, including Britain, were signatories to the Montreal Protocol in which they agreed to specific reductions in the production and use of five of the most harmful CFC ozone-depleting gases known at the time. These CFCs are used in many aerosol sprays, refrigerants, solvents, and plastic foams.

The leading nations agreed to attempt to find substitute chemicals to replace the most dangerous CFCs, one of which is called CFC-11. A possible substitute for CFC-11 is CFC-22 which, although less toxic, still represents a considerable threat. For example, it has been calculated that CFC-22 is about 20 times less harmful than CFC-11. CFC-11 has a half-life of about 75 years compared to 20 years for CFC-22. Over the next 20–30 years, however, when

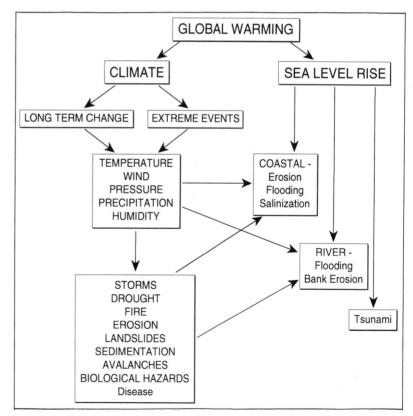

Figure 3.11 *Global warming may trigger a cascade of hazard effects – both directly through the mechanism of climate change and indirectly via sea level rise. After Campbell and Ericksen (1990).*

the ozone layer may be most rapidly depleted, it is estimated that CFC-22 is likely to be only 20 per cent as damaging as CFC-11. In fact, Peter Fabian of the Max Planck Institute for Aeronomy, Gottingen, believes that if CFC-22 is released into the atmosphere at the present rates, then by the end of this century, it will have destroyed as much ozone as the two most common CFCs, CFC-11 and CFC-12.

In 1987 CFC-22 was not in the Montreal Protocol. It may indeed be safer than CFC-11, but it is still harmful. The wisdom of replacing one dangerous CFC with another that is perhaps cumulatively just as harmful should be questioned seriously – something that indeed transpired.

In early March 1989, the British government organized an international conference on atmospheric ozone depletion, at which representatives attended from 124 other nations. The conference was in essence a stage-managed affair. Even before the conference opened, the EC, the USA, and Canada had agreed to a complete phasing out of the five CFCs and three halons covered by the Montreal Protocol. The Protocol had stipulated a 50 per cent reduction in the production and consumption of these materials by the year 2000 and so a total ban by the EC, USA, and Canada was justifiably seen as a resounding success story by the media and general public.

One of the outstanding problems was succinctly put by former Prime Minister Margaret Thatcher at the conference when she stated that the actions of the developed, industrialized world would not be sufficient by themselves to reverse the deterioration of the ozone layer. Many underdeveloped and developing nations are only just beginning to industrialize and wish to import and manufacture refrigerants, refrigerators, plastics, air-conditioning plants, and electronic goods, all of which involve the use of ozone-depleting CFCs and halons. These nations cannot, and should not, have to await the development and widespread installation of safer chemicals to replace the harmful CFCs and halons in the products they urgently need for development. This point was made at the conference by President Daniel Arap Moi of Kenya. Moi went on to ask that the developed, industrialized world take a moral and scientific responsibility for providing the technology and economic circumstances which would allow the developing nations effectively to leap-frog the CFC and halon technology.

At the London conference, another optimistic sign for the future was the decision by 20 additional countries to sign the Montreal Protocol, taking the total number of signatories to 59, although at the time of this book going to press, the Protocol has been ratified in only 32 of the signatory countries. Representatives from India and China said that their nations would be amenable to signing the Protocol if the developed, industrialized countries made a clear and attractive financial commitment to setting up a fund for adequate research and development to produce alternative chemical substances, and then transfer this technology free of charge to those nations.

In Helsinki, in early May 1989, 81 nations agreed in principle to ban eight industrial chemicals by the year 2000 which cause damage to the stratospheric ozone layer. These chemicals include five CFCs and three halon gases. At the Helsinki meeting, Eileen Claussen, of the USA's Environmental Protection Agency, predicted that even if CFCs are totally phased out and banned by the turn of the twenty-first century, chlorine concentrations will still treble from existing levels of 2.7 to 8.9 ppb by the year 2010, mainly because of the time taken for CFCs to break down in the atmosphere. A spokesperson for Greenpeace claimed that another greenhouse gas, methyl chloroform (a solvent and cleaning fluid), is increasing in the atmosphere at the rate of 7 per cent per annum. Many of the nations that met in Helsinki and agreed in principle to ban eight ozone-depleting chemicals hoped to include their proposals in the 1990 London amendments to the Montreal Protocol.

At an international meeting of the parties to the Montreal Protocol, in November 1992 in Copenhagen, revised controls on ozone-depleting substances were agreed. The deadlines for the global phase-out for most chemicals was brought forward. However, a controversial part of the Copenhagen package is the allowance of a sharp increase in the use of HCFCs, as a transitional measure to substitute for CFCs while they are phased out. The Copenhagen package consists of the following agreements:

• For *CFCs*, the phase-out date was brought

forward from January 2000 to 1 January 1996, with a 75 per cent reduction, based on 1986 levels, by 1 January 1994. The EC has proposed an interim 85 per cent reduction by 1 January 1994.

- *Carbon tetrachloride* should be phased out by January 1996 rather than 2000 as originally proposed, with an 85 per cent reduction, based on 1989 levels, by 1 January 1995. The EC has proposed an interim 85 per cent reduction by 1 January 1994.
- *Halons* should be phased out by January 1996, again brought forward from the January 2000 target date. The EC has proposed a phase-out by January 1994.
- *Methyl chloroform* should be phased out by January 1996, brought forward from January 2005, with a 50 per cent reduction on 1989 levels by January 1994. However, the 'essential use' of halons may continue beyond the new proposed phase-out dates, and a UNEP panel will prepare an assessment of such essential usage, with the announcement of a decision scheduled for 1994.
- *Methyl bromide* consumption is to be pegged at 1991 levels by 1995. More stringent controls must await scientific evaluation by two UNEP panels which are scheduled to report in 1995.
- *HBFCs*, although not in general use, should be phased out by January 1996. This is the first time that HBFCs have come under any control.
- *HCFC* use is to be capped in January 1996 at a level amounting to the sum of their consumption in 1989 and 3.1 per cent of the level of consumption of CFCs in 1989. This formula arose in order to take into account existing consumption of HCFCs which were already high in some countries in 1989, and also in recognition of their role as transitional substitutes for CFCs. The Copenhagen amendments incorporated controls for the first time on HCFC use, which is to be phased down to a 35 per cent reduction of the 1996 consumption level by 2004, a 65 per cent reduction by 2010, and a total ban by 2030.

Consumption levels of CFCs are calculated in ODP-weighted tonnes. The Copenhagen amendments include the wording that party nations 'shall endeavour to' achieve the various targets and, therefore, they cannot be considered as absolutely binding commitments, but rather as statements of intent.

The amounts of CFCs produced and consumed have decreased by 40 per cent since 1986 and, under the 1990 London and 1992 Copenhagen amendments to the Montreal Protocol on protecting the ozone layer, will continue to fall, but the release of CFCs already in use means that atmospheric concentrations of CFCs will continue to rise for some years. At least the problem has been tackled at an international level with considerable effects.

Greenhouse gas emissions

In February 1991, the US government proposed a strategy to limit global greenhouse warming by suggesting that nations should seek a comprehensive framework for the emission of greenhouse gases in preference to focusing on a single gas. In many respects, although this is a self-evident truth, one has to ask whether this proposal by the US was designed to frustrate efforts at obtaining a phased reduction in the emissions of CFCs and CO_2, or a genuine attempt to encourage a more balanced perspective on the task at hand.

A thorny issue concerning the production of greenhouse gases is the ways in which industry can substitute one form of energy generation with another. Many environmentalists would not regard nuclear energy as an alternative to conventional fossil fuel power stations and the solution to reducing emissions of CO_2. In the UK in 1989 at the public inquiry into plans for a nuclear power station at Hinkley, Somerset, David Fisk, the chief scientist at the UK Department of the Environment, gave written evidence which stated that 'nuclear power alone will not provide the solution to the greenhouse problem but the continuation of a nuclear contribution to the electricity supply will provide a diversity of fuel sources which will play a part, together with energy efficiency and renewable resources' (Pain 1989).

Much more research is needed to assess the impact on the global climate of human activities. There is a need to quantify cause and effect, and what courses of action will maximize the chances of ameliorating or reducing the destructive potential of human activities on the world climate. Predictions about economic growth and trends in fuel prices, in the context of any overall energy strategy or policy, will have a major impact on determining the acceptability to individual countries of any proposed greenhouse gas emissions targets and the means to achieve these. Many people, including environmental groups, support energy conservation and reducing energy demand as a means of controlling the emission of greenhouse gases, with cost-effective energy efficiency measures. There is also considerable debate as to the relative merits of using regulations versus economic instruments to control greenhouse gas emissions.

The issues raised by the greenhouse effect are inherently very political. There are scientists interested in exploiting the current level of media interest in order to give their 'climatic' research greater status and to persuade funding bodies to direct more research grants their way. There are environmentalists who are only too willing to believe every gloomy prediction that an anthropogenically created enhanced greenhouse effect may have, and who use the information to scare people into donating money to their cause. They are all too often unable to judge the scientific content of any argument about global warming, but simply react in a visceral manner. In essence, anyone who gets involved in the issue of global warming due to an anthropogenically enhanced greenhouse effect reacts in a political way. It is impossible to be objective, but we owe it to others, as well as ourselves, to try to become as informed as is possible about the ramifications of the greenhouse issue. Opinions expressed with some measure of knowledge allow for constructive debate and thoughtful action. But, naturally, there are vested economic interests that can so quickly come to dominate the issues raised by the greenhouse effect.

Even the United Kingdom Atomic Energy Authority (UKAEA) has seen an 'angle' in the greenhouse debate. In January 1990, the UKAEA published a glossy brochure entitled *Nuclear Power and the Greenhouse Effect*. Its 30 pages set out to persuade people that nuclear power, with the tarnished image it seems to have acquired, is the answer to human energy needs and offers an environmentally friendly alternative to the dirty conventional fossil fuel-burning power stations that emit such large quantities of greenhouse gases. In fact, the last paragraph states: 'The use of nuclear power could be readily expanded. By 2020 it could be reducing energy–CO_2 emissions by 30 per cent of what they would otherwise have been, effectively reducing global warming by 15 per cent. In the longer term it could offer much deeper cuts in greenhouse gas emissions from energy.'

The Gulf War of 1991 was probably the first war in history which really made people consider the impact on the environment that mass destruction can wreak along with the human injury and death toll. The Kuwait oil fires released approximately 240 million tonnes of CO_2, equivalent to about 1 per cent of the annual global emissions of CO_2 (Plates 9 and 10). The last oil well-head fire of about 750 was finally extinguished in Kuwait on 6th November 1991, bringing the total cost to an estimated US $1 billion. But the effect of these fires is an enormous regional pollution event in the area of the Persian Gulf. Many environmentalists predicted that the Iraqi fire-raising in Kuwait would precipitate a catastrophic effect on global climate, including a substantial cooling of the Earth's surface as solar radiation was blocked by the soot clouds, akin to a nuclear winter scenario, with the massive failure of the Asian monsoons, crop failure, and accelerated global warming. Whilst it is true that the deliberate burning of Kuwait's oil wells by the Iraqis was an act of gross environmental vandalism, the global climatic impact was exaggerated. Studies now show that the smoke from the oil wells and refineries was not injected high enough into the atmosphere to cover large parts of the northern hemisphere, nor was it produced in sufficient amounts to cause a measurable temperature change, or a failure of the monsoons, and only a small increase in global CO_2 budget resulted (Small 1991, Johnson *et al.* 1991).

By late 1992 and early 1993, however, it had become apparent that the local effects of this eco-terrorism are severe. A study commissioned by the Saudi government, published in an EC report in 1993, has shown that earlier estimates of 1–2 million barrels of oil being dumped in the Gulf have been upwardly revised to about 11 million barrels, and the Saudi Meteorology and

Environment Protection Administration claim that 2.7 million barrels are still clogging the beaches of Saudi Arabia, Iran, and Qatar, with an additional 2.6 million barrels possibly trapped beneath the sea surface. The smoke clouds that blotted out the sunlight caused a 10°C fall in sea-surface temperatures, and with it the destruction of plankton and fish larvae in the Gulf, which are at the bottom of the food chain. This has led to a fish famine, and cannibalism amongst the tens of thousands of starving seabirds. The Gulf War should have taught people to respect the planet more, even at times when human self-respect is at an all-time low. It is possible to stop wars, but their effects could be far harder to control in the ensuing peace.

In fact, the eruption of Mount Pinatubo in the Philippines more than offset any warming effects that may be caused by the Kuwait oil fires, when it ejected 3–5 km^3 of magma and 20 million tonnes of sulphur into the atmosphere. Although most of the ash and dust settled rapidly, small sulphate particles, or aerosols, remain in the atmosphere where they reflect back into space incoming solar radiation. Using simple climate models, it has been predicted that the effects of this eruption will be a global cooling of up to 0.5°C over the next few years.

An international treaty on pollution in the atmosphere is overdue. For a number of years throughout the 1980s, negotiations along this road were bogged down. A round of negotiations in Nairobi in late September 1991 ended in deadlock because of the continued US refusal to set targets and schedules concerning the emission of greenhouse gases. The argument put forward by the delegates from the USA was that setting targets is too costly and would discourage other countries from signing a climate convention. Japan also did not want targets, but rather was in favour of a 'pledge-and-review' treaty, whereby nations would agree to cut their emissions of greenhouse gases and then monitor the results. The Japanese proposed that nations stabilize their CO_2 emissions to 1990 levels by the year 2000, and were supported by Britain, Canada, and the USA. The European Commission, however, insisted on more stringent measures. So an impasse developed without any agreement on mutually acceptable emission targets and schedules. Once the storm clouds had passed on the negotiations over a climate treaty,

a much more optimistic note was struck in 1992 in Rio.

In June 1992, at the United Nations Earth Summit in Rio de Janeiro, many countries supported the UN Framework Convention on Climate Change, committing them to reducing emissions of the greenhouse gases CO_2, CH_4, and nitrous oxide to their 1990 levels by the year 2000. CFCs, which are covered by the Montreal Protocol, are not part of the Earth Summit Convention. To date, 160 countries have signed, and 14 countries ratified the Convention. Many nations still have to ratify the Convention (50 are needed for it to come into force), for example the UK plans to do so by the end of 1993 with the proviso that other industrialized signatories do likewise. Indeed, the UK's commitment to reducing its CO_2 emissions is set out in a Department of the Environment consultative document, *Climate Change: Our National Programme for CO$_2$ Emissions* (1992). Also, in the UK's 1992 March Finance Bill (the 'Budget'), higher excise duties were implemented on petrol and diesel prices, and a value-added tax (VAT) was imposed on domestic fuel and power, to come into effect in 1994 at 8 per cent, rising to 17.5 per cent in 1995, which the Government claimed was part of a broader policy on controlling environmental pollution, although in this particular case, cynics interpret this move as a straightforward revenue-raising means of direct taxation.

Conclusions

Governments are beginning to take action to control the amount of global warming caused by anthropogenically created greenhouse gases. In the USA, the US Office of Technology Assessment (OTA) has developed two scenarios assuming specified action levels for energy supply, energy conservation, and forest management, which would limit CO_2 increases to 15 per cent by the year 2015 ('moderate'), and a 20 per cent reduction in CO_2 emissions over 1987 levels ('tough'). The OTA evaluation of these contrasting scenarios is that the 'moderate' option is achievable with net savings to the economy, whereas the 'tough' option would range from a small overall saving to a cost equivalent to 1.8 per cent of the Gross National Product (GNP).

As a comparison, less detailed UK studies suggest that an 88 per cent reduction in CO_2 emissions relative to their 1987 levels, mainly through energy efficiency measures, is achievable at no net cost.

As for the future control on the emission of greenhouse gases, international agencies are beginning to debate the need for severe energy–carbon taxes. On 25th September 1991, the EC commissioners proposed an energy and carbon tax that would help reduce the emission of greenhouse gases, but raise some fuel costs by up to 60 per cent. The proposal would involve the imposition of a tax of $10 a barrel of oil over seven years. By the year 2000 this would raise industrial coal prices by 60.6 per cent, natural gas by 31 per cent, heating oil by 40 per cent, and industrial electricity by 16 per cent. Renewable energy resources, such as wind and tidal power, would be exempt from the tax, the status of nuclear power remaining undecided. Inefficient household appliances, such as washing machines and cookers, could also attract a new tax.

The debate on these proposed new taxes is just heating up, in an attempt to cool the planet down. The verdict of the jury will affect all of humanity. It is a debate that runs to the very heart of the need to reduce the emission of greenhouse gases. Is the only real way to curb these emissions by severe taxes on the production and consumption of energy? How will individual governments in Europe react to the imposition of energy taxes by the EC in terms of their perceived sovereignty? But worldwide, proposals for an energy–carbon tax appear to be one of the most practical ways forward on controlling the emissions of greenhouse gases. Finally, the 1992 UN Framework Convention on Climate Change contains no specific commitments from the signatory nations to control the emissions of greenhouse gases after the year 2000. And without any post-2000 commitments, in countries such as the UK, greenhouse gas emissions are projected to rise steeply because the savings in emissions gained from changing to gas-fired (from coal- and oil-fired) electricity-generating power stations will have been exhausted. Also, global transport-related emissions will continue to rise. The issues surrounding the emissions of greenhouse gases are going to be at the fore of both national and international discussions, certainly well into the next century.

Key points

CHAPTER 3: KEY POINTS

1 Stratospheric ozone forms part of an important natural shield around the Earth which absorbs large amounts of UV-B radiation from the Sun, thereby reducing the harmful effects of too much radiation reaching the Earth's surface. Reducing stratospheric O_3 levels results in a cooling of the stratosphere and warming of the troposphere, and the net result is a cooling of Earth's surface, but the balance is strongly influenced by latitude, altitude, and seasons. Significant depletions in stratospheric ozone concentrations have been recorded since 1977 and are, at least in part, the result of anthropogenic emissions of CFCs and other chemicals, as well as natural processes such as volcanic eruptions which produce aerosols and acids. Depletions of stratospheric ozone are now common over the poles following the Antarctic spring, when supercooled clouds act as a catalyst for the heterogeneous reactions which destroy the ozone. An increase in atmospheric CO_2 could lead to tropospheric warming and is likely to result in increased cloud cover which may activate O_3-depleting species and thereby increase the rate of O_3 depletion, but this sequence of events remains controversial. Tropospheric ozone concentrations have been increasing in polluted regions, where it is an important greenhouse gas, and contributes to photochemical smogs.

2 The Earth's atmosphere is warmed by the so-called 'greenhouse effect' where short wavelength (ultraviolet) solar radiation reaches the Earth's surface and is re-radiated as long wavelength (infrared) radiation back up into the atmosphere where it is absorbed by greenhouse gases which warm the Earth. The most important greenhouse gases include water vapour, carbon dioxide, methane, tropospheric O_3, nitrous oxide, ammonia, CFCs, and the halons. These have different global warming potentials (GWPs) and varying residence times in the atmosphere. An increase in these atmospheric gases may lead to global warming. However, there are many negative feedbacks which counteract the effects of any potential global warming caused by these gases, and the real nature of increased concentrations of the greenhouse gases remains debatable. Sophisticated 'general circulation models' (GCMs) for the interaction between the ocean–atmosphere–land systems have been developed by various organizations, including the Intergovernmental Panel on Climatic Change (IPCC), as a means of assessing the climatic sensitivity to changing atmospheric concentrations of greenhouse gases. The IPCC (1990) has suggested that a doubling of atmospheric CO_2 is likely to increase the global mean surface temperature by 1.5–4.5°C.

Evidence for human-induced global warming is fiercely debated in some quarters. It is difficult to compare historical and modern meteorological data, and although snow and Arctic sea ice cover have decreased over the last decade, it is not known if this is due to natural perturbations in the Earth's atmosphere and hydrosphere or anthropogenic causes. Arctic geotherms, however, provide evidence in the form of thermal anomalies to suggest a warming in the order of 2–4°C over the last century (ibid.). If greenhouse gas emissions continue, global warming may cause changes in global weather patterns, amongst which there could be an increased severity of storms and droughts, and an overall rise in sea level in the order of 50–60 cm by the year 2100, caused by the thermal expansion of the oceans and the melting of ice caps and glaciers.

3   Natural phenomena such as volcanic eruptions and El Nino events also cause global climate change. Volcanic eruptions emit huge quantities of aerosols which reduce incoming solar radiation and lead to global cooling usually in the order of 0.2–0.5°C for short durations (< 5 years). Greenhouse gases (CO_2), ozone-depleting gases (Cl_2, F_2) and gases which may act as catalysts for ozone-depleting reactions (SO_2 which forms H_2SO_4) are also emitted by volcanic activity, and cause either negative or positive feedback mechanisms for global climate change.

El Nino events produce relatively small changes in the Earth's climate, due to variations in the atmosphere–ocean system. Variations in oceanic circulation and upwelling of cold, nutrient-rich, deep-ocean currents affect biomass production (e.g., algal blooms) and atmospheric CO_2 concentrations. During El Nino events, more CO_2 is released into the atmosphere, and heavy rainfall occurs in the southern Pacific Ocean, and causes droughts in Indonesia and Australia and floods in South America.

4   The Earth has maintained a stable climate as far back as 3,800 Ma, even though the solar flux has increased. The Earth's self-regulation mechanisms have kept global climate stable over geological time, particularly through the 'carbonate–silicate geochemical cycle'. Clouds and cloud cover (albedo) also play an important role in stabilizing global climate, helping to reduce incoming solar radiation. Anthropogenic emissions of greenhouse gases, such as CO_2 and CH_4, may have serious implications for the long-term survival of human beings, although not for life on Earth.

5   International action on global atmospheric pollution has resulted in attempts to reduce ozone depletion (Montreal Protocol, 1987, with amendments in Helsinki, 1989 and Copenhagen, 1992) and reductions in greenhouse gas emissions (UN Framework Convention on Climate Change, 1992). It remains to be seen whether the developed and developing nations can meet the targets set by the various international agreements on atmospheric pollution.

Chapter 3: Further reading

Boyle, S. and Ardill, J. 1989. *The Greenhouse Effect.* London: Hodder & Stoughton.
A highly readable, if slightly dated, introductory book explaining the nature of the greenhouse effect. Although written for the lay-person with little or no scientific background, it may also be useful for social science students wishing to gain a reasonable understanding of the causes and effects of any global warming caused by a greenhouse effect.

IPCC (Intergovernmental Panel on Climatic Change) 1990. *Climate Change: The IPCC Scientific Assess-* ment, ed. Houghton, J.T., Jenkins, G.J. and Ephraums, J.J. Cambridge: Cambridge University Press.

—— 1992. *Climate Change 1992: The Supplementary Report to the IPCC Scientific Assessment,* ed. Houghton, J.T., Callander, B.A. and Varney, S.K. Cambridge: Cambridge University Press.
These two reports are the result of Working Group 1 of the Intergovernmental Panel on Climatic Change, set up by the World Meteorological Organization and the United Nations Environment Programme. They are essential reading and reference material for anyone interested in global climate change. The reports assess

the potential effects that human activity may have on the Earth's climate, and include sections on changes in the concentrations of atmospheric greenhouse gases; modelling of the global climate system; computer prediction of climate change; observed climate change over the last century; the detection of climate change due to human activities; changes in global sea levels due to global warming; the response of ecosystems to global climate change; and the research required to narrow the uncertainties in future predictions of global climate change. The reports consider worst-case, business-as-usual, and best-case scenarios for global climate change using various assumptions about the levels of anthropogenic emissions of greenhouse gases.

Kemp, D.D. 1990. *Global Environmental Issues: A Climatological Approach.* London: Routledge.
A climatological approach to global problems which includes good sections on the greenhouse effect, acid rain, ozone depletion, drought, and the possible effects of a nuclear winter. Appropriate for students in environmental studies and geography wishing to analyse the role of climatology in environmental change.

Nilsson, S. and Pitt, D. 1991. *Mountain World in Danger: Climate Change in the Forests and Mountains of Europe.* London: Earthscan Publications.
An account of the possible changes in the forests and mountain environments of Europe as a result of global warming and acid rain. Possible strategies in response to these changes are discussed in the hope that steps may be undertaken by policy makers to retard the potentially harmful impact. Good summaries of the IPCC's conclusions and international conventions and declarations are provided in the text and in a series of appendices.

Parry, M. 1990. *Climatic Change and World Agriculture.* London: Earthscan Publications.
Written by the chief scientist on the IPCC, and concerned with the potential impacts of climatic change on agriculture. It provides a good account of the likely patterns of change in climate and world agriculture as a consequence of global warming. Emphasis is placed on the uncertainties associated with this issue, the sensitivity of the world food system, vegetational patterns, and animal life to global climate change, the geographical limits of different types of farming, and the range of possible ways to adapt agriculture to mitigate any potential hazards caused by global warming.

I'm a goin' back out 'fore the rain starts a fallin',
I'll walk to the depth of the deepest black forest,
Where the people are many and their hands are all empty,
Where the pellets of poison are flooding their waters,
Where the home in the valley meets the damp dirty prison,
Where the executioner's face is always well hidden,
Where hunger is ugly, where souls are forgotten,
Where black is the colour, where none is the number,
And I'll tell it and think it and speak it and breathe it,
And reflect it from the mountains so all souls can see it,
Then I'll stand on the ocean until I start sinkin'.

But I'll know my song well before I start singin',
And it's a hard, it's a hard, it's a hard, it's a hard,
It's a hard rain's a gonna fall.

Bob Dylan, 'A Hard Rain's a Gonna Fall'

CHAPTER 4
Acid Rain

The Earth's atmosphere supports life, yet the skies over much of Europe and North America are rapidly becoming poisoned by toxins from power stations, factories, and vehicles. The catalogue of atmospheric pollution still seems to be growing at an alarming rate. 'Choking Greeks look to the gods for help' was one of the front-page leaders in Britain in *The Times* on 2nd October 1991. As temperatures rose to an unusual 36°C in Athens on 1st October, levels of the main pollutant gas, nitrogen dioxide, reached an average of 561 micrograms per cubic metre in the city centre, well above the 500 microgram limit, above which emergency measures have to be introduced. Indeed, in parts of Athens, the levels of nitrogen dioxide reached 696 micrograms per cubic metre, the previous record being 683 micrograms on 10th June 1991. As a result of the cocktail of toxic gases over Athens on 1st October, locally named *nephos*, more than 200 people went to hospital with respiratory and cardiac problems, and the government banned the use of private cars in the city centre between 6 a.m. and 5 p.m. Scenarios such as this one are becoming more common in major cities throughout the world because of atmospheric pollution, especially through the indiscriminate use of private motor vehicles pumping out atmospheric pollutants, some of which, such as nitrogen dioxide, contribute to acid rain.

The natural balance of gases and reactions in the atmosphere sustains not only human beings but also all the other life on Earth. Organisms, including humans, use and produce many gases, including oxygen, carbon dioxide, nitrogen, and sulphur compounds. Human activities have drastically altered the natural balance of such gases and have contributed new and extremely harmful toxic gases and other substances to the atmosphere (Figure 4.1). While rain is naturally slightly acidic, human activities can increase the acidity and push natural systems over a critical threshold level and bring about atmospheric–climatic changes which have adverse environmental effects.

Two end-member philosophies to pollution may be applied. One is to concentrate and contain, while the other is to dilute and disperse. Both perspectives have pros and cons. Traditionally, industry is the main contributor to acid rain, which is exacerbated by its common concentration in industrialized areas resulting in maximum pollution tending to be geographically contained. Arguably, by sharing the misery of pollution through geographically more distributed factories and energy plants, the effects of atmospheric pollution could actually be reduced with the more efficient dispersal of pollutants before they tended to reach threshold concentrations. In reality, commercial and demographic considerations mean that belts or regions of industrial activity are typical of any country. One of the main causes of atmospheric pollution is that polluters try to dispose of gaseous pollutants into the atmosphere, in an attempt to dilute and disperse.

ACID RAIN

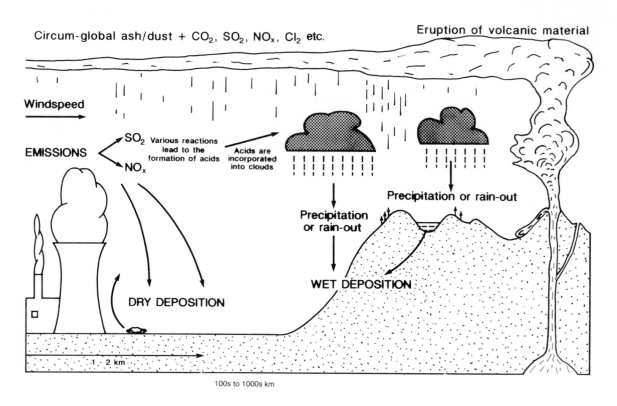

Figure 4.1 *Processes involved in the formation and deposition of acid rain. After Mannion (1991).*

The long- and intermediate-term consequences of artificially changing the composition of the atmosphere are not fully understood. The principal chemicals which produce acid rain are sulphur dioxide (SO_2), and nitrogen oxides (NO_x), and hydrocarbons. Collectively, nitric oxide (NO) and nitrogen dioxide (NO_2) are referred to as NO_x.

Human awareness tends only to surface as necessity dictates, as the effects of pollution are experienced, and not through some sort of spontaneous environmental consciousness which is separated from human exploitation of the Earth's resources. Acidic rain, however, was appreciated in the middle of the 1850s, for example in 1852 the Scottish chemist Robert Angus Smith published a paper in the *Memoirs and Proceedings of the Manchester Literary and Philosophical Society* entitled 'On the air and rain of Manchester'. The increasing acidity of the precipitation is referred to as acid rain.

The noxious effects of pollution caused by the burning of coal have been known about for centuries, though they were not identified then as acid rain. In England, during the reign of

Edward I (1272–1307), the use of sea coal (with its high levels of sulphur and trace elements), washed ashore from exposed deposits of coal, was prohibited by Royal proclamation. A third offence was punishable by execution!

The phrase 'acid rain' covers a multitude of human sins against the environment. It is not only rain but also acid mists and acid fog that have to be contended with, together with **photochemical smogs** which may or may not be acidic. Perhaps one of the most infamous smogs of recent times was the 'pea-souper' smog of December 1952 in London, when about 4,000 people were killed by inhaling water droplets that were almost as acidic as the water in a car battery! The smog lasted five days from 5th December and was caused by warm air holding down a vast acidic cloud, with no wind to disperse it. By 9th December 1952 the acid smog extended to a radius of 30 km from the centre of London. The acid in the smog was probably mainly sulphuric acid (H_2SO_4). Current estimates of the pH of the smog are that it was between 1.4 and 1.9 – more acidic than lemon juice. Even today, London is not immune

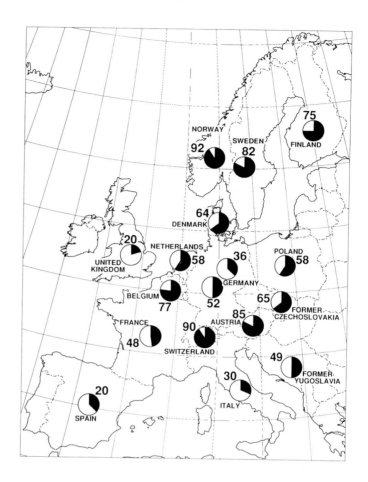

Figure 4.2 *The contribution, shown as a percentage, of external sources (solid portions) to the amounts of sulphur deposited in some European countries. Percentages are based on EMEP model calculations for 1978–82. After National Environment Protection Board (1987).*

from acidic smogs. Over the weekend 13th–15th December 1991, severe smog conditions returned to London, when on Friday 13th December levels of nitrogen dioxide at street level reached 293 parts per billion, well above the recommended EC maximum values, and the highest since 1976. The conditions arose because of a high-pressure system which brought settled weather and created warm dry air at higher altitudes. The air at ground level cooled and cold moist air started to rise, but was trapped below a layer of warmer air. The cold air stagnated because of a lack of wind to mix the air masses. A density inversion was therefore created over London and the pollutants remained trapped near street level to cause a smog. The heavy use of vehicles in the city only served to exacerbate the situation.

Other examples of serious incidents such as

the 1952 London smog include the two days in August 1984 when Athens suffered an acid smog that led to the hospitalization of 500 people and the Greek government declaring an emergency situation. Other cities around the globe, such as Mexico City, Tokyo, and Los Angeles, suffer severe air pollution and acid precipitation on a fairly regular basis (Plate 11).

Measuring the effects of acidic deposition

Acid rain has already produced many visible and measurable effects. This century, the first areas to be recognized as showing serious damage as a result of acid rain were found downwind of the major industrial centres of Britain, mainland

Europe and in North America. Scandinavia and Canada were identified earlier in the 1980s as having suffered severe damage from acid rain. Much larger areas of Europe are now known to be affected as well as less industrialized nations such as China, India, and South Africa. Although sulphur is transported from eastern North America to Europe, the amounts are small compared to the European sources. Sulphur pollution is the result of sources both within and outside individual countries. A study by the National Environment Protection Board (1987) has revealed the contribution, expressed as percentages, of external and national sources of sulphur throughout western Europe (Figure 4.2). From this study, covering the years 1978–82, it can be seen that Norway received 92 per cent of its sulphur deposition from external sources, compared to only 20 per cent from outside sources for the UK. Data such as these reveal the case for stringent international control on sulphur emissions, because of the large amounts of pollution that are exported from heavily industrialized regions to other parts of the world.

In eastern North America, anthropogenic sources of sulphur overwhelm natural sources (Galloway and Rodhe 1991). Figure 4.3 shows the estimated amounts of sulphur ($Tg\ a^{-1}$) advected (transported) eastwards from eastern North America (USA and Canada), and Figure 4.4 shows a schematic sulphur budget for the western North Atlantic Ocean atmosphere. Sulphur is removed from the atmosphere both by **wet deposition** and by **dry deposition**. Any uncertainties in the fluxes are due, at least in part, to a lack of direct measurements and because of remaining ignorance about the processes leading to dry deposition.

Throughout many parts of the northern hemisphere, trees are being poisoned and killed, and soils are becoming too acidic to support plants (Plates 4.1 and 12). It is estimated that more than 65 per cent of the trees in the UK, and more than 50 per cent of the forests in the pre-1990 West Germany, the Netherlands, and Switzerland are damaged. Acid rain affects the processes of respiration and photosynthesis in plants by damaging the breathing pores or stomata through cell damage, and also because it partly removes and dissolves the wax coatings that allow plants to regulate their growing conditions within the cells (Wolfenden and Mansfield 1991). Crops are also being affected and producing lower yields. Others, however, would argue that vegetation degradation may be due to the increase of poisonous anthropogenically created tropospheric O_3.

Lakes are becoming very acidic in North America and Scandinavia. An estimated 20,000 lakes are acidified in Sweden, 20 per cent of

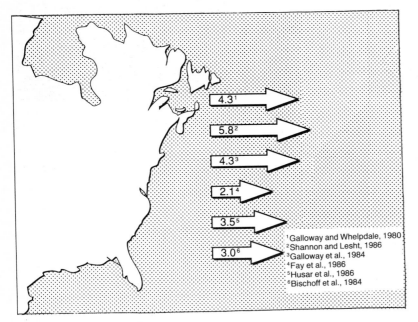

Figure 4.3 *Estimates of the absolute amounts (teragrams per annum, $Tg\ a^{-1}$) of sulphur advected (transported) eastward from eastern North America (USA and Canada) to the western North Atlantic Ocean. After Galloway (1990), in Last and Whatling (1991).*

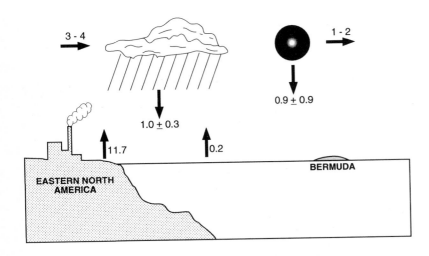

Figure 4.4 *Sulphur budget ($Tg\ a^{-1}$) for the western North Atlantic Ocean atmosphere. After Galloway and Whelpdale (1987), in Last and Whatling (1991).*

Plate 4.1 *(A) Nickel smelting plants at Nikel in north-western Russia, near Murmansk, emitting large amounts of SO$_2$ and other environmentally harmful gases. (B) Dying forests and dead silver birch trees on the hillsides surrounding Nikel – the result of acidic deposition. The tall chimney stacks actually carry the noxious emissions away from the town of Nikel in the valley around the smelters, with the result that trees within the town remain relatively unaffected unlike those of the surrounding forests. Photos taken in August 1993.*

which support no fish life. Stone buildings are being eaten away and badly discoloured.

The actual effects of acid rain on any environment vary according to whether the ground on which it falls is itself wet or dry, local climatic conditions, the way and rate at which groundwater run-off occurs, and the type of vegetation. In urbanized areas, the degree of acidic corrosion to structures will depend upon the types of materials, for example limestone buildings will suffer much more than sandstone buildings in a given area, because limestone unlike most sandstones is readily soluble in weak acidic solutions. This point is well illustrated by the weathering of

many historic monuments, such as St Paul's Cathedral in London, which has suffered considerable damage over the past few hundred years. This is constructed out of Portland limestone, a brilliant white limestone, which prior to recent industrial cleaning was discoloured to a sooty black, and preferentially hollowed out where the flow of acid rain waters are channelled across the surface of the facing stones; along these areas of stone, bulbous precipitates of gypsum ($CaSO_4$) form beneath anvils and gargoyles as a result of the reaction of acid rain (particularly sulphuric acid) and the limestone ($CaCO_3$). Scientists monitoring the degree of

Box 4.1
pH

Acidity and alkalinity are measured on a logarithmic pH scale defined as the negative logarithm of the hydrogen ion (H^+) concentration, where pure water is neutral and has a pH of 7, but natural rainwater is slightly acidic with a pH of about 6.5 because of the carbonic acid (H_2CO_3) caused by the dissolved atmospheric carbon dioxide within the rainwater. Lower values occur in acids, while pH values greater than 7 indicate alkaline substances. The further a pH value is away from 7 (towards 0 or 14 at both extremes), then the more acidic or alkaline that substance.

In the eastern USA, during the summer, water sampled near the base of clouds (where acidity is greatest) reveals typical pH values of 3.6, although pH 2.6 has been recorded. Also, in the Los Angeles area fog with a pH of 2 has been measured, equivalent to the acidity of lemon juice. Vegetated areas on mountains and hillsides are easily exposed to these very acidic situations during low cloud conditions. During rain, the acid levels are reduced through dilution typically by one-half to one pH unit. Thus, the average pH value of rain during the summer months in the north-east USA is about 4.2. In Britain, in Yorkshire and the East Midlands, the average acidity of rainwater is below a pH of 4.3 (Simpson 1990).

weathering suggest that annually as much as 0.62 microns of the surface of the limestone are lost, which over the period since the construction of St Paul's amounts to about 1.5 cm (Sharp *et al.* 1982).

To sum up briefly, acid rain has so far affected the aquatic life in rivers and lakes, the trees and other flora in the countryside and towns, contaminated groundwater and deteriorated coastal waters. Now that scientists are measuring the effects of acid rain on the various ecosystems on Earth, human concern has mushroomed, and with it the means of tackling the effects.

What is acid rain?

Acid rain occurs because of the atmosphere's continual effort to cleanse itself of various pollutants that are introduced into the air. The water droplets in clouds absorb and adsorb particulate matter and gas molecules out of the air. Not all such substances are removed by rain or precipitation, but instead remain suspended in clouds and moisture.

There are essentially two kinds of pollutants,

gaseous and particulate, which can be defined on the basis of their modes of formation as primary and secondary pollutants. Primary pollutants are produced directly from industrial and domestic activities, whereas secondary pollutants are created in the atmosphere by chemical processes acting on primary pollutants. On this basis, Last (1991) defines the four groups of pollutants as shown in Table 4.1.

The industrially produced sulphur dioxide, together with nitrogen oxides, are easily converted into sulphuric and nitric acid, respectively, and they fall as acid rain (see Tables 4.2 and 4.3). Sunlight, free oxygen, and water are the abundant ingredients that allow these reactions to take place. The reactions occur mainly in the lowest 10–12 km of the atmosphere (the troposphere). The acids, although dilute, have considerable corrosive capability. They are changing and polluting the natural environment over years rather than decades or centuries.

The chemical reactions which lead to acid rain begin as a **photon** hits an ozone molecule (O_3) to form free oxygen (O_2), and a single atom of oxygen which is very reactive. The single oxygen atom reacts with a water molecule (H_2O) to produce two electrically charged, negative hydroxyl radicals (OH^-).

Although these hydroxyl ions constitute less than one part per trillion in the atmosphere, they are effectively inexhaustible and the oxidation reactions they trigger actually regenerate more hydroxyl ions as a byproduct of the reactions. For example, the oxidation of sulphur dioxide produces hydroperoxyl radicals (HO_2^{3-}) that react with nitric oxide to produce nitrogen dioxide and a new hydroxyl radical.

Individual hydroxyl radicals are therefore capable of oxidizing millions of sulphur-containing molecules; it is only the amount of pollutant in the atmosphere which determines the amount of acid ultimately generated. The relatively scarce nitrogen dioxide in the atmosphere reacts with the hydroxyl ions to form nitric acid (HNO_3). The nitric acid then acts as a trigger for the reactions that produce sulphuric acid from sulphur dioxide. Nitrous oxides and nitric acid are produced during lightning storms, and from the exhaust fumes from supersonic aircraft; they were also formed in thermonuclear explosions during the 1950s and 1960s.

The usual way the nitric and sulphuric acids

Table 4.1 *Principal chemical pollutants.*

Gaseous, primary
  Sulphur dioxide (SO_2)
  Nitric oxide (NO) ⎫
  Nitrogen dioxide (NO_2) ⎬ Collectively referred to as oxides of nitrogen (NO_x)
  Hydrocarbons (HC)
  Ammonia (NH_3)
  Carbon dioxide (CO_2)
Gaseous, secondary
  NO_2 from oxidation of NO
  Ozone (O_3) and other photochemical oxidants formed in the lower
    atmosphere by the action of sunlight on mixtures of NO_x and
    hydrocarbons
  Nitric acid from the oxidation of NO_x
Particulate, primary
  Fuel ash
  Metallic particles
Particulate, secondary
  The reaction products of sulphuric acid and nitric acid with other
    atmospheric constituents, notably ammonia (NH_4HSO_4,
    NH_4NO_3, etc.)
  Sulphuric and nitric acids formed by the oxidation of SO_2 and NO_x
    respectively

Source: Last and Watling (1991).

Table 4.2 *Estimated sources and sinks of short-lived sulphur gases per annum ($Tg\,S\,a^{-1}$).*

Anthropogenic emissions (mainly SO_2)	70–80
Biomass burning (SO_2)	0.8–2.5
Oceans (dimethyl sulphide)	10–50
Soils and plants (DMS and SO_2)	0.2–4.0
Volcanic emissions (mainly SO_2)	7–10

Source: IPCC (1992).

reach the ground is in water droplets from clouds as rain. This is not the only way though: for example, sulphuric acid generated in gas phase reactions can condense to form microscopic droplets (0.1–2 millionths of a metre, or micrometre, in diameter) that constitute part of the haze seen over the eastern USA during the summer months. Some of these particles settle to the ground in dry deposition. Vegetation can absorb sulphur dioxide gas directly from the atmosphere, and this is also known as dry deposition.

In the stratosphere, the rate-limiting step for the oxidation of SO_2 to H_2SO_4 is believed to be the reaction of SO_2 with the hydroxyl ion, OH^-, leading through a sequence of reactions to the production of new sulphate particles, or condensation on pre-existing particles. SO_2 also acts as a catalyst in the production of O_3, but SO_2 can inhibit ozone production by absorbing incoming solar radiation, since it absorbs radiation strongly in the range 235–180 nm, weakly in the range 340–260 nm, and very weakly in the range 390–340 nm.

The chemical reactions which produce acid ground can occur in two ways, dry phase and wet

Table 4.3 *Estimated sources of nitrogen oxides per annum ($Tg\,N\,a^{-1}$).*

Natural	
Soils	5–20[1]
Lightning	2–20[2]
Transport and stratosphere	1
Anthropogenic	
Fossil fuel combustion	24[3]
Biomass burning	2.5–13[4]
Tropospheric aircraft	0.6

1 Dignon *et al.* (1991)
2 Atherton *et al.* (1991)
3 Hameed and Dignon (1992)
4 Dignon and Penner (1991)
Source: IPCC (1992), references cited therein.

phase reactions. Dry or gas phase reactions typically occur near the pollution source, such as a power station, whereas wet or aqueous phase reactions are characteristic of more distant areas and involve chemical reactions within water droplets, as in clouds, smogs, mists, sleet, and snow. Dry deposition generally occurs within 300 km as opposed to wet deposition which takes place up to the order of 1,000 km from the pollution source.

Acid groundwaters are caused by a combination of high precipitation from polluted air and clouds, rapid surface-water run-off (for example via streams), low temperatures, thin soil cover low in base metals such as calcium (as in limestones) which may reduce or neutralize the acidity, and, generally, steep slopes. In temperate climates, such as in Britain, the greatest stream run-off occurs during the spring, as snow melts from upland areas, and in the autumn storms. Typically, more than 70 per cent of the precipitation ends up as stream run-off, although this figure is extremely variable from year to year, and within a single catchment area.

The polluters

Natural processes introduce pollutants into the atmosphere which, amongst other results, may also produce acid rain. These surface processes operate over time scales much longer than human lifespans and include volcanic eruptions, while bacterial action in soil may also contribute to the atmospheric substances that produce acid rain. Human activities, however, have accelerated the release of such harmful chemicals into the atmosphere and hydrosphere and at levels that make it very hard for the Earth to cope. Over the past century, human activities have increased global emissions of sulphur gases by a factor of about three, leading to increased sulphate aerosol concentrations, mainly in the northern hemisphere (Langer *et al.* 1992).

The combustion of fossil fuels releases toxins into the atmosphere. Coal, for example, may contain up to about 5 per cent sulphur, compared to oil with typical values up to roughly 3 per cent. Approximately 120 million tonnes of SO_2 are emitted annually across the world from fossil fuel power stations, metal-smelting

ACID RAIN

Plate 4.2 *(A) The processing plant and smelter at Chuquicamata in the Atacama desert of northern Chile. Chuquicamata is the world's largest copper mine, and produces 150,000 tonnes per day of copper sulphide ore. (B) Sulphur dioxide and other fumes from the smelter are often carried by prevailing winds into the giant (4 × 2.4 km wide, 0.7 km deep) open pit copper mine, where they combine with diesel fumes from heavy machinery to produce a thick, acrid, smog, even on the brightest sunny days. The desert environment means that the pollution effects of the acid deposition are not monitored and, therefore, the environmental damage remains unassessed.*

Courtesy of Jeremy P. Richards, University of Leicester.

factories, other industries utilizing fossil fuels, and oil refineries (Plates 4.1, 4.2 and 13).

The main countries responsible for this pollution are: the former Soviet Union, the USA, China, Poland, Germany (especially what was until 1990 East Germany), Canada, the UK, Spain, Italy, and Czechoslovakia, in decreasing order of amount of SO_2 emitted. The largest single anthropogenic source of SO_2 emission in the world comes from the copper- and nickel-smelting plants in Sudbury, Ontario, which produced an annual discharge of about 630,000 tonnes of SO_2 in the early 1980s.

Conventional power stations burning fossil fuels are amongst the worst culprits for causing acid rain by the emission of SO_2 and NO_x. Oil refineries are another significant source of SO_2 emissions. In Britain along roughly 20 km of the Aire valley, Yorkshire, there is one of the largest concentrations of power stations in the UK belching out gaseous toxins. Three of these conventional power stations, Ferrybridge, Drax, and Eggborough, burn coal from the nearby new Selby coalfield and generate 20 per cent of Britain's electricity and almost an equivalent percentage of the country's total sulphur emissions. Their annual emission of SO_2 is about 600,000 tonnes, excluding other gaseous pollutants. The power stations in valleys such as the Aire have created the acid rain which has begun

to poison lakes in Scandinavia.

The former Soviet Union is the world's main producer of SO_2. A total of 60 per cent of Soviet electricity is generated from conventional fossil fuel-burning power stations that contribute 40 per cent of the SO_2 emissions. In addition, the countries to the west of the former Soviet Union contributed an estimated 5 million tonnes of sulphur compounds annually from 1979 to 1981. The secrecy of information in the old Soviet Union made it difficult to assess the real amount and extent of acid pollution, but in 1984 at the Munich conference the Soviet delegation admitted damage was occurring. In 1984, the newspaper *Pravda* reported that forests were dying from atmospheric pollution in the neighbourhood of the automobile factory city of Togliatti, near Kuibyshev in the Ural Mountains, and that lakes in the Kola peninsula and Karelia were suffering acidification.

North America is second only to the former USSR in its total emissions of SO_2. The principal offending states, in decreasing order, are Ohio, Pennsylvania, Indiana, Illinois, Missouri, Wisconsin, Kentucky, Florida, West Virginia, and Tennessee. Clearly, the most polluted states are in the heavily industrialized north-eastern sector of the USA. As in Europe where the polluters are polluting neighbouring countries, the same is true for the USA. The United States is polluting

Mexico with acids from the copper smelters in New Mexico and southern Arizona. Other parts of the USA are also guilty. In California, in the town of Corona del Mar south of Los Angeles, a winter fog in 1984 was recorded as having a pH of 1.69!

The eastern parts of Germany are the most heavily industrialized, with the old East Germany emitting an estimated annual per capita output of 235 kg of SO_2 – the annual total emissions being in the region of 4 million tonnes. As much as 90–95 per cent of the SO_2 comes from the combustion of lignite coal. In fact, this region is responsible for extracting more than 25 per cent of the world total of lignite and, therefore, not surprisingly lignite constitutes its principal fossil fuel. In the old East Germany, acid damage is reported to affect adversely more than 10 per cent of the forests, with soil acidification being severe on the Czechoslovakian border in the Erzebirge mountains. As is the case for the old Soviet Union, figures are not widely publicized and it is hard to substantiate any available data (Plate 13). Czechoslovakia itself produces an estimated annual SO_2 emission of more than 3 million tonnes.

Poland is probably the country most affected by acid pollution due to the combination of its internally generated industrial emissions and because of its proximity to other major industrialized nations whose gaseous toxins contribute to the atmospheric pollution. At the 1984 Munich conference, the Poles admitted that their annual SO_2 emission was running at 4.3 million tonnes, with a projected figure of 4.9 million tonnes in 1990. Their problem is that Poland tends to extract and burn coals and lignite with high sulphur values: it has, however, planned an emission control programme from 1991 which includes the use of fluidized bed combustion furnaces to reduce pollution levels.

The industrially developing countries, such as those in Latin America, South-East Asia, India and southern Africa, all have growing problems of acid rain and acidified soils due to industrial pollution. China, for example, has an annual SO_2 emission of more than 18 million tonnes, with only the USA and the old Soviet Union having greater levels. Of this SO_2, 85 per cent comes from coal-burning power stations and industries.

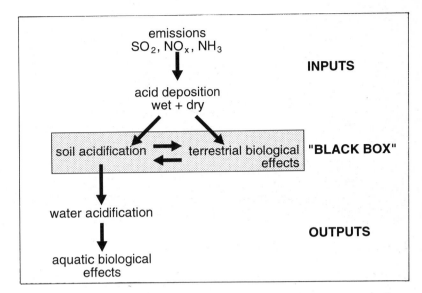

Figure 4.5 *Causal chain linking emissions of SO_2 and NO_x to soil acidification, forest effects, and aquatic effects. Reversibility of acidification may be delayed at one or more links. Some of the effects caused by long-term acidic deposition may be irreversible. A traditional approach to the study of cause–effect relationships has been to treat the terrestrial ecosystems as a 'black box'. Inputs and outputs are measured and used to deduce processes occurring within the box. After Hauhs and Wright (1987), in Last and Whatling (1991).*

Britain produces an annual SO_2 emission of about 3.5 million tonnes, with approximately 1.7 million tonnes of NO_x (1984 figures) (Plate 4.1). Rain in Britain is typically 100–150 per cent more acidic than unpolluted rain, with figures from Bush in Scotland showing the rain as more than 600 times as acidic! Acid pollution in western Europe is also a major problem, with the 19 constituent countries being in close proximity and, generally, heavily industrialized.

In western Europe, another source of acid rain, NO_x pollution, is regularly discharged into the atmosphere, with about 30–50 per cent from motor vehicles, and 30–40 per cent from fossil fuel-burning power stations. Evaporation from nitrate-based fertilizers used in agriculture can contribute significant proportions. Where fossil fuels are burned, the greater the combustion temperature, the larger the amount of nitrogen oxide given off. Where light oils and gas are combusted, almost 100 per cent of the nitrogen is oxidized, 40–50 per cent in heavy oils and 5–40 per cent in coal.

While the role of SO_4^{2-} ions in acidification of soils and waters, and the reversibility of SO_4^{2-} induced acidification is now well understood, the effect of increased NO_3^- leaching from terrestrial ecosystems and its influence on acidification is more speculative (Wright and Hauhs 1991). Figure 4.5 shows, schematically, the causal chain linking SO_2 and NO_x emissions to soil acidification, forest, and aquatic effects.

In temperate coniferous forests, nitrogen acts as a growth-limiting nutrient, therefore nitrogen inputs associated with acid deposition will probably be sequestered by the biomass and thus kept in terrestrial ecosystems. Evidence appears to suggest that areas subject to prolonged heavy loads of SO_2 and nitrogen compounds become nitrogen-saturated, resulting in greater amounts of the input nitrogen being leached into surface and groundwater run-off rather than being retained in terrestrial ecosystems (ibid.). Nitrogen saturation can result from both natural (e.g., fires) and anthropogenic causes (e.g., clearing vegetation cover). A study from the Hartz Mountains, Germany, measuring nitrate concentrations in run-off at two catchments, Dicke Bramke and Lange Bramke, shows the steadily increasing levels since the late 1970s and mid-1980s, respectively, despite constant nitrogen deposition (Figure 4.6), the increased leaching of NO_3^- apparently being associated with forest decline (ibid.). Similar results have been reported from other regions, for example Norway.

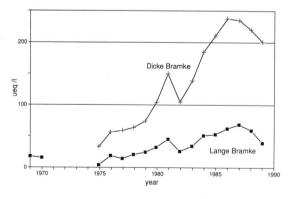

Figure 4.6 *Nitrate concentrations in run-off at two catchments: Lange Bramke and Dicke Bramke in the Harz Mountains, Germany. At Lange Bramke, NO_3^- levels have steadily increased since the mid-1980s despite constant nitrogen deposition. At the adjacent catchment, Dicke Bramke, the increase began several years earlier, and concentrations are now about twice those at Lange Bramke. Increased leaching of NO_3^- is apparently associated with forest decline. After Hauhs (1990), in Last and Whatling (1991).*

Volcanoes as natural contributors to acid rain

Volcanoes are the major natural contributory source to acid rain, often outweighing present anthropogenic emissions of sulphur dioxide. Indeed, in the geological past, major phases of volcanic activity on Earth have injected enormous quantities of SO_2 into the atmosphere, affecting global climatic and oceanographic conditions.

The eruption of Mount Pinatubo in the Philippines, on 15th June 1991, injected between 15–30 million tonnes (Mt) of SO_2 into the atmosphere, leading to the oxidation of the SO_2 to sulphate aerosols within one or two months (McCormick and Velga 1992). Other estimates of the Pinatubo eruption are that it was associated with the explosive eruption of 3–5 km^3 of total dense-rock equivalent, with sufficient sulphur to form 20 Mt of SO_2 cloud in the stratosphere (Westrich and Gerlach 1992).

Acid ground

Acid rain may fall hundreds of miles away from the source of the pollution. In extreme cases, the acidity can be reduced or neutralized by falling on alkaline soils, for example those associated with limestone areas. It is also possible to reduce the acidity by various immobilizing and buffering chemical reactions, even in slightly acidic soils such as are typical of the evergreen forests of northern Europe, the USA, and Canada.

One such buffering reaction is referred to as cation exchange, in which **cations** substitute for the hydrogen ions in the acids. Such substitution reactions involve the replacement of hydrogen ions typically by calcium, magnesium, and various other metals. These substitute metal ions are released from rocks and minerals during chemical weathering, and so are available to replace the hydrogen ions in the acids. All these reactions occur in an aqueous phase, that is, rainwater or groundwater. Clay minerals in the soil can release electrically charged particles or ions, such as the acid-producing hydrogen (H^+) and alkali-producing hydroxyl ions (OH^-). These ions can then react with other minerals and ions in the soil or bedrock to release particles that can cause either a positive or negative feedback. As an example, if clay minerals release calcium ions (Ca^{2+}), they can reduce the acidity of the pore waters in chemical reactions that tend to remove the excess of H^+ ions.

Carbon dioxide is dissolved in groundwater and rainwater which upon reaction with rocks and minerals during chemical weathering releases metal ions. During chemical weathering of rocks, negatively charged radicals, called anions, are also released, such as bicarbonate ions (HCO_3^-). The addition of sulphuric acid to soil allows the sulphate radicals (SO_4^{2-}) to displace and substitute for the calcium and magnesium ions in rocks and minerals, thereby freeing them for other chemical reactions.

The rate of these substitution (or cation exchange) reactions depends upon many inter-related factors, the principal ones being the bedrock geology, soil, vegetation, and the chemistry and characteristics of the precipitation. In certain climatic belts and soil types, the groundwater run-off will not lose its acidity because either the appropriate chemical reactions are inhibited or the environment is already too acidic. **Tundra** landscapes are just such regions. Earth scientists are beginning to develop quantitative methods for assessing the susceptibility of a soil to acidification, called its acid susceptibility, which is based on the chemistry of that soil. Clearly, this approach will allow much better modelling of cause and effect for predicting the consequences of acid rainfall upon different soils or bedrock.

There are two main sources of the hydrogen ions that cause acidity. One is external, that is from the atmosphere, and it deposits sulphuric acid (H_2SO_4), hydrochloric acid (HCl), carbonic acid (H_2CO_3), and nitric acid (HNO_3). The other is from within the soil itself, resulting from microbial activity releasing CO_2 which then reacts with water to produce carbonic acid and organic acids. Much more research is needed to evaluate the relative importance of both sources of acidity in groundwater.

The bedrock geology may be very important in inhibiting cation exchange reactions. Areas of thin and immature soil cover above quartz-rich sandstones and granites are good examples. In such soils and bedrock, not only is the quartz (SiO_2) very resistant to weathering, but also it does not contain the all-important metal ions which are necessary for cation exchange. Thus, groundwater run-off through such sandy soil or bedrock is unable to buffer the acidity of the water. Groundwater run-off in these regions, therefore, will reach rivers and lakes essentially in its original acidic state without the benefit of buffering reactions to reduce the acidity.

Countries such as the UK can be divided up into various areas with different degrees of susceptibility to developing acidic ground and groundwaters, the so-called hard-water and soft-water areas. The large parts of the UK underlain by limestone are unlikely ever to become seriously affected by acidic groundwaters because the limestone minerals (such as calcium carbonate) should be able to buffer any percolating acidic waters. Thus, the large areas where groundwaters are abstracted for public use which are underlain by the Chalk (e.g., southern England), the Jurassic limestones and other carbonate **aquifers** (throughout much of central and eastern England), and the Carboniferous limestones (mainly of central and northern England), are unlikely to develop serious problems with acidic waters. By the same token, these areas have groundwaters with a lot of dissolved mineral salts such as calcium carbonate, and it is these minerals that end up being precipitated out of solution in water pipes, kettles, and other domestic appliances to form 'scum' or 'scale'. We use the common term hard water for such water. Soft water comes from areas where there are relatively few dissolved mineral salts, where the water tends to be more acidic, such as upland regions dominated by rocks such as slate and granite.

Where thick, well-developed or mature soil profiles exist, as in many temperate latitudes, the soils may contain large amounts of electrically charged cations that can react with the nitrate or sulphate ions to **buffer** the acidity of the precipitation and percolating groundwater. As long as the buffering capacity of these soils is not exhausted, then there is a good chance that the groundwater run-off reaching rivers and lakes will have lost at least some of its acidity.

Areas in Britain where the bedrock geology is mainly granite or sandstone, however, do contain 'low alkalinity' or acidic waters. Examples include the groundwater associated with Quaternary glacial and fluvial sands and gravels, Tertiary sands, Carboniferous Coal Measures, sandstones and Millstone Grit, Cretaceous Greensand strata, Devonian sandstones (especially in south-west England), the Middle Jurassic sands of North Yorkshire, Lower Palaeozoic and Precambrian strata (apart from much of the Silurian), and the granites in Britain. Such areas tend to be characterized by acid surface waters.

An important aspect of areas with acid groundwater is that corrosion is more pronounced in metal well linings and pipes. This is because these regions are associated with the enhanced mobilization of toxic metal or base elements from any bedrock geology with which the acidic waters come into contact. Lead (Pb), copper (Cu), cadmium (Cd), and aluminium (Al) are some of the better known of these toxic metals.

In addition to industrial pollution of the atmosphere, the planting of large areas of coniferous forest in Scotland, Wales and western England is causing abnormally high levels of acidity in nearby lakes. Forests with a closed canopy are capable of reducing acid decomposition in water by as much as 30 per cent in

sensitive moorland areas. A closed canopy reduces the available sunlight reaching the ground and therefore causes a decrease in surface temperatures.

Many forest trees accelerate the rate of release of acids and potentially harmful aluminium into the soils, which is then washed into rivers and lakes. Also, the increased acidity of lake waters encourages the release of poisonous heavy metals into the waters. This is because the acidic waters allow more effective leaching of soils such that the exchangeable cations (such as Ca^{2+} and Mg^{2+}), which may help to reduce the acidity or neutralize the waters, are quickly released into groundwater and stream run-off. In short, these exchangeable cations are not allowed to stay around long enough to do the job of reacting with the acids and reducing their potency. At present, about 8 per cent of Britain is afforested, with most new or proposed plantations being in upland areas. As afforestation continues, so the problem of acid soils in such areas can only worsen.

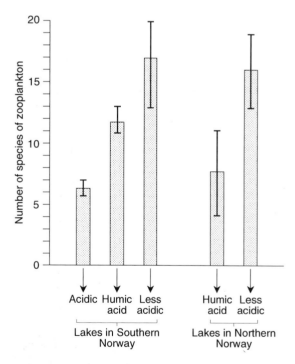

Figure 4.7 *Association between acidity and numbers of species of zooplankton, including Crustacea, Rotatoria, and Chaoborus larvae, in lakes in southern and western Norway. After Raddum et al. (1980), in Last and Whatling (1991).*

Acid rivers and lakes

Groundwater may lose its acidity upon entering lakes with the right sort of chemistry. If the water body contains bicarbonate and other negatively charged, basic ions from the chemical weathering of rocks and minerals, then they can neutralize the incoming acidic water. Earth scientists refer to this as water's acid-neutralizing capacity (ANC). The ANC value of a lake is therefore a measure of its susceptibility to becoming acidic. Rivers and lakes with a high ANC may be immune to acidification and neutralize the impact of acid rain. If the ANC value is zero, then such water bodies are very susceptible to becoming acidified as a result of acid rain, especially if they are sited near to the pollution sources.

Acidified rivers and lakes have tell-tale chemical signatures: the water tends to contain large amounts of sulphates and aluminium, the indigenous organisms may well show signs of alteration in terms of physical characteristics, population density and types of communities, and there may even be little or no aquatic life in the case of very polluted water systems. Some seasonality may be present in a lake's chemistry in that at certain times of the year it has a high ANC value that is reduced by buffering reactions during other months. These lakes are clearly potentially on the brink of becoming acidified on a permanent basis and need careful monitoring and precautionary measures to stop their deterioration.

High levels of aluminium in lakes may not always be due to acid rain releasing this element. Fluorine released into water from the bedrock geology, such as granitic rock, for example, can increase the ability for aluminium to be dissolved into the water.

Acidification of freshwater has deleterious effects on the activities of microbes, plants, and animals in poorly buffered ecosystems (Muniz 1991). The biota can be influenced directly, or in more subtle ways, for example causing a change in the balance between acid-sensitive and acid-tolerant species at different trophic levels. Figure 4.7 shows just one example of how acidity affects the number of species of zooplankton in lakes in southern and western Norway.

Results from North American lakes

In the United States in the mid-1980s, a committee of the National Academy of Sciences (NAS) reported on the pH and alkalinity of the water in hundreds of lakes in New York, New Hampshire, and Wisconsin, which roughly equates with the ANC and is a measure of the buffering ability of the lake water. The committee compared data from the 1920s to 1940s with those of the present day (NAS 1986, Mohnen 1988). The committee found that, on average, pH and alkalinity have increased in the Wisconsin lakes while remaining essentially the same in New York and New Hampshire. In New York State (including the Adirondack Mountains), however, a trend towards acidification of many lakes was recorded. Acidified lakes were also found in states such as eastern Pennsylvania and Michigan.

The NAS committee was also able to make some further disturbing observations concerning the aquatic life in these lakes. In particular, in 11 of the sampled lakes in the Adirondack Mountains, a change in pH was accompanied by a change in the assemblages of brown algae and

micro-organisms called diatoms. Six of these 11 lakes have demonstrably become more acidic since the 1930s, with present pH values of less than 5.2. The most rapid acidification had occurred in the period immediately prior to the 1980s. Acid rain was the only plausible explanation that the committee could find for this acidification.

In the Adirondack Mountains, the lakes have low pH values, and are therefore very acidic, because of the extremely acid rain in western New York State (as low as pH 4.1) and the quartz-rich, granite-floored lakes and soils with their poor buffering ability. This is a good example of how the bedrock geology conspires to make the groundwater maintain its acidity after the acid rain has fallen.

There are, however, large numbers of acidic lakes, such as in Florida and southern New England, which are believed to owe their acidity not to acid rain but to organic acids produced as vegetation decays. In part, the acidity of these lakes is also as a result of pollution from agricultural fertilizers finding their way into the lake waters.

In the eastern USA, geochemical investigations of selected watersheds, hydrological flow-paths (e.g., rivers and subsurface seepage), and lakes suggests that the bedrock geology is probably the most important factor in controlling the chemistry of the surface waters and, therefore, water acidity. Rainfall can be very acidic. Individual storms with a pH as low as 3.4 and up to 5.6 have been recorded for this area.

Threatened forests and woodlands

Acid rain is responsible not only for adversely affecting many of the world's rivers, lakes, and groundwater, but also apparently for threatening large parts of the flora, particularly trees in Europe and North America. In former West Germany where the effect of acid rain on trees has been so dramatic in the past few decades, they have coined the term 'forest death' or *Waldsterben*.

False-colour satellite imagery is used to detect damaged crops and forests in various parts of the world. This is useful because in the infrared part of the spectrum, it is most sensitive to varying amounts of chlorophyll, which is produced during photosynthesis in plants. The satellite pictures have revealed large areas of dying and diseased woodland and forest, probably caused by acid rain.

In North America, statistics on trees in the Adirondack Mountains, the White Mountains of New Hampshire, and the Green Mountains of Vermont have shown a death rate exceeding 50 per cent in the last 25 years of the red spruce, which grow above 850 m altitude where they are most susceptible to the effects of acid rain. Below this altitude, there has also been a deterioration in the health of both softwood and hardwood trees.

Earth scientists do not really understand the chain of chemical reactions leading to the death or loss of vitality in many parts of the world's woodlands and forests. Possible reasons include the leaching of important potassium, calcium, magnesium, and other chemical elements from the soil by percolating acidic groundwater derived from acid rain. Without such chemical nutrients, the health of trees suffers and they die. Acid rain can also upset the ecosystems in soil by causing the impoverishment and death of micro-organisms which help to release the all-important nutrients. Percolating acidic groundwater may also release large quantities of aluminium, which then competes successfully with other elements such as calcium for binding sites in the roots of trees. Tree growth then slows down. High nitrate levels in soils, for example from nitric acid, can affect the **symbiotic** fungi colonizing the roots of conifer trees which provide nutrients to the trees and help to protect them against disease. Acid rain also affects the leaves of trees and other plants by damaging the plant cells. Acid rain or acidic clouds and fog will leach out nutrients such as calcium, potassium, and magnesium from conifer needles at a faster rate than they can be replenished through the tree's root system.

Finally, Pleistocene lake sediments can be accurately dated, and analysed for many different components which can provide a palaeoclimatic signature and also indications of past lake acidity levels, for example by using diatoms and fly-ash particles (Battarbee 1984, 1986, 1992, Charles *et al.* 1989).

Recovery and sensitivity of ecosystems

Studies, particularly from North American (Canadian) and Norwegian lakes, are now showing that recovery or reversal of acidified ground and groundwater/run-off can occur following reductions in acid deposition, the rate depending on the sensitivity and degree to which ecosystems have been affected (Wright and Hauhs 1991). Table 4.4 shows examples of surface

Table 4.4 *Summary of examples of surface water acidification and its reversal. The examples are grouped by recent trends in the amounts of acidic deposition. The data suggest continued acidification with constant and slightly decreased acidic deposition, but reversal and recovery in response to large decreases in deposition.*

Acid deposition trend	Start year	End year	No. years	SO_4^* deposition (meq m^{-2}a^{-1}) start	end	change (%)	Surface water ueq l^{-1} (start year) Ca*+Mg*	SO$_4$*	alk	(end year) Ca*+Mg*	SO$_4$*	alk	F obs.	Trend, surface water acidification
Constant														
White Oak Run	1980	1987	8	100	115	15	63	77	24	67	94	12	0.24	Acidifying
Slight decrease														
Plastic Lake	1979	1989	11	77	62	−19	146	137	21	141	135	7	2.50	Acidifying
Large decrease														
Clearwater Lake	1973	1989	17	140	100	−29	460	575	−49	365	313	−22	0.36	Recovering
Hubbard Brook	1965	1986	22	135	90	−33	76	124	−25	63	102	−23	0.59	No change
Loch Enoch	1978	1988	11	75	50	−33	19	62	−79	16	27	−25	0.09	Recovering
RAIN KIM	1984	1989	6	58	16	−72	22	125	−108	1	35	−33	0.23	Recovering

Source: Wright and Hauhs (1991).

water acidification and its reversal, the results suggesting continued acidification and slightly decreased acidic deposition, but reversal and recovery in response to large decreases in deposition.

It is possible to construct geographic maps to indicate the sensitivity of different areas to acidic deposition, for example Figure 4.8 shows the relative sensitivity of ecosystems in Europe to acidic deposition.

The five sensitivity classes of ecosystems to acidic deposition proposed by Hornung *et al.* (1986) are shown in Table 4.5. More sophisticated attributes of these five sensitivity classes have been added by Chadwick and Kuylenstierna (1990), by focusing on four groups of factors, each of which is subdivided into two or four categories: bedrock geology, soil type, land-use, and amounts (not quality) of rain. Arguably, these refinements actually add more subjectivity to the sensitivity classification (Last 1991).

Another facet of sensitivity studies is the use of models to predict the likely future trends in acidic deposition. For example, Figure 4.9 shows current estimated annual deposition of sulphur resulting from natural (58 Tg S a^{-1}) and anthropogenic (70 Tg S a^{-1}) sources. Using predictions that by the year 2020, rates of sulphur emissions

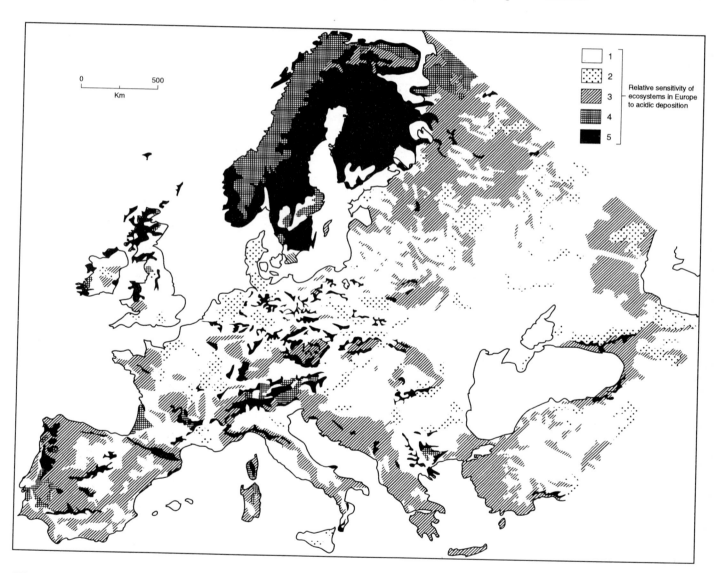

Figure 4.8 *Relative sensitivity of ecosystems in Europe to acidic deposition (1–5, on an increasing scale of sensitivity). A preliminary assessment of the sensitivities of aquatic and terrestrial ecosystems. After Chadwick and Kuylenstierna (1990).*

Table 4.5 *Sensitivity classes for ecosystems.*

	Soil/rock combination	Occurrence of acidic waters
Class 1	Acid soils over rocks with little or no buffering capacity	Acid waters will occur at all flow levels
Class 2	Acid soils over rocks with low buffering capacity	Acid waters likely at all flow levels
Class 3	Acid soils over rocks with moderate buffering capacity	Acid waters may occur at high flow levels
Class 4	Acid soils over rocks with infinite buffering capacity	Acid waters could occur at very high flows
Class 5	Non-acid soils over any rock type	Acid waters will not occur

Source: Hornung *et al.* (1986).

due solely to population increases will be about 30 Tg S a^{-1} (Galloway 1989, cited in Last and Whatling 1991), and assuming a modest increase in per capita energy consumption, Galloway and Rodhe (1991) predict that global sulphur emissions will increase to about 100 Tg S a^{-1}, with an environmental impact covering a much greater area than the regions which will have most of the increased population, namely Asia, South America, and Africa. A realistic increase in sulphur emissions of 15 Tg S a^{-1} in South-East Asia, for example, could alter global sulphur deposition. The value of 15 Tg S a^{-1} was chosen because of the intense convective vertical mixing of air currents over South-East Asia, resulting in the upward transport to the free troposphere of substantial amounts of sulphur which then, due to the stronger and more prolonged winds, can ensure the horizontal dispersal over considerable distances across the low-latitude belts compared to emissions in more temperate, higher-latitude regions.

Figure 4.10 summarizes the global problem of acid emissions whilst Figure 4.11 shows the variation in the acidity of rain.

Mopping up the mess: clean technologies

Appreciating the problem of acid rain and researching the damage wrought is certainly a first necessary step toward alleviating the effects of acid rain. As with so many environmental issues, the main obstacles to becoming more ecologically sensitive are ignorance and greed. Ignorance, perhaps, is the easier hurdle to overcome. Greed is more difficult, because enlight-

ened people may choose not to listen if they see profit margins decrease and even their livelihood being threatened. The philosophy behind national approaches to reducing acid deposition, together with other forms of environmental damage, varies considerably. In Germany, for example, the clean-up philosophy is to use the 'best available technology', whereas in the UK it is the 'best available techniques not entailing excessive cost' (or BATNEEC).

So, what can be done, and what is being done, to reduce the effects of acid rain? First, ways must be found to reduce the harmful chemical emissions from the traditional power stations which run on fossil fuels such as coal. Second, industry and domestic consumption must burn 'cleaner' coals such as those with low levels of sulphur and other very harmful chemicals. These coals may be naturally low in sulphur or have been treated by various washing processes. Such approaches have already produced lower levels of sulphur emissions from conventional fossil-fuelled power stations in the former West Germany, North America, and Japan.

In former West Germany, for example, wet limestone is introduced into the hot gaseous emissions of some power stations where it can scavenge up to 90 per cent of the sulphur dioxide. This process, unfortunately, does not reduce the amount of nitrogen oxide (NO$_x$) given off and it leads to a slightly reduced efficiency for the power station; the sulphur-rich waste also presents a disposal problem. The present solution is a classic case of swings and roundabouts. Society must, nevertheless, weigh up the benefits and disadvantages of cleaning up the environment. Furthermore, research and development should continue to seek better,

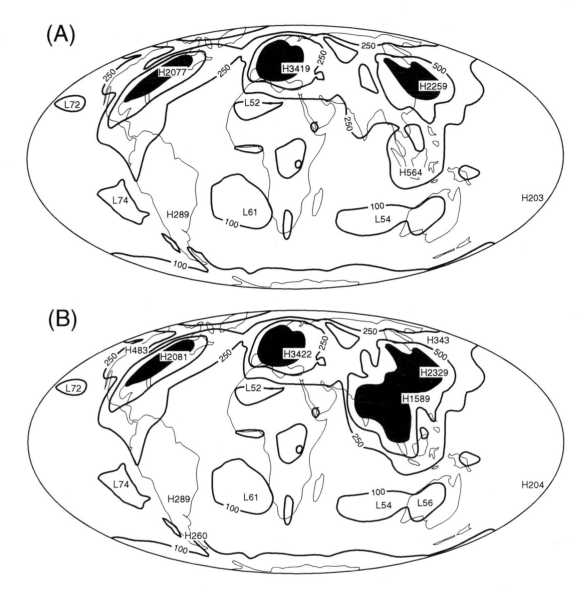

Figure 4.9 (A) Estimated annual deposition of sulphur resulting from natural (58 Tg S a^{-1}) and anthropogenic (70 Tg S a^{-1}) sources. Units are in mg S m^{-2} a^{-1}. (B) Estimated annual deposition of sulphur following an assumed increased emission of 15 Tg S a^{-1} in South-East Asia. Other emissions are as in (A). 'H' and 'L' deposition refer to the maximum and minimum deposition amounts, respectively. After Rodhe et al. (1991), in Last and Whatling (1991).

more environmentally sound means of cleaning up the atmosphere.

Other new technologies for reducing the noxious emissions from coal-fired power stations include the process called **atmospheric fluidized-bed combustion**. In this process, a turbulent bed of coal and limestone particles is suspended (or fluidized) by high-velocity, upward-streaming air with combustion occurring at a steady and lower temperature than in more conventional burners. As a result of this process, the amount of harmful nitrogen oxides is reduced and much of the sulphur dioxide reacts with the limestone.

In Germany, a new filtering process is being developed to remove both the sulphur dioxide and nitrogen oxide from coal in the conventional power stations. The technology appears capable of removing these pollutants from the cheapest and dirtiest lignite coals. The cleaning process uses very porous pellets of **active coke** (about 5 mm in diameter) that catalyse reactions to

123

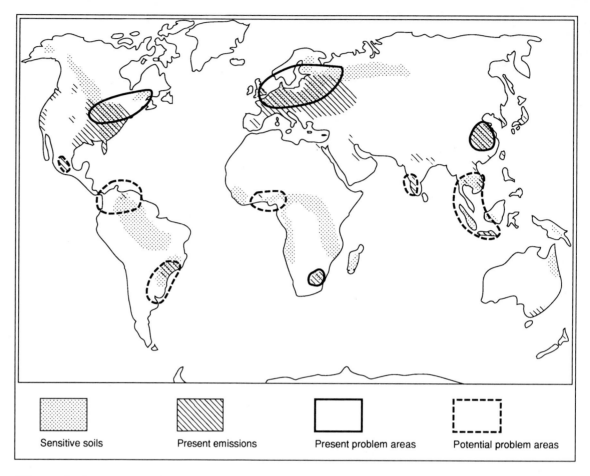

Figure 4.10 *The global problem of acid emissions and precipitation in the late 1980s. The map shows regions of acid emissions, areas where soils are likely to suffer damage by acid rain, and present and future problem zones. After Rodhe et al. (1988).*

remove the pollutants. The most impressive performance of this new technology was at a power station near Arzberg in north-eastern Bavaria, which burns brown lignite coal from Czechoslovakia with 4,000 mg of sulphur m⁻³ of coal, where the filter process only left 10 mg m⁻³ of sulphur.

In Britain, the electricity-generating companies are currently evaluating the use of sea-water to remove the sulphurous gases emitted by conventional fossil fuel-burning power stations. Basically, the polluted flue gas goes through a pre-scrubber process to remove any residual ash, along with the chlorides and fluorides. It then enters the main scrubber or absorber where the counterflow of seawater removes about 90 per cent of the sulphur dioxide from the gas. The acidified water is then funnelled into a tank where it is aerated to convert the sulphite to

sulphate, neutralize the pH and drive off the carbon dioxide and oxygenated water. The cleaned effluent is then pumped into the sea, and the clean gas discharged. Although seawater could be successfully used in any desulphurization process, the water would also pick up any heavy metals such as mercury and selenium from the coal, which could then find their way into the effluent discharged back into the sea. The disadvantages could outweigh the benefits. If these potential problems can be overcome, then the seawater method for desulphurization could provide an alternative to the use of limestone which generates mountains of gypsum.

On the domestic front, the noxious emissions from cars can be reduced by conservation, for example through limiting private travel by car to journeys necessitated by the absence of public transport. Naturally, this means some sacrifices in

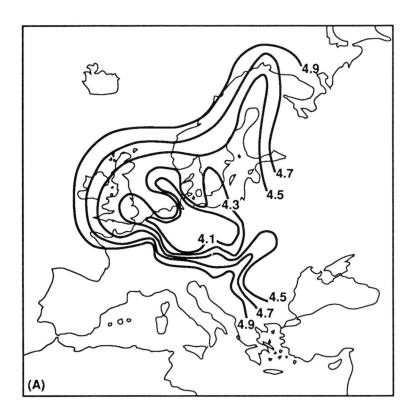

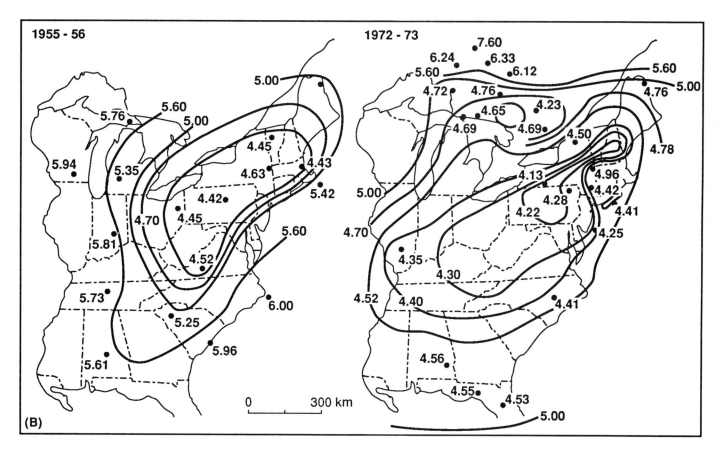

Figure 4.11 *Variation in the acidity of rain in (A) Europe, and (B) eastern North America. Numbers refer to pH. After Blunden and Reddish (1991).*

terms of convenience, but unless a longer-term perspective on atmospheric pollution and acid rain is taken, then the countermeasures that eventually have to be taken will prove all the more austere. Car manufacturers should continually strive, via research and development, to produce 'cleaner' internal combustion engines.

Governments should both enact restrictive legislation on tough acceptable standards of emissions and through financial incentives encourage vehicle manufacturers to market 'cleaner' transport. It is also the responsibility of governments to make public transport convenient and economically attractive to the populace. While governments undeniably must take a considerable share in the responsibility for a less polluted atmosphere, individual members of society should not shirk the responsibility to conserve energy and, thereby, reduce the causes of atmospheric pollution and acid rain.

Major oil companies are now becoming much more environmentally aware. In April 1991, BP Chemicals Nitriles and Nitrogen division, for example, announced in its company newsletter *BP Chemicals World* a US $17 million plan to build a new nitric acid plant to replace the 35-year-old unit at the Lima Complex, Cleveland, in the USA. It is envisaged that the new plant, which will boost Lima's nitric acid production capacity by 27 per cent to 90,000 tonnes a year, will result in a 95 per cent reduction in the nitrogen oxide emissions that cause acid rain. Clearly, if industrialists are really capable of tying increased production to significantly reduced environmental risk, then these are the sorts of developments to welcome cautiously.

In 1989, Britain committed itself to spending £1.8 billion in a programme to reduce acid rain by utilizing a cleaning-up process in its conventional, fossil fuel power stations. The problem is that this new process may actually pose a different threat to some of the country's most beautiful countryside. Millions of tonnes of limestone will be required in order to extract the sulphur dioxide from the combustion furnaces. This limestone will have to come from areas of natural beauty, including national parks in the Yorkshire Dales and the Derbyshire Peak District where the geology is most suitable for obtaining the relatively pure limestone needed in the power stations.

Environmental groups are understandably concerned that a substantial expansion in the quarrying operations in these beauty spots will both desecrate such areas and lead to a drastic reduction in the ecological niches available for certain wildlife, particularly birds. Another major problem is that the chemical reaction of limestone and sulphur dioxide produces calcium sulphate ($CaSO_4$), more commonly known as gypsum. Although gypsum has industrial uses, for example in the production of plasterboard and as a road-fill material, large amounts have to be dumped as industrial waste.

In order to assuage any public outcry over the sulphate-scrubbing process, the former Central Electricity Generating Board (CEGB) and the Department of the Environment maintained that they would develop a limestone-recycling process to avoid producing massive and ever-growing industrial waste tips of gypsum, a process that requires only 8 per cent of the limestone used in the gypsum-generating method. This policy was announced by Lord Belstead in 1987 when he was Junior Environment Minister. Also in 1987, the former CEGB announced that the first limestone-recycling plant was to be installed on Merseyside in the Fiddler's Ferry power station, while the first gypsum–limestone plant was to operate near Selby, Yorkshire, in the Drax power station.

Following the launch of the £500 million Drax gypsum–limestone programme, in February 1989 the former CEGB announced that it would not continue with the limestone-recycling plant at Fiddler's Ferry power station or at any of its other power stations. In public, the former CEGB board claimed that no firm decisions for or against the installation of limestone-recycling plants had been made at the other power stations in Britain after the back-sliding at Fiddler's Ferry. In addition, sceptics maintain that the privatization of the electricity industry will, in any event, change the financial situation so that the incentives for installing limestone-recycling plants will be considerably reduced as this cleaning operation erodes profitability.

Funding research into acid rain

In an article in *New Scientist*, Pearce (1990) asked why media interest and research funding seem to have forgotten acid rain. The answer is that interest has shifted to global warming. To some extent, the effects of acid rain are now more widely appreciated, and some of the problems of acid rain are being tackled. Yet, despite the move away from research into acid rain, with its attendant media interest on a very low back-burner, studies are showing that acid pollution in the coal-mining provinces of south-west China are now approaching levels found only in the most polluted parts of the USA. David Schindler, from the Canadian government's Freshwater Laboratory, believes that the acidification of large areas of the temperate lands, with far fewer diversity of species than in the tropical regions, make them much more vulnerable to catastrophic disruption with the loss of a few species. He also argues that acid rain may have made the ecosystems, such as those in Europe's forests, more vulnerable to global climatic change associated with the greenhouse effect: 'Greenhouse warming may enhance the effects of acid precipitation on boreal lakes and streams. If greenhouse droughts cause water tables to rise, then a sudden rainstorm could release many years of accumulated sulphuric acid in a single deadly acid pulse.'

The fallout of nitrogen in acid rain might also lead to the emission of greenhouse gases such as nitrous oxides from acidified soils. Scientists such as Melillo, from the Woods Hole Marine Biological Laboratory, Massachusetts, USA, have estimated that air pollution deposits about 18 million tonnes of nitrogen per year in the temperate northern hemisphere, which has increased the natural emission of the greenhouse gas, nitrous oxide. Excess nitrogen in acidified soils might reduce their ability to absorb one of the other important greenhouse gases, methane (CH_4). Such a reduction in the soil 'sink' for CH_4 could have been an important, yet little appreciated factor in the doubling of atmospheric concentrations of methane since pre-industrial times (Pearce 1990).

A countervailing argument is that acid rain could act as a negative feedback mechanism to reduce the potency of the greenhouse effect. Nitrogen fallout in the oceans might allow increased biological activity so that the oceans could absorb greater amounts of atmospheric CO_2, the main greenhouse gas.

Wigley, Head of the Climatic Research Unit at the University of East Anglia, UK, has suggested that sulphate in acid rain might act as a 'seed' for the formation of clouds that protect the planet from the Sun's rays, and so reduce any greenhouse effect. James Lovelock, in his book *The Ages of Gaia* (1988), describes how marine plankton give off sulphurous particles, called dimethylsulphide, which also act as seeds in the formation of clouds. So, acid rain may actually exert some negative feedback mechanisms that reduce the amount of global warming.

Acid rain needs continued study and monitoring if the prospects for the future are to be predicted accurately. Its harmful, polluting effects need to be minimized. As the study of the greenhouse effect and global warming gathers pace, so too should research into any positive or negative feedback mechanisms that might enhance or reduce acidic deposition. From a brief look at research into acid rain, the following section deals with international action on acid rain.

International action on acid rain and photochemical smogs

Although acid rain has been known about for decades, its impact on the environment was consistently denied by politicians until the 1980s. President Reagan spent much of his first term in the White House denying Canadian claims that acid rain caused by emissions from American power stations was killing fish north of the 49th Parallel in Canada and destroying maple trees. It was not until 19th March 1986 that the British government, under Prime Minister Thatcher, freely accepted the link between acid rain, water pollution, and the death of fish in Norwegian and Swedish lakes. About 17 per cent of the acid which falls over Norway comes from Britain.

In the USA, the main legislation against air pollution is the 1963 Clean Air Act, considerably amended in 1970. In 1970, a 10-year plan was conceived to set standards of national air quality, including SO_2, NO_x, O_3, and hydrocarbons, all of which can contribute to acid rain. The required standards were never met, and in 1977 there were amendments to the Act in order to set less stringent targets. Despite verbal pledges to meet the issue of acid deposition head-on, President Reagan stalled and by the time George Bush assumed the presidency, very little had changed regarding controls on the polluters and levels of acceptable air quality.

Europe is the most heavily populated and industrialized continent; it is also the most polluted. More than 50 per cent of the world-wide global sulphur and nitrogen pollution comes from Europe. With statistics such as these, people have to ask whether or not European politicians are showing the appropriate level of concern by encouraging investment in research into the problem of acid rain and stimulating debate over the best ways of tackling the problem. What has been done and what is being done to counter the effects of acid rain and acid deposition?

There have been various international meetings to address the problems associated with air pollution and acid rain. Following is a brief summary of just some of the most significant.

In the early 1970s, the Scandinavian countries seemed to be a lone voice in the wilderness crying out for greater environmental awareness. However, the wind of change blew with the 1975 Helsinki Conference on Security and Co-operation at which the Soviet Premier, Leonid Brezhnev, proposed that attempts be made on the three pan-European issues, energy, the environment, and transport. The Convention on Long-Range Transboundary Air Pollution arose from this. The convention was signed in November 1979 in Geneva by 35 countries, including the USA and Britain. This was the first environmental agreement to involve North America, western Europe, and the Eastern Bloc. The treaty, since ratified by most of the participating nations (35 by March 1985 and also the EC), dealt with voluntary controls not only on levels of SO_2 emissions, but also other atmospheric pollutants including various sulphur, nitrogen and chlorine compounds, polycyclic aromatic hydrocarbons, and heavy metals.

The Conference on the Acidification of the Environment took place in Stockholm in 1982. It resulted in international agreement to endeavour to reduce acid pollution to the environment. The conference came about at a time when the first reports of serious acid rain damage to trees in the then West Germany were becoming common public knowledge – the West German

Green Party won its first *Landtag* seats in 1981, and in March 1983 their first seats in the *Bundestag*. Amongst the agreed measures, a threshold value of 0.5 g of sulphur per square metre per annum was accepted as the level above which sensitive aquatic ecosystems would begin to suffer adverse effects. The conference was divided into two separate sessions, an initial scientific and technical session, followed by a ministerial meeting. At the latter meeting, ministers from many of the major developed, industrialized nations accepted the seriousness of the acidification of the environment and agreed to make concerted efforts to reduce pollutants such as SO_2 and NO_x emissions using available technologies.

At a meeting of the 1982 Stockholm Convention in 1983, a Nordic resolution was proposed for the mutual reduction in emissions of SO_2 by 30 per cent in a decade, from 1983 to 1993, with the 1980 levels of emission taken as the baseline. The proposal was rejected by the Executive Body of the Convention in Geneva during the 1983 meeting. Amongst the members of the Executive Body who voted against the Nordic proposal were the USA, Britain, France, and the Eastern Bloc countries. Frustrated by the lack of progress on this issue, the nations supporting the Nordic proposal met in Ottawa in March 1984 to sign their own agreement. In 1984, members of the 30% Club pledged themselves to a 30 per cent reduction in SO_2 emissions over a decade. They were Sweden, Finland, Switzerland, the former Soviet Union, Luxembourg, Belgium, Bulgaria, the former East Germany, the Ukraine, Byelorussia, Liechtenstein, Italy, and Czechoslovakia (Hungary joined in 1985). The Netherlands and Denmark agreed to a 40 per cent reduction by 1995, while a 50 per cent reduction in SO_2 emissions was pledged by France by 1990, West Germany and Austria by 1993, Norway and Canada by 1994, and Denmark by 1995.

The Multilateral Conference on the Environment held in Munich in June 1984 was attended by the environment ministers from the 1979 Convention countries with the two-fold aim of encouraging additional nations to join the 30% Club, and exploring further ways and means of reducing both SO_2 and NO_x emission levels. Environment ministers from the countries within the Stockholm Convention of 1983 also attended. At the conference, it was requested as a matter of some urgency that, by 1993, the Executive Body of the United Nations Economic Commission for Europe convention adopt a specific agreement for a reduction in the annual sulphur emissions, or their transboundary levels (fluxes). It was also recommended that the total annual emissions, or trans-boundary fluxes, of NO_x be reduced by 1995. It was at this meeting that Patrick Jenkin, the British Environment Minister, declared the 'indivisibility of environmental and economic policies'.

The Helsinki Meeting, in July 1985 was the third meeting of the Executive Body (United Nations Economic Commission for Europe convention) of the Stockholm 1982 Convention, which drew up a protocol for reducing sulphur emissions. The Protocol bound its signatories to a 30 per cent reduction of SO_2 emissions at source by 1993 on the 1980 levels – in other words, on the same terms as those embodied in the 30% Club. Thus this was a Protocol to the 1979 Convention on the Reduction of Sulphur Emissions or their Transboundary Fluxes by at least 30 per cent.

By September 1987, 16 nations had become signatories to the agreement which embodied most of the sentiments of the original Nordic proposal that led to the creation of the 30% Club. Countries such as Britain, Poland, and the USA did not become signatories to the agreement. Poland, whose sulphur emissions had actually been increasing since 1980, made the excuse that it would not have the technology to control its emissions until the 1990s.

The International Conference on Acidification and its Policy Implications, held in Amsterdam in May 1986, focused on a number of issues, perhaps the most notable of which concerned NO_x emissions. It was not until 1988 that a draft agreement was reached for the fixing of emission levels for NO_x at 1987 levels to be reached by 1994.

At the Sofia Meeting in 1988, 25 nations signed the protocol that was formulated as a result of the Amsterdam conference in 1986. Twelve EC countries agreed to even more stringent targets to reduce NO_x emission levels by 30 per cent by 1998. The USA did not agree to the 30 per cent reduction but signed the protocol. This was a Protocol to the 1979 Convention Concerning the Control and Emission of Nitrogen or its Transboundary Fluxes.

On 1st March 1993, negotiations opened in Geneva under the auspices of the UN Economic Commission for Europe (ECE) for a new 'sulphur protocol' aimed at reducing SO_2 emissions. It is intended that this protocol should be signed within the near future, and that it should replace the 30% Club which, with notable exceptions such as the UK, was signed by most Europeans. The new protocol will establish different clean-up targets for individual countries; the date by which these targets should be met is under negotiation. The targets or thresholds are based on the concept of perceived '**critical loads**' of acid rain that any sensitive ecosystem (e.g., lakes, forests, etc.) can tolerate before being harmed. For many countries in the EC, it appears that on economic grounds it will prove too costly to achieve the critical load targets, so it has been proposed that states should reduce the gap between the critical load and existing acid rain fallout by 50 per cent – the so-called '50% gap reduction'. Since 1980, European SO_2 emissions from fossil fuel-burning (coal and oil) power stations have been considerably reduced, partly because of the installation of flue-gas desulphurization equipment in power stations, but also as a result of the changeover to other energy sources such as natural gas and nuclear power. Examples of reduced SO_2 emissions since 1980 are 50 per cent in the Netherlands and Belgium, 70 per cent in the former West Germany and Austria, 60 per cent in France, but only 25 per cent in Britain – virtually all due to the closure of small coal-fired power stations (Pearce 1993). Along with other members of the EC, Britain was a signatory to the 1987 Large Combustion Plant Directive requiring countries to reduce SO_2 emissions, for example for the UK to 60 per cent of its 1980 levels by the year 2003. Scientists at the UK government's Institute of Terrestrial Ecology, Cambridgeshire, however, argue that even these reductions are insufficient to get many of Europe's most threatened environments below the critical loads. The Large Combustion Plant Directive set different levels of reduction of NO_x and SO_2 for each EC country.

Conclusions

In preparation for the UN negotiations, European countries have been compiling their critical load maps for soil and freshwater, but there are clear signs of some countries attempting to 'adjust' the figures. Two years ago Britain, for example, drew up a critical load map for soil which the British government claimed showed that only 8 per cent of the soil would experience greater levels of acid deposition than the critical load. A 1992 report prepared by the environmental pressure group Friends of the Earth, however, claims that had the UK adopted the mapping methods employed by other European countries, then this 8 per cent figure would actually increase to 47 per cent. In the first critical load maps, the UK was divided into 1 km squares and assigned a critical load for its soils. Later maps by government scientists lumped these small areas into 100 km² boxes and, as in previous maps, assigned a critical load based on the dominant, rather than most sensitive soil type. Many west European countries, however, have assigned critical load values as those which are necessary to protect 95 per cent of the soils in each box (equating to 70 per cent or less for the UK method). Clearly, if countries are going to 'cheat' or simply use statistical methods that massage any potential problem areas in evaluating the critical load maps, then this is a rather depressing outlook for any concerted international efforts at cleaning up the natural environment.

Key points

CHAPTER 4: KEY POINTS

1 Human activities result in the emission of various pollutants (principally SO_2, NO_x, NH_3, hydrocarbons, and particulate matter) to the atmosphere, which lead to a number of environmental problems such as poor air quality (e.g., the *nephos* over Athens on 1st October 1991) and acidic deposition or acid rain with very low pHs. Acidic deposition is produced by natural or human activities, amongst which the principal acids produced are carbonic acid (H_2CO_3), formed by the reaction of carbon dioxide and water, sulphuric acid (H_2SO_4), formed by the reaction of SO_2 with H_2O, and nitric acid (HNO_3), which is formed by the reaction of nitrogen oxides with water. The pH of some human-induced acidic deposition has been recorded as low as 2. Volcanoes which emit NO_x, SO_2, and Cl_2 also produce acidic deposition.

2 Acidic deposition is a particular problem in industrialized regions and countries where the combustion of fossil fuels releases large quantities of SO_2 into the atmosphere. Some regions and countries are polluted because of acidic deposition caused by industrialized parts of other countries up-wind. Acidic deposition results in acidification of groundwaters, surface water, damage to life (particularly forests and aquatic life), and building decay. Buffering reactions due to the presence of certain clay minerals, and because of cations within water in some soils and lakes, may reduce the immediate effects of the acidic deposition on those environments. The susceptibility of a soil to acidification is quantified as its acid susceptibility.

Regions can be classified on this basis. Areas where the soils are calcareous and/or the bedrock is limestone have a low susceptibility compared to areas where clastic and acidic igneous rocks predominate. Acidic groundwater may cause corrosion of water pipelines and the mobilization of toxic metals such as lead (Pb), copper (Cu), cadmium (Cd), and aluminium (Al). The susceptibility of lakes to acidification is measured by the acid-neutralizing capacity (ANC). Acidification of lakes has deleterious effects on their ecology. 'Forest death' is also a consequence of acidic deposition, which may be caused by direct cell damage or by a depletion from within the soil of important nutrients which support the plants.

3 Recovery from acidification of the environment can occur and is dependent on the sensitivity of the ecosystem. Reducing acidification involves cutting down on anthropogenic emissions, mainly from fossil fuel-burning power stations, for example by using appropriate clean technologies such as atmospheric fluidized-bed combustion, and active coke, and cleaning combustion engines in both domestic and commercial vehicles. Research and development of existing and new clean technologies are happening, but without clear incentives for those who use and develop such technologies, and penalties for polluters, so it appears that acidic deposition will remain as a major regional environmental problem.

4 At a series of gatherings, international conventions and agreements have been signed during the last few decades in order to reduce poor air quality

*Key points
continued*

and the emissions of SO_2. These have included the Convention on Long-Range Transboundary Air Pollution in the 1970s; the Conference on the Acidification of the Environment, Stockholm 1982; the 30% Club, introduced in Ottawa in 1984; the Multilateral Conference on the Environment, in Munich in 1984; the Helsinki Meeting in 1985; the International Conference on Acidification and its Policy Implications, held in Amsterdam in 1986; and the Sofia Meeting of 1988.

Chapter 4: Further reading

Journal of the Geological Society of London 1986. *Geochemical Aspects of Acid Rain.*
A useful, if somewhat specialized, thematic set of research-level scientific papers on acid deposition. Recommended reading for courses with an in-depth appreciation of this topic.

Last, F.T. and Watling, R. (eds) 1991. *Acid Deposition: Its Nature and Impacts.* Edinburgh: The Royal Society of Edinburgh.
An excellent volume on acid deposition, produced as an edited conference proceedings, aimed at students and teachers, including researchers, with a good scientific background. This book is a reference work which should be recommended to science-based students undertaking courses in environmental science.

McCormack, J. 1989. *Acid Earth: The Global Threat of Acid Pollution.* London: Earthscan Publications.
This is a highly readable book written by a freelance environmental writer. It should appeal to those wishing to understand the nature of acid deposition and the international context in which nations have sought to reduce the associated problems.

Pearce, F. 1987. *Acid Rain.* Harmondsworth/New York: Penguin Books.
A readable book by a senior editor at *New Scientist*, summarizing the results of work undertaken by scientists into the problems associated with acidic deposition. The book examines acidic deposition and its effects on the natural environment and people's health, the corrosion of building materials, the acidification of water sources, the damage to parts of the biosphere and environmental degradation, and policy issues relating to national and international efforts aimed at reducing the emissions that cause acidic deposition.

Water, water, everywhere,
And all the boards did shrink;
Water, water, everywhere
Nor any drop to drink.

The very deep did rot; O Christ!
That ever this should be!
Yea, slimy things did crawl with legs
Upon the slimy sea.

About, about, in reel and rout
The death-fires danced all night;
The water, like a witch's oils,
Burnt green, and blue and white.

Samuel Taylor Coleridge,
'The Rime of the Ancient Mariner'

Water resources and pollution

A blue planet

The Apollo space missions in the 1960s revealed the Earth to countless millions as the 'Blue Planet'. Everyone's imagination was captured by this new vision of the Earth from out in Space (Plate 1). The Earth appears as a blue planet from Space, not because of its watery surface (hydrosphere) and atmosphere rich in water vapour, but because the atmosphere scatters the blue part of the electromagnetic spectrum. The seas and oceans cover about 71 per cent of the Earth's surface (Plate 5.1), and contain more than 361 million cubic km of water. Not only did life on Earth begin in water and emerge from water, but animals themselves are made up of more than 70 per cent water with plants containing over 90 per cent water by weight.

Water as a resource

Water is essential for life on Earth. Within organisms, water provides the medium within which the complex metabolic processes necessary for life take place. Organisms simply cannot function without water and if deprived will rapidly die. In addition, the water must be clean. Human beings, the most complex of organisms, are affected by the most subtle variations in water chemistry and supply. According to the World Health Organization (WHO), an esti-

mated 1,200 million people lack a satisfactory or safe water supply. There are clear disparities between domestic and municipal, and industrial, water consumption in the developing compared with the developed, industrialized world (see Tables 5.1 and 5.2). Also, the greatest amount of waste water occurs in the developed and industrialized nations, for example Europe and North America.

People do not take enough care of this essential resource, however, and often appear not to value it. Water is polluted directly or

Plate 5.1 *Part of the hydrologic cycle – the coupled ocean–atmosphere system.*

WATER RESOURCES AND POLLUTION

Table 5.1 *Domestic and municipal water consumption.*

	1980s				Year 2000 projection			
Region	Population (millions)	Water with-drawal (km³)	Consump-tive use (km³)	Waste water (km³)	Population (millions)	Water with-drawal (km³)	Consump-tive use (km³)	Waste water (km³)
Europe	496	48	10	38	512	56	8	48
Asia	2,932	88	53	35	3,612	200	100	100
Africa	589	10	7	3	853	30	18	12
North America	411	66	20	46	489	90	22	68
South America	279	24	14	10	367	40	20	20
Australia and Oceania	26	4.1	1.2	2.9	30	5.5	1.5	4
Former USSR	282	23	5	18	310	35	5	30
World total	5,015	263.1	110.2	152.9	6,173	456.5	174.5	282

Source: World Resources Institute (1990).

indirectly by introducing substances that are a hazard to human health, which lead to a reduction in amenities, and prevent water activities, such as swimming and fishing. Figure 5.1 shows the world consumption of water expressed as the world average annual consumption per person. These figures show that the industrialized, developed countries use far more water per capita than the non-industrialized, developing world. Examples of per capita consumption of water for 1988/9, in litres per day, are as follows: Austria, 145; Belgium, 108; Denmark, 190; Finland, 151; France 159; Hungary, 205; Italy (1984), 220; Luxembourg, 176; Netherlands, 167; Spain, 126; Switzerland, 264; Sweden, 194; UK,

136; West Germany, 145 (Water Services Association 1992, Table 9A).

Water supply is a problem not just for the developing world. In California, for example, there is a serious looming water supply crisis. The Sacramento river index which measures the water flow in five northern Californian rivers that feed the state's largest reservoirs, is at one of its lowest levels in history. The water-parched city of Santa Barbara on the Pacific coastline is leading the way after being afflicted by a five-year drought by turning to the ocean. In May 1991, the city council approved the immediate construction of a US $37.4 million **desalination plant** intended to provide 9.1 billion litres of

Table 5.2 *Water use in industry.*

	1980s				Year 2000 projection			
Region	Population (millions)	Water with-drawal (km³)	Consump-tive use (km³)	Waste water (km³)	Population (millions)	Water with-drawal (km³)	Consump-tive use (km³)	Waste water (km³)
Europe	496	193	19	174	512	200–300	30–35	170–175
Asia	2,932	118	30	88	3,612	320–340	65–70	255–270
Africa	589	6.5	2	4.5	853	30–35	5–10	25
North America	411	294	29	265	489	360–370	50–60	310
South America	279	30	6	24	367	100–110	20–25	80–85
Australia and Oceania	26	1.4	0.1	1.3	30	3.0–3.5	0.5	2.5–3.0
Former USSR	282	117	12	105	310	140–150	20–25	120–125
World total	5,015	759.9	98.1	661.8	6,173	1,153–1,308.5	190–225.5	962.5–993

Source: World Resources Institute (1990).

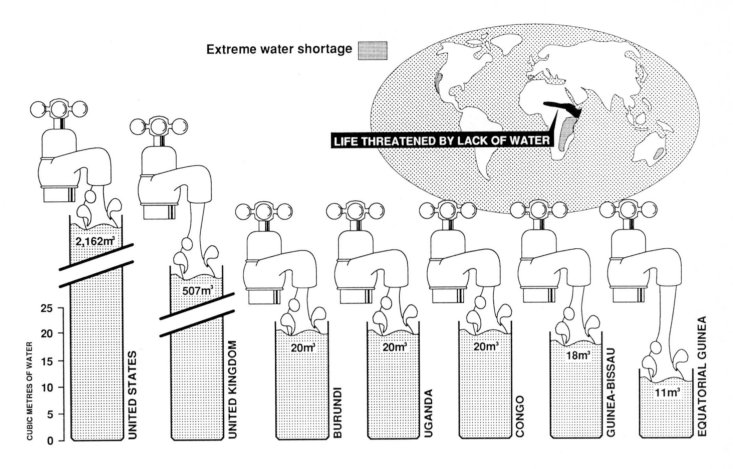

Figure 5.1 *Average annual consumption of water per capita in selected countries. Redrawn after World Resources Institute (1990).*

water per year for Santa Barbara and two neighbouring communities; when complete, this will be the largest municipal desalination plant. Until this scheme was approved, some of the Californian communities, such as the agricultural community of Goleta, considered importing fresh water using ocean-going tankers from Canada. Other communities in California, such as San Luis Obispo, are also now on the verge of approving a desalination plant.

In developed countries, droughts (see Chapter 8) are usually manifest in the issuing of drought orders and the imposition of temporary bans on the excessive use of water, for example, on the use of hosepipes. In some countries water meters are being introduced to charge by volume, and to influence demand – already the case in most European countries. It appears that metering water can reduce demand by between 7 and 55 per cent (OECD 1987), but typically between 10 to 15 per cent in regions that have

similar climates and consumption patterns to Britain (OFWAT 1990). In a detailed 1993 report, POST (the UK Parliamentary Office of Science and Technology has indicated that leakage control programmes appear to give higher savings than a universal metering programme, even where external meters with an ensuing reduction in supply pipe leakage are used. In England and Wales, about 22 per cent of the water input to the supply system is lost from the water companies' distribution systems, with a further 8 per cent lost from customers' supply pipes (ibid.).

Rivers

Humankind has been polluting water since the early days of civilization. This is because we have always been attracted to live by river environ-

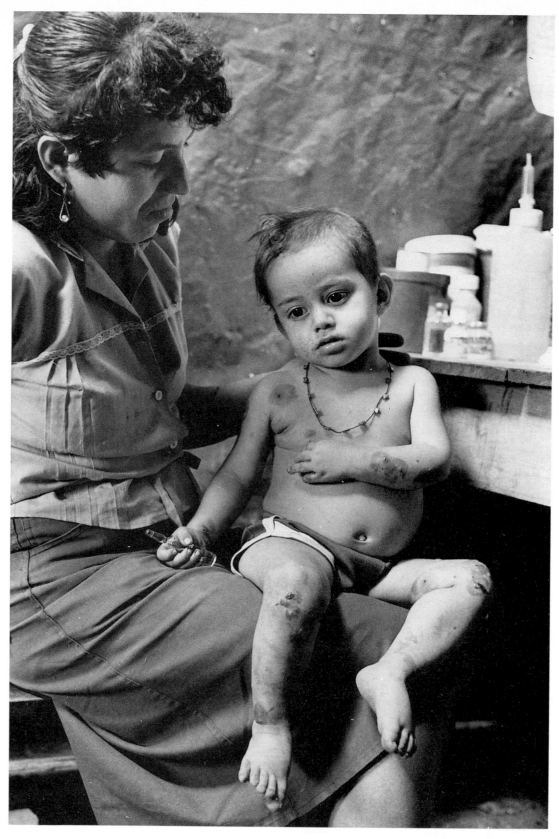

Plate 5.2   *Skin infection exacerbated by inadequate nutrition, dirty water, and poor sanitation, Communidad Santa Martha, El Salvador. These problems are being rectified by local health programmes.*
Courtesy of Rhodri Jones/Oxfam.

136

ments where a continuous supply of water for drinking and farming activities, such as irrigation and watering animals, is available. Consequently, the early civilizations, such as those in Mesopotamia (the land between the Tigris and Euphrates rivers), the Indus civilizations in Pakistan, and the great dynasties of China (Yellow and Yangste rivers), developed along big rivers.

Rivers are polluted by adding organic waste, including human and animal excreta and agricultural fibrous waste from harvested plants. Rivers have been regarded as sewers for centuries. For example, accounts of Elizabethan London describe people taking their sewage down to the Thames where the bucket was emptied and refilled with river water for household use! Much of this waste is degraded by microbes in the water through a natural process called **self-purification**. When populations are large, and excessive amounts of waste are produced which end up in the water supplies, the natural process of self-purification cannot keep pace with the input of pollutants, so the water quality rapidly deteriorates. Such organic waste is responsible for the transmission of infectious diseases such as cholera, typhoid, dysentery, and diarrhoea. Inadequate nutrition and poor medical care in many developing countries exacerbate the health problems associated with polluted waters (Plate 5.2). As long ago as 1389, Richard I of England outlawed the dumping of dung, filth, and garbage into streams and rivers near cities, but humans continued to discharge their waste into water supplies. The results have been frequent cholera epidemics, such as those experienced in Victorian London which killed thousands of people.

It was not until the turn of the century that people such as Dr John Snow in London made the connection between bad water quality and cholera epidemics. Laws were passed in the British parliament which included the Public Health Acts of 1848, 1872, and 1875, the Rivers Pollution Prevention Act of 1876, and the Land Drainage Act of 1930. These restricted the type and quality of pollutant that could be disposed of within sewerage systems, and made pollution a prosecutable offence. More recently, the EC has introduced regulations, directives, and new standards for water quality, such as the 1973 Community's First Action Programme on the Environment. Stronger legislation was enforced

in the USA, especially during Richard Nixon's presidency in the early 1970s, which paid particular regard to the dumping of waste at sea. Similarly, the Canadians introduced the Arctic Water Pollution Bill in 1970, restricting waste disposal and shipping activities in Arctic waters.

Despite such legislation, pollution by organic waste is still a major problem throughout the world. For example, in many countries raw sewage is flushed directly into the sea via outfalls. Today, around the coastlines of the Mediterranean, it is a common sight to see human excrement, toilet paper, sanitary towels, disposable nappies, and condoms on many beaches, something that the new EC legislation is designed to stop.

Additionally, rivers are still being polluted, not only by traditional organic wastes, but also by industrial wastes that now form the most common and hazardous pollutants. Industrial wastes include radioactive chemicals, dangerous organic chemicals, nitrates, heavy metals, and oil. In the UK, in 1987–9, the number of reported pollution incidents in rivers in England and Wales was 23,253, almost double the number for 1980-1 when 12,500 incidents were recorded. The situation has not improved. In 1991 the independent consumer magazine *Which?* examined river pollution in Britain with more disappointing statistics. *Which?* claimed that during the last decade there has been a significant increase in reported river pollution, with around 10 per cent of the river network in England and Wales being so badly polluted that fish cannot survive in the water. In 1989–90, there were more than 25,000 recorded pollution incidents in England and Wales, about 1,000 more than the preceding year and approximately twice that of 1982. Of course, the anti-alarmist, business-as-usual lobby would put this increase in reported pollution incidents down to enhanced public awareness of environmental issues. It seems reasonable to assume that at least some of these statistics are indeed due to increased public concern, but it would also be foolish to dismiss the figures on this basis. On 11th December 1991, the NRA (National Rivers Authority) in England and Wales released a report which showed that overall water standards have been dropping for the past decade. At the same time as these results were released, the NRA announced a five-year plan to force industrialists,

Plate 5.3 *Washing clothes in a stream – a feature of everyday life, Communidad Santa Martha, El Salvador.*
Courtesy of Rhodri Jones/Oxfam.

water companies and farmers to improve the quality of water in rivers, canals, and estuaries. Legally-binding standards would be introduced for water quality.

The increase in pollution is disturbing. Even when an incident is reported in the UK little action is taken against the polluter. For instance, in 1986 only 254 prosecutions were made out of a total of more than 20,000 incidents. The maximum penalty that a magistrate can impose is a mere £2,000, a very small slice of an industrialist's profit when compared to the expense of treating waste before it is discharged. In the USA, the situation is better because individuals can take out law suits against polluters, which may result in penalties of up to US $10,000 per day for violations, criminal penalties of from US $25,000 to US $42,500 for a first offence, and

US $50,000 for a second offence, and/or imprisonment.

More disturbing is the fact that the less-developed countries may have even greater problems than the developed countries, since government legislation and implementation is crude due to poor resources. There are still areas, such as in the Indian subcontinent and Latin America, where streams are used as open sewers, as well as a source of water for household use (Plate 5.3). James Bedding (1989) graphically described such a situation in Cairo and drew attention to the high infant mortality rate (131 in 1,000 children die before the age of 5). This is mostly as a result of gastroenteritis, cholera, and typhoid related to poor sanitation and water quality. A new network of sewers, pumping stations and treatment works for Cairo is scheduled for completion in the near future, the 'Greater Cairo Wastewater Project'. Raising finance for such schemes, estimated at well over £200 million for the Cairo project, is not easy in a developing country.

Oceans and seas

The oceans and seas have acted as the dumping ground for human waste for many centuries, but the problem has only become very serious in the industrialized and highly militarized world of the twentieth century. The waste ranges from domestic sewage, oil, detergents, pesticides, toxic metals, and PCBs, to radioactive waste. The persistence of such pollutants in the oceans varies considerably (Figure 5.2). Much of this chapter is devoted to some of these pollutants, such as oil.

Every year since 1986, 8 million tonnes of raw sewage have been dumped into the ocean at the Mid-Atlantic Bight, an area about 100 miles off the coast of New York and New Jersey (American Geophysical Union 1992b). When sewage was first dumped here, it was believed to be a safe and ideal site because of the strong ocean currents and deep water. More recently, scientific studies co-ordinated by the US NOAA's National Undersea Research Program have revealed the extent of sewage sludge accumulation at the site.

Sediment traps were suspended between 5 and 100 m above the ocean bottom and the trapped material tested for silver, the trace metal which is one of the most sensitive inorganic tracers of increasing amounts of sewage. Background levels in the oceans are believed to be approximately 0.1 ppm, and elevated amounts would be regarded as around 5 ppm. The sediment traps contained material with silver levels up to 16 ppm, an order of magnitude greater than the background level. At the site, sea-urchins were found to contain as much as 25 per cent sewage-derived organic material in their body tissue, showing that they are actively feeding on the sewage. The implications of this type of study have yet to be evaluated fully but, whatever the official government's views on the dumping of raw sewage at sea, it is clear that the sludge is not removed as fast as it is discharged.

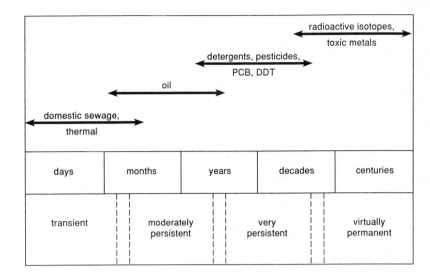

Figure 5.2 *The persistence of pollutants in the oceans. After Smith and Warr (1991).*

The seawater is becoming polluted by potentially toxic waste. At present, the UK dumps sewage sludge in the North Sea, both from coastal outflow pipes and from ships, but under an EC directive this will cease in 1998.

The water cycle

In order to understand what happens to pollutants in water, it is necessary to examine the global water cycle, or what is known as the **hydrological cycle**. Water is present on Planet Earth as oceans and seas (97.41 per cent of the total water), on land as rivers, lakes, within soil, animals, and plants, and in the atmosphere as water vapour (0.014 per cent). The remaining water is stored as ice within the ice caps and glaciers, and in the ground as groundwater (2.576 per cent). Figure 5.3 shows the ways in which water is transferred continuously from these main components by evaporation, vapour transport, precipitation, and flow across the surface of the land.

Water is precipitated from the atmosphere as rain or snow, falling on the land and the oceans. Some of this water will percolate into the soil and bedrock and flow as groundwater, often towards the sea, and some of the water will flow via rivers to lakes and oceans. The amount of precipitation is counterbalanced by evaporation of water from

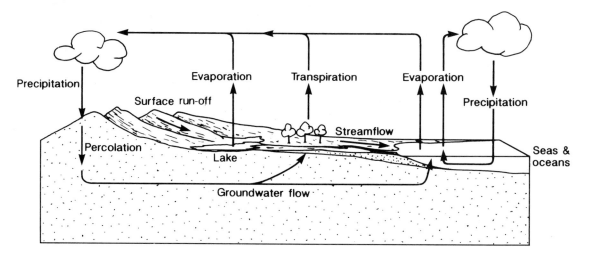

Figure 5.3 *The hydrological cycle showing the movement of water through the atmosphere, lithosphere, hydrosphere, and biosphere.*

the seas and lakes, from the soil by direct evaporation or drawn up by plants and then released during **transpiration**. Once the water has been evaporated it is transported in the atmosphere as vapour, until it condenses and returns to the Earth in the form of precipitation. This is the hydrological cycle.

The precipitation of water may introduce pollutants such as sulphur dioxide which is partly responsible for the acidification of water in acid lakes (see Chapter 4), airborne lead, and many radioactive substances (see Chapter 6). Precipitation, however, may have the positive effect of diluting polluted surface waters, and it can then be regarded as essentially pure water. The precipitation may then be stored for long periods as ice within mountain glaciers or ice caps, or it may flow over the surface of the Earth, transporting any pollutants, perhaps picked up on its journey towards the sea. The water that percolates into the ground may also carry pollutants with it to cause polluted groundwater.

Groundwater in many parts of the world is a vital water resource. In deserts and semi-arid regions it may be the only source of water. The flow of surface water and groundwater to the sea carries the pollutants, which become concentrated as the seawater evaporates to leave the pollutants behind in still greater concentrations. Evaporation from the soil may also lead to a form of water pollution known as **salinization**. This involves the drawing of salts and pollutants towards the surface, as soil water evaporates or is

taken up by plants and is replaced by water which comes from depth. This, in turn, is drawn up to the surface, concentrating salts as the water evaporates. This causes great problems in semi-arid regions, especially where irrigation is practised.

In addition to understanding the path of pollutants in the hydrological cycle, it is important to consider the fate of pollutants when organisms eat, drink, or inhale polluted water or water vapour. Animals form part of a hierarchical system known as a **food chain**. Put simply, organisms at the bottom of this chain obtain their food from plants, while predatory creatures further up the chain consume those below. Humans constitute the top of many of these chains. For example, plants may be eaten by small fish, which in turn are preyed on by larger fish, and fish are then consumed by humans. In reality the different chains interlink in a complex way to form **food webs** (Figure 5.4).

There is a relative accumulation of pollutants up the food chain in more complex organisms, at least for toxins that are stored. Plants that are polluted, for example, may be eaten by the small carnivores with little adverse effects, but as the relative pollution levels rise, toxins become concentrated higher up the food chain where the effects can be magnified. It is not quite this simple because higher up a food chain, larger creatures may have a greater resistance or ability to deal with poisoning.

The knock-on effect in a polluted food chain

was illustrated vividly in Rachel Carson's famous book, *Silent Spring* (1962). The book begins by describing a picturesque town in the heart of America renowned for its bird and fish life. It then goes on to recount a strange plight that swept the area. The birds and fish began to die until, one spring, there was no more dawn chorus and no fish in the rivers. What Rachel Carson was describing were the effects of pesticide poisoning on the area and how the poisoned insects, which formed part of the food chain, transferred this poison further up the food chain to the higher organisms, such as birds and fish. She paid particular attention to the pesticide **dichloro-diphenyl-trichloro-ethane (DDT)** that was being widely used throughout the world at that time. DDT was commonly washed off the crops into streams and into the ground. The plants absorbed this water containing DDT, and small creatures assimilated the poisons into their bodies by drinking the polluted water. The affected organisms then passed the pollutant up the food chain. Later, this pesticide was banned in the USA, but it is still used today in some less-developed countries. There are other pollutants widely used today that can have similar effects and can be transmitted just as easily up the food chain, yet because their effects are either less publicized or less well understood, they continue to be used. Multinational corporations continue to market products in the developing countries which are banned in the developed countries because they are deleterious to the natural ecosystem. Profit is the only motive.

To understand the causes and effects of pollutants in the main groups of toxic chemicals, they are considered in more detail in the following sections.

Sewerage, sewage and sludge disposal

Sewage is a cocktail of dissolved and suspended materials, and its quality is defined in terms of the suspended solids (SS), **biochemical oxygen demand (BOD)**, and ammonia content. Raw, untreated sewage typically contains 1–7 per cent solids. A general formula for the dry solids of sewage sludge is $C_5H_7NO_2$. SS is determined by weighing after filtration of a known volume of sample through a standard glassfibre filter paper, expressed as mg l^{-1}. BOD is determined by sample incubation at 20°C for five days, with the amount of oxygen consumed also being given in mg l^{-1}. Domestic sewage, with about 1,000 mg l^{-1} of impurities, comprises about two-thirds organic material, with the main constituents being nitrogen compounds. Between countries, standards vary considerably, and within the EC minimum standards and controls for municipal wastewater and the disposal of sewage sludge are being introduced through the Urban Waste Water Treatment Directive (91/271/EEC). This requires secondary treatment of municipal wastewater as a minimum, except in areas designated as less sensitive where primary treatment alone may be permitted. It also requires the cessation of sludge dumping at sea by 1998.

Disposal of sewage sludge, either to land or sea, is frequently based on economic rather than environmental considerations, although in many countries this is changing. Land and sea disposal both have their pros and cons. Application of sludge to agricultural land, either in liquid or dried form and commonly as soil/subsoil injection, has the advantage of resource recovery but may cause more immediate environmental damage than disposal at sea. Land disposal can lead to contamination of groundwater and surface streams by viruses, bacteria, protozoans and other pathogens, persistent toxic organic compounds, and toxic heavy metals and, therefore, requires careful monitoring. On the credit side, sludge with its high nitrogen and phosphorous content provides a valuable fertilizer.

The balance between various sewage sludge disposal routes varies between countries, being dependent on factors such as appropriate technology, perceived acceptable standards, and purely economic considerations. In England and Wales, sewage sludge disposal for 1991/2 was as follows: 51 per cent farmland, 22 per cent sea, 11 per cent landfill, 7 per cent incineration, and 9 per cent other (Water Services Association 1992). As the dumping of sludge at sea is reduced and finally ceases, other forms of disposal are becoming more important, such as incineration. There are plans to build Europe's largest waste incinerator in south-east London, projected to generate 100 MW electricity from 1.5 Mt of waste per annum. Thames Water plans to build two incinerators capable of burning 0.5 Mt of sewage sludge per annum. Throughout

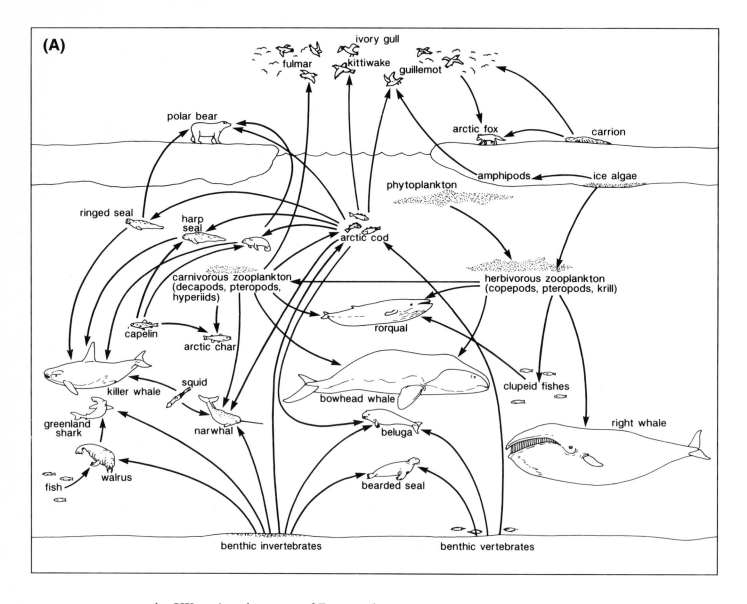

(A)

ivory gull

fulmar kittiwake

guillemot

arctic fox carrion

polar bear

amphipods ice algae

phytoplankton

ringed seal

harp seal

arctic cod

carnivorous zooplankton (decapods, pteropods, hyperiids)

herbivorous zooplankton (copepods, pteropods, krill)

capelin

rorqual

arctic char

squid

clupeid fishes

killer whale

bowhead whale

right whale

greenland shark

narwhal

beluga

walrus

bearded seal

fish

benthic invertebrates benthic vertebrates

the UK, as in other parts of Europe, there are other schemes under way for the construction of incinerators to burn domestic waste and sewage sludge. These changes have implications for local air quality and broader environmental change. Incineration produces more immediate air pollution and the release of CO_2 to the atmosphere, whereas sea dumping leads to the release of more CH_4. Incinerating dry sludge releases about 53 per cent of its weight in carbon, equivalent to approximately 195 per cent CO_2. In calculating CO_2 budgets, however, there is little to no net contribution to atmospheric CO_2 levels because life-cycle analysis shows that it represents a short-cycle sequestration and release of CO_2.

In countries such as England and Wales,

sewerage systems carry the sewage to the treatment site or discharge point via 'foul sewers' (carrying only domestic and industrial effluent) and 'stormwater systems' (discharged directly to natural water courses). In older towns and cities, however, combined foul and stormwater systems are used, leading to large differences in the flow of sewage during storms. In Britain, water pollution was a recognized problem even in 1850, and in 1876 Parliament passed the first Act to control water pollution; between 1898 and 1915, the Royal Commission on Sewage Disposal outlined the objectives of, and requirements for, sewage treatment. The main purpose of sewage treatment is to destroy disease-carrying organisms and to remove, or at least dilute, toxic chemicals;

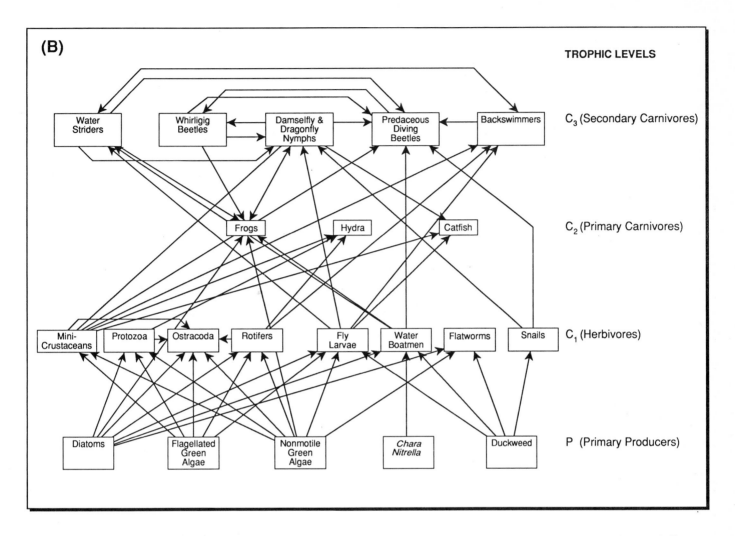

Figure 5.4 *(A) Simplified food web for the Arctic Ocean. After Open University (1991). (B) Simplified food web of a small meadow pond. The arrows show the direction of energy flow. Not all species are confined to one level, especially carnivores which may use all trophic levels. Some animals, particularly insects, may occupy different levels at different times of their life cycle. After Clapham (1973).*

Box 5.1

Sewage treatment

Sewage treatment usually involves a preliminary screening stage to remove larger suspended and floating material, followed by primary sedimentation to separate out solids, and then a secondary, biological treatment of the sludge, commonly associated with an anaerobic stage in a digester and an aerobic stage in settlement tanks and/or lagoons, the latter processes being referred to as digestion. Sludge treatment and disposal can account for 40 per cent of the operating costs in a wastewater treatment plant. An important part of the treatment of sewage sludge is the reduction of the water content, or dewatering, thereby allowing easier handling and reduced transport costs since the volume of sludge is much less: 30 million tonnes of wet solid corresponds to 1.25 million tonnes of dry solid. Dewatering processes vary considerably in both duration and effect, for example it may take more than two months using drying beds (giving about 25 per cent solid content of the sludge), or 2–18 hours under pressure filtration at 700 kPa in cloth-lined chambers to form cakes typically of 25–50 per cent solids. Alternatively, vacuum filtration, operating with a pressure difference of about 70 kPa, produces sludge with 15–20 per cent solids.

other aspects include the reduction of unpleasant odours (see Box 5.1). The indirect reuse of water is common in some countries and tends to be on the increase, such as in the UK where it is predicted that by the year 2000 the amount of reuse will have approximately doubled from 1990 levels. Where reuse is practised, nitrate concentrations in river water are an area of concern and subject to monitoring; high levels of nitrate in potable (drinking) water can cause methaemoglobinaemia or 'blue-baby' syndrome.

Traditional organic waste, microbes, bacteria, and viruses

Human excrement discharged in an untreated form into rivers and seas constitutes a major pollutant. Such discharges are frequent in developed countries as well as in less-developed countries. In England and Wales, many of the 6,500 sewage treatment plants discharge sewage effluent into the rivers, rather than the open sea or estuaries, and are old and cannot handle the increased amounts of sewage. About 20 per cent of the UK sewers were laid before 1914, many of which were designed to carry rainwater as well as sewage (*Which?* 1991). A major programme to improve and upgrade these plants is currently under way.

The result of using manure from animal husbandry can be considered as equally detrimental to the environment, a point emphasized in an article on the production of manure by piggeries in Denmark (*New Scientist*, November 1988). These piggeries produce around 94 million tonnes of manure each year, much of which cannot be treated or disposed of adequately. This manure produces not only ammonia gas (NH_3) to pollute the atmosphere and damage vegetation, but also nitrates that are washed into streams. The manure concentrates heavy metals into the soils at toxic levels. These heavy metals, such as copper, cadmium, and zinc, which in high enough concentrations become poisonous, were originally added to the pigs' feed to increase their growth. These metals, however, are discharged by the pigs in their excrement. When the manure is used as fertilizer or stockpiled, the metals may then leach out of it and end up in streams and groundwater. Large amounts of these heavy metals may be retained in the soil and can lead to the death of earthworms that are so essential for the breakdown and aeration of the soil. The knock-on effect of decreased soil breakdown is soil erosion, which then accelerates the input of the heavy metals to streamwater. A vicious circle results. The heavy metals are then washed into the drainage systems to find their way into the water used for human consumption, and can ultimately cause serious illnesses.

Excrement is often associated with particular microbes, bacteria, and viruses, among which cholera and typhoid are numbered as particularly rampant and dangerous. In recent years, there have been several outbreaks of animal parasites, such as cryptosporidia. For example, affected water supplies in the Swindon area of the UK during February 1989 threatened 20,000 people and caused many people to suffer diarrhoea and vomiting. Though these effects were mild, there should be concern for those people with suppressed immune systems, such as babies, old people on particular immune-suppressive drugs, and other vulnerable people. They may be more adversely affected by such disorders and could die.

Other cases of cryptosporidia have been reported, including a similar outbreak in Ayrshire in 1988, and a case of giardiasis in Leeds in 1980, which infected 3,000 people. Giardia is a particularly unpleasant organism which results in dysentery and may cause death in weak individuals. It affixes itself to the gut lining, multiplies extremely rapidly and is very difficult to eradicate without strong antibiotics. It is prevalent in tropical developing countries, but the number of cases of giardiasis in Europe, even amongst people who have not travelled abroad, has been increasing in recent years. Another disease that is on the increase is leptospirosis, commonly known as 'sewerman's disease', which can be caught from contact with rats' urine. Fifteen people died in the UK from this disease in 1989. Some experts believe that modern farm slurry discharges are responsible for the rising infection rates.

The Scottish Home and Health Department has been carrying out investigations with universities and laboratories in Scotland and is particularly concerned about the parasitic organisms in water supplies. The results indicate that

Plate 12 *Dead spruce trees, the result of acidic deposition in the Ore Mountains, Czechoslovakia.*
Courtesy of Greenpeace/Tickle.

Plate 13 *Coal-fired electric power station emitting smoke near Midland, Pennsylvania, USA.*
Courtesy of Greenpeace/Visser.

Plate 16   *Attempt to clean up the oil pollution in Alaska after the 1989 Exxon Valdez disaster, using high-powered water jets.*
Courtesy of Greenpeace/ Merjenburgh.

Plate 14   *Clean-up of an algal bloom that washed up on an Adriatic beach in Italy in July 1989.*
Courtesy of Greenpeace/ Ferraris.

Plate 15   *The grounded Braer oil tanker off Sumburgh Head in the Shetland Isles, Scotland, leaking oil into the sea in January 1993.*
Courtesy of Greenpeace/ Hodson.

Plate 17 *Hydropolitics create unrest and threaten the socio-political and cultural stability, and health, of water-deficient regions. The Jordan basin has one of the most acute water problems in the Middle East, and waters are shared between three Arab countries – Jordan, Syria, and Lebanon – and their traditional enemy, Israel. In accordance with the peace settlement for the Lebanon, a US marine stands guard over Beirut.*
Courtesy of James Nachtwey/Magnum.

within many of the large water bodies, especially recreational lakes, river water, sewage, and even in treated drinking water, there are viable parasitic cysts. This poses a particularly daunting threat to water supplies.

Nitrates

Pollution of rivers, seas, and drinking water by nitrates from fertilizers and traditional organic waste has become a serious problem. Nitrates, however, are essential chemicals for the metabolic functions of plants. They are naturally produced by nitrofixing bacteria which fix nitrogen from the atmosphere in the decomposing organic matter and form nitrates. Nitrates are also formed by the weathering of rocks during the formation of soil.

In areas where intensive agriculture is practised, the soil may become depleted in nitrates with resultant poor crop yields. To remedy this, artificial fertilizers are added to the soil, or muck naturally rich in nitrates is spread over agricultural land. This effectively makes the soil more fertile. If there are excessive quantities of nitrates, however, they percolate into the groundwater supplies, flow into streams and rivers and eventually reach the sea or lakes where they become concentrated. Nitrates may also be concentrated in groundwater, which tends to be as a very long-lived effect because of the slow turnover of groundwater reservoirs. This may have detrimental effects on aquatic and marine ecosystems, and may lead to public health problems when drinking water is contaminated.

The quantity of artificial nitrates used on arable land in attempts to improve agricultural yields has increased considerably over the last 20 years. In the UK, for example, 1.3 million tonnes of nitrates were used in 1986, twice that of 1966. Natural fertilizers or manure also produce large quantities of nitrates. In addition, animal husbandry produces huge amounts of excrement, both manure and urine, which cannot be used as fertilizers. This excrement is often leaked or intentionally released into water supplies in order to dispose of the large quantities produced by animal husbandry. Soil itself will naturally produce nitrates through the microbial activity of nitrofixing bacteria by supplying nitrogen as

nitrates directly to the plant. When the plant decays nitrates are deposited in the soil.

The standards set by the EC put an acceptable limit of 50 mg per litre (mg l^{-1}) of nitrate concentration in public water supplies. The World Health Organization (WHO) recommends a limit of 100 mg of nitrates per litre. We do not really know if these levels are safe, but they are commonly exceeded. In 1984, the WHO produced evidence to show that nitrate pollution was responsible for 'blue-baby' births (methaemoglobinaemia). In this disease, nitrates react with bacteria in the gut and deplete the blood oxygen levels which then affect the brain and heart muscles. It has even been suggested that some forms of stomach cancer result from nitrate poisoning. In Britain, the EC identified 52 areas where concentrations exceeded permitted nitrate levels and it was estimated that some 1.3 million people were affected. Mothers were told to use bottled water for their babies because boiling water does not remove nitrates.

At Rothamsted Experimental Station in Hertfordshire, UK, research into the effects of nitrates in the soils has identified the main conditions under which nitrates leach into water supplies. Suggestions have been made for the improved use of nitrate-rich fertilizers (Addiscott 1988). Little, however, is being done to reduce the levels of nitrates in contaminated water.

It is possible to remove nitrates from water chemically. Various chemical techniques that are technically available to transfer nitrates from one body of water to another, but are not used on any significant commercial basis, include selective ion exchange, reverse osmosis, electrodialysis, and distillation. Biological processes are also available. One of the new nitrate-removal processes involves using aluminium powder to reduce the nitrate to the less harmful ammonia, nitrogen, and nitrite (Murphy 1991). With such a range of nitrate-removal processes available to us, there really is little reason for the current complacency.

A blooming problem

A superabundance of nitrates in lakes and seas, introduced by groundwater and streams rich in nitrates, causes water quality problems. Often, other compounds such as phosphates will also be present with the nitrates derived from detergents washed or flushed into the water sources. Within lakes, the great increase of these chemicals, either naturally or accelerated by pollution, and many of which are nutrients for some organism, will initially produce prolific growth of algae. As the algae grow, they reduce, and may eventually cut off, the light to other plants which are so necessary in the processes that lead to oxygenating bottom waters. The larger amounts of decaying plant material being rotted by bacteria on the bottom stimulate the production of greater numbers of bacteria which, in turn, further deplete the oxygen in the water. The low free oxygen levels can rapidly cause the fish to die, and before long, a diverse living ecosystem may become lifeless apart from bacteria. Scientists call this process **eutrophication**, and describe lakes as being eutrophic when this process has occurred.

Many of the Great Lakes in the USA suffer to varying extents from this condition. Similar conditions may occur in seawater, where green and red algae become very prolific when nitrates and phosphates are introduced into the water above threshold levels. Such phenomena are called **algal blooms**. Red blooms, produced by red algae, are a common sight in tropical waters, but they may also occur naturally along coasts where cold nutrient-rich bottom waters rise to the surface by a process known as upwelling. These blooms are associated with cycles of increased algal growth and large fish populations living on the algae and the new supply of nutrients. The consequent over-productivity can result in a depletion of free oxygen in the waters and eventually lead to large-scale suffocation and death of marine life. Similar cycles are experienced in areas where pollutants such as nitrates and phosphates are introduced into the sea by waste disposal. Up to 50 per cent of the global 'new' production of biomass in the oceans caused by upwelling occurs in the eastern equatorial Pacific, a natural process of high productivity.

Recently, blooms of algae have begun to threaten the waters of the North Sea and the Mediterranean. Graphic descriptions appeared in newspapers during the summer of 1989, such as in *The Independent* (27 July 1989) entitled

'Slime and Scum lap the moonlit beaches' (Plate 14). The article described the problems facing waters and resorts in northern Italy due to extraordinarily large blooms of green algae in the heavily polluted waters. In the Mediterranean, beaches that are now greatly infected include those between Trieste and Rimini, and around the coasts of Malta, Marseille, Barcelona, Algiers, and Alexandria. Milne (1989) described especially alarming blooms of green algae in the seas adjacent to the major estuaries of the North Sea. He illustrated the widespread extent of elevated nutrient levels and how they combine to create oxygen-deficient waters. He identified four main estuaries in the UK that are experiencing toxic algal blooms: the Thames, Humber, Tyne, and Forth. The biggest problem, however, is in areas along the eastern regions of the North Sea, such as the Rhine, Rhem, Weser, and some of the Norwegian fjords. Fortunately, he reported that at the 1987 North Sea ministerial conference held in London, most countries agreed to try to cut the input of nutrients by 50 per cent by 1995 to help reduce the problem.

Dangerous organic chemicals

Amongst the class of dangerous organic chemicals, one of the principal chemical elements is chlorine. Organic chemicals containing chlorine are commonly referred to as 'chlorinated hydrocarbons'. The most notorious are the **polychlorinated biphenyls (PCBs)** and, to a lesser extent now, DDT (dichloro-diphenyl-trichloroethane).

PCBs are used mainly in the manufacture of paints, plastics, adhesives, hydraulic fluids, and electrical components such as generators. Most of the PCBs which eventually enter the world's rivers and seas do so from the atmosphere having been transported as aerosols. The pesticide DDT is air-sprayed on to crops, and can be transported considerable distances in the atmosphere as well as through rivers and lakes to the sea.

DDT is essentially non-biodegradable because, when it does eventually break down, it forms equally harmful chemicals with a lifespan measured in decades. PCBs are biodegradable by the action of micro-organisms, but the process takes a long time. PCBs are non-flammable, unless incinerated at very high temperatures (up to 1,200°C) to stop the formation of very toxic chemicals called furan and dioxin. They are therefore very difficult to break down. PCBs are actually more persistent than DDT. Both DDT and PCBs concentrate in the fatty tissues of living organisms and may become concentrated along the food chain as one animal eats another. In many developed nations, legislation has made the use of DDT illegal but it is still widely used in the agricultural economy of developing countries.

The adverse effects of PCBs and DDT are most pronounced in animals with a backbone (vertebrates). For example, higher than expected levels of PCBs were found in the 12,000 fish-eating seabirds washed up on the shores of Britain in 1969. These chemicals inhibit normal growth in animal populations and cause the bones of seabirds to be abnormally thin. Thinner bones are weaker and so the affected birds are unable to sustain their flight in the normal way which makes them more vulnerable to exhaustion, injuries and predation. Ironically, DDT has brought many environmental problems, which include the development of DDT-resistant strains of malarial mosquitoes.

PCBs can pose a threat to ocean mammals, such as seals, polar bears, and whales, which in an extreme case could face extinction if no action is taken to stop the release of PCBs into the atmosphere and seas. Recent analyses of the blubber of dead carnivorous marine mammals show a substantial increase in the quantity of absorbed PCBs. PCB levels in these animals cause infertility. At a level of 50 ppm of PCB in the male blubber, there is a sharp cut-off in the production of sperm. In the summer of 1991, the Ministry of Agriculture in the UK discovered PCB levels up to 320 ppm in a dead baby bottlenose dolphin found off Dyfed, Wales, and 93 ppm in a dead porpoise in Cardigan Bay, Wales. Twenty-two bottlenose dolphins are known to have died in Cardigan Bay in 1989. Deep-ocean killer whales have been shown to contain 410 ppm PCBs, with values of 833 ppm in dolphins off the coast of Europe, considerable distances away from sources of PCB pollution. Even polar bears in the Arctic have been found to have levels of 10 ppm PCBs. Cummins, a geneticist at the University of Western Ontario, Canada, has suggested that 'should only 15 per cent of the world's PCBs at present in use, storage, or

simply dumped in developing countries, ever enter the oceans they would be sufficient to cause the extinction of most, if not all, marine mammals and the chemical fouling of ocean fisheries, rendering them unsuitable for use by humans'. This view, although at the most alarmist end of the spectrum, is shared to varying degrees by many other scientists. The toxicity of PCBs is unquestionable. The issue is whether or not the environmental consequences of dumping PCBs in the seas will be appreciated at an international level in time to safeguard the survival of marine and other life.

Tributyn (TBT), a type of **organotin** used as an anti-fouling paint on boats, has also been shown to cause serious damage to marine life such as shellfish. France banned its use in 1982, whereas the British government set limits on the amount of TBT in anti-fouling paints. In 1986, scientists showed that these limits were still too high and shellfish were still being seriously affected. Estuaries where pleasure craft are common still show levels of poisonous chemicals, including TBT, that are at unacceptable concentrations. As an example, during the summer of 1986, the Crouch river estuary in Essex, UK, showed concentrations of organotins varying from 50 to 130 nanogrammes per litre. The target set by the British government for water quality is 20 nanogrammes per litre. In that same year, five out of the eight estuaries investigated by the Ministry of Agriculture, Fisheries and Food (MAFF) did not meet the target.

In Los Angeles, California, the 15,000-strong Surfrider Foundation has brought a successful lawsuit against the Louisiana-Pacific Corporation and the Simpson Paper Company for polluting the ocean with effluents including dioxins and other toxic chemicals from their paper mills near Eureka. The Louisiana-Pacific Corporation and the Simpson Paper Company have to pay US $5.8 million in fines and invest more than US $50 million in reducing the effects of the dangerous effluent. Part of the financial package, negotiated with the help of the US Environment Protection Agency (EPA), also includes provision that the pulp companies do not harm the abalone mollusc, two species of echinoid, sea-urchins, and sand dollars, and the seaweed, kelp. Settlements like this are unusual at the moment, but pave the way for the possible pattern of future lawsuits and ensuing agreements.

WATER RESOURCES AND POLLUTION

Heavy metals

Mercury, lead, arsenic, tin, cadmium, cobalt, selenium, manganese, and copper are examples of what are called **heavy metals**. The natural weathering of rocks on land and volcanic activity introduce these heavy metals into the rivers and seas, but in many instances they occur at unnaturally high concentrations because of mining activities and other industrial processes. Rivers may transport large amounts of these heavy metals to the seas and oceans where they become even more concentrated. In large enough doses, these metals can prove lethal to organisms, including humans. Heavy metals, therefore, constitute another major category of pollutants in rivers, coastal waters, and seas.

Figure 5.5 shows the increase in toxic metals since the beginning of this century in the River Rhine, as a result of the increased industrialization. Notice the steady increase of contaminants up until the late 1960s and early 1970s when levels began to decrease with the increase in environmental awareness and legislation.

The most common toxic pollutants are lead salts (e.g., methyl lead), mercury, and arsenic. If these heavy metals accumulate in the body they can cause brain damage and even death. A particularly serious case that has been well documented was the poisoning of people from the fishing community of Minamata in Japan in the 1950s. The symptoms were numbness of the limbs, speech impairment, and loss of co-ordination. The cause was eventually traced to methyl mercury introduced into the area from a nearby factory, but it was some time before the factory admitted liability.

Sceptics argue that there is little direct proof of sea pollution by heavy metals. Sophisticated analytical techniques available today, however, can show the source of pollutants. By measuring the ratio of lead $^{204}Pb/^{206}Pb$, and $^{207}Pb/^{208}Pb$ isotopes (an element whose atom may have different numbers of neutrons in its nucleus to give several atomic varieties of different masses) in near-surface waters of the Pacific Ocean off San Francisco, scientists have been able to trace local pollution to ^{204}Pb, ^{206}Pb, ^{207}Pb, and ^{208}Pb industrial lead and polluting waters thousands of kilometres west across the other side of the ocean in Asia. These contaminated ocean currents come to the surface, or upwell, as cold-water current off California (Fanning 1989). The lead was introduced to the water column from industrial lead aerosols and advected across the Pacific under the influence of the prevalent westerly wind system.

Lead in old piping for water supply can cause serious health problems. In March 1989, the EC published a report showing lead levels in drinking water exceeded those set by the EC in 1985 in every Scottish region. The principal culprit was lead piping. The maximum acceptable lead level set by the EC is 50 micrograms of lead per litre. The British government has set its own standard at double this figure, that is, 100 micrograms per litre. Excessive lead levels, even at low dosages, are known to cause hyperactivity in children and to affect their performance in attainment and ability tests. Some experts believe that there is no safe level of lead in drinking water and that the government should provide grants to all householders to remove any existing lead fittings in their water systems.

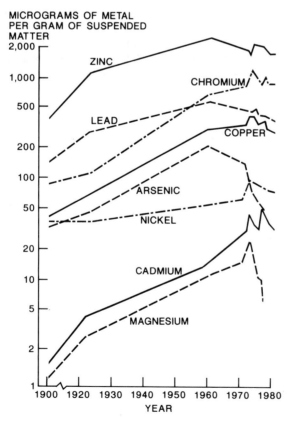

Figure 5.5 *Changes in metal contamination associated with suspended matter in the River Rhine. After World Resources Institute (1990).*

The single biggest cause of lead poisoning, however, probably comes from car fumes. This lead is introduced into the air by engine exhaust and is blown or washed by rain into water sources. As long as people choose to use leaded rather than unleaded petrol, there is little that can be done to reduce this form of pollution to the environment. Taxation may be used to discourage the use of leaded petrol. Fortunately, with the cheaper price of unleaded petrol and people's increased awareness of environmental issues the amount of leaded petrol used is slowly being reduced.

The element aluminium is also a health hazard and is a relatively new addition to the list of chemical elements that are frequently cited in the media as being at unacceptably high levels in some sources of drinking water. In January 1989, an article appeared in the medical journal *Lancet* which linked high levels of aluminium in tapwater to Alzheimer's disease, the most common cause of senile dementia. It is estimated that perhaps one in twenty people over 60 years of age may be affected by this disease. High levels of aluminium tend to occur in acidified natural water, not only in areas of pollution, but also where certain geological rock and soil types exist, for example above average levels have been measured in northern England especially around the River Tyne.

The only known cause of Alzheimer's disease is a genetic defect. In approximately one-twentieth of the people diagnosed as having Alzheimer's disease, a genetic defect in chromosome 21 has been identified by John Hardy of St Mary's Hospital Medical School in London (Ferry 1989). Chris Martyn from the Epidemiology Unit of the Medical Research Council, USA, has found that in districts where drinking water contains more than 0.01 mg of aluminium per litre, there was 1.3–1.5 times the incidence of Alzheimer's disease compared to areas with lower aluminium levels.

Using a laser microprobe mass analysis technique, Dan Perl of Mount Sinai Medical School, New York, has identified significantly higher levels of aluminium in the brain cells of people who suffered from 'Guam disease' (named after the Pacific island of Guam where it is endemic), allied to Parkinson's disease which causes senile dementia. Perl believes that there are probably close links between these various diseases that cause dementia, and elements such as aluminium. The present EC standard for aluminium in drinking water is 0.2 mg per litre. There is little available evidence to say whether even this permitted amount is too much. Statistics from the Ministry of Agriculture, Fisheries and Food in the UK show that wheat contains 2 mg of aluminium per kg, and dry tea leaves have as much as 1,000 mg per kg. The Ministry also calculates that the average person has a daily intake of about 6 mg, 10 per cent coming from drinking water and the remaining 90 per cent from food.

The effects of increased aluminium levels in drinking water were dramatically reported in the UK on the BBC television programme *Panorama* on 20th March 1989. The programme revealed what has become known as the 'Camelford Incident'. The incident occurred because a lorry driver delivering aluminium sulphate to the Lowermoor water treatment works in north Cornwall, UK, which was to be used as a water purifier, dumped 20 tonnes of the hazardous chemical into a purified water tank by mistake. This water was then released to the main supply and delivered to 7,000 households and 20,000 people in the Camelford area. The result was a discoloration of household water and an increase in its acidity, which helped dissolve copper, lead, and zinc, further adding to the metal pollutants in the water.

The people of Camelford developed sickness, diarrhoea, mouth and nose ulcers, blistered tongues, bloody urine, and aches and pains in the muscles. Those suffering from arthritis found themselves in excruciating pain and some people began to show signs of Alzheimer's disease with a loss of memory. Later, farmers reported that calves were born at only half their normal weight, breeding rabbits became sterile and piglets were born with both male and female parts. The water was found to contain 6,000 times the maximum levels considered safe by the EC and WHO. The situation was exacerbated by the denial that the water had been polluted by the South West Water Authority (SSWA) who were responsible for the treatment plant. The denials continued despite the complaints by local people and an independent report, by a local biologist and expert on water pollution (Cross 1989). In fact, the *Panorama* report revealed that the SSWA had known about the pollutant since 8th January, but continued to be economical with the truth about the incident and it even put an advertisement in the local paper addressed to the residents in the Camelford area assuring them that the water was fit to drink. The long-term effects have yet to be evaluated.

A new generation of phosphate-eliminating sewage treatment plants, and the increasing prohibition of phosphates in detergents, have led to considerable reductions in phosphate concentrations in surface waters, but this has created new concerns. Replacing phosphates in detergents with complexing agents could cause the increased mobilization of various toxic heavy metals which result in serious pollution to surface waters. However, as a result of a 12-year analysis of phosphate, manganese, and cadmium levels in the River Glatt, Switzerland, and the adjacent aquifer infiltrated by the river water, von Gunten and Lienert (1993) have postulated that the lower phosphate levels have decreased the amount of oxidizable organic carbon, thereby creating less reducing conditions in the infiltrating water. Consequently, there is decreased reductive dissolution of Mn and Cd in the groundwater and, therefore, actually an unexpected improvement in drinking water quality with respect to some toxic metals.

Radioactive waste

Radioactive waste forms one of the most frightening pollutants. Much of the waste has a whole range of short- to long-term disastrous effects, from the mutation of genes and the birth of deformed organisms to the development of cancer and death in humans. Worst of all is the fact that radioactive waste is a legacy to future generations. The nature and effects of radioactive waste are dealt with in Chapter 6 on nuclear issues.

Oil pollution

Worldwide, oil pollution has steadily grown with the increased transport and use of oil. It is estimated that there are over 3.6 million tonnes of oil spilt into the sea every year, mainly as a result of shipping accidents involving oil tankers (Figure 5.6). There are many other sources of oil spills, including natural seeps from the sea bed, for example from oil fields via fissures in rocks. Other seeps include off-shore oil seeps from drilling and oil exploration, airborne oil droplets from unburnt fuel and combustion engines on land, and discharges from shipping, either because of washing out of engines or tanks, or water used as ballast from oil tankers or deliberate discharges into the sea as ecological terrorism.

In 1967, the *Torrey Canyon* ran aground off south-west England, leaking 118,000 tonnes of crude oil that affected 20 miles of coastline and killed 40,000–100,000 sea birds. One-third of this oil was burnt using bombs, napalm, and rockets fired from fighter-bombers and thousands of gallons of oil were removed using oil-dispersing detergents. On 4th February 1970, the supertanker *Arrow* ran aground in Chedabucto Bay, Nova Scotia and released 10,000 tonnes of oil, contaminating 300 km of coastline. On 6th August 1983 off the Cape coast of South Africa, the Spanish supertanker *Castillo de Belver* caught fire and split in two while fully laden with 200,000 tonnes of crude oil. A slick developed that was about 20 miles long and up to 3 miles wide. On 16th March 1987, the *Amoco Cadiz* ran aground on rocks near Portsall, France, releasing 223,000 tonnes of crude oil and contaminating a 300 km stretch of the Brittany coast. In April 1991, a blazing oil tanker still laden with 100,000 tonnes of oil sank to the seabed off the Italian Riviera. Of the total 140,000 tonnes, 40,000 were burned in the fire. The initial oil spills are often exacerbated by the rapid spread of oil across the surface of water aided by waves, wind, tides, and ocean currents. This can make even a relatively small spill a widespread problem, but in certain circumstances it actually helps to disperse the thick oil and break it up into less serious patches.

With the exception of the Gulf oil slick, the five largest recorded oil spills were all the result of oil tanker accidents. One of the most recent of these occurred on Tuesday 5th January 1993, when the 17-year-old single-hulled oil tanker *Braer*, operating on a Liberian flag of convenience for the American oil company Ultramar, ran aground on the southern tip of the Shetland Isles, Scotland, after losing all engine power in

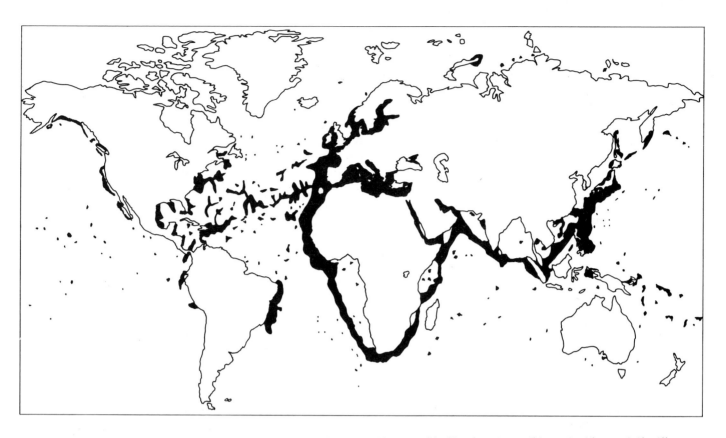

Figure 5.6 *Distribution of major oil slicks throughout the oceans (shown in black). There is a striking coincidence of oil spillage with major shipping routes, for example off the west coast of Africa. After Mysak and Lin (1990).*

storm conditions (Plate 15). By the following Wednesday, 12th January, all the 84,500 tonnes of light crude oil that the 89,730 tonne *Braer* was carrying had escaped into the stormy seas that had not abated since the disaster first happened. Hurricane-force winds prevented a full-scale clean-up operation from getting under way (although detergents were sprayed from six DC3 aircraft fitted with infrared tracking equipment in an attempt to disperse the oil slick), and also stopped any attempts at salvaging the stricken tanker before it broke up on 11th January. This oil spill affected important, internationally-recognized bird and otter communities around the coast of the Shetlands, and raised a number of important issues about the transport of oil and the risk of oil pollution. Amongst these issues, the principal concerns are: (1) the whole issue of ageing oil tankers, commonly only single-hulled, sailing under flags of convenience with inexperienced international crews that may have had problems communicating with each other in an emergency, all done in

order to save money at the expense of the environmental and safety factors; (2) the routes taken by oil tankers, travelling close to ecologically sensitive and important coastal habitats for birds and other marine life; (3) the procedures to be adopted by a ship's crew when a tanker appears to be out of control, and the responsibility of captains to take the decision to radio a May-day call for assistance without awaiting a response from the owners; and (4) clean-up procedures and technology following a major oil pollution event, for example the desirability of using detergents, which also cause marine pollution, in order to break up the oil.

The second largest oil spill occurred on 24th March 1989, when the *Exxon Valdez* ran aground at Bligh Island, Alaska, leaking 10 million gallons of crude oil over an area of more than 900 square miles of water and affecting hundreds of miles of shoreline. The spilt oil was often over 10 cm thick along the shore and it is estimated that less than 4 per cent of it has been removed so far. Amongst the dead were an estimated 3,500–5,500 otters, 200 harbour seals and almost 400,000 birds (Plate 16). More than one year after the catastrophe, scientists studying the effects of the pollution have found themselves increasingly embroiled in the litigation as lawsuits are filed against Exxon. Researchers have been hired by Exxon, the State of Alaska, and the US government to assess the extent of the damage. Beside the two lawsuits with the highest profile, those of the State of Alaska and the US Department of Justice which each seek millions of US dollars in compensation, there are about another 150 lawsuits by environmental groups, local organizations, and individuals. So, with all this litigation under way, many of the professional assessments of the environmental damage remain unpublished and sensitive. And with the information remaining secret, the wider scientific community cannot consider the actual damage and begin to use the data to develop strategies to reduce any environmental impact that a future major oil spill might have.

In January and February 1991, during the Gulf War between Iraq and the American-led, United Nations backed Allied Forces, Iraq's President Saddam Hussein resorted to ecological terrorism which resulted in oil pollution to the Gulf. In the Gulf War, oil was discharged into the Gulf of Arabia off the coast of Kuwait by the Iraqi invasion forces and spread along the coast of Kuwait and Saudi Arabia. Additionally, so bad was the pollution caused by the huge plumes of smoke from more than 600 oilwell heads deliberately set ablaze by the Iraqis, that thick layers of soot and sludge have been found more than 2,000 miles away in the Himalayas. Saddam Hussein's scorched-earth policy in Kuwait may have far-reaching consequences for the environment and it will be many years before these can be assessed fully as part of the cost of the Gulf War.

The deliberate sabotage and discharge of vast quantities of oil from oil refineries by the Iraqi army created one of the biggest oil slicks ever known to civilization, though the exact quantity of oil released remains uncertain. Initial estimates of 10–11 million barrels, used by environmental groups and politicians, were subsequently downwardly revised to about 2 million barrels. In fact, 50 per cent of the oil spills evaporated within the first 24 hours. The oil slick spread for over 150 km along the coasts of

Kuwait and Saudi Arabia, and had catastrophic effects for the ecosystems of Abu Ali Bay, but did not affect the large oil refinery port of Jubail further south. The effects of pollution were enhanced due to the low tidal range in the Gulf over the extensive intertidal zones or flats, commonly being 2 km wide. It is in these zones, which support vast quantities of life, that much of the early devastation occurred.

Abu Ali Bay has become the focal point for environmental groups. This has been the Mecca for conservationists and environmentalists to wade in oily waters and to present the aftermath of the Gulf War as an ecosystem in crisis. The seabed is smothered with tar that has been raining down through the water column as balls of tar, and has been sterilized of life below the slick, apart from the proliferation of oil-consuming bacteria. The darkest nightmares of pollution have indeed been realized here, but further offshore the coral reefs and fish life, together with the rest of the marine life and fishing catches seem to have survived unscathed. Despite this dark deed, was this really the environmental catastrophe that the media and environmentalists offered to the world? The question was discussed in many political and scientific arenas across the world (e.g., Browning *et al.* 1991 and Bakan *et al.* 1991). After assessing all the evidence, the verdict was that the Gulf War was more of a human than an ecological disaster. Yes, environmental damage has been done, but the amount is, thankfully, much smaller than originally feared. In fact, the Gulf is being polluted, not so much by Saddam Hussein's environmental terrorism, but more by the slow and inexorable long-term spillage of oil through accidents and deliberate acts. In the Gulf, about 1 million barrels of oil per year find their way into the seawater, and three-quarters of Saudi Arabia's beaches were polluted before the Gulf War.

The effect of oil pollution can be disastrous for many ecosystems. Birds, fish, invertebrates, and plant life can, and often do, die in large quantities. The behaviour of oil when spilled into the sea, and its dispersal and degradation, are shown in Figure 5.7. Oil, especially gasoline pollution, affects fish because it is toxic, and it causes suffocation, either by coating the gills, or by coating the surface of the sea and inhibiting the diffusion of oxygen into the water. Oil sinks

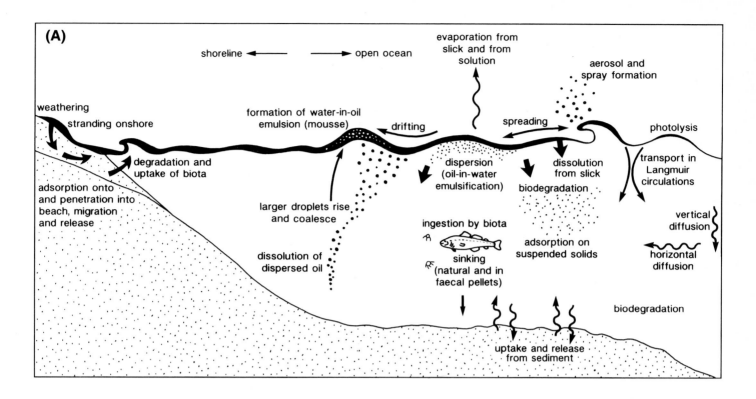

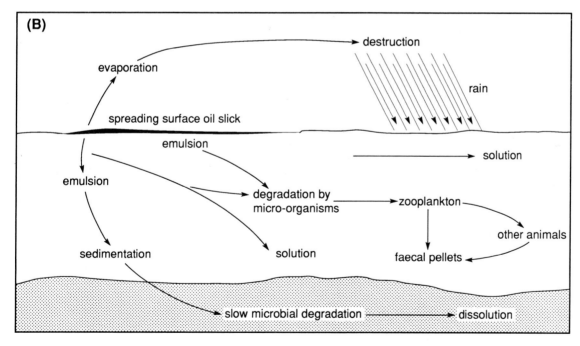

Figure 5.7 *(A) The behaviour of oil released into the sea. Redrawn after Open University (1991). (B) The persistence of pollution in the oceans, its dispersal, and degradation. After Smith and Warr (1991).*

to the bottom of the sea and may immobilize sperm and reduce fertilization rates. Similarly, oil affects birds by coating their feathers which then reduces buoyancy and insulation. Lipid pneumonia and intestinal irritation can result. It may also affect birds' eggs, reducing the permeability of the egg, and cut down the food supply as the fish life dies.

Oil pollution also affects many other organisms, such as plants and invertebrates, all an essential part of food webs and the ecosystem. In the Gulf, oil represents a major threat with the destruction of vast quantities of sea grass in the subtidal zone (only a few metres deep) and algae in the intertidal zone. These plants form the base of important food webs, providing the food for many invertebrates, which in turn provide food for higher organisms. Destruction of these food webs means the death of vast quantities of marine animals, many of which are already under threat, such as the sea cow which grazes on sea grass.

The obvious immediate effects of oil spillage on humans is to poison the marine life, and reduce the consumption of fish and fish yields. Oil is inflammable and explosive. Beaches contaminated with oil are unattractive for recreation, to such an extent that tourism in polluted areas may decline along with investment from tourist revenue. Inland, oil can pollute local water supplies and cause the build-up of various toxins. As an example, the extraction of oil from the oil shales of Estonia is causing serious pollution problems, with local water wells being polluted and large concentrations of heavy metals carried to the Baltic Sea. The technology for this oil production is decades old and needs updating to clean up the area.

Various methods are available to try to constrain the spread of an oil spill and to disperse the oil. These include burning the oil, and using high-capacity pumps to draw off the surface oil, adhesion collectors (mechanical devices that scrape the oil up from the surface of the water), sinking agents (e.g., carbosand to settle the oil on to the sea bed), and emulsifiers and dispersants to break the oil into droplets to increase its surface area and aid bacterial degradation of the oil (Plate 16). Biological agents such as **desulfovibrio** and **desulfomaculum** are also employed, and act by using the oil as a food source.

The short-term effects of oil pollution are easy to observe. The long-term effects on the ecosystems, however, are still not known. Particularly disturbing may be the effect on the benthonic fauna (seafloor-dwelling animals) as oil settles on the ocean floors. These latter risks have not really been assessed.

Beaches

Many of the world's beaches, both in well-known coastal resorts and those which are more remote, are suffering from serious water pollution. Untreated sewage, oil, industrial chemicals, and radioactive waste are all contributors to the pollution.

Untreated sewage is the most common culprit. Many of the beaches in the western Mediterranean are unsightly and represent a health hazard because of sewage. This problem is not confined to Europe. Industrial waste and raw sewage threaten the famous Guanabara Bay beaches beneath the statue of Christ which towers over Rio de Janeiro. The whales which once inhabited the bay have gone and the dolphin population has dwindled.

In Australia, the third largest tourist attraction in the country, Sydney's Bondi Beach, is unsafe for bathing on two out of every five days because of the deposits of raw sewage and industrial waste (Beder 1990). Sydney has three main outfalls and four minor ones for the discharge of sewage. The largest discharges 640 million litres of sewage per day in dry weather, while the outfall on the northern headland of Bondi provides 165 million litres of sewage a day. Immediately north of Sydney harbour, the third outfall contributes 325 million litres of sewage a day. All of this sewage only undergoes a primary treatment process to remove most of the solids. In the US, secondary treatment is mandatory. A 1987 study of the levels of heavy metals and pesticides in fish caught near one of the sewage outfalls (Malabar), and leaked to the *Sydney Morning Herald* (January 1989), found dangerously high levels of organochlorines (such as benzene hexachloride, DDT, and heptachlor epoxide). In eight red morwong that were caught 300 m from the outfall, the average level of benzene hexachloride was 122 times greater than the recommended limit set by the Australian National Health and Medical Research Council.

In Britain, around the nuclear power station of Sellafield, the once-popular beaches on the Irish Sea coast west of the Lake District are contaminated by seawater with high levels of radioactive materials.

Hydropolitics

Water is the most precious resource on Earth, more valuable than gold and diamonds, oil or land. Humans have controlled the course of rivers for thousands of years in order to irrigate land and make it more fertile (Plate 5.4). Its superabundance, however, commonly makes it an undervalued resource, although this is not the case in every country. In countries and regions where there is a scarcity of clean water, water management is a potential source of co-operation or conflict. Water is a commodity to be bargained with, bought and sold, and used to influence policy in neighbouring countries.

In the Middle East, where the climate is arid, rivers are few, population growth is rapid (only exceeded by Africa), nations are already hostile towards one another, and most of the significant rivers flow through more than one country, the conditions are ripe for political tension and military conflagrations over water. Examples of tension over water resources are many. Israel currently taps an aquifer in the West Bank, and has done so since the Six-Day War in 1967. The Palestinians claim that much of this water is transferred to the rest of Israel or used by Israeli settlers in the occupied territory.

In the 1950s, Israel diverted waters from the Sea of Galilee, or Lake Tiberias, to the densely populated coastal strip and further south into the Negev Desert. The result was to divert saline springs into Jordan and so degrade the quality and quantity of useful water flowing into that country. In the 1960s, Jordan planned to dam a tributary of the Jordan river, the Yarmuk, which rises in Syria and flows along part of the border between the two countries, but also forms a border between Israel and Jordan. Israel would not entertain the project without Jordan's guarantee of its right to Yarmuk waters. Shortly afterwards, the Six-Day War ensued, Israel occupied the Golan Heights and the site of the proposed dam, and the Jordanians shelved the scheme. Jordan and Syria now have a new site for a dam and reservoir, but to date international funding has not been forthcoming in the absence of agreement of all the involved parties. Israel has so far declined to support this scheme. Actually, Syria already uses so much of the Yarmuk waters that the Jordanians have a considerably reduced potential water resource (Plate 17).

The Yarmuk-type saga does not end with Israel, Jordan, and Syria. Syria relies upon the waters of the Euphrates river which passes upstream into Turkey. In order to irrigate south-eastern Anatolia, Turkey is currently constructing dams along the course of the Euphrates (such as the Ataturk Dam), and the Tigris. The Euphrates and the Tigris flow through Iraq to the Persian Gulf, so like other countries in the Middle East, it is also concerned about any upstream activities to tap, divert, and store water.

In wars, enormous damage to water supply is a common occurrence. For example, in January 1993 in the current Balkan conflict, the retreating Serbian soldiers blew up the reservoir behind the Peruca Dam in Krajina. Fortunately, the reservoir has drained safely, removing the risk to

Plate 5.4 *Artificial irrigation of the upper reaches of the Indus River in Ladakh, north of the high Himalaya, has created a fertile valley capable of supporting many villages and small towns.*

about 20,000 people inhabiting the valley below. Destroying the enemies' water supplies has a notorious pedigree. About 4,500 years ago in ancient Mesopotamia, along the Euphrates, the king of Umma destroyed the banks of the irrigation canals, unleashing torrents of water on his downstream neighbours at Girsu. In the sixteenth century, William of Orange deliberately flooded large parts of the southern Netherlands to impede the Spanish, and in the seventeenth century the French under Louis XIV were defeated by the Dutch using the same tactics. Today, the World Bank and other funding agencies are increasingly refusing to provide investment for dams in regions of the world where there are extreme political tensions, such as war zones.

On a more optimistic note, some governments are becoming more aware of the need to balance the human demand for water resources with a concern for the effects on the environment. In France, the demand for more water from the Loire is posing a threat to the fragile marshes or wetlands along the course of the river and its tributaries. Along the Mediterranean coastline, since 1945 the Camargue has lost roughly half its wetlands because of drainage and development associated with tourism. In 1980, the French wetlands, covering an area exceeding more than 1.2 million hectares, were designated as being of international importance for wildlife, yet by 1990 more than half this area had been drained or was at risk (Purseglove 1991). There are now more than 20 hydroelectric dams and other barriers along the Rhône between Geneva and Arles, something that has radically altered the vegetation and ecosystems along this mighty river. In France, the growing environmental awareness, and the lessons which have been learnt from the over-zealous development of the major water courses, have actually led to the scrapping of various plans for new dams, for example pressure from conservationists helped to reverse a decision to construct two new dams in the upper reaches of the River Loire. To date, four major projects for the construction of dams along the Loire and its tributaries have been abandoned, partly as a result of lobbying by environmentalists (ibid.).

In England and Wales, the privatization of water for party political reasons, through the Water Act 1989, consolidated into the Water Resources Act 1991 and the Water Industry Act 1991, led to the establishment of 10 water and sewerage companies having responsibility to provide clean water and sewage treatment, together with its disposal. Additionally, there are another 22 companies who provide water only. The Office of Water Services (OFWAT) and the National Rivers Authority (NRA) were created to regulate and monitor the economic and environmental aspects of water, respectively. This privatization has raised many issues, including the fundamental question of whether the responsibility for managing and distributing something as necessary and basic to life should be left in the hands of companies whose first concern is profit, and their shareholders.

These examples of the politics of water management serve to show just how precious and potentially explosive are the issues concerning water. Co-operation and sensible planning between countries can enhance the living standards and quality of life for a whole region. Selfish and thoughtless water management, driven by greed and power-politics, can create regional tensions and even wars.

Conclusions

In conclusion, it can be seen that there are many different ways that water becomes polluted and many types of pollutant. Polluted waters not only affect delicately balanced or fragile ecosystems, but can also have detrimental or catastrophic effects on human health. Today, environmental pollution is becoming more prevalent as populations rise and humans put greater and greater stress on the environment through intensive agricultural and industrial processes that produce large amounts of waste as by-products. Unless such activities are carefully controlled and pollutants more effectively dealt with, the consequences may be disastrous for the whole planet, let alone specific geographical areas.

National and international legislation and public awareness are paramount in the prevention and control of water pollution. It is discouraging to see so many governments neglecting their responsibilities, for example the privatization of the Water Boards in Great

Britain principally for commercial gain, without proper policeable water standards. Judith Cook (1989) discussed the ethics and politics of privatizing the Water Boards in the UK in her book *Dirty Water*. She came to the same conclusion that the Mayor of Birmingham, Joseph Chamberlain, made almost 100 years ago when trying to obtain clean water and sanitation for his city:

'it is difficult and indeed almost impossible to reconcile the rights and interests of the public with the claims of an individual company seeking as its natural and legitimate object, the largest private gain'. On an international scale, it is even more disconcerting to witness just how little effort is being made to combat water pollution.

Key points

CHAPTER 5: KEY POINTS

1 Clean and abundant water is an important resource essential for life. There are water shortages in many countries and human activities have polluted water resources throughout history. Many of the world's beaches are suffering severe pollution, some with dangerously high levels of toxic chemicals. An understanding of the hydrological cycle is important in understanding the routes that pollutants can travel through the ecosphere, particularly through food chains and webs. There are many types of pollutants, both natural and created by human activities.

2 Sewage and sludge is one of the most common pollutants, often being disposed of in an untreated form or raw state in streams and seas. Treatment involves filtering, biological digestion, and drying to produce sewage sludge which is easier to handle and dispose of. Sewage treatment also aims to destroy disease-carrying organisms, such as cholera, typhoid, cryptosporidia, leptospirosis, and giardia. Sewage sludge properly treated can provide a valuable fertilizer, or energy source through incineration to generate electricity. The decay of organic matter produces NH_3 and nitrates which, if in sufficient concentrations, may contaminate the atmosphere and water sources. Manure or treated and diluted sewage sludge can be added to soils as a fertilizer.

3 Nitrates are also produced by traditional and artificial fertilizers, and may be washed into water sources. Nitrates are believed to be responsible for diseases such as methaemoglobinaemia and stomach cancer, and are responsible for the eutrophication of lakes and algal blooms which cause the depletion of free oxygen in standing bodies of water, such as lakes, and the consequent suffocation of aquatic animals. Nitrates may be chemically removed from water by cation exchange, reverse osmosis, electrodialysis, distillation, and biological processes, but these are expensive to implement.

4 Dangerous organic chemicals (chlorinated hydrocarbons, notably PCBs, DDT, and TBT) are manufactured for paints, plastics, adhesives, hydraulic fluids, electrical components, defoliants, pesticides, and anti-fouling paints, and the waste from such industrial processes is commonly dumped in water courses. Such chemicals take a long time to biodegrade or break down, so they concentrate in food chains and poison animals and plants.

5 Heavy metals (principally mercury, lead, arsenic, selenium, cobalt, copper, and magnesium) are naturally produced by weathering, but industrial processes discharge large quantities into water courses which may reach toxic levels. The metals become concentrated at high trophic levels within food webs, and may cause serious effects to life, such as brain damage and even death in animals. Lead piping for water resources, and the use of aluminium in cooking utensils, may further concentrate these metals in the human body. Aluminium may be a contributory factor in causing Alzheimer's disease and Guam disease. The sources of heavy metal pollutants can be traced using isotopic methods, and this may provide an effective means of tracing metal pollution to specific polluters.

6 Radioactive waste poses a very serious threat to health, and bequeaths a long-term legacy.

7 Oil pollution is a major environmental problem, and results mainly from shipping accidents, offshore oil exploration, oil droplets from unburnt fuel, combustion engines on land, deliberate discharges from ships (particularly during tank cleaning operations), natural seepages from the seabed, and ecological terrorism. Oil pollution affects many ecosystems, poisoning organisms by coating fish gills leading to suffocation, immobilizing fish sperm, coating birds' feathers and thus impeding flight and reducing insulation, leading to the death of birds and aquatic mammals such as otters,

reducing the permeability of birds' eggs, causing lipid pneumonia in mammals, seriously disrupting food webs, and devastating local ecosystems. Clean-up technologies include containing the spread of oil by using booms, the dispersal and break-up of oil slicks using detergents, combus- tion, collecting the oil, and bacterial degradation.

8 The importance of water as a valuable resource causes international and regional conflicts. Governments are becoming increasingly aware of the need for long-term agreements between nations that must share common water resources.

Chapter 5: Further reading

Gourlay, K.A. 1988. Poisoners of the Seas. London/ New Jersey: Zed Books Ltd.
An interesting book about the pollution of the seas in eight chapters: the oceans and how they work; humanity and the seas; the black death – oil; in the shit – sewage; toxic technology; hazardous chemicals and heavy metals; the future in danger – radioactivity; portents and possibilities.

Hinrichsen, D. 1990. *Our Common Seas: Coasts in Crisis.* London: Earthscan Publications.
This book is based on UNEP data and describes the growing pressures on coastal ecosystems throughout the world. Case studies are provided to illustrate the local successes in protecting marine and coastal environments.

Open University 1991. *Case Studies in Oceanography and Marine Affairs.* Oxford: Pergamon Press.
A well-illustrated textbook produced to support an Open University course on oceanography. It examines marine resources and activities, and the development and nature of international conventions on the sea, and provides case studies on the Arctic Ocean and the Galapagos Islands. This book emphasizes the complex interactions between the political, economic, and environmental aspects of development within the marine realm.

Price, M. 1985. *Introducing Groundwater.* London: Chapman & Hall.
An excellent introductory text to hydrogeological/ hydrological aspects of groundwater. The book is written for non-specialists and uses minimal technical and mathematical formulae. There are thirteen chapters including: water underground; water circulation; caverns and capillaries; soil water; groundwater in motion; more about aquifers; springs and rivers; deserts and droughts; water wells; measurements and models; water quality; groundwater, friend or foe?; some current problems.

Be advis'd;
Heat not a furnace for your foe so hot
That it do singe yourself. We may outrun
By violent swiftness that which we run at,
And lose by overrunning.

Spoken by Duke of Norfolk.
William Shakespeare,
All is True (Henry VIII), Act I, Scene i

Death or salvation?

The nuclear age has promised both death and salvation. Today, in the Nuclear and Space Age, humankind has come a long way, technologically, from the heady early days when the atom was split. Many of the physicists and other scientists involved in the early atomic experiments wanted to believe in 'atoms for peace'. At its inception, nuclear power offered an endless supply of clean and cheap energy, but also destruction and fear.

Military strategists, however, saw things differently. Nuclear weapons would be the ultimate deterrent in a brave new world of **Mutually Assured Destruction (MAD)**. Nuclear weapons would keep the peace and if conflagrations occurred, then they would ensure the speedy restoration of the military superiority of the superpowers. Thus, through the late 1940s and 1950s, post-War reconstructionist, 'we've never had it so good' USA and Europe, the early days of atomic research seemed to offer something for everyone.

Despite the mood of optimism about the dawning of a Nuclear Age, the horrors of Hiroshima and Nagasaki had made their mark. A vocal minority of public opinion foresaw the problems that a nuclear future would bring to humankind. They believed that nuclear weapons should never be used, that the consequences of using nuclear weapons could never be justified,

and that we should 'Ban the Bomb'. So, by the start of the 1960s, there was open public debate over the morality of nuclear weapons, and today there is also the ongoing debate over the financial issues of a nuclear arms programme. Table 6.1 gives the military and education expenditures for selected countries in 1987, and Figure 6.1 shows the money spent in 1990 per person on defence in NATO countries. There are many people who would argue that in the face of a world recession and the need for greater social provision and life chances for a country's citizens, these levels of expenditure are far too high. The debate is complicated and there are many vested interests involved in the arguments for or against high levels of expenditure on nuclear arms.

In the early days, nuclear waste disposal and accidents at nuclear power stations were perceived as much less of an issue. It only became common currency in the media in the 1970s with the dawning of the 'environmentally conscious era'. Now, the two big nuclear issues are whether or not nuclear energy should be used, and concern over the production of nuclear waste and its disposal, as well as concern over the morality of nuclear weapons, the arms race and the verification of treaties to police nuclear arms. These issues are explored in this chapter, although nuclear energy as a fuel is dealt with in Chapter 7.

Table 6.1 *Military and education expenditure in selected countries.*

	Military expenditure (1987)		Educational expenditure (latest year obtainable: % GNP or GDP)	Ratio of military to educational expenditures	Literacy rate (%)
	$ billion	% GNP or GDP			
Australia	4.99	2.5	5.6	0.45	99.5
Bangladesh	0.32	1.8	2.1	0.86	33.1
Botswana	0.02	2.2	6.0	0.37	70.8
Bulgaria	6.66	10.3	7.1	1.45	95.5
Burkina Faso	0.05	3.1	2.5	1.24	13.2
Canada	8.84	2.2	7.4	0.30	95.6
China	20.66	4.4	2.7	1.63	72.6
Costa Rica	0.02	0.6	5.2	0.11	92.6
Cuba	1.60	5.4	6.3	0.86	96.0
Egypt	6.53	9.2	5.5	1.67	44.9
Ethiopia	0.44	8.5	3.9	2.18	3.7
France	34.83	4.0	6.1	0.66	98.8
Guyana	0.34	8.9	10.1	0.88	95.9
Iran	21.12	7.9	3.8	2.08	61.8
Iraq	16.70	30.7	3.8	8.08	45.9
Japan	24.32	1.0	5.1	0.20	100.0
Lesotho	0.01	2.3	3.5	0.66	73.6
Mauritania	0.04	4.2	7.9	0.53	28.0
Mozambique	0.10	8.4	1.2	7.00	16.6
Nigeria	0.18	0.8	1.8	0.44	42.4
Pakistan	2.23	6.5	2.1	3.10	25.6
Peru	2.20	4.9	2.9	1.69	87.0
Senegal	0.10	2.2	4.7	0.47	22.5
South Africa	3.40	4.4	3.8	1.15	79.3
Tanzania	0.08	3.3	4.3	0.77	85.0
USSR	303.00	12.3	7.0	1.76	99.0
UK	31.58	4.7	5.2	0.90	100.0
United States	396.20	6.5	7.5	0.87	95.5
Venezuela	1.38	3.6	6.8	0.53	89.6
Zambia	0.17	6.6	5.4	1.22	68.6
Zimbabwe	0.28	5.0	7.9	0.63	76.0

Literacy is differently defined in various countries. Thus UK and Japanese figures reflect all considered capable of reading, Canadian figure excludes functionally illiterate.
GNP, GDP = gross national/domestic product
Source: Encyclopedia Britannica Yearbook (1990).

Historical perspective

The dawning of the Nuclear Age can be considered to have occurred in 1919 when Ernest Rutherford and his team, at Cambridge University in the UK, smashed the atom, to convert a nucleus of nitrogen into a nucleus of hydrogen (eight years after Rutherford had discovered the atomic nucleus). On 19th December 1938 in Berlin, at the Kaiser Wilhelm Institute, two nuclear chemists, Otto Hahn and Fritz Strassman, split the uranium atom into roughly two equal parts and recognized that **nuclear fission** had occurred. Actually in 1934, two years earlier, Fermi, working with Frederic and Irene Joilet-Curie, had created artificial radioactivity by bombarding a target with **alpha particles**, but they were not aware of the full significance of their discovery.

Research into nuclear physics and chemistry was given its real momentum on 9th October 1941, when President Roosevelt sanctioned the US construction of an atomic bomb. In 1945, it was President Harry S. Truman who made the fateful decision to use the atomic bomb against Japan. Little Boy was detonated at an altitude of just under 2,000 ft over Hiroshima on 6th August and over 70,000 people died, whilst Fat Man was dropped on Nagasaki on 9th August 1945 causing around 35,000 citizen deaths. Only weeks earlier, on 16th July 1945 in the New Mexico desert near Alamogordo, the first American test of a nuclear bomb, code-named Trinity, took place under the direction of Robert Oppenheimer (see Newhouse 1989).

Before considering various other aspects of nuclear issues, it is worth briefly considering the history of nuclear arms treaties between the superpowers. The progenitor of the succession of treaties was the 'Atoms for Peace' speech by the US President Eisenhower on 8th December 1953, and the growth in 1957 of the International Atomic Energy Agency (IAEA) committed to promoting the peaceful use of nuclear energy and preventing its destructive use. Box 6.1 summarizes the principal international treaties which chart the progress of nuclear arms negotiations.

To date, six nations have openly tested nuclear weapons – the USA, the former Soviet Union, Britain, China, France, and India (Plates 18 and 19) – and they are currently observing a moratorium on nuclear weapons with the exception of China. Of these nations, the USA, Britain, China, France, and the Russian states of Belarus, Kazakhstan, and the Ukraine are declared nuclear weapon states with the capability of delivering nuclear weapons, whereas of the undeclared nuclear weapon states Israel, India, and Pakistan, only Israel is known to possess deliverable weapons. Algeria, Iran, Iraq, Libya, North Korea, and Syria are known to be working on obtaining nuclear weapons, whilst Argentina, Brazil, South Africa, South Korea, and Taiwan are believed to have ceased developing a nuclear capability. States who probably possess or will soon have a nuclear capability are often referred to as 'threshold nations'. While the nuclear nations promote non-proliferation of nuclear weapons, many of the threshold nations are still moving to join the nuclear club, mainly because of the perceived threat to national security from neighbouring states. The disparity in nuclear capability between nations has created two diametrically opposed philosophies. One is that peace will be achieved through the containment

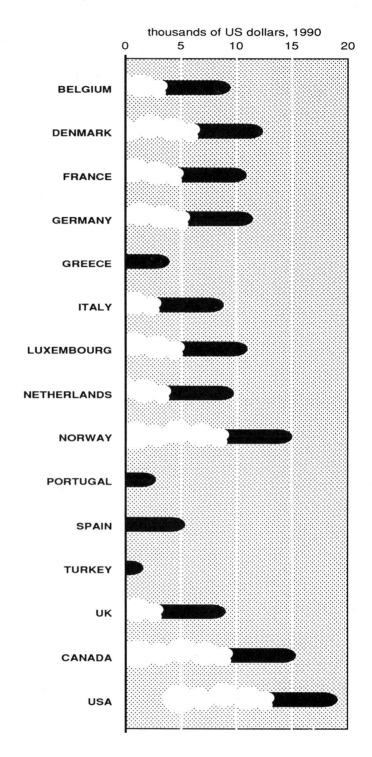

Figure 6.1 *Comparisons between the military expenditure per capita of the top 15 developed countries. From the* Environmental Guardian *(11 February 1992).*

163

of nuclear weapons and non-proliferation. The other is that peace will come first through nuclear proliferation in the developing countries and later through denuclearization by every nation.

Threshold nations complain that the IAEA treats them differently from the nuclear nations by imposing stricter verification procedures such as on-site inspections, seismic and satellite monitoring, perimeter monitoring, and checking for the diversification of nuclear fuels. Most inspectors come from the nuclear nations, and it is alleged that they may connive to cheat by turning a blind eye to the movement of prohibited nuclear materials between nuclear states. More than 4 million tonnes of **plutonium** appear to have gone missing from the British nuclear industry, apparently to the US weapons programme (Hassard 1992). It is issues such as these that generate a distrust by the threshold nations of the nuclear states such as India who have refused to sign the NPT because they perceive it as merely maintaining the nuclear *status quo*. There is little indication that the

world is addressing these issues head on, and without clear international agreements between nuclear and non-nuclear nations, the problems are being shelved for the near and intermediate-term future. In its attempts to contain the threat to world peace caused by a spread in the number of states with a nuclear weapons capability, the USA and its allies cannot guarantee the current national integrity of Iran against Iraq, India against China, the Arab states against Israel, North Korea against South Korea, or Pakistan against India, and it is tensions such as these which fuel the scramble to acquire nuclear weapons. Proliferation cannot be stopped, but international treaties can slow the pace of growth and contain the time when many more nations will have joined the nuclear club.

Radioactive fallout from nuclear tests

Atmospheric nuclear testing between 1945 and

Box 6.1

International nuclear arms agreements

In 1963, the Limited Test Ban Treaty (LTBT) was signed by President John F. Kennedy for the USA and the Soviet Premier Nikita Khrushchev for the Soviet Union, and banned nuclear tests in the atmosphere, under water or in space. All tests have been under ground since 1972. Kennedy had ideally wanted a Comprehensive Test Ban Treaty (CTBT) to stop the testing of all nuclear weapons, but the LTBT represented the best compromise. Indeed, Kennedy believed it was the greatest achievement of his Presidency.

In 1968, the Non-Proliferation Treaty (NPT) was proposed on 1st July in Geneva and signed by more than 100 states (notably, unsigned at the time by the French). Those states with nuclear weapons undertook to make every effort to rid themselves of their nuclear weapons and to provide access to peaceful nuclear energy for the non-nuclear signatory states, the latter agreeing not to attempt to acquire nuclear weapons. The NPT also banned the superpowers from helping others to acquire a nuclear capability. This was a somewhat asymmetric agreement between the nuclear and non-nuclear states. China refused to sign, and is believed subsequently to have traded nuclear secrets with South Africa, Pakistan, Iraq, North Korea, and Algeria. The NPT was ratified

in 1970 by Britain, the USA, the Soviet Union, and 40 non-nuclear states. Today, more than 140 nations have signed the NPT, with recent signatories being France, China, and South Africa, and significant non-signers including Argentina, Brazil, India, Iran, and Pakistan.

In 1972, the Strategic Arms Limitation Treaty I (SALT I) was signed by the US President Richard Nixon and the Soviet Premier Leonid Brezhnev in Moscow on 26th May. The treaty, in two parts, comprised an Anti-Ballistic Missile (ABM) treaty limiting both superpowers to just two ABM sites, further reduced to a single site each under an agreement in 1974, together with an interim offensive arms agreement to set ceilings on the deployment of certain nuclear weapons. Amongst these weapons, it was agreed that the Soviet Union could deploy 740 submarine-launched ballistic missiles (SLBMs) on 62 submarines, with the US being restricted to 710 SLBMs on 44 submarines. It was also agreed that no additional intercontinental ballistic missile (ICBM) launchers could be added to those that were operational by 1st July 1972, SLBMs were limited to those operational in May 1972, new 'heavy' ICBMs could not be deployed, and future modernization and replacement was

1980 was responsible for putting a large, but still unquantified amount of nuclear fallout into the atmosphere (Plate 18). The Partial Test Ban Treaty of 1963 led to the cessation of most of the testing, but some non-signatory countries, such as France, continued testing until 1980 and beyond. These continued tests deposited large quantities of additional **radionucleides** such as strontium-90 (^{90}Sr) and caesium-137 (^{137}Cs) which found their way, via precipitation, into the hydrosphere (see Chapter 5).

Radioactive waste is very much a legacy of the twentieth century bequeathed to future generations. Table 6.2 lists the typical yields of the so-called 'actinide group' of chemical elements, in curies, together with their typical yields of fission products. The main radioactive elements that are involved in polluting the environment and causing ill health and death are the isotopes of iodine, strontium, caesium, and ruthenium. Iodine accumulates in the thyroid gland. Strontium, which is chemically similar to calcium, is absorbed through the walls of the intestine and collects in the bones. Caesium behaves chem-

ically in a similar manner to potassium and, therefore, can be distributed throughout the body in much the same way. Ruthenium has no chemical analogue with a biological function.

The effects from radioactive iodine can be ameliorated by saturating the thyroid gland with safe iodine through iodine tablets. Radioactive iodine has a short **half-life** and will quickly decay to safe levels soon after the event. Isotopes of strontium and caesium, however, are widely distributed throughout the body and their accumulation cannot be controlled by the aid of tablets. Also their half-life is long (40 days to 30 years) which results in prolonged effects.

Several sources are responsible for generating radioactive pollutants. One natural source of radiation which leaks into the atmosphere is from rocks and sediments rich in radioactive elements. The following section deals in particular with one of the most common naturally occurring radioactive gases, radon.

---

restricted to a one-for-one basis. According to the memoirs of Dr Henry Kissinger (1982), the then Secretary of State under Nixon, it was the land-based ICBMs, known as the SS-7s, that the US was most eager to get rid of. The SS-7s had a range of over 6,000 miles and carried warheads of 6 megatons, 70 of which were reasonably invulnerable in hardened underground silos.

Also, in 1972, a Biological Weapons Convention was signed by the superpowers to ban biological weapons and order the destruction of existing stockpiles.

In 1974, the Threshold Test Ban Treaty (TTBT) was signed as an agreement between the USA and the former Soviet Union not to carry out nuclear tests of more than 150 kilotons explosive yield, equivalent to 150 tonnes of TNT or 10 times the Hiroshima bomb. Despite the treaty, both sides continued to deploy warheads with greater kilotonage, such as the US Minuteman 3 and Peacekeeper missiles with 350 kiloton warheads and the former Soviet ICBMs such as the 550 kiloton SS-19s and the 3.6 megaton SS-17s.

In 1976, the Peaceful Nuclear Explosions Treaty (PNET) was formulated between the former Soviet Union and the USA. In essence, this treaty came about because the superpowers did not

want a loophole in any existing international nuclear arms control agreements, whereby it would be possible for either side to make a military device as a peaceful explosive, for example the theoretical use of nuclear devices in the construction of canals or to divert waterways (ibid.). Detailed verification procedures formed an important part of this treaty, for example, under this treaty for the first time the Soviets allowed on-site inspection. Because the treaty was more concerned with technical details than earlier treaties, it generated relatively little debate and rather than attract the bitter animosity and discussion that was associated with the SALT agreement, it attracted media indifference. In fact, the treaty was never ratified by the US Senate.

In 1979, the Strategic Arms Limitation Treaty II (SALT II) was signed by the US President Jimmy Carter and the Soviet Premier Leonid Brezhnev to limit the strategic missiles and bombers deployed by both parties to a maximum of 2,400. Certain restrictions were to apply within this broad framework. Weapons equipped with multiple independently targeted re-entry vehicle (MIRV) warheads or air-launched cruise missiles were limited to a maximum of 1,320. Either side could deploy no more than one new ICBM before

Box 6.1

International nuclear arms agreements

Plate 6.1 *The 1986 summit meeting between the US and Soviet leaders, Reagan and Gorbachev, took place in this building, Höfdi, in Reykjavik, Iceland.*

Radon gas

Radon gas (see Box 6.2) is a major component of the **background radiation** dose received by populations in certain geographic areas, for example it is the biggest contributor to radiation exposure in Britain where, from the total average annual radiation dose to the population of 2.5 millisieverts (mSv), approximately 1.2 mSv comes from radon. In total, radon contributes approximately 50 per cent of the environmental radiation compared to 12 per cent from medical sources and only 0.1 per cent due to anthropogenic nuclear discharges. Since the 1960s, studies such as that undertaken by the National Radiological Protection Board (NRPB) in the UK have drawn tentative links between high incidences of cancers (e.g., lung cancer and leukaemia) and high concentrations of household radon levels. The NRPB has calculated that as many as 2,000 deaths per year in the UK may be the result of the radiation produced by radon,

Box 6.1

International nuclear arms agreements

December 1985. ICBMs could not carry more than 10 warheads per missile and a ceiling of 1,200 ballistic missiles carrying MIRV warheads was imposed.

On 8th December 1986, the Intermediate Nuclear Forces (INF) Treaty was signed by the US President Ronald Reagan and the Soviet Premier Mikhail Gorbachev to eliminate an entire class of intermediate-range nuclear weapons (Plate 6.1). It was agreed to remove and destroy 1,286 nuclear missiles from Europe and Asia, containing more than 2,000 warheads. The treaty involved the most stringent and comprehensive verification procedures to date in any arms control agreement.

In 1991, the START Treaty (Strategic Arms Reduction Treaty) was signed on 31st July in Moscow, by Presidents George Bush for the USA and Mikhail Gorbachev for the Soviet Union. This treaty was the result of nine years of diplomatic groundwork. The treaty was designed to reduce the Soviet and US strategic nuclear weapons by around 35 per cent on each side over a seven-year period. It would still leave the Soviets with 7,000 warheads and the Americans with 9,000, and the treaty did not address the ground-based multiple-warhead missiles or sea-launched cruise missiles. So, while undeniably releasing some of the pressure for over-arming on the part of the superpowers, the treaty may well be viewed in the fullness of time as a major PR exercise rather than a significant and irreversible reduction in the nuclear capability of either side. On 27th September 1991, President George Bush announced that the US would destroy a further 3,050 nuclear weapons, many of which are currently based in Europe or at sea.

Also in 1991, on 5th October President Mikhail Gorbachev announced the Soviet Union's offer on disarmament, in which he offered to match the sweeping cuts in nuclear weapons announced by President George Bush for the USA the previous week. The Soviet offer amounted to the most far-reaching reductions in the nuclear arsenal since the end of the Cold War, and went much further than originally embodied in the START Treaty signed in June of 1991. The October proposal envisaged a 1998 target figure of 5,000 long-range strategic nuclear warheads instead of the 6,000 proposed in the June treaty. Gorbachev also proposed that the USA and the Soviet Union enter immediate negotiations on additional radical cuts of around 50 per cent in strategic offensive weapons. As with the US proposal, the Soviets would effect a complete elimination of all former Soviet nuclear artillery shells and warheads for tactical missiles, and remove all tactical nuclear weapons from submarines and ships. In addition, the Soviets would remove all heavy bombers armed with

and under certain circumstances the increased risk of contracting lung cancer may be comparable to those risks faced by heavy smokers. Epidemiologists, however, argue that a direct link between radon and an increased risk of contracting cancer cannot be proved statistically.

There are many factors controlling the rate of migration and the concentration of radon in the human environment. Concentrations of atmospheric radon are generally highest in regions comprising rocks which contain the parent sources, namely uranium and thorium, and have effective pathways for the migration of the gas, for example granites and some sedimentary rocks such as certain types of shales and ironstones, particularly where there are **phosphate nodules**. The concentrations, however, cannot be directly correlated with the concentrations of uranium and thorium in rock because the uranium is present in different host minerals, some of which are more easily weathered, and so release higher concentrations of radon more effectively. The

rate at which radon passes through the ground is important in controlling the concentration of radon at the surface or within the soil. Rocks with higher permeability (interconnectedness of cavities), such as highly fissured and faulted rocks, allow radon, in the form of gas or dissolved in groundwater, to migrate at a faster rate into the human environment. Some of the highest concentrations of radon are often associated with springs rather than known uranium-rich rocks. This is because groundwater pressure decreases when it comes to the surface allowing the release of dissolved gases such as radon from the contaminated water.

During rain storms, pore spaces within soils and rocks fill with water and prohibit the migration of radon and allow its build-up within the soil, and when the ground dries radon is released in high concentrations. Diurnal variations may occur because dew inhibits the release of radon from the soil at night and in the early morning, while during the day radon is released as the soil

Box 6.1
International nuclear arms agreements

nuclear weapons from alert status, order a one-year moratorium on nuclear tests, and cut 700,000 jobs in their 4 million-strong army. Gorbachev pledged that all the nuclear artillery munitions removed from tactical missiles would be destroyed, together with at least some from the submarines and ships. He also announced that there would be moves to stop work on modified short-range missiles for heavy bombers and a mobile small ICBM; six nuclear missile submarines, with a total of 92 launchers, would be removed from active service; more than 500 ICBMs, including 134 with MIRVs, would be stepped down from day-to-day alert, and the plans to construct new launchers for ICBMs on railcars would be scrapped, with the missiles being put into cold storage. The offer remains under discussion by the superpowers.

In 1992, the START 2 Treaty was signed on Sunday 3rd January in Moscow, between Presidents George Bush for the USA and Boris Yeltsin for Russia. This treaty would in effect deny both sides a first-strike capability by eliminating about 15,000 warheads by the year 2003, leaving America with 3,500 and Russia with 3,000 warheads. Amongst the terms of the treaty, Russian SS-18 missiles are to be scrapped, and American B1 and B52 bombers converted to conventional use. The goals to be reached by the year 2003 include the following: bomber

warheads: 750 Russian and 1,250 American; submarine-launched warheads: 1,750 Russian and 1,750 American; land-based missile warheads: 500 Russian and 500 American; the elimination of all land-based multiple-warhead ICBMs. The implementation of START 2 is to be achieved in two stages over ten years.

START 2 cannot take effect until START 1 has been ratified by the three former Soviet republics of Ukraine, Belorussia, and Kazakhstan (Ukraine has also not yet signed the non-proliferation treaty because it claims that the western funds towards helping to destroy their 176 long-range missiles are inadequate).

Postscript to Box 6.1

In January 1994, the Ukraine entered into negotiations with the US to destroy its nuclear arsenal in exchange for US $12 billion. The deal, signed in Moscow by the US President Bill Clinton and the Ukrainian President Leonid Kravchuk, means that the Ukraine will dismantle the 1,200 nuclear warheads on its SS-24 and SS-19 missiles that were left in its territory when the Soviet Union collapsed. The missiles will be shipped to Russia for destruction, and the enriched uranium reprocessed in the USA. The deal requires ratification by the Ukrainian Parliament which has yet to be done.

Table 6.2 *Typical yields of (A) actinides,* (B) nuclear fission products†.*

(A)

Element	Half-life (years)	Decay interval**									
		0.3 yr		10 yr		500 yr		10,000 yr		100,000 yr	
		(g)	(Ci)	(g)	(Ci)	(g)	(Ci)	(g)	(Ci)	(g)	(Ci)
Neptunium-237 (^{237}Np)	2.1×10^6	760	0.59	760	0.59	786	0.61	810	0.63	790	0.61
Plutonium-238 (^{238}Pu)	86	5.8	105	5.5	100	0.1	1.8	–	–	–	–
Plutonium-239 (^{239}Pu)	24,400	27.5	1.7	27.5	1.7	32	2.0	59	3.6	4.5	0.3
Plutonium-240 (^{240}Pu)	6,580	8.5	2.0	19.2	4.5	38.4	8.8	13.9	3.2	–	–
Plutonium-241 (^{241}Pu)	13.2	4	464	2.4	273	–	–	–	–	–	–
Plutonium-242 (^{242}Pu)	379,000	2	0.009	2	0.009	2	0.009	2	0.009	1.7	0.007
Americium-241 (^{241}Am)	462	54	189	55.6	198	29.5	103	–	–	–	–
Americium-243 (^{243}Am)	7,370	82	17.0	8.2	17.0	77	16.0	31	6.5	–	0.001
Curium-244 (^{244}Cm)	17.6	30	2,570	19.7	1,700	–	–	–	–	–	–
Total grams/tonne fuel		974		974		965		916	100	796	50
Actinides,* including daughter products (approx.)			200,000		10,000		900				

*Actinides = series of chemical elements beginning with actinium-89, and continuing to lawrencium-103. Includes intermediate and long half-lives in the waste stream from processing 1 tonne of light water reactor fuel irradiated to 33×10^9 watt-days (thermal) per tonne of fuel
**g = grams per tonne of fuel; Ci = curies

(B)

Element	Half-life	Curies remaining			
		10 yr	100 yr	500 yr	1,000 yr
^{144}Ce/^{144}Pr	285 days	300	–	–	–
^{106}Ru/^{106}Rh	367 days	1,100	–	–	–
^{155}Eu	1.8 yr	160	–	–	–
^{134}Cs	2.1 yr	8,300	–	–	–
^{125}Sb/^{125}Te	2.7 yr	980	–	–	–
^{90}Sr/^{90}Y	28.1 yr	1.2×10^5	1.32×10^4	0.6	–
^{137}Cs/^{137}Ba	30 yr	1.6×10^5	2.1×10^4	2	–
^{151}Sm	90 yr	1,100	520	30	0.4
^{99}Tc	2.1×10^5 yr	15	15	15	15
^{93}Zr	9×10^5 yr	3.7	3.7	3.7	3.7
^{135}Cs	2×10^6 yr	1.7	1.7	1.7	1.7
^{107}Pd	7×10^6 yr	0.013	0.013	0.013	0.013
^{128}I	17×10^6 yr	0.025	0.025	0.025	0.025
Total curies (approx.)		300,000	35,000	53	22

†Includes intermediate and long half-lives from processing 1 tonne of light water reactor fuel irradiated to 33×10^9 watt-days (thermal) per tonne of fuel

Chemical element symbols: Ce = cerium; Pr = protactinium; Ru = ruthenium; Rh = rhodium; Eu = europium; Cs = caesium; Sb = antimony; Te = tellurium; Sr = strontium; Y = yttrium; Ba = barium; Sm = samarium; Tc = technetium; Zr = zirconium; Pd = palladium; I = iodine
Source: Cassedy and Grossman (1990).

Plate 18 *The nuclear mushroom cloud produced during a US nuclear test on Mururoa atoll, Pacific Ocean.*
Courtesy of Jean Gaumy/ Magnum.

Plate 19   *Craters produced by underground nuclear explosions at the Nevada Test Sites, USA.*
Courtesy of Rex Features.

Plate 20   *The damaged reactor number 4, the Sarcophagus, at Chernobyl following the accident in 1986.*
Courtesy of Rex Features.

Plate 21 *A US Trident II missile being fired from a submarine.* Courtesy of Rex Features.

dries out. Also, reduced atmospheric pressure during a cyclonic depression increases the soil–atmosphere pressure gradient which may result in an increase in the rate of release of radon from the soil and help concentrate it in the atmosphere (Miller and Ball 1969, Miller and Ostle 1973). These factors contribute to complex variations in atmospheric radon concentrations.

Atmospheric radon, however, is easily dispersed by the wind; it is only when its dispersal is inhibited that it becomes a hazard. The most common way in which it is concentrated is by becoming trapped within a building, particularly houses. Figure 6.2 shows the main ways that radon can enter and is concentrated in the home. The majority of the radon enters through the floor from the ground and accumulates beneath houses in cavities where there are gaps between the floor and the ground. The concentration of radon within a building is aided by the indoor lower pressures compared to the ambient atmospheric pressure, creating a pressure gradient which drags air into the building and is greatest when buildings have chimneys. Radon may also enter the house by the release of dissolved radon in shower and bath water, and from walls which have been constructed from rocks such as granite or gypsum board which have high uranium concentrations.

Insulated houses (i.e., double-glazed, draught-proofed) impede the escape of radon and hence may have relatively high concentrations. The variation of radon concentrations depends on human activity. Levels of radon vary annually, the highest concentrations being between November and March when the home is kept warm and ventilation is reduced. Daily variations are very common, with the highest concentrations at night when people are in bed and the radon is trapped, and lower concentrations in the morning and evening when people open and close doors and windows thereby releasing the radon into the atmosphere.

The average amount of radon in household air in Britain is 20.5 Bq/m^3 (1 becquerel or Bq is equivalent to 1 atomic disintegration per second), but the NRPB found levels more than 200 Bq/m^3. In the UK, the Radiation Regulations 1985 dictate that annual radon doses should not exceed 50 mSv, equivalent to a radon level of 1,000 Bq/m^3. Areas where there is a probability of more than 1 per cent of the households having levels above 200 Bq/m^3 have been designated as 'Affected Areas'. Radiation levels in some Affected Areas may exceed 8,000 Bq/m^3. The acceptable safe concentrations of radon are difficult to assess because the relationships between radon exposure and the incidence of lung cancer are based on studies of workers in uranium mines who have received their doses in a totally different environmental setting. Also, it is difficult to assess and measure the concentra-

---

Radon (Rn) is a naturally occurring colourless, odourless, and tasteless radioactive gas produced by the decay of uranium (U) and thorium (Th) in rocks and soils. There are three main isotopes (radon-219, radon-220, and the most important in terms of its effects on humans, radon-222). Exposure to radon gas increases the risk of lung cancer because once inhaled the gas sits in the respiratory tract and lungs. ^{222}Rn has a half-life of 3.82 days and is produced in the decay chain starting with uranium-238 (^{238}U). ^{220}Rn (thoron) which has a half-life of 55.3 seconds is produced by a decay series of ^{232}Th; and ^{219}Rn (actinon) with a half-life of 3.92 seconds is formed by the decay of ^{235}U. With a longer half-life, ^{222}Rn has more opportunity to migrate from rocks and soils into the human environment, in homes and places of work. Alpha particles are formed and released into the environment during many of the disintegrations involved in the decay series of ^{238}U

through radium-226 and finally to the stable daughter isotope lead-206 (^{206}Pb). Alpha particles can cause tissue damage, but as these particles are relatively large and possess a high electrical charge, they do not easily travel through clothing or skin. They may, however, enter the body in drinking water and can be inhaled during respiration. Since ^{206}Rn has a half-life which is appreciably longer than the time required for clearance of gas from the respiratory tract, few alpha particles are produced while the radon is in the lungs and thus few alpha particles enter the body. The daughter products, however, which include polonium-219 (half-life 3.05 minutes), lead-214 (half-life 26.8 minutes), and bismuth-214 (half-life 19.7 minutes) have shorter half-lives and are more important in irradiating the lungs and contaminating drinking water, thereby increasing the risk of lung cancer.

Box 6.2
Radon

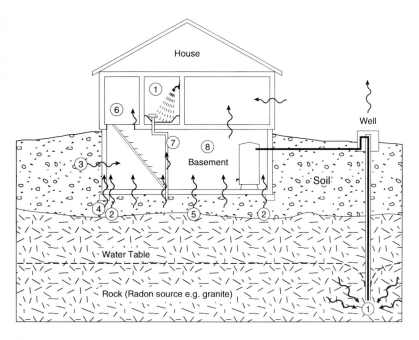

Figure 6.2 *Ways in which radon enters homes: 1 = radon enters groundwater and is used in the house for bathing and drinking; 2 = through construction joints; 3 = through cracks in the walls; 4 = through cavities in the walls; 5 = through cracks in solid floors; 6 = through cracks in suspended floors; 7 = through service pipes; 8 = radon concentrating in basement.*

campaign to encourage householders to reduce radon levels in their homes by altering their construction and design, part of which included a document produced by the Department of the Environment entitled *The Householders' Guide to Radon* (1990). Recommendations to help reduce radon levels include regularly ventilating the home, sealing the floor and depressurizing spaces beneath ground or basement floors. Governments are becoming more concerned with the likelihood of the build-up of radon within households, and surveys are under way in many developed countries to identify areas of high radon concentrations and to assess the acceptable levels of radon for households.

tions of household radon because of the annual and diurnal variations, discussed above. In Britain, the NRPB have measured radon levels using a track-etch method comprising a plastic such as cellulose nitrate which is sensitive to the passage of alpha particles. With the aid of such measurements and good public co-operation, the NRPB published its first assessment of a potentially Affected Area in 1990 which included the counties of Devon and Cornwall. No areas in Cornwall and Devon had concentrations below the 1 per cent probability of homes being above the recommended level (also referred to as the 'Action Level') of 200 Bq/m^3, giving a 3 per cent lifetime risk factor of a fatal cancer. This Action Level represented a compromise by the British government between setting very low risk levels for cancer versus costing the remedial work which would be necessary to modify existing houses and buildings if an even safer radiation dose level were fixed. At the set level, the NRPB estimated that up to 100,000 homes in Britain (60,000 in south-west England alone where there are large areas of granite basement) may exceed the safety limits. From the results of this study, the British government has launched a

Radioactive waste

Radioactive or nuclear waste is produced from the production of nuclear weapons, in electricity generation using nuclear power, and in smaller quantities from medical practice, some industrial activities, and certain types of scientific research (obviously, mainly but not exclusively from nuclear research).

The range of activities associated with the operation of civil nuclear reactors, from uranium extraction, enrichment, fuel fabrication, reactor operation, spent-fuel storage, spent-fuel reprocessing, and other aspects of waste management, are referred to as the nuclear fuel cycle (NFC). The decommissioning of commercial nuclear reactors is not usually assumed to be part of the NFC (Berkhout 1991).

The disposal of radioactive waste is one of the most sensitive of environmental issues. Nobody, whether pro- or anti- the creation of radioactive waste through nuclear energy and other means, wants to see this waste dumped in their proverbial backyard. So, issues involving the disposal of radioactive waste include a very large measure of selfish motives mixed with the altruistic.

Radioactive substances dumped at sea can pollute water as low-level radioactive waste. In the USA dumping was terminated in 1970, and in European countries in 1982, but many other countries continue to dump their waste in the oceans. Since 1975, all radioactive waste disposal at sea has been conducted in accordance with the London Dumping Convention (1975).

Many radioactive wastes are extremely toxic, with their radioactivity exponentially decaying through time – the half-life being the time taken for half the original amount of material to decay. Half-lives and toxicity vary greatly amongst the radioactive chemical elements. Table 6.2 shows the fission products and actinides (group of 14 intensely radioactive elements, with increasingly short half-lives for those with the highest atomic numbers) which would be typical for 1 tonne of spent liquid reactor fuel (LWR). From these tables, it can be seen that the half-lives vary up to millions of years, and typically range over at least tens to thousands of years, far longer than human recorded history. Thus any nuclear accidents can contaminate the natural environment into the foreseeable future, and beyond!

Radioactive waste is introduced into water as liquid effluent discharged from two commercial nuclear-fuel reprocessing plants (one in the UK and the other in France). Long-lived nucleides such as strontium-90 (^{90}Sr), caesium-137 (^{137}Cs), and ruthenium-106 (^{106}Ru) are frequently discharged. An example of such an incident occurred on 11th November 1983 when Greenpeace informed the public that the beach below Sellafield nuclear plant in Cumbria, UK, had become highly contaminated by radioactive waters being washed up from the Irish Sea where it had been discharged from Sellafield. On 7th December the Government issued a warning to the public that a 25-mile stretch of beach had become contaminated to levels of over 1,000 times the normal background radiation. Ironically, independent studies of the same beach showed that the beach had begun to be contaminated as far back as 1979, but little was done to counteract the problem.

In addition to the beach pollution at Sellafield, the environmental pressure group, Friends of the Earth, revealed new high levels of an isotope of caesium (^{137}Cs) and also Americium (^{241}Am), a particularly dangerous isotope associated with plutonium in the River Esk that flows into the sea at Ravenglass in Cumbria. Friends of the Earth were particularly concerned for the health of anyone spending more than 20 hours a week along the river bank because they would be at risk from inhaling radioactive dust blown up from the drying sediment along the river banks in quantities which could exceed the recommended safety levels.

The transport of radioactive substances is a hazardous process. In the North Sea, on 28th August 1984, for example, the French cargo ship, the *Mont Louis*, carrying 450 tonnes of uranium-rich material, collided with a ferry and capsized off Belgium. Disasters of this type which involve radioactive material in transit have happened and, undoubtedly, will occur again. They give cause for concern and demand more stringent international legislation to minimize the dangers of handling and transporting such waste.

Scientists have also shown concern about the risks of dumping radioactive waste in salt mines. In 1991, however, in a US $800 million project, the American government planned to store 4,250 barrels of radioactive waste in underground caverns in New Mexico. The salt could collapse to make access impossible, or at least extremely difficult, should the containers leak.

Accidents at nuclear power stations can cause serious contamination of the environment. Particularly memorable examples of this type of pollution occurred at Chernobyl in 1986, Windscale (now called Sellafield) in 1957, and Three Mile Island Nuclear Power Station in 1979. The effects of polluted rainwater from the Chernobyl accident in the former Soviet Union in April 1984 have still to be assessed fully. The Soviets stated that there was little effect on the waters around Chernobyl with water supplies to the main city of the province, Cave, remaining uninterrupted after the accident. There has been no real assessment of how the fallout polluted water supplies, let alone how waters might be treated if contaminated. Some authorities, such as Heinz Hansen (Riso National Laboratory, Denmark), believe the concentration of radioactive nuclei in contaminated waters supplied to humans is small compared to other sources that concentrate the nuclei, such as dairy products, fish, fruits, and vegetables. Many of the food sources, however, which concentrate these nuclei may derive them from water sources (O'Neill 1989).

Underground and subsea repositories for radioactive waste storage are being adopted by many governments and/or waste-disposal companies as the most acceptable option. This is exemplified by the US and British decisions. For example, in Britain, on 11th December 1991, the company responsible for the disposal of radioactive waste from reprocessing spent reactor fuel rods, NIREX, announced that the chosen site for the disposal of low-level and intermediate-level radioactive waste (LLW and ILW) is to be an onland underground site near Sellafield in Cumbria. There are, however, still proponents of an alternative undersea site below the Irish Sea off the coast near Sellafield. Both options have serious drawbacks, but a subsea site would have been the worse choice, principally because it would have posed the greatest difficulties in monitoring the radioactive waste. Furthermore, the retrieval of such waste from below the seabed would be logistically much more difficult than if it were stored under ground.

Following exploratory drilling, NIREX decided to dispose of the ILW and LLW within the deeply buried volcanic 'basement' to this area known as the Borrowdale Volcanic Series. These rocks are extensively exposed at the surface throughout much of the Lake District and are therefore well known to geologists. Where the Borrowdale Volcanic Series is exposed at 'outcrop', the rocks are invariably highly folded, fractured, jointed, and dislocated by geological faults. In the absence of readily available detailed data on the equivalent deeply buried Borrowdale Volcanic Series rocks offshore from Sellafield, there is no reason to assume that they are not similarly fractured, jointed, and faulted. Many of the geological faults are vertical to near-vertical and, therefore, not easy to detect in vertical boreholes. Such deformation of the rocks has created fluid-filled and gas-filled cavities at all scales, from the microscopic to the macroscopic, which is known as **porosity**. The interconnectedness of these spaces is known as permeability. Should radioactive waste escape following some unforeseen accident, it is through this porosity and permeability that it will move, possibly towards the surface in groundwater systems. The highly fractured rocks associated with geological faults are probably the principal pathways for fluid movement. This could occur for both the underground and subsea sites. Sealing the fracture porosity and the destruction of the associated permeability to acceptable minimum values in the Borrowdale Volcanic Series will present major scientific and technical problems.

Prior to making a firm decision of where to bury the radioactive waste, it may well have been much better to await a detailed quantitative understanding of the fracture porosity in the proposed host rocks designated to contain the LLW and ILW, together with an open and public debate, with all the relevant information freely available for inspection by independent bodies. Supporters of storage of ILW and LLW in a site below the seafloor must accept that such an option has all the potential problems associated with onland underground storage, but with the additional complications imposed by operating in the sea where there are also possibilities for shipping accidents and bad weather conditions hampering storage and/or retrieval.

To minimize the long-term risk to the environment caused by the disposal of radioactive waste, and until the undersea repository option has been fully evaluated, the storage of LLW and ILW perhaps should be in surface sites where the waste would be most readily retrievable and accessible for ease of monitoring in preference to its disposal in the poorly understood deep repository in the Irish Sea. Onland underground storage would be preferable to subsea storage, but neither option is satisfactory given the sites from which NIREX must choose. In order to gain more information, additional boreholes are needed to supplement the very few drilled to date.

Nuclear accidents and fallout

One of the ways in which past thermonuclear explosions can be detected is from suitably located high-latitude ice cores. Atmospheric thermonuclear weapons testing (ATWT) generates large amounts of **Beta activity**, a radionucleid, and associated nitrous oxide (NO), approximately 10^{10} g per megaton of NO, equivalent to about 2×10^{10} g nitric acid (HNO_3), commonly measured as nitrate (NO_3^-). The mechanism by which this occurs is the fixation of nitrogen by nuclear fireballs which generate atmospheric nitrogen oxides (NO_x), and ultimately nitric acid that falls as rain and snow, or precipitation. A study of ice and snow cores from Mount Logan (60°36'N, 140°30'W), in the Yukon Territory, Canada, collected in 1980 at an altitude of 5,340 m and calibrated to an accuracy of less than half a year, revealed that during the era of intense atmospheric thermonuclear weapons testing, between

1952 and 1980, the observed all-time peak pulse in gross Beta activity coincided with the largest nitrate kick (Holdsworth 1986). After taking account of other factors that might have led to these peaks, including volcanic eruptions, it was concluded that ATWT was indeed responsible for a significant part of the nitrate content.

The following sections explore some of the consequences of the most well-documented accidents at nuclear power stations.

Three Mile Island

On 26th March 1979, the Three Mile Island nuclear power plant near Harrisburg, Pennsylvania in the USA, underwent an accident sequence that almost led to a meltdown of the reactor core. The accident was part human error and part machine failure. The initial action in the sequence of events on that day was human error, because a major feedwater valve was erroneously left closed at the end of a routine maintenance exercise the day before the accident. The plant operators were unaware of the closed valve for several critical minutes. They also received false information on the correct setting of another valve. Both of these faults led to the operators taking counterproductive corrective action in shutting down the cooling water pumps which resulted in a drop in the coolant level in the reactor core. The reactor underwent a near-**meltdown**, in which the reactor overheated towards a critical state, before these errors were observed and remedial action taken.

The consequences of this accident were that the reactor Number 2 was seriously damaged and within the containment building, the entire area became prohibitively radioactive. Fortunately, nobody was killed.

The ensuing clean-up operation, still unfinished more than a decade after the event, nearly bankrupted the operating company, General Public Utilities, which was left with an estimated bill for US $1 billion to make the area safe and repair the damage. The significance of this accident at Three Mile Island is that it represents a turning point in public awareness about the costs of nuclear power and nuclear technology. Since the incident, no more US-commissioned nuclear power stations have been constructed. Even those people in favour of nuclear power stations tend to favour the option of 'not in my back yard'.

Chernobyl

The Chernobyl accident must rank as the worst in the history of nuclear power. On 26th April 1986, reactor Number 4, now christened The Sarcophagus, at the nuclear power plant of Chernobyl in the Soviet Ukraine, suffered a meltdown and released large amounts of nuclear fallout that spread across much of northern Europe and the former Soviet Union (Plates 6.2 and 20). The accident was part human error and part design flaw. In the explosion, the 2,000 tonne lid of the reactor was dislodged and fell back into the reactor to become precariously balanced at an angle. The floor of the reactor dropped by about 4 m in the explosion. Radioactive elements were released into the atmosphere, especially iodine-131, strontium-90, and caesium-137. The nuclear fuel in the reactor overheated and mixed with the sand used to line the reactor, then flowed like lava into the rooms below the reactor to solidify as a highly radioactive glass. It was only when the Soviet clean-up team, known as the 'liquidators' or 'bio-robots', came to examine the reactor core that, to their consternation, all they found in the reactor core were the twisted remains of the cooling rods.

Plate 6.2 *Reactor number 4, The Sarcophagus, at Chernobyl.*
Courtesy of Rex Features.

The clean-up operations at Chernobyl have been a catalogue of disasters, mismanagement, and underfunding. Even today, after the Soviets have requested international aid, there has been little help and equipment forthcoming. Western governments seem to see Chernobyl as a Soviet disaster and not a global catastrophe. Of the estimated 600,000 people who have been involved in the clean-up, about 250,000 have already received their lifetime's safe dosage of radiation. Many have died from **Acute Radiation Syndrome (ARS)** and suffered radiation burns from the intense fall-out at Chernobyl, but the exact numbers are unknown. Vladimir Chernousenko, the scientific director of the 30 km radius exclusion zone around Chernobyl, has estimated that 7,000–10,000 people could have died in the clean-up, but the Soviet authorities officially gave a figure of 250–350 dead.

In this and the next century, The Sarcophagus will remain unsafe. Amongst the plans that have been proposed to contain the further release of radiation at Chernobyl, have been suggestions to bury the reactor in sand (but the reactor fuel might overheat and cause another nuclear explosion), embed it in concrete (making it difficult to monitor the state of the solidified nuclear fuel), or erect a second sarcophagus over the damaged reactor to seal it hermetically for at least a few hundred years.

It is obvious that the results of this single nuclear accident at Chernobyl cannot easily be contained, yet governments are busy planning and constructing more nuclear power stations which will increase the risks of still further unforeseen disasters. No reactor design, however well engineered, can be totally safe. It is a question of acceptable levels of risk measured against the perceived benefits of nuclear energy which have to be weighed in the balance. Over and above these arguments, there is the often unstated military requirement for nuclear power stations to process and reprocess radioactive material for weapons. Governments certainly find it easier to hide these military motives behind the smokescreen of a debate about nuclear energy.

Tomsk-7

In early April 1993, an explosion and fire at a plutonium plant east of Moscow, the Tomsk-7 military complex in Siberia, caused by an explosion in a tank of uranium, led to radioactivity covering the entire complex 14 miles outside the town of Tomsk. The explosion occurred under ground when nitric acid was being added to uranium in a stainless steel tank which caused the temperature to rise rapidly. At **nuclear reprocessing plants** such as Tomsk-7, the spent nuclear fuel is separated by dissolving it in acid to allow the recovery of the uranium and plutonium. In the case of the Tomsk-7 plant, waste gases in the uranium tank were ignited and blew the concrete shield off the tank in a large explosion.

This type of accident is amongst the most serious that can occur at a reprocessing plant. On the Russian scale of nuclear accidents, from one to seven, the latter being the most serious, the Tomsk-7 incident was initially rated by the Russians as level 3 to 4, or possibly higher. Later analysis revised the seriousness downwards to a relatively minor incident. However, radiation levels in the contaminated area were measured at between several milliröentgens to several **röentgens**, substantially more than the maximum permitted annual dosage for nuclear workers. Although prevailing winds caused the radioactive cloud to move away from populated areas and drift north-eastwards, the radioactivity poses an environmental hazard, and to date has resulted in a large-scale clean-up operation with extensive top-soil removal, and raised concerns about the melting snow finding its way into water courses to affect human health.

Nuclear war and a nuclear winter

The view that the consequences of a nuclear war could provoke a 'nuclear winter' was first presented in 1982. In their book, *A Path Where No Man Thought*, Sagan and Turco (1990) portray the possible effects of the detonation of a 1 megaton nuclear explosion a few kilometres above New York city (also, see Turco *et al.* 1984):

The fireball radiation, traveling at the speed of light, has already ignited flammable structures ten miles and more from the city center. The shock wave, traveling at the speed of sound, has not yet reached the city. . . . As the nuclear shock wave is leaving the city, skyscrapers and most buildings have been blown down. Fires are momentarily extinguished by the blast wave, and smoke is propelled away from the city. Looming over the scene is the mushroom cloud, which sucks debris up to high altitudes – into the lower stratosphere for a ground-burst of yield greater than about 200 kilotons. The shock wave has passed. Many fires ignited by the fireball, and others – set, for example, from broken or demolished gas mains – begin to rage. The fires spread and merge over an area of 100 square miles or more. Great clouds of rolling black smoke rise above the fires. The inferno becomes a firestorm like a roaring fire in a fireplace with the flue open, but on a vastly larger scale, a huge column of convective air establishes itself, sucking up flames and carrying smoke to high altitudes. Winds in the firestorm can exceed hurricane force. Many days later, hovering over the flattened city is a vast smoke pall extending into the stratosphere. Simultaneous development, and subsequent spreading and merging, of many such soot clouds at many altitudes can lead to a nuclear winter.

The cloud of soot injected into the stratosphere could blanket the Earth, and block the normal passage of sunlight to the Earth's surface. Consequently, there would be a drop in mean air temperature of more than a few degrees Centigrade and, because the soot cloud would remain for at least many months, this temperature fall could seriously affect many food chains. The knock-on effect could affect the very survival of many species, including human beings. This bleak scenario of freezing temperatures and decimated food chains does not even take into account the radiation fallout, and the effects of any pyrotoxins (poisons from the fires). The idea of a nuclear winter remains controversial, not as a chain of events, but rather the precise effects which would result from a given set of starting conditions, taking account of the magnitude and location of the initial explosion(s), the prevailing climatic conditions (wind direction, etc.), and the amount of explosive matter ejected to various altitudes.

The concept of a nuclear winter has even been applied to the geological column, for the impact of a 10 km diameter meteorite impacting the Earth at the end of the Cretaceous Period, approximately 65 million years ago, causing the extinction of many species including the dinosaurs (see Chapter 2).

Nuclear arms and verification

Nuclear weapons pose one of the most threatening forces to the future survival of humans and the entire ecosystems of the planet. We cannot 'uninvent' these ominous weapons which possess a combined power capable of destroying the world many times over. Lord Zuckerman, a former UK Chief Scientific Adviser, commented that 'No one can ban what is not yet invented.' Treaties will inevitably lag behind scientific and technological advances in any field, including that associated with nuclear weapons. The USA and the former USSR have had this destructive nuclear capability of overkill, or Mutually Assured Destruction, since the 1950s.

The exclusive and highly sought-after membership of the world club of nations with a nuclear weapons capability is slowly growing. It appears that secret trading of nuclear information between China and Algeria is leading to Algeria becoming the first Arab nation to possess the nuclear bomb. Within two years, the nuclear plant in the foothills of the Atlas Mountains, 165 miles south of Algiers within a military exclusion zone, is expected to start producing 8 kg of plutonium annually.

Whatever people's personal views about the build-up and ever-increasing sophistication of nuclear weapons, concerned scientists have a duty to seek ways and means by which they can reduce the threat of a nuclear holocaust. This statement instantly begs the question of just what is the concerned scientist's role in such issues? Scientists invented nuclear weapons in the first place because they were given the necessary financial support and environment in which to satisfy the perceived needs of politicians and military strategists. Once invented, the decisions about nuclear arms deployment and development, or reductions, became the responsibility, primarily, of politicians and the armed forces. Nuclear arms issues are essentially moral and political, but the scientist in her or his own right can still make a valuable contribution.

The verification arena is where scientists can contribute to increasing the level of confidence in nuclear treaties which may be signed between nuclear powers. Generally, the confidence that signatories have in their ability to monitor the compliance by other signatories to a nuclear test ban treaty, partial test ban, threshold treaty, or any other treaty, will tend to determine the levels at which targets or limits are set. It is scientifically possible, for example, to detect and monitor underground explosions, and to discriminate between earthquakes and nuclear explosions.

By carefully analysing the seismic signals that emanate from earthquakes and underground explosions, their characteristic fingerprints can be identified. It has been argued by the British Seismic Verification Research Project group that about 15 monitoring stations outside the old Soviet Union could detect an underground explosion of 1 kiloton anywhere within the former USSR, assuming that the nuclear test was not set off during an earthquake, and that there was 'efficient coupling' (BSVRP 1989). Efficient coupling basically means that the explosion is not set off say within a large underground cavity where a significant amount of the energy waves from the test are absorbed by the air. Naturally, there are ways of trying to cheat on treaties, for example by attempting to 'decouple' nuclear explosions in underground cavities, or by trying to set them off during a cycle of earthquake activity. As long as nuclear explosions are above a certain very low threshold, or yield, and as long as the superpowers allow a certain amount of monitoring on their territory, then the technology is available to discriminate between earthquakes and explosions, even if both take place over the same time period. In other words, the science of **verification** is effective as long as there is good will on the part of signatories to nuclear treaties – that is, if there is adequate verification.

The need for verification

In the arms arena, verification processes are discussed and built into many international agreements in order to provide the means by which partner states to any treaty can attempt to satisfy themselves that the other signatories are indeed abiding by the terms of the agreement. The verification process involves both an initial phase of monitoring and data acquisition, followed by an evaluation of the gathered intelligence data. Most disputes centre on the monitoring processes rather than the evaluation of such information. Agreements and deals can only be struck where parties to a treaty believe that the terms of agreement are fair and equitable, where the signatories feel that what they bargain away at least matches (or is perceived as exceeding) the accrued benefits and, perhaps most importantly, that the conditions stipulated in the agreement are verifiable. It is in this last arena of verification that scientists can make a valuable contribution.

There are all sorts of people involved in the verification arena. Amongst the more scientifically based is the British Seismic Verification Research Project (BSVRP), co-ordinated by scientists in the Geology Department at Leicester University in the UK. It was only as recently as the mid-1980s that foreign nationals were allowed to set up seismic monitoring stations within the Soviet Union with the express purpose of monitoring nuclear tests. This agreement came about after the US non-governmental Natural Resources Defense Council proposed a scientific exchange between the USA and former USSR to demonstrate that verification need not be an obstacle to a Comprehensive Test Ban Treaty. Three stations were installed at Bayanul, Karasu, and Karkaralinsk, less than 250 km from the test site near Semipalatinsk in east Kazakhstan. In February 1987, however, when the former USSR resumed nuclear tests, the American equipment was switched off by the Soviet military. After fresh negotiations, the US scientists agreed to move their equipment from within 250 km to more than 1,000 km from the test sites. It was at this stage that the BSVRP group was invited to join them, with the number of monitoring stations increasing to five.

It has been argued that any strategic arms system that can be monitored for arms control purposes must be vulnerable to a first strike, simply because of the accuracy of modern weapons. There is little point in spending vast sums of money in knowing the exact location of sophisticated weapons if you lack the capacity to neutralize them in the early stages of a conflict. So, first-strike survivability can create problems for arms control negotiations. The development of Midgetman or SS-25 mobile ICBMs, which were more survivable than their predecessors, created a need for even more sophisticated anti-missile missiles. And so the arms race spirals upwards in a vicious circle of ever-increasing costs as the stakes get higher. At what stage can we create a breathing space in the arms race and a chance for more considered judgement? Any development in the nuclear arms race has an associated, limited 'window of opportunity' for constraint to slow or halt the technological advances.

Ironically, it is the same technological advances which drive the arms race that can improve the verifiability of treaties. Through the 1960s, the improved missile targeting and battle management capabilities facilitated the remote-sensing (satellite) verification of nuclear weapons installations. Technology is very much a double-edged sword in the arena of arms control.

Policing missile deployment

One of the most difficult aspects of verifying nuclear arms agreements is the policing of the nature and number of warheads on any missile. Short-range missiles (i.e., with a strike capability up to 300 miles or 482 km) provide part of NATO's flexible response which represents an option intermediate between the long-range missiles and conventional weapons. Of the Allied powers, Britain maintains that this class of nuclear missiles must remain (Plate 21). The so-called third zero option would be the abolition of this entire class of weapons, but Britain argues that such a ban would prove extremely difficult to verify.

One of the most controversial short-range missiles is the supersonic Lance missile that can carry either nuclear or conventional warheads. Furthermore, both types of warhead are easily interchangeable and it is this aspect that is at the heart of a current controversy between the USA and Britain on one side, against the old West Germany. Germany wishes to negotiate a reduction in the number of short-range missiles that are deployed, whereas the USA and Britain are not in favour of such action. In fact, the USA actually wishes to update the Lance missile that was originally deployed in the 1960s, in a programme called the 'follow-on-to-Lance' (FOTL) which is still at an early stage. If FOTL proceeds, then it is intended to increase the range of the Lance missile to 450 km from the present 110 km with a conventional warhead or 135 km with the lighter nuclear warhead. Part of the reason for updating the Lance missile is to increase its accuracy which, it is argued by Bill Arkin of the Institute of Policy Studies, Washington DC, should allow the fitting of a warhead with a smaller yield, that is, 1–10 kilotons and an accuracy that is two or three times better than the current accuracy of 300 m at full range where some of the warheads have a yield of 100 kilotons.

Improving the accuracy of a missile is extremely costly. NATO possesses 692 Lance missiles and 88 launchers, whereas the former Soviet Union has 1,600 launchers for short-range missiles. Both superpowers tend to argue that the number of launchers is more critical than the total count of warheads. The inherent problems in verifying any reductions in short-range missiles which might be negotiated, because of the ease of substituting conventional for nuclear warheads, has led some military analysts to suggest that a possible solution to the problem is to abolish this entire class of weapons. Without the weapons themselves, any launchers detected by either superpower would be a clear violation of a treaty. Other viable solutions include sealing the warheads into the missiles and 'tagging' them in some way at the assembly stage or specifying the number of allowable launchers without specifying the type of warhead.

The main ways in which military installations and missile launching sites can be monitored is through the use of 'spy' satellites. Such remote sensing of the Earth has generally been jealously guarded by the military, until the recent deployment of commercial reconnaissance satellites. The intelligence business became public on 22nd February 1986, when an Ariane rocket was launched from French Guiana with a SPOT satellite (SPOT 1) into an orbit 832 km above the Earth. The satellite, jointly owned by French, Swedish, and Belgian interests, and managed through the French Space Agency (CNES), is available to the public and media.

SPOT, with two cameras and sensors, can deliver a resolution of 10 m in panchromatic or 20 m in the green, red, and near-infrared wavebands (Zimmerman 1989). This resolution, at down to 10 m, represents about a three-fold improvement on the Landsat 4 and 5 satellites. It was the SPOT satellite that gathered detailed pictures from the former USSR of the burning reactor at Chernobyl, the missile early warning radar station at Krasnoyarsk, the Soviet naval base at Severomorsk on the Kola peninsula, the nuclear testing site at Semipalatinsk, the laser laboratories of Sary Shagan, and the ballistic missile base at Yurya.

Whilst surveillance experts can argue about the quality of the remotely sensed images of military sites, the advent of commercial reconnaissance satellites opens up a whole new proverbial ball game in which non-military (third-party) organizations can gain access, at a price, to verifying at least certain aspects of arms control treaties and agreements. The very fact that third-party verification of nuclear arms treaties is possible means that there is a useful deterrence function which could be served by independent organizations, including the United Nations, to discourage false allegations being made by any party to an agreement in which they may have a vested interest in non-compliance. Hopefully, in the years ahead, the prevarication over signing nuclear arms treaties which is based upon arguments over the non-verifiability of agreements, together with the attendant suspicions of cheating, can be largely allayed by designating a non-partisan, third-party, satellite-based verification. The United Nations may well prove to be the most acceptable organization to assume the mantle of this responsibility.

On 20th March 1992 came the first arms control agreement of the post-Cold War period, in which NATO and the former Warsaw Pact countries announced completion of a treaty covering aerial surveillance of all territory from Vancouver to Vladivostok. The treaty sets out conditions under which spy flights can be made, guaranteeing that any information received must be disseminated to any of the signatory countries to the treaty. It is this kind of open skies treaty which has the potential for removing any climate of fear, and substituting it with mutual trust.

The death of the Soviet Union: a new world order

The dramatic overthrow of the Soviet President Mikhail Gorbachev in a military *coup* in the early hours of Monday morning, 19th August 1991, and his replacement by a right-wing, senior hardline, so-called Emergency Committee of eight recidivists created international fear. The West anticipated a return to a more isolationist Soviet Union with a rejection of the policies of perestroika and glasnost. The status of the various arms treaties and the future continuation of moves towards significant reductions in the superpowers' nuclear arsenals became unclear.

Fortunately, the abject failure of the military *coup d'état*, because of the stance of President Boris Yeltsin and his supporters in the Russian parliament building, the show of popular support from the Soviet citizens themselves, the indecision and disunity amongst the 'gang of eight' with their conflicting military demands that they were never able to meet, the violence through the night of 20th and into the early hours of 21st August with the aborted military action, the defection of key personnel on the morning after, and the exertion of international pressure and condemnation, allowed the return of Gorbachev on the evening of Wednesday 21st August 1991.

When Gorbachev resigned from his position in the Communist Party on Saturday 24th August, he signed the death knell for the party that had terrorized Soviet citizens for more than 70 years. On Christmas Day 1991, at around 7 p.m. in the Soviet Union (5 p.m. GMT), Gorbachev stepped down as President of the Soviet Union and handed power over to Boris Yeltsin who became the President of the Russian Federation. With this transfer of power went the control of the nuclear arsenals. Shortly before 6 p.m. GMT, the Red Flag was lowered over the Kremlin, symbolizing the demise of the Soviet Union and the birth of the commonwealth of independent states. In his resignation speech, Gorbachev spoke of the changes which had happened since he assumed power in 1985, and of the new order where free elections, a free press and a multiparty political system had become realistic aspirations.

This string of momentous events has served to show the often fragile nature of peace, upon which so much confidence and trust throughout the rest of the world depends. Perhaps, with the new mood of optimism sweeping the dismembered Soviet Union for a more democratic future, there will be greater opportunities for significant arms reductions, particularly in the nuclear sphere. As a cautionary counterbalance, the breakup of the Soviet Union may lead to the new sovereign states claiming control of the nuclear weapons on their territory, as the Ukraine did in late October 1991 when it announced its independence. There is, then, the potential for the proliferation of nuclear weapons, not a reduction, as a number of small states overnight become independent nuclear powers. It appears, however, that Russia may be able to assume full control of the nuclear arsenals and in any nuclear arms talks become the superpower that will deal directly with the USA. As this book was written, the full implications of this second revolution in the Soviet Union this century remain uncertain. A new world order is being born out of the ashes of the fire of Soviet communism, with the potential for more superpower co-operation on issues that affect the global environment.

A major outcome of the decreased international tension between the so-called Free World and the former Soviet Union is the decommissioning or scaling down of certain nuclear capabilities. For example, in Britain, the Royal Navy expects to decommission about 10 nuclear submarines by the year 2000, all of which will be contaminated with radioactive waste. If the submarines are dumped at sea, as planned, then although the nuclear fuel will be removed, there will be marine pollution by at least low-level radioactive waste.

Conclusions

The world political arena is entering into a new and uncertain age of American and Russian co-operation. The Cold War of fear is being replaced by an economic and commercial conflict where the technology of the nuclear arms race is being used to earn foreign currency. The vast military industries and infrastructure of the East and West are looking for work. There is now talk of a joint defence programme between America and Russia. Areas of possible co-operation include sharing satellite information about missile attacks and other sensitive military information. The form of any response to an attack on either nation, or a third party, is controversial, because the order to intercept would have to be taken jointly and the mechanism for such a course of action remains unclear. Sharing sensitive military information is by no means an inevitable consequence of the present *rapprochement* between the USA and the former Soviet Union. The American early warning system deep within Cheyenne Mountain, Colorado, is unlikely ever to route unfiltered military intelligence direct to the Russians, and the converse is true; there will always be a reason to withhold or filter the intelligence. In the Gulf War, intelligence information was passed to the Israelis from Cheyenne Mountain, but even this was filtered. Meanwhile, the second-league nuclear countries, such as Britain, look on with interest as spectators. A consequence of real co-operation between Russia and America could be to render the British nuclear capability obsolete. There are those who caution that greater co-operation between the USA and Russia might accelerate rather than reduce the international nuclear arms race. Countries such as China, Japan, North Korea, and India might feel the need to arm more heavily in order to combat any superpower alliance. Perhaps we should be cautiously optimistic about a joint defence programme between America and Russia, because the international consequences are yet to be assessed fully.

Key points

CHAPTER 6: KEY POINTS

1 Nuclear technology and nuclear weapons are accepted as a necessary evil by some, and unacceptable by others. Whatever a nation's viewpoint, the exploitation of nuclear energy and possessing a nuclear arsenal have their associated risks from accidents, together with problems for the disposal of radioactive waste.

2 The Nuclear Age began in Cambridge University, UK, in 1919, with Rutherford's

NUCLEAR ISSUES

Key points continued

experiments on the structure of the atom. Nuclear fission was first achieved by Hahn and Strassman in 1934. The development of the atomic bomb was sanctioned by Roosevelt in 1941, and the first two bombs were used against Japan in August 1945. Treaties signed to limit arms proliferation and nuclear testing include: the 1963 Limited Test Ban Treaty (LTBT); the 1968 Non-Proliferation Treaty (NPT); the 1972 Strategic Arms Limitation Treaty I (SALT I); the 1974 Threshold Test Ban Treaty (TTBT); the 1976 Peaceful Nuclear Explosions Treaty (PNET); the 1979 Strategic Arms Limitation Treaty II (SALT II); the 1987 Intermediate Nuclear Forces (INF) Treaty; the 1991 Strategic Arms Reduction Treaty (START); and the 1992 START 2 Treaty.

3   Policing nuclear missile deployment is difficult, and effective and adequate verification mechanisms are necessary in order to instill confidence in a treaty, and in order to check that no nation is cheating. Mechanisms for verification include seismic verification to monitor underground nuclear testing, remote sensing, scientific exchange programmes, and on-site inspections of nuclear installations.

The collapse of the Soviet Union and the birth of the Commonwealth of Independent States of the Russian Federation may result in greater opportunities for significant arms reductions, although real concerns remain over the proliferation of nuclear weapons, and their pre-emptive use by politically unstable regimes in various war zones.

4   Radon is a major component of the natural background radiation dose received by many people, and results from the decay of radioactive minerals within rocks. Tentative links have been drawn between cancer in humans and areas of high radon concentrations. Factors which concentrate radon in the environment include: the geological setting and the bedrock type; building design and the materials used; water sources; and atmospheric conditions. Governments have commissioned surveys to identify areas of high risk and have provided recommendations to help reduce the effects within the home.

5   Nuclear waste results from the production of nuclear weapons and energy, medical products, and scientific research. Waste management in the nuclear power industry includes the nuclear fuel cycle. Prior to 1970 for the USA and 1982 for Europe, waste could be dumped at sea, but since 1975 all dumping has had to comply with the London Dumping Convention. The half-life and toxicity of radioactive chemicals vary greatly, a factor that determines the risk and magnitude of any potential contamination. The transport of nuclear waste is hazardous, and accidents may affect very large areas for extremely long time periods. Acceptable underground nuclear waste disposal relies on many factors, including the geological setting and bedrock geology, groundwater movement, the nature of the containers in which the waste is sealed, and its monitorability and retrievability.

6   Nuclear accidents at power stations have occurred, for a variety of reasons, but all involving some degree of human error, and include Three Mile Island in 1979, Chernobyl in 1986, and Tomsk-7 in 1993. These accidents serve to highlight the potential dangers, clean-up problems and costs, and the environmental damage.

Chapter 6: Further reading

Berkhout, F. 1991. *Radioactive Waste: Politics and Technology.* London: Routledge.
This is a companion book for students in environmental studies, geography, and public administration. The book focuses on radioactive waste management and disposal policies in three European countries – the UK, Germany, and Sweden. A detailed historical account of the policy processes in these three countries is presented, and the theoretical and public policy implications are evaluated. A particular strength of this book is in its comparative approach, and the way in which Berkhout sets the issue of radwaste management at the centre of the current debate about nuclear power, the environment, and society.

Newhouse, J. 1989. *The Nuclear Age: From Hiroshima to Star Wars.* London: Michael Joseph.
An excellent history of the post-Second World War events and personalities behind the single most important issue of the past 50 years – nuclear weapons and the nuclear arms race. This book is well researched and highly readable. It tells a dramatic story of confrontation and *rapprochement*, of scientific and technological advance, of diplomatic wrangling, and blatant shows of military strength, through the Cold War and the Cuban Missile Crisis, to the Star Wars programme, and the more recent US–former Soviet Union arms

agreements. It is an invaluable account of the Nuclear Age.

Pasqualetti, M.J. (ed.) 1990 *Nuclear Decommissioning and Society: Public Links to a New Technology.* London: Routledge.

This edited collection considers the impact of the decommissioning of nuclear power plants on the environment. The authors look at problems that have already been recognized, those that will arise through the decommissioning task itself, and the likely implications for the future. Key issues such as nuclear energy policy, jobs, waste, economics, the law, public opinion and public policy, siting, land use, and legacies are examined in detail.

Sheehan, M.J. 1988. *Arms Control: Theory and Practice.* Oxford: Blackwell.

A useful analysis of the origins and development of arms control, and the issues which underpin arms control. The problems of verifying treaties, and the political context in which arms control negotiations, both domestic and international, are considered. This is a useful supplementary book for any student interested in understanding arms control issues.

O LORD, why grievest Thou? –
    Since Life has ceased to be
Upon this globe, now cold
    As lunar land and sea,
And Humankind, and fowl, and fur
    Are gone eternally,
All is the same to Thee as ere
    They knew mortality.

Thomas Hardy,
'By the Earth's Corpse'

A crisis of conscience

Most of the energy needs for domestic and industrial consumption come from only a few precious natural resources. They may seem limitless in the context of the human short lifespan, something to squander and let future generations replace. Never has the preciousness of this resource been appreciated as much as during the world energy crisis of October 1973. The crisis was precipitated following the Arab–Israeli war when the Arab oil-exporting nations cut the export of oil to Israel's ally, the USA. American citizens suddenly became aware of their extreme dependence on oil as a commodity. Following the oil embargo by the Arab nations, the crisis was lessened but did not abate. The 1970s were dogged by successive energy crises. At the end of the 1970s, oil was more than US $30 a barrel, compared to US $3 a barrel at the start of the decade. The Organization of Petroleum Exporting Countries (OPEC) cartel seemed determined to raise the price of oil *ad infinitum*. Domestic bills for oil, gas, and electricity rose sharply, fuelling the inflationary spiral of the 1970s throughout the developed world.

A useful outcome from the energy crisis was the realization that the Earth's fossil fuels are a limited, finite resource. There also came the awareness that individuals and countries have a very real responsibility, not only to future generations, but also to themselves as concerned and responsible tenants of this fragile planet to husband these resources in a prudent manner. Whatever is taken from the Earth has a price and a cost for the environment. No energy source is free, environmentally completely safe or limitless. Energy must be used more wisely, in order to minimize the environmental hazards and optimize the efficiency with which it is produced.

Energy use and energy reserves

It has been estimated that humankind annually expends an amount of fossil fuel that it took nature, on average, about 1 million years to produce. In 1990, global energy expenditure amounted to an annual 1.3 billion tonnes of coal equivalent, four times greater than in 1950, and 20 times more than in 1850. In 1990, the global energy supply breakdown was 77 per cent from fossil fuels, 18 per cent from 'renewables' (mainly hydropower, wood, crop wastes, dung, and wind), and 5 per cent from nuclear energy (Holdren and Pachauri 1992). Approximately 1,200 million people living in developed, industrialized countries consumed over two-thirds of this total energy supply, while less than one-third went to the 4,100 million people in the developing world (ibid.).

The consumption of energy in developing countries is rising rapidly, and by the end of this

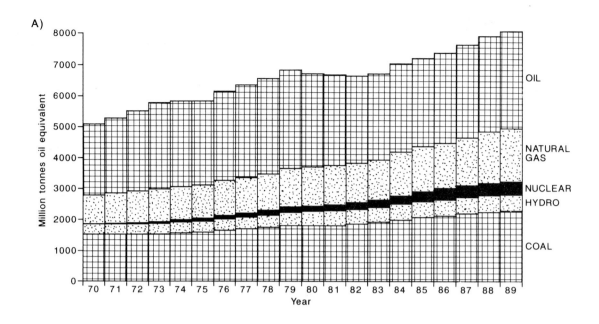

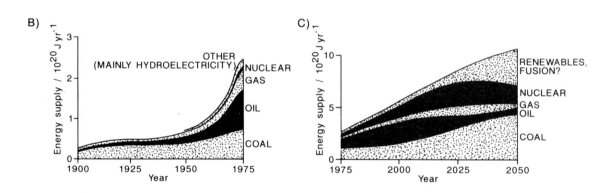

Figure 7.1 *(A) World energy use 1970–90. (B) World energy supplies 1900–75. (C) Projected demand in energy requirement for the period 1975–2050. After Blunden and Reddish (1991).*

Table 7.1 *Global primary energy supplies for 1987 (expressed in exajoules).*

	Coal	Oil	Gas	Primary electricity	Total commercial	Biomass	Total energy
World	88.7	104.6	58.2	33.0	284.5	36.9	321.3
Industrial	63.5	77.0	51.7	26.6	218.7	5.5	224.2
Developing	25.2	27.7	6.5	6.4	65.7	31.3	97.1
Share of industrial countries	72%	74%	89%	81%	77%	15%	70%
Share of developing countries	28%	26%	11%	19%	23%	85%	30%

Biomass figures for developing countries are underestimates
Source: US Congress Office of Technology Assessment (1992b).

century will dominate energy markets worldwide. Even allowing for a growth rate in the demand for energy in the developing countries 1–2 per cent lower than the present trend, the demand is likely to exceed 100 million barrels a day of oil equivalent (mbdoe) by 2010, and possibly 200 mbdoe (World Bank 1992).

Coal, oil, and natural gas account for 88 per cent of the global energy used, with nuclear fuel supplying most of the remaining needs. Underdeveloped and developing nations, however, still tend to rely heavily upon other fuel sources such as wood, crop waste, and dung. Oil accounts for roughly 38 per cent of commercial energy consumption, with natural gas contributing about 20 per cent. Figure 7.1 shows the world energy use, supplies, and projected demand for the major current fuels until the year 2050. World **primary energy** supplies for 1987 are shown in Table 7.1, from which it can be seen that the developed, industrialized countries utilize 70 per cent of the total energy. Energy data for 1989, from the United Nations Statistical Office, provides a useful breakdown of the energy production and consumption by region and fuel type, and is given in Table 7.2. Again, the indus-

Table 7.2 *Commercial energy production and consumption by region and fuel in 1989 (expressed in petajoules[a]).*

Region	Liquids[b] Production	Liquids[b] Consumption	Gases[c] Production	Gases[c] Consumption	Solids[d] Production	Solids[d] Consumption	Primary electricity[e] Production Nuclear	Production Hydro[f]	Consumption	Total Production	Total Consumption[g]
World	130,299	116,573	70,497	70,144	95,713	97,019	6,783	7,680	14,522	310,972	298,258
Developing countries	72,732	29,789	13,033	10,391	34,618	34,963	340	2,593	2,992	123,316	78,135
Oil-exporting developing	59,773	11,325	9,531	6,907	460	447	0	365	358	70,129	19,037
OPEC[h]	47,384	6,771	7,242	5,020	234	187	0	210	210	55,070	12,188
Non-OPEC[i]	12,389	4,554	2,290	1,887	226	260	0	155	148	15,060	6,849
Oil-importing[j]	12,959	18,464	3,502	3,484	34,158	34,516	340	2,228	2,634	53,187	59,098
Africa	621	1,434	16	46	4,068	2,880	14	120	126	4,840	4,486
Asia and Oceania	8,764	12,332	2,305	2,171	29,374	30,904	299	1,030	1,330	41,773	46,737
Latin America	3,574	4,698	1,180	1,267	716	732	27	1,078	1,178	6,575	7,875
Industrialized countries	57,567	86,784	57,464	59,753	61,095	62,056	6,443	5,087	11,530	187,656	220,123
OECD industrialized	31,286	67,606	28,905	33,719	37,322	39,036	5,511	4,090	9,643	107,114	150,004
North America[k]	22,043	35,321	21,158	21,403	23,595	21,117	2,193	2,090	4,292	71,080	82,133
Western Europe[l]	8,108	22,486	6,859	9,621	9,457	13,000	2,660	1,505	4,197	28,589	49,304
Pacific[m]	1,134	9,799	888	2,695	4,270	4,919	658	495	1,154	7,445	18,567
Non-OECD industrialized	26,281	19,178	28,559	26,034	23,773	23,020	932	997	1,887	80,543	70,119
Central Europe	813	3,129	1,611	3,063	8,687	8,510	165	195	458	11,471	15,160
Former USSR	25,468	16,049	26,948	22,971	15,086	14,510	767	802	1,429	69,071	54,959

a 1 petajoule = 10^{15} joules = 947.8×10^9 Btus
b Includes crude petroleum and natural gas liquids
c Includes natural gas and other petroleum gases
d Includes bituminous coal, lignite, peat, and oil shale burned directly
e Production and consumption of electricity assessed at the heat value of electricity (1 kilowatt hour = 3.6 million joules), the equivalent of assuming a 100% efficiency
f Includes geothermal and wind
g World electricity production shown as less than consumption because of incomplete trade data
h Algeria, Ecuador, Gabon, Indonesia, Iran, Iraq, Kuwait, Libya, Nigeria, Qatar, Saudi Arabia, United Arab Emirates, and Venezuela
i Developing countries whose exports of petroleum and gas including re-exports account for at least 30% of merchandise exports: Afghanistan, Angola, Bahrain, Bolivia, Brunei, the Congo, Egypt, Mexico, the Netherlands Antilles, Oman, Syria, Trinidad & Tobago, and Yemen
j Also includes countries that are self-sufficient in oil whose oil exports are less than 30% of merchandise exports
k Canada and United States
l Does not include Turkey, which is included in total Asia and Oceania oil-importing
m Australia, Japan, and New Zealand
Source: UN Statistical Office data from World Resources Institute (1992).

trialized countries, especially the OECD industrialized countries, consume substantially more liquid fuels than they produce, and have a smaller but nevertheless negative balance of gas and solid fuels, a situation which is less common in developing countries.

Perhaps the biggest challenge for developing countries in relation to energy consumption is to develop and implement technologies which help reduce the emissions of gases and particulate matter (dust and smoke), which have both local and possible global environmental impacts. In order to be more environmentally conscious, we need to endeavour to use energy resources preferentially that create fewer pollutants as by-products. Natural gas, for example, is 'cleaner' than oil and other fossil fuels. The combustion of natural gas releases 14 kg of CO_2 for every billion joules of energy produced, compared to oil at 20 kg and coal at 24 kg.

The way in which developed, industrialized countries provide their energy services to the developing world is important for the following reasons (outlined in the US Congress, Office of Technology Assessment 1992b): (1) international political stability, through steady, broad-based economic growth; (2) humanitarian concerns, helping developing countries to meet their energy requirements; (3) trade and competitiveness, facilitating the internationalization of economic growth; (4) global environmental issues will only be tackled on an international basis if there is co-operation between the developed and developing countries; (5) global oil markets, if World Energy Conference predictions are accurate, will change such that developing countries will account for 90 per cent of the increased world oil consumption between 1985 and 2020, and this may cause both higher prices and greater price instability, with impacts for all countries in terms of inflation, balance of trade, and overall economic performance; and (6) global financial markets are affected by the indebtedness of the developing countries, a large part of which is incurred through building their energy sectors, and unless debt repayments are restructured then they will contribute to international instability in the world's financial institutions.

Before going on to look at an historical perspective of energy, and dealing with energy resources, it is important to mention ways in which power output and energy consumption are measured. Power output is measured in megawatts (MW), electrical output as megawatts electrical (MWe); 1 megawatt is 1,000 kilowatts (kW). There are various ways of measuring energy consumption, for example, as the amount of energy used to produce a given amount of economic output (the energy/gross domestic product or GDP ratio). As an illustration, in the UK the shift from coal to gas has helped to improve the energy/GDP ratio, because the use of gas involves cleaner technologies and cleaner fuels, for example, differences in the balance of fuel used to generate electricity; per unit of electricity generated from gas, there is approximately a 50 per cent reduction in the CO_2 emissions compared to that derived from coal.

This chapter explores the various energy resources from fossil fuels to nuclear energy, and alternative energy resources, all of which are considered in the context of projected energy requirements into the twenty-first century.

An historical perspective

Until the Industrial Revolution began in England last century, demand for energy resources was relatively modest. Water or hydropower and fire power from wood and peat were the principal means of obtaining energy. Archaeological evidence in caves in the Peking area of China shows that humans were utilizing fire power in at least 400,000 BC.

The natural energy resources of wood, fossil fuels such as peat and coal, water power from streams and rivers, and wind energy harnessed in windmills, seemed adequate and did not pose any long-term environmental problems. Any atmospheric pollution seemed to be merely a local phenomenon associated with certain areas such as the 'Potteries' in the north-west English Midlands. Nobody really thought that there was any serious threat to the environment, much less to Planet Earth. How could they? Science and research had not advanced to the sophistication of this age where it was capable of establishing complex cause-and-effect relationships about the global balance of atmospheric gases.

By the early 1970s, natural gas was considered a fossil fuel for premium use such as in domestic heating and cooking, with oil as the major intermediate and, perhaps, long-term energy resource. In the US, the Gulf of Mexico, and in Britain, the northern North Sea, were fast becoming the largest areas of high-tech exploration and discovery, respectively. Coal seemed like yesterday's fuel: expensive to mine and dirty to use.

In 1974, OPEC abruptly quadrupled the world oil price. It is probably fair to say that this dramatic increase in the price of crude oil precipitated the world economic recession that became evident in the late 1970s and early 1980s. Politicians and industrialists suddenly became obsessed with a need for alternative, viable energy resources and nuclear energy appeared to provide the answer.

The 1980s saw the groundswell of environmentalists lobbying for safer, 'greener' energy resources, and a less wasteful use of these resources, together with an increasingly shaken faith in nuclear energy, especially after Chernobyl. It is within this new-found green political climate that energy resources and resource planning must be placed. While issues such as the greenhouse effect, acid rain, and water pollution tend to dominate the headlines (with the exception in 1992 of the debate over energy policy associated with the British government's proposed closure of more than 30 coal mines), energy issues tend to take second place – even though acid rain, for example, is caused by the emissions particularly from coal-fuelled power stations, and greenhouse gas emissions are heavily linked to fossil fuel use.

Conventional fossil fuels

In order to discuss fossil fuels, it is necessary to define what is meant by the term mineral/fossil fuel 'proven reserves'. Proven reserves do not represent the total amount of reserves that are estimated to be ultimately recoverable, but rather that part of the reserves judged to be recoverable under the extant economic and operating conditions. Proven reserves, therefore, are dependent upon world commodity prices, the state of exploration, and recovery technology. As an example, due to changes in the economic conditions, and because of explora-

tion, over the past 20 years the proven reserves of oil and gas have shown an increase. Reserve lifetime (reserve/production ratio) is a measure of the duration of the reserves at the existing rates of extraction and demand. At current consumption rates, world coal reserves have a predicted lifespan of 200–400 years, whereas for oil the projected longevity of supply is very dependent on region, for example, more than 100 years for some Middle Eastern states, but with a global lifetime of about 56 years, averaging 13 years for the USA and UK (Blunden and Reddish 1991). Any figures such as these must be treated with circumspection as their derivation involves considerable uncertainties, and political–commercial calculations.

Estimating the reserves of fossil fuels is difficult. In an attempt to do this, scientists and economists use a variety of techniques which include using the physical evidence for known locations and predictive or statistical methods that make assumptions about the size and type of reserves which may be discovered. The 'estimated resources' available to humankind will always outstrip the 'actual recoverable reserves'. Fossil fuels are being depleted by human activities at a rate 100,000 times faster than they are being formed (Davis 1990). Extractive technology can change to such an extent, however, that reserves which seemed inaccessible become within human reach. So, the figures that follow should not be taken as gospel but rather as a reasonable estimation of what is available.

The three main conventional fossil fuels are coal, oil, and natural gas. These also constitute the largest source of greenhouse gases.

World coal reserves have been estimated at 3,160 billion barrels of oil-equivalent (BBOE) compared to a world natural gas figure of 425 BBOE, and an estimate for oil of 700×10^9 barrels (Cassedy and Grossman 1990). Of course, reserves are not the same as recoverable resources. The estimated recoverable resources are 25,600 BBOE for coal, 1,200 BBOE for natural gas, and 1,863 billion barrels for oil (ibid.). In 1950, the world's proven reserves of oil and gas stood at 30 billion tonnes of oil equivalent (btoe), whereas today they exceed 250 btoe, including a total world consumption over the past 40 years of 100 btoe (World Bank 1992).

Coal

Between 1950 and the present, proven coal reserves have risen from 450 btoe to 570 btoe (World Bank 1992), updated to 694 btoe (Bowler 1993). Global coal resources are equivalent to at least 10,000 billion tonnes of oil. Coal, formed by biological and geological processes acting on the buried remains of plant material from ancient peatlands and swamplands, is the principal conventional fossil fuel. It remains the largest fossil fuel resource, but coal extraction and combustion in power stations has quite a high economic and social cost – acid rain, for example, is caused (but not solely) by coal-burning power plants emitting sulphur dioxide. Underground mining can be dangerous because of immediate accidents, such as gas explosions and roof collapse. Long-term illnesses and fatalities such as *pneumoconiosis* are even more commonplace than these immediate accidents.

Coal is formed from the remains of vegetation that grew in wetlands and swamps (collectively called mires). Under favourable climatic and burial conditions, the decaying plants form peat, which is then converted into various ranks of coal. About 1 m of peat typically forms in time periods between 4,000 and 100,000 years, with 10 m of peat compacting and transforming into approximately 1 m of coal. The ideal environments in which peat, and subsequently coal, can form are associated with major river deltas where the rate of subsidence and burial occurs at about the same rate as the vegetation is being established. There are different types, or ranks, of coal – **anthracite**, **bituminous**, sub-bituminous, and lignite – all of which have different calorific (energy) values, and produce varying amounts of polluting gases. Anthracite has the highest calorific value, with lignite having the least.

In the nineteenth century, the ascendancy of coal as the major fossil fuel was closely tied to the Industrial Revolution which began at Ironbridge in the English Midlands. Coal has been used mainly to generate steam to drive turbine engines and make electricity. The importance of coal as a global energy resource peaked in the 1920s, when it accounted for roughly 70 per cent of fuel use, but today coal only supplies 26 per cent of the world's energy needs. Oil and gas have replaced coal as the main fossil fuel, a trend that began after the Second World War, but which really accelerated in the 1960s and 1970s. For comparative purposes, quantities of coal are

Box 7.1

Clean coal technologies

Coal combustion is associated with relatively high emissions of environmentally harmful gases (such as the greenhouse gas CO_2, and SO_2 which is a cause of acid rain) and particulate matter, and for this reason there has been considerable research to develop 'clean coal' technologies that might provide both increased energy efficiency and reduced atmospheric pollution (Table 7.3). In the USA, there is growing pressure on the electricity-generating industry to use the cleanest coals, that is, the low-ash, low-sulphur coals. The choice is between the black coals with their higher calorific value, but also higher sulphur levels, which occur in relatively thin seams in the Appalachian coalfields, and the younger coals with lower calorific values, less sulphur, and which occur in thicker seams further west. In order to compensate for the shift towards increased coal mining in the more westerly coalfields, the coal producers of the eastern USA are looking to greater exports to European markets.

'Clean coal' technologies (CCTs) are seen by many as providing improved energy efficiencies and lower emissions (see POST 1992a), but because of coal's high carbon content, CO_2 emissions from coal will always be higher than gas – with the carbon/energy ratio of gas being half that of coal.

In Europe, SO_2 and NO_x emissions limits are given in the EC Large Combustion Plant Directive. For the UK, this Directive stipulates a 20 per cent reduction in SO_2 emissions by 1993, from the levels in 1980, and to 40 per cent by 1998, and 60 per cent by the year 2003. Also, by 1993, UK NO^x emissions should be reduced by 15 per cent, and 30 per cent by 1998. As for CO_2 emissions, the UK is committed to no more than its 1990 levels by the year 2000.

In many industrialized countries, there are ongoing changes in the balance of primary energy sources, because of both economic and political considerations. In the UK, for example, in 1990, about 40,000 MW of coal-fired generating capacity supplied 68 per cent of the nation's electricity, with nuclear energy supplying about 21 per cent and oil approximately 9 per cent. The proposed introduction of 10,000 and 20,000 MW gas-fired combined cycle gas turbines

expressed as billions of barrels of oil equivalent (BBOE). By doing so, it is easy to appreciate that coal constitutes the largest fossil fuel resource available.

In the West, many of the world's most accessible and easily extractable coal reserves are now nearing depletion or are worked out. The remaining coal seams are often thinner and broken by geological **faults**, which makes the prediction of their extent and extraction more difficult.

The old mining communities that were so commonplace in the US Appalachian Mountains (and epitomized in films such as *Coalminer's Daughter*), or in central and northern Britain and throughout large parts of Belgium, France, and Germany, are in decline. The decline of coal mining and the associated redundancies is an emotive issue. Countries such as Germany provide heavy subsidies for their coal production, because of the social aspects of providing employment and for reasons of energy security. Nations such as the UK actively encouraged the development of competition in power generation following the privatization of the Central Electricity Generating Board (split into two generator companies, PowerGen and National

Power), by promoting the so-called 'dash-for-gas', involving the construction of new combined cycle gas turbine stations (CCGTs). In Britain, opencast mines are common, but their output is falling, not because of a lack of exploitable sites, but because of local opposition to their development on the basis of environmental considerations: there are total proven reserves of 300 million tonnes, 25 mt more are found each year which would require deep mining to extract, and British Coal estimates coal resources at 130 billion tonnes.

In Britain, on 13th October 1992, the announcement by the government of the closure of 31 of the country's remaining 50 deep coal mines, and 30,000 job losses from the industry, caused a public outcry. The result of protests from many trade unionists, economists, and the general public, resulted in the President of the Board of Trade giving a stay of execution for 21 of the 31 pits, with the government promising a wide-ranging review, although at the time of this book going to press, it seems that far fewer coal mines will remain open – probably around 10–12, based on a UK government White paper published in March 1993. The political/energy policy, and socio-economic issues lying behind

---

(CCGTs), seen as having the advantage of contributing less CO_2, NO_x, and SO_2 than fossil fuel-burning power stations, will change this balance.

In conventional pulverized fuel power stations, coal is ground or powdered, and then undergoes combustion to give high-pressure steam which drives turbines. Flue gas desulphurization removes up to 90 per cent of the SO_2 produced. A by-product from this process is large amounts of the calcium-sulphate mineral, gypsum, with industrial uses such as in the manufacture of plasterboard. In some industrialized countries such as Germany, flue gas desulphurization has been routinely fitted to coal-fired power stations. As part of the UK's package of measures to reduce SO_2 emissions, flue gas desulphurization is being fitted to some of the existing coal-fired power stations, such as at the new Drax power station projected to go into operation in 1994, and possibly at Ferrybridge. NO_x emissions can be reduced by between 30 and 50 per cent, depending on the type and age of the power station, by installing 'low NO_x' burners, and 50–

80 per cent reductions are possible through flue gas cleaning techniques, such as selective non-catalytic reduction, or the more costly selective catalytic reduction.

Commercially available clean coal technologies include fluidized bed combustion (FBC), and gasification, together with hybrid systems such as British Coal's Topping Cycle, which is at a developmental stage. In fluidized bed combustion, either under pressure (pressurized fluidized bed combustion), or at normal atmospheric pressure (circulating fluidized bed combustion), a mixture of coal and crushed limestone is partially suspended by an upward-moving stream of air or oxygen from the bottom of the combustion chamber. The limestone can absorb around 90 per cent of the SO_2, and NO_x production is inhibited because of the relatively low operating temperatures. Worldwide, there are approximately 200 operational circulating fluidized bed combustion plants, with four in the UK. The technology is commercially proven for plants having up to 150 MWe capacity. Larger-scale demonstration plants, however, are under

Box 7.1
Clean coal technologies

the pit closures in the UK, due to changing power-generating technologies and a rethinking of the energy strategy, illustrate the deep-seated concerns of ordinary working people at the prospect of drastically scaling down a traditional energy industry with very strong community bases. The effects of such closures will have far-reaching socio-economic impacts in a society where the high rate of unemployment makes finding alternative employment very unlikely. The issues surrounding the pit closures in the UK are discussed in some depth in a special issue of the journal *New Scientist* (Charles 1993, Ridley 1993, Cross 1993, Bowler 1993, Maitlis and Rourke 1993).

While countries such as the UK may be able to meet their environmental obligations through turning to natural gas, other nations such as India and China have a rapidly rising energy demand which they will need to meet, and very large indigenous coal reserves. If such fossil fuel reserves are to be used in an environmentally acceptable way, clean coal technologies (see Box 7.1) are very important to these nations. Unfortunately, the widespread adoption of clean coal technologies is unlikely to happen unless developed countries implement the technology first, and offer technology transfer capabilities.

Table 7.3 *Comparison of energy efficiencies and emissions.*

Technology	Generating efficiency*	SO$_2$ removal (%)	NO$_x$ emissions (mg m^{-3})
Pulverized fuel + flue gas desulphurization + low NO$_x$ burners	38–39	90	500–650
Circulating fluidized bed combustion	39–40	90	100–300
Pressurized fluidized bed combustion	41–43	90	150–300
Integrated gasification combined cycle	43–44	99	120–300
Hybrid combined cycle (British Coal Topping cycle)	46–47	90	150–300
EC Large Combustion Plant Directive (for high-sulphur fuel)		90	650

*Based on lower heating value
Source: POST (1992).

Oil

The past few decades may well be seen retrospectively as the 'Age of Oil', to the extent that Daniel Yergin (1991), refers to the present

Box 7.1

Clean coal technologies

way, such as in France where the 250 MWe Electricité de France plant is under construction. In pressurized fluidized bed combustion plants, coal is combusted at pressures of 10–20 atmospheres, with additional energy efficiency being achieved through allowing the hot pressurized combustion gas into the gas turbine along with the steam. Countries where pressurized fluidized bed combustion plants are operational or planned include the USA, Japan, Britain, Sweden, Spain, and Eastern Europe. Integrated gasification combined cycle plants produce gas from coal by reacting the coal with steam and air/oxygen in a gasifier, with about 99 per cent of the SO$_2$ removed from the gas, along with other impurities, and the gas is used to drive a gas turbine. Worldwide, gasifier plants are operational or planned in the USA (with strong government support as part of its clean coal strategy), Britain,

Germany, the Netherlands, and Spain.

The most desirable technical aspects of combustion and gasification are synthesized in hybrid combined cycle systems. In such power stations, some coal is used to produce steam and drive steam turbines by combustion in a fluidized bed, while the rest is converted into hot gas to drive gas turbines. An example of a hybrid system is in the UK, at Grimethorpe, where British Coal has pioneered the Topping Cycle, and other pilot plants have been constructed in the USA and Germany.

For future energy production, there are other clean coal technologies being developed that may provide more than 50 per cent energy efficiency, for example those that combine coal gasifiers and fuel cells, and also magnetohydrodynamics, but these are not expected to become commercial plants for at least 20 years.

generations as the 'Hydrocarbon Society'.

Oil is the world's largest and most pervasive business, with commercial and political influence transcending national frontiers. Indeed, the American company Standard Oil became one of the first truly multinational enterprises (MNEs). The first oilwell, which was drilled by Edwin Drake in Pennsylvania, near Titusville, struck oil on 27th August 1859, at about 69 ft below the ground. The advent of the internal combustion engine ensured the supremacy of oil as a fuel. Virtually all international conflicts this century have involved a struggle for military and political control over oil fields or oil supplies, for example, the Suez Crisis of 1956 or the invasion of Kuwait by Iraq in 1991. Many household names are associated with the history of oil, such as the entrepreneurs Rockefeller, J. Paul Getty, and Armand Hammer, but civilizations have been aware of oil for thousands of years – bitumen seepages were tapped around 3,000 BC in Mesopotamia, and in the first century AD, bitumen's cauterizing capacity in medicine was commented upon by the ancient Roman naturalist Pliny. Oil, such as tar or bitumen, found an early use in warfare. Homer, in the *Iliad*, describes the Trojans using simple firebombs, probably of tar set alight. Bitumen also had a more constructive use, for example, it was used as a building mortar throughout the ancient Middle East. Today, up to thousands of years later, civilizations' uses of oil and its derivatives remains essentially unaltered, merely more sophisticated.

Most oil and natural gas is formed from the remains of marine micro-organisms that died and accumulated on the seafloor. Under favourable burial temperature (**geothermal gradient**) conditions, the organic chemical compounds are transformed into relatively short-chain hydrocarbon molecules. If the temperature is too high, the oil is broken down into volatiles or natural gas, of which methane (CH_4) is the principal constituent. Over time, the oil and gas migrate through the microscopic cavities, or pores, of a sediment or rock and become trapped in certain geological situations to form hydrocarbon reservoirs. Many of the best sites of major oil and gas fields occur in ancient sedimentary environments associated with continental margins that were created by the breakup and rifting apart of land masses in what were equatorial to temperate latitudes, because it is in such sites that the lake and marine conditions are favourable to high organic productivity, fast burial of organic carbon, and appropriate geothermal gradients and sediment burial histories.

In the early 1970s, about 40 per cent of global fossil fuel use came from oil, whereas today the figure has dropped to around 38 per cent. In 1988, oil consumption rose by 3.1 per cent globally, with oil production averaging 8.8 million metric tonnes per day (World Resources Institute 1990). Of this quantity, US consumption is about 25 per cent of the total global figure and constitutes approximately 50 per cent of the oil used by all the OECD countries, who together account for more than half the world's oil demand. In 1988, the global production of oil was 5,313,303 million tonnes, with consumption being slightly greater at 5,326,785 million tonnes.

The developed, industrialized nations use far more oil than they produce. For example, in 1988 North America and Western Europe produced 22,857 and 8,290 but consumed 36,171 and 24,875 petajoules of oil, respectively (1 million tonnes of oil equivalent = 41.87 petajoules). In contrast, in the same year Africa produced 10,991 and consumed 3,609 petajoules of oil, and the Middle East produced 30,954 and consumed 5,669 petajoules of oil.

In 1984, in an analysis of the likely effects of an oil import crisis, the US Congress Office of Technology Assessment concluded that the USA then had the technical capability to replace 3.6 MMB/D (million barrels per day) of oil imports, equivalent to a curtailment of 70 per cent of US net imports and a loss of 20 per cent of the US oil supplies, which was regarded as a comfortable margin in any realistically projected oil crisis. The Gulf War in 1991 refocused American concern about the security of energy supply. By 1990, US petroleum consumption had risen from 15 to 17 MMB/D and domestic production had decreased from 10.3 to 9.2 MMB/D, oil imports had increased from 5 to 8 MMB/D, and the share of US oil needs supplied by imports had risen from 33 per cent to more than 40 per cent. In a 1991 report, the US Congress Office of Technology Assessment concluded that if a 70 per cent curtailment in oil supply were to affect the USA, believed by some experts to be an unlikely but not impossible situation, and assuming a similar scenario to that in the 1984 report, then using all currently available oil replacement technologies they could only displace 2.9 MMB/D of 1989 oil use within five years. This replacement potential has to be offset against an ongoing decline in domestic US oil production, which gives an effective replacement capability of between 1.7 and 2.8 MMB/D. All this adds up to the fact that between 1984 and 1991 the USA has decreased its ability to respond effectively to a serious and prolonged oil shortage, and in such an event would fall some 5 MMB/D short of the 1984 capability. It is perhaps authoritative reports such as these that are providing at least some of the impetus for nations investing in alternative renewable energy resources.

In many developed, industrialized countries oil consumption is falling because of improved energy efficiency, decreased oil intensity (oil used per dollar of gross national product, GNP) associated with energy diversification, with the shift to other fuels such as natural gas, nuclear energy, etc., increased strategic petroleum reserves, and in some regions environmental considerations. Finally, it is important to distinguish oil import dependence from oil import vulnerability. Import dependence, measured as a percentage of domestic consumption met by foreign imported oil, can contribute to import vulnerability but does not in itself necessarily cause import dependence. 'Import vulnerability arises out of the degree and nature of import dependence, the potential harm to the economic and social welfare from a severe disruption in physical supplies or prices, or its duration, and the likelihood of such a disruption occurring' (ibid.). However, the magnitude of many nation's dependence on oil imports remains a source of deep concern and a central part of any considerations that help formulate energy policy, and provides one of the main reasons for energy diversification programmes.

Natural gas

From 1965 to the present, proven reserves of natural gas have increased five-fold to 100 btoe (World Bank 1992). Natural gas will probably be the fastest-growing energy resource. The present global use of natural gas stands at roughly 19 per cent and this figure is predicted to rise. Natural gas provides an alternative to energy requirements that are dependent on oil or coal, and it is a cleaner fuel in terms of atmospheric pollution.

Although the exploitation of natural gas as an efficient energy source is well advanced, current predictions are that global reserves will not last beyond about 120 years, whereas economically recoverable coal reserves might last for another 1,500 years (Fulkerson *et al.* 1990). Natural gas could last up to three times as long as the 120-year figure if unconventional sources were to be tapped, or if it becomes economically viable to recover the less accessible reserves.

Leakage and losses of CH_4 from distribution systems, including pipelines, oil/gas wells, and domestic use, provide a contribution to anthropogenic emissions of greenhouse gases (see Chapter 3). Estimates for the annual leakage rates range between 1 and 5 per cent of the total amount used each year in Europe and North America, and may be as high as 10 per cent in countries such as the former Soviet Union. The total global quantity could reach as much as 50 million tonnes per year, but more accurate estimates need to be made.

Nuclear energy

In 1953, 'Atoms for Peace' was the slogan and programme launched by President Dwight D. Eisenhower at the United Nations. Commercial nuclear energy was portrayed as the panacea – a limitless, inexpensive and clean energy resource. Many people perceive the underlying motive force behind the development of nuclear energy as a military need for still greater resources, encapsulated in the phrases of the day such as 'A Bigger Bang for a Buck' and 'Massive Retaliation'.

Conventional nuclear power uses mainly uranium in the process of nuclear fission, in which atomic particles are split and large amounts of energy released (see Box 7.2). The nuclear particles released from a fission event cause further fission of neighbouring uranium atoms in 'chain reactions'. The heat which is generated by these chain reactions is harnessed to raise steam to drive electricity-generating turbines.

There are a number of different types of

Plate 22 *The Hoover Dam in Nevada famed for the earthquakes it initiated.*
Courtesy of Comstock.

Plate 24 *Arco Corporation's photovoltaic power plant at Carrisa Plains, California, USA.* Courtesy of Comstock.

Plate 23 *Wind farms at Altamont Pass, California, USA. California produces over 80 per cent of the planet's wind-generated power, but this is still only a mere 1 per cent of the USA's yearly energy needs.* Courtesy of Comstock.

Plate 25 *Geothermal pumping plant near Grindakiv in Iceland which supplies hot water around the Keflavik Peninsula for domestic use. Cold fresh water is pumped into the ground at the plant and reaches temperatures of 95–125°C before being pumped via a system of pipes, which have a total length of 300 km, to the user who eventually consumes the water at about 80°C. (A) The plant and one of the distributary pipes. Evidence for the relatively recent volcanicity and high heat flow in this area is provided by the fresh poorly vegetated lava flow in the foreground and the young volcanic cones in the distance. (B) The 'Blue Lagoon' at the plant which provides hot baths for tourists and earth scientists.*

nuclear reactor, such as Magnox stations, pressurized water reactors (PWRs), and advanced gas-cooled reactors (AGRs). Nuclear power has been a priority for many post-Second World War governments.

The British civil nuclear energy programme provides a good example of one country's approach to the issue of a changing energy policy. The Magnox programme was initiated in the 1950s, even though the Conservative government of the day did not expect this programme to be economic (House of Commons Energy Committee 1990). In 1964, the Labour government authorized a second reactor programme, the AGRs, involving a twenty-fold scaling-up from a small prototype. By the early 1970s, the AGR programme had run into serious difficulties, both economic and technological. Another reactor type, the steam-generating heavy water reactor, was also briefly considered, but shelved in favour of continuing with the AGR programme. In 1979, the newly elected Conservative government announced a third programme, the PWRs with an intended electricity-generating capacity of 15 gigawatts and costing (at 1979 prices) £15 billion over 10 years from 1982. In Britain in 1987, the first PWR, Sizewell B, was formally authorized by the government, and is scheduled for completion in 1994 – the other three identical PWRs which were approved have effectively been abandoned until the 1994 review of Britain's nuclear programme. One of the common themes running throughout the nuclear programme has been the escalating costs of reprocessing spent fuel, and the estimated costs of decommissioning reactors.

Nuclear energy was seen as the sensible, modern answer to the future energy requirements until nuclear accidents such as Three Mile Island and Chernobyl (see Chapter 6). Environmentalists then began to emphasize the negative factors such as safety and the issue of the dumping of nuclear waste. In the USA, no new nuclear power station has been commissioned since 1978, at least partly as a consequence of the nuclear accident at Three Mile Island in 1979. The other factor espoused by environmentalists is that it is simply uneconomic to order any new nuclear power plants. There are currently moves by the nuclear industry to persuade Congress to simplify the way in which operating licences are

obtained, something that has already been started by the US Nuclear Regulatory Commission (NRC). The principal obstacle is that two hearings are required to make operational any nuclear power station in the USA – one prior to construction, and the second following construction but before the NRC grants an operating licence. Investors are disinclined to put their money behind such a lengthy process from commissioning to operation.

Critics of nuclear power argue that, in the final analysis, the lower limit of accident probability may be taken to an irreducible minimum level of human error, and that however technically safe a nuclear reactor may be, the level of risk is still unacceptably high. This is, of course, a value judgement. Many environmentalists will accept a level of accident risk which is generally considerably lower than that which is acceptable to industrialists. Politicians, depending upon their allegiances, will express the complete spectrum of opinions.

The governments of some countries have decided to reduce drastically, or eliminate, the use of nuclear energy. In January 1991, the Swedish government committed itself to phasing out the use of nuclear power by the year 2010. Sweden actually generates more nuclear power per capita than any other country – 50 per cent of its energy comes from 12 nuclear power plants. At the same time, Sweden has allocated 3.8 billion krona (£352 million) over the next five years to developing alternative energy resources and to improving energy efficiency.

These resolutions by Sweden also include a pledge to maintain high levels of employment and an annual economic growth rate of 1.9 per cent. By the same token, it is committed to not building any more new **hydroelectric** dams – the source of the remaining energy needs. The final sting in the tail for Sweden is that it has also agreed that by the year 2010, its emissions of CO_2 will be reduced to the 1986 levels. Can these lofty goals be realized? The balance in this tricky equation seems to depend on improved energy efficiency (MacKenzie 1991).

All of the foregoing discussion has concerned nuclear fission – energy released by splitting heavy atoms such as uranium. The goal of energy from **nuclear fusion**, released by the bringing together of the light atoms of hydrogen through the fusion of the isotopes of hydrogen known as deuterium and tritium, is still a long way off. On the evening of Saturday 9th November 1991 there was a false alarm, when at the Culham laboratories in Oxfordshire, England, in a £75 million per year project jointly funded by 14 European countries including Britain, nuclear scientists believed that they had made the breakthrough in an experiment lasting only two minutes, in which their furnace named Torus achieved temperatures of 300 million °C, 20 times hotter than in the core of the Sun. They thought that, for the first time, they had succeeded in creating a controlled fusion experiment which released nuclear power, other than in a nuclear bomb. Later attempts at repeating this experiment failed, and there is a consensus that the experiment never generated power by nuclear fusion.

Hydrogen energy

One of the exciting new energy resources is the use of energy from sunlight to produce chemical fuels, for example hydrogen energy (see Box 7.3), supported by some as an easily transported and readily storable fuel, two attributes which have advantages over the limitations associated with converting solar energy directly into electricity. Since most of the regions of the Earth where there is ample sunlight for efficient solar energy plants to be constructed tend to be in remote areas away from large population centres, transportable chemical fuels may well provide an attractive energy proposition. Many early attempts to generate chemical fuels involved producing the light and combustible gas hydrogen by the electrolysis of water. Upon combustion of hydrogen, water is produced. If the electricity used to burn the hydrogen is gen-

erated from a non-fossil fuel such as wind, hydro-, or solar power, then this process can be seen as an environmentally benign fuel.

Renewable energy

Over the next decades, energy based on fossil fuels will continue to provide most of the world's energy requirements, and it is unlikely that present overall energy systems will change radically. It is likely, however, that within the second quarter of the next century the extent and depletion of certain oil and gas reserves will have begun to reflect on energy prices (World Energy Commission 1992). Global CO_2 emissions will rise throughout the 1990s, and Western Europe will not achieve the target 20 per cent reduction by 2005 as set out in Toronto. The World Energy Commission believes that the target is achievable by 2020, provided that an appropriate mix of energy is available and that effective incentives are given to promote energy saving. With these factors in mind, there is a need to consider new technologies for transport fuels, and for government/EC assistance in achieving their technological and commercial viability.

Renewable energy resources are becoming more attractive because, unlike conventional fossil fuels, they offer 'cleaner' technologies associated with lower emissions of greenhouse gases and do not contribute to acid rain. Furthermore, they offer energy diversification and, therefore, greater energy security. Many developed countries are encouraging programmes of utilizing renewables, such as the Non-Fossil Fuel Obligation in the UK which requires regional electricity-generating companies to provide customers with a certain proportion of their elec-

Box 7.2

Obtaining nuclear energy

Nuclear energy is currently obtained from splitting atoms, termed nuclear fission. Typical yields of fission products are shown in Table 6.2, giving some indication of the longevity of many of these radioactive substances. Intensive research is currently under way to harness this energy by combining subatomic particles in the process of nuclear fusion. Fusion is seen as having an enormous advantage over fission because it should produce much less radioactive waste as by-products, and the principal raw material is heavy water or deuterium, something that is effectively in limitless supply.

In order to develop and install nuclear fusion power stations successfully, it is still going to take enormous amounts of investment and time. Nuclear fusion is still not technically, let alone commercially, viable.

tricity supply from nuclear and renewable sources. The Non-Fossil Fuel Obligation was introduced to subsidize nuclear power, and the support of renewables was only added to give it an acceptable environmental gloss.

Some countries obtain extra revenue for the development of alternative energy sources from the conventional fossil fuel sector. For example, in the UK, the Fossil Fuel Levy is used to impose an 11 per cent surcharge on all electricity generated from fossil fuels, and the revenue raised (£1.25 billion in 1990/1) subsidizes electricity produced from non-fossil fuel sources – mainly nuclear energy and renewables, such as tidal and wind power. Clearly, this type of taxation carries many implications for any energy policy, not least of which is the positive discrimination against fossil fuels in favour of alternatives and renewables.

The following sections examine the principal renewables.

Hydroelectric energy

Water power is amongst the oldest energy resources. Before the Industrial Revolution in the nineteenth century, water mills were a common sight. Hydroelectric power (hydropower) is widely utilized around the world. It is a clean energy source but has certain environmental disadvantages such as the destruction of large areas of natural vegetation and/or farmland. Capital investment costs are normally very high for the construction of dams and reservoirs. Nevertheless, hydroelectric power may prove to be one of the most acceptable 'compromise' energy sources.

Hydropower is currently the world's largest renewable energy resource. Since there is so much information available on regional watersheds, hydro-resources tend to be the best audited of all the renewable energy options. These resources are defined as potential hydroelectric capacity, measured for example in kilowatts or megawatts. It has been estimated that the global potential hydroelectric capacity is enormous at 2.2 million megawatts, 'double the present installed world generating capacity for power plants of all types and sizes' (Cassedy and Grossman 1990).

Environmental problems associated with diverting water courses and the construction of dams include the loss of vegetation and agricultural land, and displacement of the population.

One of the ways to produce hydrogen is by the high-temperature decomposition of sulphuric acid (H_2SO_4), which releases water and sulphur dioxide (SO_2). The SO_2 is then reacted with iodine and water to produce hydrogen iodide (HI) and H_2SO_4, followed by the thermal decomposition of the HI molecules to yield free hydrogen (H_2) and iodine (I_2) gas. More recently, methods for producing hydrogen gas have included the electrolytic process where SO_2 and H_2O react to release H_2SO_4 and H_2, the advantage being that only 0.29 volts are required, compared to about 2 volts for the high-temperature decomposition of H_2SO_4. Another electrolytic process involves the reaction of bromine (Br_2), SO_2, and H_2O to produce hydrogen bromide (HBr) and H_2SO_4, followed by the application of 0.62 volts to the HBr molecules to release H_2 and bromine gas (Br_2). Today, the efficiency of producing electricity commercially from solar sources is around 12 per cent and 70 per cent for electrolysing water, with an overall efficiency value of about 8 per cent (Dostrovsky 1991).

Although it is theoretically possible to generate hydrogen from water by heating it to more than 2,000°C (obtainable from concentrated solar energy), the technology is not yet available to stop the hydrogen and oxygen from recombining to form water vapour as the gases cool. Other sources of hydrogen include fossil fuels and plant material, the latter being known as biomass. The biomass option only requires temperatures of 700–900°C and steam in the absence of air to produce a gaseous cocktail of H_2 and carbon monoxide (CO). It is feasible to ensure a mix of H_2 and CO in the ratio of 2:1 and 3:1 and to use this so-called syngas to synthesize liquid fuels such as pure hydrogen, methanol, and gasoline. This biomass technology is still in its infancy, but deserves sustained industrial and government funding because it provides the potential for the production of energy using renewable resources that are environmentally more friendly.

Box 7.3
Producing hydrogen energy

Changes in river courses may affect the animal, plant, and fish life.

Apart from the obvious problems of dam failure and the resultant catastrophe of flooding causing loss of life, there are many additional problems associated with dam construction and reservoirs. These include the flooding of ecologically important areas, for example the Kariba Dam in Zimbabwe floods 5,100 km² of land and is destroying one of the natural habitats of the rhinoceros and elephant. A further example is the Tucurui Dam in the Amazon which is causing the destruction of virgin rainforest and displacing thousands of people. The Volta Dam in Ghana displaced 78,000 people from 700 towns and villages, and was associated with a string of resettlement problems and increased pressures on neighbouring areas. Such flooding causes the decay of drowned vegetation leading to the acidification of the lake waters and the anaerobic conversion of organic matter to greenhouse gases such as methane.

Other problems include modification to surface and groundwater hydrology. In northern Quebec, for example, recently constructed hydroelectric power stations have altered the hydrology of an area comparable in size to Switzerland, much of this region originally having been forest. Since the construction of the Hoover Dam, which impounded Lake Mead on the Colorado river in 1935, earthquake activity appears to have increased, and this has been linked with the increased height of groundwaters and loading of the underlying rocks due to reservoir construction (Plate 22). Prior to the construction of the Hoover Dam, there were no earthquakes noted in that area: since 1935, more than 1,000 earthquakes strong enough to be felt by the local population have been recorded. Worldwide since then, more than 15 other reservoir constructions are known to have generated large earthquakes, including the magnitude 5–6.5 earthquake associated with the Koyna Dam in India which caused the loss of 177 lives.

Despite some of the negative aspects of using water power to generate electricity, hydroelectric schemes are seen by many people as one of the most acceptable, least environmentally damaging energy resources. Clearly, the careful and sensitive selection of sites for hydroelectric power plants and dams can ameliorate many of the potential problems. Very small-scale hydroelectric ('micro-hydro' and 'mini-hydro') power plants, however, can be much more appropriate in many rural situations, and do not suffer from most of the side-effects associated with large-scale dam construction.

Wind power

Wind power is one of the most underutilized renewable energy resources. Wind farms are capital-intensive but cheap to run, with greater wind speeds providing more power, and hence more cost-efficient plants (Plate 23). While many countries still seem uninterested in a serious commitment to wind power, countries such as Denmark appear to be blowing full-steam ahead. Table 7.4 shows the national targets for wind power development for selected countries.

The largest wind farm in Denmark is sited in the west of Jutland at Velling, where there are 100 wind turbines with a capacity of about 13 megawatts. Denmark also has the world's first offshore wind farm which will help supply the targeted 10 per cent of its energy requirement by the year 2000. The farm, which is expected to generate 12 million kilowatt-hours of electricity, has 11 wind turbines, sited in up to 5 m of water 1.5–3 km offshore from Vindeby on Lolland Island in the Baltic Sea. The co-operative company, Elkraft, which runs the wind farm, estimates that if the turbines have a 20-year lifespan, then the electricity will cost about 0.63 Danish krone (£0.058) per kilowatt hour. The turbines are controlled and monitored on land through optical-fibre cables built into the 12-kilovolt power cable on the seafloor that transfers the wind energy from the offshore turbines to land.

In the Netherlands, the government has committed itself to the largest programme of windmill construction since the seventeenth century. It wishes to encourage the construction of 2,000 new windmills and increase wind power by an equivalent 2,000 per cent by the end of this century. Their intention is to generate more than 1,000 megawatts of wind power. Currently, there are about 1,000 windmills in the country. In order to encourage this growth, the government will provide subsidies of up to 40 per cent to build the windmills, which will make the cost

Table 7.4 *National targets for wind power development in selected countries.*

Region		Output at 25% load factor
Denmark	c. 1,200 MW of wind power by 2000	2.6 TWh a^{-1}
Germany	250 MW of wind power by 1995	0.55
Greece	400 MW of wind power by 2000	0.88
Italy	600 MW of wind power by 2000	1.3
Netherlands	400 MW of wind power by 1995	0.88
	1,000 MW of wind power by 2000	2.2
California, USA	10% of electricity from wind by 2000	c. 30.0
India	5,000 MW of wind power by 2000	11.0

1 TW = 1 million kW
Source: House of Commons Energy Committee (1992).

of wind power about equal to that of nuclear or conventional fossil fuel-generated power. But, of course, this energy resource is far cleaner. By developing more energy-efficient windmills, the government hopes to increase the power output and so reduce the unit costs.

Those who are sympathetic to the technology of renewable energy believe that Britain is squandering the best wind in Europe. In the UK in 1990, only six wind projects were included in the list of renewable energy projects covered by the first Non-Fossil Fuel Obligation (NFFO), which was set by the Department of the Environment in Britain to allow renewable energy projects to establish themselves. Subsequently, in the second NFFO, this was increased to 220 MW at 58 sites. In a recent UK government White Paper (Department of the Environment 1988), the contribution from wind power has been set at 1,500 MW by the turn of the century. As a comparison, countries such as Denmark have set their sights on much higher targets, with a projected 1.5 per cent of its electricity coming solely from wind power, and 10 per cent by the year 2000 (see above). Even India has set a target of 5,000 megawatts by the beginning of the next century. Environmentalists in Britain believe that the low target (less than 2 per cent) in the NFFO is because the Department of the Environment is only offering very short-term contracts of up to eight years. This time limit arose because the EC would not allow a subsidy on nuclear power to go beyond 1998, but the UK government is seeking agreement from the EC to extend support for renewables. Yet the House of Commons Energy Committee (1992) has stated that 'the wind energy resource in the United Kingdom is

particularly rich', with the technical potential of onshore wind energy as 45 TWh/year, and offshore being as high as 140 TWh/year. The British Wind Association estimates that the UK has 40 per cent of Europe's total realizable wind energy potential. During the past decade or more, there have been considerable technical and economic improvements in wind energy technology (see Figure 7.2), all of which have served to make this renewable more viable.

Hopefully, governments will encourage the harnessing of wind energy through attractive contracts to prospective companies, and by allocating a specified proportion of their energy requirements to wind resources. The environmental impacts of wind farms are that they can cause local noise pollution and electromagnetic interference. Perhaps the major problem is that they are seen by many as unsightly. This has been a particular problem in the UK, where many of the windiest sites are in areas of outstanding natural beauty, national parks, and other scenic sites.

Tidal energy

Tidal power, which is inexhaustible and intermittent (although absolutely predictable and reliable), utilizes the generally semi-diurnal (twice-daily) rise and fall of the tides to generate electricity. Coastlines around the world are divided into: microtidal, where the tidal range is less than 2 m; mesotidal, with 2–4 m tidal ranges; and macrotidal, where the range is greater than 4 m. Macrotidal coasts, with tidal ranges some-

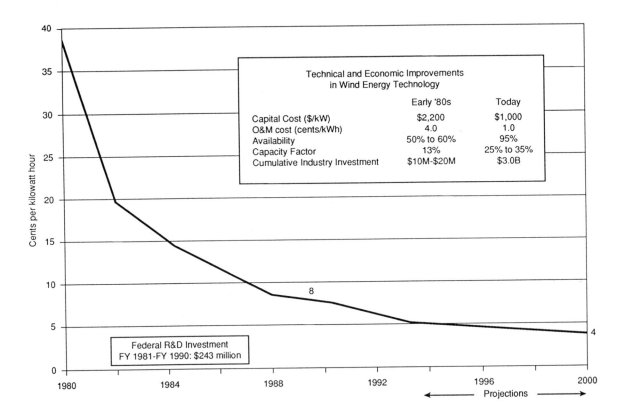

Figure 7.2 *Cost of electricity from wind in the USA.*
Source: House of Commons Energy Committee (1992).

times greater than 15 m (as in the Bay of Fundy, Nova Scotia), provide the best conditions to harness tidal energy.

In the UK, where the EC estimates that the technical potential of tidal power is 53 TWh, with about 90 per cent of this figure in eight estuaries, or about 20 per cent of current electricity demand in England and Wales (House of Commons Energy Committee 1992), there are major tidal energy schemes for the Severn, Mersey, Wye, Conway, and Humber river estuaries, along with other relatively small sites. The technology is largely proven, but for such major engineering projects the capital costs are high, although offset against this is that the lifetime of the barrages is expected to be around 120 years. Cost estimates are in the range of £10.2 billion for the Severn barrage (17 TWh or 8,640 MW annual output), £966 million for the Mersey barrage (1.4 TWh or 700 MW annual output), and £72.5 million for the Conway barrage, with annual running costs at £86 million, £17.6 million, and £600,000, respectively. Estimated electricity prices for the Severn barrage, at a

16.5-year payback period, are £0.06 per kWh, and at 20 years, £0.05 per kWh (ibid.). Broadly comparable prices are estimated for the other tidal barrages. A £4.4 million preliminary feasibility study for erecting a tidal power barrage across the Severn estuary has recently been completed with the conclusion that there are no overriding environmental or technical barriers to the project. This study was financed by the Department of the Environment, the former Central Electricity Generating Board (CEGB), and a consortium of construction companies named the Severn Tidal Power Group. The 10-mile barrage would be erected from Bream Head in Somerset, to Lavernock in South Wales, with a projected cost of about £9 billion.

The Severn tidal barrage would have a power output equivalent to five nuclear reactors, but costing six times the price of a nuclear pressurized water reactor (PWR) such as that being constructed at Sizewell, Suffolk (£1.5 billion). The tidal power station would house 200 turbines and generate more than 7,000 megawatts of electrical energy at a unit cost of £0.0379 per kilowatt hour (PWRs produce electricity at £0.0224 per kilowatt hour). Although the superficial economic comparisons between the tidal power station and a PWR suggest that the latter is better value, the working life of the PWR at Sizewell will be 35 years whilst it will be 120 years for the Severn barrage. The barrage scheme would also employ an estimated 30,000 people over its nine-year construction. Some would argue that, on balance, the economic and environmental case seems to favour tidal power stations over nuclear schemes such as the proposed Severn estuary scheme.

Tidal barrages can have a significant environmental impact, for example by impairing fish reproduction, as spawning takes place in areas where fresh water and saline marine waters mix, which in the case of the Severn could move the zone up to 30 km seaward. Also, barrage construction would reduce the area of mudflats washed by the tides, leading to more plant growth and corresponding reductions in the amount and diversity of invertebrate animal life. The ability of estuaries to support wintering and migrating bird populations could be affected, although this is an area of dispute and in need of further research. Also, tidal barrages can adversely affect migratory and spawning fish populations.

Wave power

The energy associated with waves can be used to generate electricity. Wave energy is very much in the research and development (R&D) stage, with few operational devices, and considerable uncertainties over costs. The UK and Norway have led R&D into wave energy.

Wave energy may involve small-scale shoreline (or onshore) wave energy, and large-scale offshore wave energy devices. Offshore devices have potential energy levels of three to four times those of the nearshore devices. One example of the technology, known as the 'Edinburgh Duck', or 'Salter's Duck', utilizes a large hinged plate that moves up and down in response to passing waves. The problem with this, as is the case for other wave-energy devices, is the reliability of the technology, but an attraction is that any environmental impacts are likely to be minimal.

Solar energy

The Sun is the main source of energy – it allows photosynthesis, the process whereby plants convert solar energy into chemical energy which is then changed into fossil fuels. Every year, energy from the Sun, or solar radiation, bathes the Earth's surface in roughly 15,000 times the current global energy supply, equivalent to an estimated annual energy supply of approximately 178,000 terawatts or 5.6 million exajoules (1 terawatt-hour, TWh = 1,000 million kilowatt-hours, kWh; 1 exajoule = 10^{18} joules). Other experts, such as Dostrovsky (1991), quote the amount of solar radiation reaching the Earth's surface annually as totalling 3.9 million exajoules (equivalent to the amount of heat released by the combustion of 22 million tonnes of oil). About 30 per cent of this total solar budget, however, is reflected back into Space and 50 per cent is absorbed, converted into heat and re-radiated. The remaining 20 per cent powers the hydrological cycle, but only a mere 0.06 per cent drives photosynthesis. The Sun is directly responsible for solar power, wind power, and hydropower.

Table 7.5 shows the solar radiation in selected countries. It is clear that many developing countries have considerable potential for developing solar energy plants. Box 7.4 summarizes the process of converting solar energy into electricity.

Solar power has come a long way since 1876 when two British researchers, W.G. Adams and A.E. Day, were the first to convert sunlight into electricity using a selenium cell. Energy from the Sun, or solar energy, is being used in many countries throughout the world. Not only is

Table 7.5 *Solar radiation in selected countries.*

Region	Solar radiation (kWh/m²/year)
Mali	2,490
Niger	2,450
Mexico	2,080
Sierra Leone	2,000
Venezuela	2,000
India	1,950
Brazil	1,880
Chile	1,630

Source: US Congress Office of Technology Assessment (1992b).

electricity generated from solar energy for domestic and commercial use, but also many everyday gadgets run on solar power such as solar-powered calculators, watches, personal cassette players, and radios. Satellites also rely on solar energy.

In the USA, electricity produced by solar power costs about 30 cents per kilowatt-hour, and this cost is rapidly dropping. In February 1989, a Californian company opened a plant to convert sunlight energy into heat (Plate 24). This solar-thermal plant expects to produce power at less than 3 cents per kilowatt-hour. As a comparison, in the USA, where conventional power stations use fossil fuels to drive electricity generators, running costs are about 3 cents per kilowatt-hour.

Southern California plugs into the largest power socket ever built. Here, the Sun-drenched, arid Mojave Desert is home to 600,000 computer-driven parabolic mirrors pointed skywards to collect the energy in the Sun's rays. This power farm reaps the almost daily harvest of solar energy and testifies to the practicality and utility of natural alternative energy resources. In southern California, the Mojave Desert power farm at Kramer Junction is the largest of the three solar energy complexes operated by Luz International. The farm covers an area of about 1,000 acres and generates 90 per cent of the world's grid-connected solar energy at 275 megawatts of power. The parabolic mirrors track the Sun to focus its rays on to pipes filled with synthetic oil which is heated up to about 390°C. The super-heated oil is then used to boil water to drive steam-turbine engines. It is as simple as that. A cocktail of sunlight, air, and water demonstrates the viability of this energy resource.

In Europe, the largest solar energy plant is situated near Koblenz in Germany, at Kobern-Gondorf on the slopes of a former Moselle vineyard. It can generate 340 kilowatt-hours at peak capacity. Just outside Chur in eastern Switzerland, the motorway is lined with solar panels for many hundreds of metres. This kind of approach to supplementing other energy resources provides an unobtrusive and sensible location for such panels. Other countries should follow suit.

As to the investment in R&D for solar energy, amongst the market leaders is Japan which spends about US $50 million annually, compared to US $43 million for 1990 by the United States. In fact, in 1989 in the absence of rapid results, the Reagan administration cut the R&D figure to US $35 million from its 1981 high of US $150 million (Spinks 1990).

Box 7.4

Converting sunlight energy

The energy from the Sun strikes the Earth every day as a bombardment of packets of light energy, termed photons, the sum of which is commonly referred to as the solar flux. It is stored in photoelectric cells either for immediate or later use. Naturally, countries with the most potential for using solar energy are those which have the sunniest weather, thus this energy source is more appropriate in some countries than others.

There are two main methods of converting sunlight to electricity. One is to convert the sunlight to heat or thermal energy, and then use the heat to raise steam to generate electricity. This method has now been eclipsed by the photovoltaic cell, a device for changing solar energy directly into electricity and which uses semiconducters (Figure 7.3). Efficiency, although improving, remains a problem with these cells. The efficiency of these cells varies between 15 and 29 per cent (monocrystalline silicon cell), and 12 to 18 per cent (polycrystalline silicon cell) (Spinks 1990).

The US Department of Energy aims to improve this to 35 per cent, with 40 per cent efficiency being regarded as the theoretical maximum. There are many types of crystals that are now being used to make solar cells, such as copper indium diselenide, gallium arsenide, and gallium antimonide. Gallium arsenide, appears to be one of the most efficient semiconductor materials, but unfortunately is also one of the most costly. There is also much interest in amorphous silicon cells which have a lower efficiency than crystalline cells but are cheaper to produce. Despite all these difficulties, the increased efficiency and reduced costs of solar panels have conspired to reduce the cost of solar power to an average of about US $0.5 per kilowatt-hour from a 1980 figure of around US $1.5 per kilowatt-hour.

In the future, it may be viable to use photovoltaic panels as cladding on buildings to supply electricity during the summer months to run air-conditioning systems.

Countries where solar energy is used tend to supplement their electricity grid by solar energy rather than see it as a mainstay of their energy budget. The environmental argument, however, is that even if solar power is marginally more expensive, it should still be used since the long-term cost to the planet through continuing to burn fossil fuels at present rates will be greater.

Geothermal energy

The Earth continuously emanates heat energy from rocks and molten magmas at depth, referred to as **geothermal energy**. It is an essentially limitless resource that is presently underutilized, and is estimated worldwide at almost 5 GW (US Congress Office of Technology Assessment 1992b). The two main sources of geothermal energy are heat extraction from rocks and the tapping of hot water, including oceanic-thermal energy by heat exchange pro-

cesses (see Box 7.5). Table 7.6 shows the geothermal electricity generated in the USA and selected developing countries in 1990.

There are geothermal energy programmes in a number of countries, for example in the USA in northern California where more than 2,000 megawatts of power are produced. Some other very suitable countries, where the geothermal gradients are high, include Italy, the Philippines, Mexico, and Iceland (Plate 25). In Britain, technical problems have bedevilled the hot dry rock project that aimed to tap geothermal energy from the Cornish granites, and its future remains uncertain. In the port of Southampton, geothermal energy is tapped from hot salt waters (brines) below the city, which are used to heat public and private buildings. The brine is pumped out of the well at a rate of about 12 litres per second and a temperature of 74°C, giving a basic 1.1 megawatts of power. Additional energy sources, including a diesel generator and a gas/oil boiler, raise the maximum power output of the scheme to 14 megawatts. Clearly, although

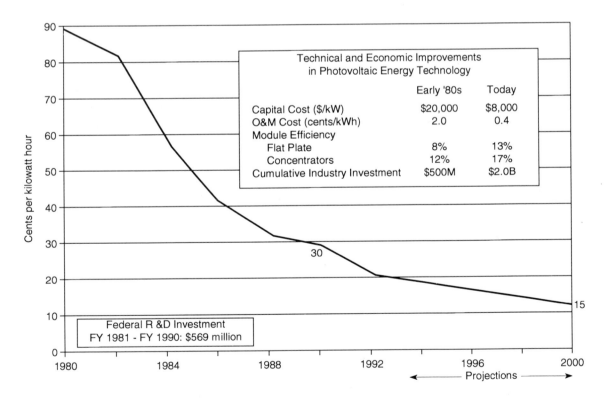

Figure 7.3 Cost of electricity from photovoltaics in the USA.
Source: House of Commons Energy Committee (1992).

ENERGY

Table 7.6 *Geothermal electricity generation in selected countries in 1990.*

Country	Geothermal capacity (MW)	Technology
USA	2,827	all technologies
Philippines	894	single flash
Mexico	725	dry steam, single flash, double flash
Indonesia	142	dry steam, single flash
El Salvador	95	single flash, double flash
Nicaragua	70	single flash
Kenya	45	single flash
Argentina	< 1	binary
Zambia	< 1	binary

Source: US Congress Office of Technology Assessment (1992b).

geothermal energy only provides a relatively small energy source, it makes a contribution to providing power and has environmental benefits.

Geothermal energy does have some adverse environmental implications. CO_2 and hydrogen sulphide (H_2S) emissions may be high, the latter being limited in the USA to 30 ppbv. Technologies exist to eliminate these emissions, for example the so-called binary technology. Water consumption is also high, with the amount required being particularly dependent on the plant design. CO_2 emissions depend on the amount of carbon contained in the particular resource, averaging about 5 per cent of the CO_2 emissions per kWh output from a coal-fired power station. There may be land subsidence above geothermal wells, and water supply can become contaminated by saline, possibly toxic waters, but these problems can be managed successfully.

Biomass energy

Biomass fuels are produced from plant and animal matter wastes and residues, and account for about 14 per cent of the current world total energy requirements, equivalent to approximately 25 million barrels of oil per day, equal to OPEC's current crude oil production (Hall *et al.* 1992).

Natural degradation of waste under controlled conditions in landfill sites releases energy which can be tapped. It is estimated that the biodegradation of 1 tonne of refuse can ideally produce 400 cubic metres of gas, equivalent to 7,500 megajoules of heat energy. Generally, less than 25 per cent of the degradable fraction of such waste will break down within the first 15 years, with most being produced in the early stages. Thus, the efficiency of energy production from decaying material will decrease rapidly with time.

Organic matter such as food products decay quite rapidly, principally through the action of bacteria. This decomposition initially uses free oxygen until the decaying matter becomes oxygen-deficient, when the natural breakdown of organic matter by organisms occurs without free oxygen as anaerobic digestion. This comprises three main stages. First, simple sugars such as glucose ($C_6H_{12}O_6$) are produced. These are then converted into fatty acids, mainly acetic acid, by acetogenic bacteria. Finally, methanogenic bacteria under anaerobic conditions break down the organic matter, the organic acids, into a **biogas** or landfill gas. All landfill sites release methane (CH_4), a greenhouse gas. By making use of it to produce heat and power, CH_4 is converted to CO_2, a less potent greenhouse gas and its use reduces the release of CO_2 from other

Box 7.5
Geothermal technologies

Technologies for geothermal heat extraction include direct steam, single-flash, double-flash, and binary systems. In direct steam, the simplest of the technologies, steam is piped directly from the subsurface reservoirs and then used to drive steam turbines to generate electricity. In the single-flash process, use is made of the underground hot water which is 'flashed' into steam to drive turbines, and in the 'double-flash' process, a second 'flashing' is undertaken to utilize up to about 20 per cent of the energy that would otherwise escape. In binary units, an intermediate working fluid is used to transfer energy from the geothermal reservoir to the electricity-generating turbines. An advantage of the binary cycle, which is especially appropriate in developing countries, is that it can use relatively low-temperature resources (e.g., 170–180°C), and this cycle is suitable for relatively small plants, typically 5–10 MWe.

fossil fuels that would otherwise have been burnt.

Biogas is a cocktail of roughly equivalent amounts of methane and carbon dioxide, a few per cent by volume of hydrogen and nitrogen, together with traces of sulphur-containing organic molecules and halogens, plus some heavier hydrocarbons. Landfill sites with household waste commonly have temperatures in the region of 35°C which is ideal for the natural biodegradation stages outlined here. The biogas or landfill gas can be burnt to generate electricity and/or heat.

The first commercial venture to exploit the production of landfill gas was about 14 years ago in the USA, in California at Palos Verdes. Countries such as Germany and Britain followed suit only nine years ago. About 70 such sites operate in the USA with the largest being at Puente Hills, California, which generates 46 megawatts that is sold to the local electricity grid. Approximately 50 production sites operate in the former West Germany.

Throughout England and Wales, 300 potential sites for the economic production of landfill gas through biodegradation were identified in 1986 by the former Department of Energy, now incorporated into the Department of Transport and Industry (DTI). The Department of Energy estimated that the total annual production of biogas from these sites alone would be equivalent to 1.3 million tonnes of coal, of which at least 1 million tonnes is economically viable to recover. In 1989, just 26 of these 300 sites were being exploited. In Bedfordshire, for example, London Brick (part of the Hanson Trust Group) sold off the large open pits created by the extraction of clay for brick making as landfill sites for the dumping of household waste from which CH_4 is marketed. In Britain, estimated savings on conventional fuels from biogas production increased from £3.5 million in 1986 to approximately £6 million in 1988. Savings for 1989 were expected to be in the region of £12 million with a further doubling in 1990.

Although the principal use of bio gas is in the generation of electricity or supply to nearby boilers and furnaces, another use is as a substitute for high-grade fuels after it has been cleaned. On the debit side, the production of bio-gas from landfill sites raises environmental issues. Biodegradation produces greenhouse gases, including small amounts of CFCs. Until more research has been conducted into the management of landfill sites and the associated production of biogas, it is not known if the benefits outweigh the disadvantages. The onus is upon the proponents of biodegradable waste and biogas to show not only that these processes provide an efficient way of converting unwanted waste into energy, but also that the process need not necessarily harm the environment. The use of this gas provides a way of utilizing something that would otherwise escape into the atmosphere as a greenhouse gas.

Not only can energy be produced from the degradation by biological means of organic matter, for instance bacterial decay in the process of biodegradation, but there is also the potential to reclaim or recycle the various metals, glass, and plastics that form part of such waste. In a typical year, Britain produces more than 80 million tonnes of mixed household, commercial, and industrial waste. An estimated 28 million tonnes of this waste is biodegradable. The problems of waste disposal are increasing, both from the point of view of available land for dumping and the danger from the by-products of the natural breakdown of the waste, let alone the intrinsic dangers from certain dumped industrial chemicals.

Typical household waste has a calorific value of about one-quarter to one-third of that of coal, equivalent to an energy yield of approximately 9 megajoules per kg. A proportion of this energy can be released simply through combustion or incineration to produce electricity. An alternative is to convert the waste into a solid, liquid, or gaseous fuel, but this is very much for the future. Combustion is easy, but it may release harmful gases into the environment, including various greenhouse gases.

Road transport and energy

Transport services are a major and integral part of economic development. Many countries have to import large quantities of transport fuels, leading to import dependence. The principal control on traffic growth is the GDP and fuel price. The two fastest-growing sectors of energy consumption are for the supply of electricity and transport. Western Europe consumes 18 per cent of world commercial primary energy supplies, but contains only 7 per cent of the global proven coal reserves, 2 per cent of proven oil reserves, and 5 per cent of proven natural gas reserves, and is therefore a net importer of energy, a situation that is likely to remain for the foreseeable future. It is factors such as these which lend credence to arguments in favour of developing alternative, renewable energy supplies (energy security), and when combined with environmental issues such as clean air requirements conspire to make liquid **biofuels** (see Box 7.6) worthy of much more attention as possible, large-scale substitutes for fossil fuels. An area of present and future concern, and fraught with powerful vested-interest lobbies in preserving the dominance of oil as a fuel in motor transport, is the use of biofuels in transport policies. This issue is dealt with at some length because of the potential for biofuels to substitute for petroleum products.

With the increased use of transport in most countries, the level of atmospheric and other environmental pollution is growing. Transport emissions account for a large part of the global emissions of air pollutants which contribute to photochemical smogs and pose health hazards. In India, for example, petrol-fuelled vehicles, mainly two- and three-wheeled, account for 85 per cent of the carbon monoxide, and 35–65 per cent of the hydrocarbons from fossil fuels, with diesel vehicles (buses and trucks) being responsible for 90 per cent of the urban nitrogen oxide (NO_x) emissions (US Congress Office of Technology Assessment 1992b).

Liquid biofuels for motor transport

Biofuels (Box 7.6) for motor transport are perceived by many people as providing a suite of alternative and renewable motor fuels which cause less environmental pollution (Box 7.7), permit greater energy diversification, decrease the dependence on petroleum products (security of energy supply), and provide a potentially profitable use for fallow agricultural land, for example set-aside land in Europe under the EC Common Agricultural Policy (CAP). The development of biofuels in motor transport may depend on favourable tax regimes (just as the launch of lead-free petrol required tax incentives), probably combined with total or partial tax exemptions for pilot studies, something that is permitted under the EC Council Directive on the harmonization of excise duties, but taken up by few member states. Given that fossil fuel oil prices are in real terms at their lowest ever, the marginal to slightly uneconomic aspects of biofuels for use in motor transport may change by the early part of the twenty-first century.

In the EC, Germany, Switzerland, France, and Britain are the biggest cultivators of oilseed rape (Plate 26). In Europe, the Italian company Novamont produces and supplies **rape methyl ester (RME)** to 17 Italian cities, the Berlin taxi association, all the taxis in Bologna, buses in Zurich, and the ferry on Lake Como. In Britain, RME-based motor fuel is used in a pilot study for three public transport buses in Reading. RME is also used in niche markets, for example as a lubricating oil in North Sea oil rigs. In 1992, Novamont produced about 50,000 tonnes of RME, and in 1993 about 100,000 tonnes using its new 60,000 tonne capacity plant in Leghorn, Italy, which is Europe's first and largest plant solely for the purpose of biodiesel production. Previously, the main customer for Novamont's RME had been the Italian domestic heating market due to environmental considerations, that is, the need to cut SO_2 emissions drastically in cities such as Milan (where La Scala is heated by biodiesel). In France, near Compiègne, a 20,000 tonne per annum factory is being built. In Austria, where Vogel and Noot are the main manufacturers of equipment for biodiesel plants, in 1991 almost 20 per cent of the 50,000 hectares of rape cultivated was used to produce RME, and a group of agricultural co-operatives ensure its distribution to more than 50 fuel stations.

In Europe, trials to evaluate bio-ethanol as a fuel have been conducted in Germany, Italy, and Sweden, with a small public distribution of blended bio-ethanol in France. The USA has major transport biofuels programmes using bio-ethanol, such as the US 'Gasahol' programme, supported for energy diversification and environmental reasons. Ethanol production in the USA began because of the Cuban embargo on sugar exports to the USA, which led to a market crisis and the decision to expand substantially the US domestic production of sugar. This led to overproduction, and the conversion of the surplus sugar into energy. In the Midwest, USA, bio-ethanol has been used for about the past 10 years as a 10 per cent blend in gasoline (gasohol). Brazil is the biggest and most successful user of bio-ethanol for motor fuel. Under the 14-year 'Proalcohol' initiative in Brazil, set up in response to the world oil crisis in early 1975, currently about one-third of Brazil's 12 million cars use pure ethanol from sugar cane, with the rest using a 20 per cent blend. The dry pulp left over from crushing the sugar cane, or 'bagasse', can be used as an animal feed or combusted in high-pressure steam turbines to generate electricity, and the potassium-rich liquid waste produced in the distillation of ethanol provides a good fertilizer. Brazil now produces annually about 10 million tonnes of bio-ethanol, giving major cost reductions through economies of scale, but bio-ethanol still costs about US $40 per barrel to produce.

Despite some commonly held beliefs about the health risks to populations living near areas of extensive oilseed rape cultivation, there is no scientific/medical evidence to support such a contention. A British study by the Department of Environmental and Occupational Medicine, Aberdeen University, Scotland, compared 2,000 people in rural areas, half of whom live in areas extensively cultivated for oilseed rape, but otherwise there were no known differences. The results show no statistical difference in allergy and asthma complaints between the populations, implying that people living in areas of extensive cultivation of oilseed rape are no more prone to asthma and other allergies than the rural population living outside such areas. There is, however, a 20 per cent greater incidence of complaints associated with allergic reactions to pollen in rural compared to other areas, but this is not specifically related to oilseed rape.

Under the Common Agricultural Policy (CAP) reform agreement reached by the EC Council in May 1992, EC growers of cereals, oilseeds, and protein crops (other than those categorized as small producers) are required to set aside 15 per cent of the land on which they are claiming arable support in relation to the 1993 harvest. The change from tonnage- to land-area-based subsidies to farmers for the cultivation of oilseed rape may alter farming practice, because prior to the CAP the higher-yield

winter varieties of rape (average 3.2 tonnes per hectare) were more popular. Although the spring crop gives lower yields (average 2.2 tonnes per hectare), its cultivation requires fewer inputs and, therefore, is cheaper. Rape tends to be grown on heavy soils more suited to winter cereals which may be unsuitable for spring varieties. If up to 5 per cent of EC gasoline were to come from biofuels, then at 1992 figures between two and five times the current area of set-aside land, depending on the selected crop, would need to be cultivated for biofuel production. In the EC, blended biodiesel may be economically more viable, at least for the short term, compared to bio-ethanol, principally because its use would involve the least disruption to the present energy production and distribution systems, and best fit the present EC farming practices and policies.

More employment in rural areas and associated industries could be created by developing a biofuels market, but with job losses in the fossil fuel and allied chemicals industries. Due to the relatively small volume of biodiesel marketed within the EC, RME producers and suppliers such as Novamont, have targeted high-volume users in niche markets, such as public transport bus and taxi operators. Other potential niche markets are where the environmental advantages of the greater biodegradability of biodiesel can be exploited, as in inland waterways and urban use. The economy of scale is helped by the simplification of the distribution system for biodiesel in being delivered direct to the high-volume user.

Oil company studies suggest that liquid biofuels will only marginally improve EC energy security, with the 5 per cent of gasoline level given over to biofuels accounting for a 1.5 per cent elimination of oil imports (equivalent to 0.5 per cent of EC primary energy consumption). The oil companies and closely allied chemicals industries believe that biofuels must compete on the present market as pure fuels or as octane enhancers with alternative oxygenates.

For the EC, a study by the Department of Land Economy at Cambridge University, UK, predicts that within the EC by the year 2000, 1–1.5 million hectares of land will be removed from food production, and that this figure will increase to an estimated 5–5.5 million hectares by 2010. If the consumption of diesel

Table 7.7 *Biofuel emissions relative to conventional fuels.*

Pollutant	Main effect	Biodiesel	Blend with petrol	
			Bio-ethanol 10%	Bio-ethanol 95%
Sulphur dioxide (SO$_2$)	acid rain	almost 100% less	10% less	60–80% less
Nitrogen oxides (NO$_x$)	acid rain, ground-level ozone	up to 20% increase	2% increase	no difference
Unburnt hydrocarbons (HC)	respiratory diseases, possible carcinogens	typically, 10–40% reduction	6% increase	up to 15% less
Carbon monoxide (CO)	respiratory diseases	up to 30% less	4% less	up to 8% increase
Particulates	respiratory diseases	up to 40% less	no significant difference	up to 5% increase

Source: After POST (1993b).

remains stable and all this land were used for the cultivation of oilseed rape for biodiesel under a rotational set-aside agricultural system, then by the year 2000 an estimated 10–15 per cent (1–1.5 million tonnes) of the diesel market could be satisfied by this fuel, growing to 51–56 per cent (5.1–5.6 million tonnes) by 2010. If such projections were to prove accurate, then it is not surprising that the oil companies and allied chemical industries feel a certain amount of discomfiture at the prospect of the increased use of biofuels in motor transport.

Environmentalists, including the producers of biofuels, would argue that even though the net volume of biofuels in motor transport is small, and is likely to remain so in the near future, the increased use of biofuels can help to achieve specified emission standards to stabilized levels and, therefore, warrant government support through encouraging further R&D, and additional pilot studies (possibly with tax concessions/exemptions). Ultimately, the use of biofuels in motor transport hinges around any perceived environmental benefits (see Table 7.7) and increased energy security over and above shorter-term economic factors.

Box 7.6

Types of liquid biofuels

Biodiesel can be produced from most edible oil crops. In Europe, oilseed rape is the principal crop used to produce biodiesel, from which the oil (called rape seed oil, oilseed rape) is converted into rape methyl ester (RME) (Plate 26). In the USA, soya oil is the main source, whereas canola oil is used in Canada. The main opportunity for future energy from edible oils could be a cost reduction through using acid oils and waste oils. As with many industrial monoculture crops, oilseed rape should not be grown continuously on one site because of the risk of disease build-up in the plants and, therefore, it should be cultivated as a rotational crop.

Oil is extracted from the rape plant by crushing, with 3 tonnes of rape yielding 1 tonne of usable RME. Most diesel engines can run unmodified on unblended rape oil but become clogged after several days, prevented through removing the glycerine from the oil by mixing 1 tonne of rape oil with 110 kg of methanol in the presence of the catalyst nitrogen hydroxide, and heated to temperatures of 40–50°C. The glycerine then settles out to leave the clear liquid RME. Glycerine, with many industrial uses, such as pharmaceuticals, cosmetics, and explosives, is a high-value co-product, giving a significant economic credit to RME production. Biodiesel and ETBE appear to have no serious technical limitations to their use in motor transport. Blends of less than 10 per cent RME with mineral diesel fuel are commonly used since this involves little, if any, modifications to existing engines. Engine performance is slightly reduced where RME is used, estimated by the Italian tractor manufacturer, Same, as less than 2 per cent compared

The use of biofuels as alternatives to conventional petroleum products requires a substantial shift in public attitudes, probably towards an acceptance of paying a premium for motor biofuels, and any new engine technology. Such changes would require a change of the fuel market through subsidies and tax incentives, and possibly a pollution or carbon tax. It could be argued that since energy markets are already regulated through government pricing policies, a shift towards energy being sold at prices that fully cover all costs – including environmental costs (often referred to as internalizing the externalities) – may lead to biofuels becoming more competitively priced compared to conventional petroleum products, at least as niche-market fuels. In the USA, tax concessions/ exemptions give a subsidy of about 60 cents per gallon of biofuel gasoline. In California, tailpipe emission control standards are more stringent than in any other state, and it is this factor rather than purely economic considerations that is stimulating the expansion of biofuels in motor transport.

Energy efficiency

The fundamental problem with the consumption of non-renewable energy resources is that individuals and countries simply use too much,

Table 7.8 *Emissions associated with production/saving of 1000 MW of electricity.*

Method	CO_2 tonnes per annum
Coal	5,912,000
Nuclear	230,000
Hydroelectric	78,000
Wind	54,000
Tidal	52,000
Loft insulation	24,000
Low-energy lighting	12,000

Source: Mortimer (1989).

and human activities are invariably wasteful of valuable energy commodities. Present attitudes tend to accept wastefulness, and encourage a business-as-usual approach. Energy conservation is popular as a buzz phrase but not in action. The growing public concern over poor air quality and the emissions of greenhouse gases associated with the production of electricity (see Table 7.8), however, has led to governments, particularly in the developed, industrialized countries, taking a more serious view of these issues, including proposals for a carbon or carbon/ energy tax in the EC (see Box 7.8). International action to reduce greenhouse gas emissions and generally improve air quality is dealt with in more detail in Chapters 3 and 4.

The 1973 oil price rises forced energy planners to rethink their strategies. Unfortunately,

---

with mineral diesel fuel. The high aniline point of RME means that it destroys rubber more easily than mineral diesel oil, a problem that is not manifest within the engine, but requires addressing for distribution systems using rubber hoses, for example in oil tankers and at petrol stations. The use of stronger rubber in the distribution system, at cost, provides a solution.

The main sources of bio-ethanol (pure ethanol is neat alcohol) are from cereals (wheat and barley), sugar beet, maize, and surplus wine, and from lignocellulose (wood and straw) using either acid or enzyme hydrolysis, the technology of which is not sufficiently advanced as yet and requires further research, for example to improve enzyme efficiency.

The technology for producing bio-ethanol (fuel ethanol) from feedstocks containing sugar

and starch is well proven and the processing equipment is commercially available on an industrial scale. Ethanol is produced by fermentation, and recovered by a distillation process. Bio-ethanol blended with petrol must have the water content removed (anhydrous ethanol) by a dehydration process after distillation, as even small amounts of water lead to phase separation. For wheat, the yield is about 6.7 tonnes of field-harvested grain per hectare, equal to about 2.7 kg of grain producing 1 litre of bio-ethanol. Bio-ethanol is used as a blend or as a pure fuel. Vehicle engines require substantial modification to run on pure bio-ethanol such that they cannot then run on conventional petrol. Blended motor fuel containing less than about 10 per cent bio-ethanol does not require such modification.

Box 7.6
Types of liquid biofuels

from the standpoint of energy-resource management, the 1980s have seen a return to relatively cheap fossil fuels and so discouraged, on purely economic considerations, further improvements in energy efficiency. The International Energy Agency (IEA) has estimated that if available, economically viable ways of conserving energy were implemented, then by the turn of the century energy efficiency could be 30 per cent greater than the present level.

The need to conserve both existing reserves of fossil fuels and the growing awareness of environmental impact have led to increased energy-efficient technologies (see Table 7.9). Necessity has been the mother of invention. Energy-efficient buildings that are better insulated, optimize available natural light and are constructed from better materials, new types of combustion furnaces (e.g., condensing furnaces that require 28 per cent less fuel to produce an equivalent amount of energy compared to conventional furnaces), the cogeneration of power (e.g., heat and electricity from the same process) and a greater reliance on public transport are but a few of the ways in which people can use energy resources more sensibly.

Cars and light commercial vehicles consume one out of every three barrels of oil produced and, in the USA, contribute 15 per cent of the total CO_2 emissions into the atmosphere. Attempts both to reduce the polluting capability of vehicles and to conserve energy (e.g., through the use of unleaded petrol and better engine design) are becoming increasingly popular throughout the developed, industrialized world.

Table 7.9 *Energy efficiencies of fossil fuel power plant systems.*

Power plant system	Current efficiency (%)
Current US power plants (overall average)	33
Oil and natural gas combustion	
Aircraft derivative turbines	40
Combined cycle systems	47
Coal combustion	
Atmospheric fluidized-bed combustion	38
Pressurized fluidized-bed combustion (using a combined cycle)	42–44
Integrated gasification combined cycle	43
Fuel cells	40–60
Cogeneration	up to 85

Source: World Resources Institute (1990).

Electric cars and trams provide cleaner and less noisy forms of public transport. In many European cities, such as Amsterdam, Oslo, Geneva, and Basle, the trams are a pleasant feature of the urban landscape.

Most housing has not been designed with energy efficiency in mind. It's a pity, because an energy-efficient home is less harmful to the environment, saves money, and can be much more pleasant to live in. Also, such homes need not be expensive to construct.

There are many ways of making homes more energy-efficient (see Oliver 1991). Turning lights off when leaving rooms empty, using

Box 7.7

Energy balance and CO_2 balance of biofuels

Most studies of biofuels indicate a positive energy balance (i.e., the energy value of the fuel obtained is greater than the total amount of energy input at all stages in its production), although the magnitude of the energy credits are a common source of dispute. The fate of the co-products from biofuel production is crucial to the overall energy and CO_2 balance. An ETSU study (Culshaw and Butler 1993) suggests that the energy ratio for biodiesel production (output/input) ranges between 1.3 and 3.8, depending on the use made of the by-products. In 1991, the British consultancy, Environmental Resources Ltd (ERL), on behalf of the European Fuels Oxygenates Association (EFOA), examined the energy inputs and associated CO_2 emissions arising from the production of fertilizers and pesticides, farming operations, crop harvesting, transport, processing and drying, and onward transport of bio-ethanol and co-products. The calculations were undertaken for a modern industrial plant producing 80,000 tonnes of fuel annually from four principal crops that are the main candidates for large-scale bio-ethanol production in Europe, that is, wheat, sugar beet, sweet sorghum, and Jerusalem artichoke. In its report, ERL showed that only the currently uneconomic option of using the co-products from bio-ethanol production as fuels would yield a reasonable energy credit and a significant reduction in CO_2 emissions.

better insulation, double-glazing, and draught-proofing around doors and windows, are but a few of the ways. New homes should be constructed to minimum standards. Windows facing south and east are the most efficient at receiving the Sun's rays – and also brighter. Depending upon the latitude and climate of a country, its housing should be designed so that windows preferentially face the appropriate directions and are of a suitable size.

In countries with a high amount of sunshine hours, solar panels can supplement or provide all the domestic energy requirements. Superinsulation, where buildings are sealed against draughts with a ventilation system that is controllable to allow fresh air in winter, saves significant amounts of energy. In Sweden, legislation now decrees that all new houses must have less than three changes of air per hour. Superinsulated homes may have fewer than one air change per hour, compared to many homes where there are typically far more. And, in another attempt at being more environmentally conscious, CFC-free plastic foams are available for cavity filling.

Some countries, such as the UK in 1993, appear to be pursuing a policy of increasing fuel and power prices through imposing a value-added tax (VAT), therefore leaving the consumer to decide priorities in terms of both expenditure and improving energy efficiency. An alternative means of encouraging energy efficiency and energy conservation is by using **demand-side management (DSM)**, first developed in California in 1975 as a response to the steep price rises following the first oil crisis in the early 1970s (see Cragg 1993).

Demand-side management is advocated by environmental groups and energy economists in many countries, because it makes electricity-supply companies into suppliers of energy services. In DSM, the supplier tries to reduce demand for energy rather than building a new power plant to meet the increased demand. DSM is a cheaper option than constructing new power stations at considerable cost, because the electricity supplier distributes energy-efficient consumer goods, such as thermal insulation and energy-efficient light bulbs, and could offer, for example, free energy audits to large energy consumers. Naturally, at least part of, if not the entire, cost of energy-efficiency, energy conservation drives is passed on to the consumer,

but this should be much less than that incurred by building new power plants. This has the added advantage that the output of pollutants from power stations is reduced. Examples of DSM in action include South California Edison, in 1988, providing 450,000 energy-efficient light bulbs to low-income customers (thereby saving an estimated 8 MW electricity capacity); in 1987, 59 US power companies offered rebates to customers if they purchased more energy-efficient refrigerators; and the Tennessee Valley Authority provided US $250 million as interest-free loans for improving domestic insulation. In Washington, DC, the environmental World-watch Institute calculated that by 1988, US measures for energy conservation instigated by the largest six power companies had saved 7.24 gigawatts of electricity-generating capacity. Also, a 1992 report by the Boston-based Goodman Group, commissioned by Greenpeace, estimated that, besides creating 80,000 new jobs to date, for every US $1 spent on DSM, US $1.5–1.75 is saved on constructing new power plants. The Goodman report also emphasized the enormous potential of DSM, since currently only about 1 per cent of the turnover of power companies is invested in energy efficiency. DSM should be an integral part of both developed and developing countries' energy policy.

Finally, an exciting future means of energy efficiency and conservation will be through the use of so-called 'smart materials', made of polymers (very long-chain molecules) which can actually sense and respond to external stimuli by changing density, colour, and other physical properties. Smart materials behave like organisms, which is in many respects the antithesis of conventional building materials such as concrete. Smart material technologies are currently being developed, with plans on the drawing table for a smart building in Tokyo. Such smart buildings will use much less energy than those constructed from conventional materials, because the outer skin will be designed to respond to changes in ambient atmospheric and interior temperature and pressure, and will be able to respond, as required, to computer-input commands to increase insulation, warm or cool parts of – or the entire – building, etc. In other words, smart materials will be alive and aware of their environment and, therefore, able to capitalize on changes in the environment.

Energy production and environmental damage

The environmental impact created by virtually any energy resource that is utilized will cause changes that have some deleterious effects. As custodians of a fragile Earth, it is the responsibility of individuals and societies to develop and use energy resources that have a minimal impact on the environment. The combustion of fossil fuels such as oil and coal cause acid rain and produce greenhouse gases. The cumulative effect of the global energy consumption releases about 5 billion tonnes of carbon into the atmosphere every year. Combined with the other greenhouse gases, by the middle of the next century the Earth's surface temperatures could be changing at a rate 10–100 times faster than that which occurred at the end of the last glaciation. Nuclear fission for energy and weapons generates long-lived radioactive waste.

At present, most electricity-generating processes use fossil fuel power stations, which emit greenhouse gases, in particular CO_2. In a Friends of the Earth submission to the UK Parliamentary Select Committee (Mortimer 1989), figures were presented for the CO_2 emissions associated with the production or saving of 1,000 MW of electricity 'taking into account related mining and fuel producing processes', expressed as CO_2 tonnes per year, and which are listed in Table 7.8.

Vehicle emissions

Motor vehicles have a major impact on the environment (see Box 7.9). Tailpipe emissions, emissions control technology, and legislation are major environmental issues associated with motor transport. Emissions include evaporative hydrocarbons occurring throughout the fuel system (dependent upon the volatility characteristics of the fuel), and combustion emissions, that is, unburnt hydrocarbons (HC), carbon monoxide (CO), and mixed nitrogen oxides (NO_x), the latter contributing to acid rain. Tailpipe combustion emissions include particulates such as carbon particles, that can be coated with various hydrocarbons. There are other emissions, such as formaldehyde which generally occurs as a trace gas, that can be particularly toxic and **carcinogenic**.

Tailpipe emissions standards between countries are very variable, with the Californian legislation and planned standards being, as has always been the case in the USA, the most stringent (Table 7.10). California has introduced a plan for the progressive reduction of vehicular emissions to help the state achieve national air quality standards by the year 2010. The plan includes the progressive introduction of Transitional Low Emission Vehicles (TLEV) Low Emission Vehicles (LEV), and eventually Ultra-Low Emission Vehicles (ULEV), and Zero Emission Vehicles (ZEV). ZEVs are vehicles that

---

Box 7.8

EC carbon or carbon/energy tax

A measure under active consideration by the European Commission to reduce the emissions of the main greenhouse gas, carbon dioxide, is the proposed carbon or carbon/energy tax. This proposed tax would be calculated on the basis of 50 per cent related to the carbon content of the fuels, and 50 per cent to their energy content, increasing gradually over seven years, from US $3 on each barrel of oil equivalent, to US $10 per barrel of oil equivalent by the year 2000, assuming the tax comes into effect in 1993/4. The tax would reduce energy consumption in general, but particularly with regard to the use of coal (House of Commons Trade and Industry Committee 1993).

From its inception, the EC proposal was consistently supported by Germany, the Nether-lands, Belgium, Denmark, Italy, and Luxembourg. The other member states were more equivocal and/or opposed the imposition of this tax. The poorer EC countries – Spain, Portugal, Greece, and Ireland – claimed it would hinder their economic growth. France claimed the tax would be unfair, because much of its electricity generation is from nuclear power and, therefore, its CO_2 emissions are low anyway. The UK claims that its domestic taxation policy (e.g., increased value-added tax, VAT, on domestic fuel and a 10 per cent increase in the duty on fuels, imposed in the 1993 March Finance Bill or Budget) represents an adequate response to seeking to control future CO_2 emissions.

It has been argued that even the possibility of a carbon/energy tax may already have had an

Table 7.10 *Vehicle tailpipe emissions legislation in California (A), and planned standards in California (B).*

(A)		1991/92	1993	TLEV	LEV	ULEV	ZEV
HC	g/mi	0.39	0.25	0.125	0.075	0.04	0
	g/km	0.24	0.15	0.078	0.047	0.025	
CO	g/mi	7.0	3.4	3.4	3.4	1.7	0
	g/km	4.375	2.13	2.13	2.13	1.06	
NO$_x$	g/mi	0.4	0.4	0.4	0.2	0.2	0
	g/km	0.25	0.25	0.25	0.125	0.125	
HCHO	g/mi		0.015	0.015	0.015	0.008	0
	g/km		0.0093	0.0093	0.0093	0.0093	

(B)	1991/92	1993	TLEV	LEV	ULEV	ZEV	NMOG (g/mi) for fleet average
MY 91/92	100%						0.390
MY 1993	60%	40%					0.334
MY 1994	10%	80%	10%				0.250
MY 1995		85%	15%				0.231
MY 1996		80%	20%				0.225
MY 1997		73%		25%	2%		0.202
MY 1998		48%		48%	2%	2%	0.157
MY 1999		23%		73%	2%	2%	0.113
MY 2000				96%	2%	2%	0.073
MY 2001				90%	5%	5%	0.070
MY 2002				85%	10%	5%	0.068
MY 2003				75%	15%	10%	0.062

TLEV	Transitional Low Emission Vehicle		LEV	Low Emission Vehicle
ULEV	Ultra-Low Emission Vehicle		ZEV	Zero Emission Vehicle
NMOG	Non-Methane Organic Gases		HCHO	Formaldehyde
g/mi	grams per mile		g/km	grams per kilometre

impact in so far as electricity-generating companies now perceive coal-fired power stations as having a much higher risk attached to them. Clearly, such an increased risk encourages the use of other fuels, for example natural gas. There are those who argue that a carbon/energy tax will help promote cost-effective energy investment, although energy-intensive industries may well be damaged. Countries such as the UK would probably find that a carbon/energy tax would have only a small impact on its CO_2 emissions, with EC estimates of UK gross energy consumption in the year 2000 being only 2.5 per cent lower and fossil fuel consumption 3.5 per cent lower compared to a situation without the proposed tax (ibid.).

A strong argument in favour of a carbon/energy tax is that it would curb the CO_2 emissions from the transport sector, which is the fastest-growing area of fossil fuel consumption. However, the transport sector is notoriously unreactive to small changes in price, therefore there is perhaps little reason to be optimistic that modest taxation changes would have much of an impact. Although the impact of a carbon/energy tax would probably be small in reducing CO_2 emissions, because gross energy demand appears to react only in a modest way to fiscal measures, it may well be that nations can only achieve any set emissions targets through a mixture of both large- and small-impact measures.

Box 7.8

EC carbon or carbon/energy tax

produce zero exhaust and evaporative emissions of any pollutant, with California committed to ZEVs accounting for a minimum 2 per cent of sales by 1998 and 10 per cent in the year 2003. In the USA, Title 2 (relating to motor vehicles, fuels, and their emissions) of the revised Clean Air Act, signed by President Bush in November 1989 following protracted negotiations between the House of Representatives and Senate, included: (1) the imposition of tighter tailpipe emissions standards and the establishment of compliance testing and maintenance programmes; (2) the establishment of a reformulated gasoline programme; (3) legislation relating to clean fuels vehicles, which could lead to the introduction of alternative fuels; (4) legislation covering operators of vehicle fleets in areas where there are specific air quality problems; and (5) re-affirmation of the rights of individual states with particular air quality problems to set more severe emissions standards, but which must not exceed the Californian standards, a restriction imposed so that vehicle manufacturers would not have to produce customized vehicles for each state, but rather two models, one to meet Federal standards and the other to comply with the laws in California.

The inconclusive, often apparently contradictory results of the tailpipe emissions tests are due to the different engine performance characteristics and varying test cycles used, causing greater differences in rates of emissions than may be attributed to the use of the biofuel itself. Without similar test data sets from different laboratories, available emissions results suffer from limitations due to the non-comparability. In Europe, a standard emissions test involves sampling emissions at various steady-state engine speeds, whereas a more representative measure of the in-service situation may be the US transient testing, since the latter includes conditions that better approximate to acceleration and deceleration under varying engine conditions. It is important that emissions tests are undertaken on in-service vehicle engines, which requires more pilot studies.

Biofuel combustion produces insignificant amounts of sulphur compounds such as SO_2 (because most biofuels contain much lower amounts of sulphur compounds compared to petroleum products) which contributes to acid rain.

The combustion of biofuels is often referred to as 'CO_2 neutral', because potentially as much CO_2 is sequestered from the atmosphere during plant growth as is released during combustion.

Box 7.9

The impact of motor vehicles on the environment

Carbon dioxide (CO_2) Vehicle impact on emission: 19 pounds into the atmosphere per gallon;[a] 300 pounds per 15-gallon fill-up; 14 per cent of the world's CO_2 emissions from fossil fuel burning from motor vehicles

Tropospheric ozone Although ozone in the lower atmosphere does not emanate directly from motor vehicles, they are the major source of the ozone precursors: hydrocarbons and nitrogen oxides

Carbon monoxide (CO) Concentrations in the lower atmosphere increase by 0.8–1.4 per cent per year.[b] 66 per cent of OECD country emissions (78 million tonnes) from motor vehicles in 1980.[c] 67 per cent of US emissions from transportation in 1988.[d]

Nitrogen oxides (NO_x) 47 per cent of OECD country emissions (36 million tonnes) from motor vehicles in 1980.[c]

Hydrocarbon compounds (HC) 39 per cent of OECD country emissions (13 million tonnes) from motor vehicles in 1980.[c]

Chlorofluorocarbons (CFCs) 54.1 tonnes consumed by US mobile air conditioners annually.[e] 35.6 tonnes consumed in the US annually through leakage, service venting, or accidents.[e]

Diesel particulate (Tiny carbon particles, hazardous to respiratory tract, visibility, and as a possible carcinogen.)[f] No overall measurements. Diesel engines emit 30–70 times more particulate than petrol-fuelled engines.

Lead 90 per cent of airborne lead from petrol vehicles.

Lead scavengers Additives to remove lead; some (notably ethylene dibromide) may be carcinogenic.[f] Significant amount emitted.

Aldehydes (incl. formaldehyde) Exhaust emissions correlate with hydrocarbon (HC) emissions. Diesel engines produce a higher percentage.

Benzene (identified as carcinogen) Present in both exhaust and evaporative emissions; 70 per cent of the total benzene emissions in the US come from vehicles.

The CO_2 budget, however, must account for the energy expended in activities associated with planting, cultivating, harvesting, transportation, and processing which, if generated from fossil fuels, will have a net contributory effect on atmospheric CO_2 levels. Even allowing for the use of substituted biofuel energy at all production stages, the use of pesticides and fertilizers manufactured from petroleum products, and other energy inputs that use fossil fuels, may result in a net input of CO_2 into the atmosphere. Increased use of pesticides and fertilizers could further reduce any energy and CO_2 credits. To achieve an overall reduction in atmospheric CO_2 levels relative to mineral (fossil) fuels, the waste vegetable matter from biofuel cultivation must be burnt as an energy source rather than used as an animal feedstuff, something that is less economically attractive. There is an argument expressed most strongly by the oil companies that CO_2 emission targets are better met through technical improvements to the diesel engine aimed at increasing engine efficiency and reducing fuel consumption, and that the use of biofuels as alternatives for fossil fuels provides, at best, a very small reduction in atmospheric CO_2 levels.

Conclusions

Energy policy, realistically, is not formulated in a political vacuum. Decisions about the energy resources used by a nation depend upon many complex economic, social, and military factors. A nation's accessible resources play a major role in determining energy policy.

The geographic distribution of energy resources is extremely unequal. Countries such as Japan have virtually no commercially viable large reserves of fossil fuels. They rely upon the import of almost all of their energy resources. This lack of natural energy resources partly explains its interest in areas such as Antarctica where the carve-up of territory for economic exploration and exploitation has fortunately been frustrated by the Antarctic Treaty Environmental Protocol 1991 which stopped the destruction of this major natural wilderness (see Chapter 10). Other nations such as the former USSR and South Africa are amongst the world's richest lands in energy resources. Estimates of total versus recoverable reserves are subject to continual reappraisal, with the result that the downwardly revised figures can lead to more aggressive economic and military policies. Examples of downwardly revised estimates

---

Non-diesel organics Smaller amount per vehicle, but more mutagenic overall, than diesel particles.

Asbestos Used in brake linings, clutch facings, and automatic transmissions. About 22 per cent of the total asbestos used in the USA in 1984 was used in motor vehicles.[f]

Metals US EPA has identified mobile sources as significant contributors to nationwide metals inventories, including 1.4 per cent of beryllium and 8.0 per cent of nickel. Arsenic, manganese, cadmium, and chromium may also be mobile source pollutants. High-risk hexavalent chromium which does not appear to be prevalent in mobile source emissions.

Notes
[a] This figure refers to direct tail-pipe emissions only. Transportation, refining, and distribution account for perhaps 15 to 20 per cent of total emissions.

[b] Khalil, M.A.K. and Rasmussen, R.A. 'Carbon Monoxide in the Earth's Atmosphere: Indications of a Global Increase.' *Nature*, 332, (245), March 1988.
[c] Organization for Economic Co-operation and Development. *OECD Environmental Data*, Paris, 1987.
[d] US Environmental Protection Agency, *National Air Quality and Emissions Trends Report 1988*, p. 56.
[e] US Environmental Protection Agency, *Regulatory Impact Analysis: Protection of Stratospheric Ozone*, Washington, DC, December 1987.
[f] Carhart, B. and Walsh, M. *Potential Contributions to Ambient Concentrations of Air Toxics by Mobile Sources. Part 1*, paper presented at the 80th Annual Meeting of Air Pollution Control Association, New York, 24 June 1987.
Source: OTA

Box 7.9
The impact of motor vehicles on the environment

include many of the non-OPEC nations such as the former USSR, which currently consumes roughly 15 per cent of the global oil production and, because of its rapidly growing consumption, needs to increase production and oil imports. In 1987, the USA imported US $40 billion of oil, equivalent to about one-third of the nation's trading deficit, and in the same year the Pentagon spent US $15 billion on protecting these supplies of oil.

OPEC controls about 75 per cent of the known crude oil reserves. The dependence on the Middle East countries for oil by the countries of the industrialized world means that their economies are directly affected by the pricing of crude oil by OPEC. About 70 per cent of natural gas reserves are held by the former USSR and the Middle East. Of the calculated 950 billion metric tonnes of global coal reserves, estimated to last another 275 years if present production rates are maintained (Kumar *et al.* 1987), the USA and former USSR each control 25 per cent. Much of the remaining reserves are in Europe, Africa, Australia, and Asia, particularly China. Approximately 20–30 per cent of the energy requirements of industrialized countries come from coal. Almost 75 per cent of China's energy is from coal.

The environmental lobby against dirty fuels

which pollute the atmosphere, produce greenhouse gases and cause acid rain, has stimulated the search for cleaner alternatives. Nuclear fuel has been seen by many people as a viable alternative, though environmentalists express concern about this because of the nuclear waste and ever-present dangers of another, perhaps far more serious Chernobyl or Three Mile Island disaster. Today, nuclear energy supplies roughly 17 per cent of the world's electricity, with France as the leading country obtaining about 70 per cent of its electricity by this method, and Japan about 35 per cent.

Although nuclear energy does not produce greenhouse gases or acid rain, the cost of nuclear reactors is very high and the public concern over the disposal of nuclear waste and nuclear accidents conspires to make it less attractive as a political option. These perceived disadvantages have meant that the USA has not ordered a new nuclear plant since 1978. Most operational nuclear reactors are of the light water type, and the pro-nuclear lobby has suggested that advanced reactor designs could restore public confidence in such energy.

The funding of research and development into energy resources can help to predetermine the outcome of decisions to adopt energy policies and to invest further in certain directions. In

Box 7.10

Biofuels and motor transport emissions

Biofuels may provide a contribution to reducing the emission of greenhouse gases, and gases which contribute to photochemical smogs and acid rain. Amongst these, bio-ethanol and biodiesel are considered here. The high volatility of ethanol produces evaporative emissions 5–220 per cent above those for conventional petrol, the effect being moderated by the presence of a co-solvent (e.g., tertiary butyl ether) used to increase the solubility of ethanol and water in petrol which decreases the possibility of phase separation. Compared with conventional diesel, bio-ethanol combustion gives reduced HC and CO emissions, variable higher or lower NO_x, and increased emissions of aldehydes (which are toxic and contribute to the development of smog and low-level ozone) and esters. Particulates (mainly carbon particles) are not significantly different, but other emissions, such as the carcinogenic organic chemicals called alkanes, alkenes, ketones, and aldehydes, are higher. These increases are offset against the greater emissions of an even

more carcinogenic group of organic chemicals from mineral diesel called polyaromatic carbon compounds (PACs). Additionally RME can cause problems with lubricants, exhaust catalyst durability, and poor consumption records. NO_x levels can be slightly reduced by altering engine configuration (e.g., timing) and ignition temperature, at a trade-off of increased particulates (unburnt and partially burnt HC) but still well below those from mineral diesel. RME combustion also produces exhaust fumes that many regard as having an unpleasant odour, but this can be rectified with additives in the biodiesel. More biodiesel is retained in the engine lubricating oil because it is less volatile than mineral diesel, a factor that can lead to greater emissions of particulates from the biodiesel when it is ignited.

Catalysts are being used to help meet the increasingly stringent emissions control regulations. It is technically easier, by the addition of oxidative catalysts to the fuel, to remove the

Britain, for example, there is an enormous difference in the R&D money committed to nuclear energy versus renewables, let alone other resources. In the period 1990–91, the R&D breakdown in £ millions was: nuclear fast reactors, 84.6; nuclear fusion, 26.9; other nuclear, 17.5; coal-based energy, 7.6; oil and gas, 10.7; energy-efficient R&D, 11; renewables, 20.3. With about two-thirds of the total financial support in R&D directed towards nuclear energy, a lot of influential people have made a large commitment to this fuel source, and vested interests would find it correspondingly hard to change direction.

As for energy efficiency, many of the measures to utilize renewable energy resources and construct more energy-efficient buildings incur additional costs to society and the individual, both in financial terms and in the way in which people live their lives. There is a price to pay for everything.

It should be the responsibility of governments to encourage a more prudent use of energy. They should provide attractive subsidies to companies that wish to construct power stations or energy farms that rely upon renewable energy, something that the UK Non-Fossil Fuel Obligation (NFFO) does. Builders who are prepared to construct houses and other buildings to improved standards should be given similar

incentives. Indeed, in the USA in the late 1970s, relatively generous research budgets were allocated for renewable energy resources, replaced by tax concession schemes in the 1980s. A spin-off from these policies was the 'Californian wind rush' in which about 16,000 privately owned wind turbines were installed, mainly as wind farms, with a generating capacity of 1,500 megawatts. After the experimental period in which many wind turbines revealed design faults and were poorly sited, the tax concessions were withdrawn in 1985 and the industry contracted. Despite the juddering start to renewable energy resources in many countries, the only way to overcome design flaws and improve energy efficiency in the new breed of power plants, governments must be prepared to invest heavily in these alternative technologies. In Britain, the government's advisory body, the Energy Technology Support Unit (ETSU), has stated that it is feasible for half the nation's energy requirements to be met from renewable energy resources by sometime early in the next century. So, it is technologically possible, but the big question is whether or not the will is there to achieve this goal. Without real government support for a programme of moving over to a much greater emphasis on renewable energy, societies may be even less prepared for forthcoming energy crises.

Box 7.10
Biofuels and motor transport emissions

increased emissions from the combustion of RME of mainly single carbon-chain aldehydes, alkenes, and ketones than the PACs that occur in larger volumes in mineral diesel or petrol. Adding such catalysts (e.g., platinum-rhodium) to biofuels entails a relatively small increase in cost.

Many of the EC countries are committed to stabilizing emissions of CO_2 at 1990 levels by the year 2000. In Britain, an Energy Technology Support Unit (ETSU) study (Culshaw and Butler 1993) calculated that for each litre of biodiesel used, 1.5 kg of CO_2 is saved (i.e., CO_2 emitted from the substituted fuel minus CO_2 emitted during biodiesel production). A 1993 report by Germany's Federal Environment Office (UBA) shows that if RME displaced 400,000 tonnes of fossil fuels a year, equivalent to 640,000 tonnes of CO_2, it would account for 0.5–0.7 per cent of Germany's total emissions, a small part of the 25 per cent target reduction. Also, the UBA report shows that the consumption of 1 kg of mineral

diesel fuel, including the preliminary chain, releases 3.40–3.49 kg CO_2, together with 0.9 kg/kg of emissions from other climate-relevant trace gases expressed in CO_2 equivalents, giving a total 3.5–3.6 kg CO_2 equivalent per kg for mineral diesel, compared with 1.9–2.3 kg CO_2/kg for substituted biodiesel which takes into account the production (agriculture), processing (rape seed oil extraction and esterification), and transport. If co-products are given CO_2 emission credits, for example glycerine and oilseed rape groats, then the CO_2 balance is 1.2 kg CO_2/kg for rapeseed oil and 0.8–1.4 kg CO_2/kg for RME, depending upon the recoverability of the glycerine, giving 3.5–4.0 kg CO_2/kg for rapeseed oil and 2.7–4.4 kg CO_2/kg for RME. For many countries, at current and realistically projected levels of oilseed rape cultivation, RME will make only a small contribution to the overall target reduction levels for CO_2.

The EC is currently undertaking a five-year (1993–97) research and development plan to promote renewable energy sources within the community, the ALTENER plan, which is expected to cost 125 million ecu. One of the EC Commission proposals is to reduce CO_2 emissions drastically by the year 2005, as a result of which it will be necessary to increase the contribution made by renewable energy resources to meet EC energy demands, triple the amount of electricity produced by renewables, and use more biofuels. The ALTENER plan makes provision to fund four categories of projects concerning: (1) technical studies and evaluations to set forth technical specifications and standards; (2) measures provided for by the member states to extend or create infrastructure for renewable energies, acquire equipment, and organize professional training; (3) measures to establish an information network enabling a better coordination among the activities of the member states; and (4) actions in the sector of biomasses, especially biofuels. It is through the prescient action of individual countries, or groups of nations as in the case of the EC, providing substantial funds for renewables that any impending energy crisis in the twenty-first century can be averted. Naturally, the fossil fuel and allied chemical industries are deploying every commercial and environmental reason at least to slow the wind of change that is blowing through the energy markets. Such delaying tactics will conserve company profits and jobs in those industries over the short term, but as pollution standards and control measures become ever more stringent, and as the price of oil goes up from its present historic low, a greater reliance on renewables will inevitably come about and with it a sharp decline in the use of fossil fuels.

On a final but by no means unimportant note, military expenditure and the military policy of the world's leading nations are strongly governed by the need to maintain supplies of various energy resources, particularly oil from the Middle East. It is for reasons such as these that any armed conflict in the Middle East, with the associated economic implications, creates such international tension with the potential for local wars to become 'flash points' for the rapid escalation into much larger-scale international conflagrations. Arguably, the Gulf War in early 1991 was actually solely about preserving American and European interests in Middle East oil supplies.

Key points

CHAPTER 7: KEY POINTS

1 Industrialization, population growth, increased living standards, concerns about pollution, and the serious depletion of many traditional energy resources such as fossil fuels, have increased the need to develop renewable and clean energy. Most of the global energy supply comes from finite resources (77 per cent from essentially fossil fuels), and from renewables (18 per cent hydropower, wood, crop waste, dung, and wind), and nuclear energy ($c.$ 5 per cent).

2 Conventional fossil fuels (coal, oil, and gas) have low reserve lifetimes (e.g., world coal = $c.$ 200–400 years, world oil = $c.$ 56 years), but these figures have a large uncertainty attached to them because proven reserve estimates are modified by improved technologies of exploration and recovery, together with fluctuations in the world commodity prices. Extraction and use of fuels causes many environmental problems including open cast mining and piling of waste which scar the landscape. Leakages of oil and gas from pipelines and installations also cause environmental pollution problems.

3 In the fossil fuel sector, environmental awareness has led to improved, cleaner technologies. Clean coal technologies, for example, are being developed which include fluidized bed combustion, flue gas desulphurization, gas-fired combined cycle gas turbines, gasification, and British Coal's Topping Cycle. To reduce atmospheric pollution, many governments in the developed countries have agreed to reduce their emissions by target dates, but such agreements have yet to make any headway in some of the major polluters, such as China and the former Soviet Union, two of the largest users of coal.

4 Nuclear energy, produced by nuclear fission, has been developed since the Second World War. The industry has encountered many problems during this time, including several nuclear accidents and the problems associated with the safe disposal of nuclear waste. These problems have resulted in some countries deciding to reduce their

Plate 26 *Crops of oilseed rape in southern England. The oil from these plants can be converted into rape methyl ester (RME) for use as biodiesel, either as a substitute for conventional diesel or as a blend.*

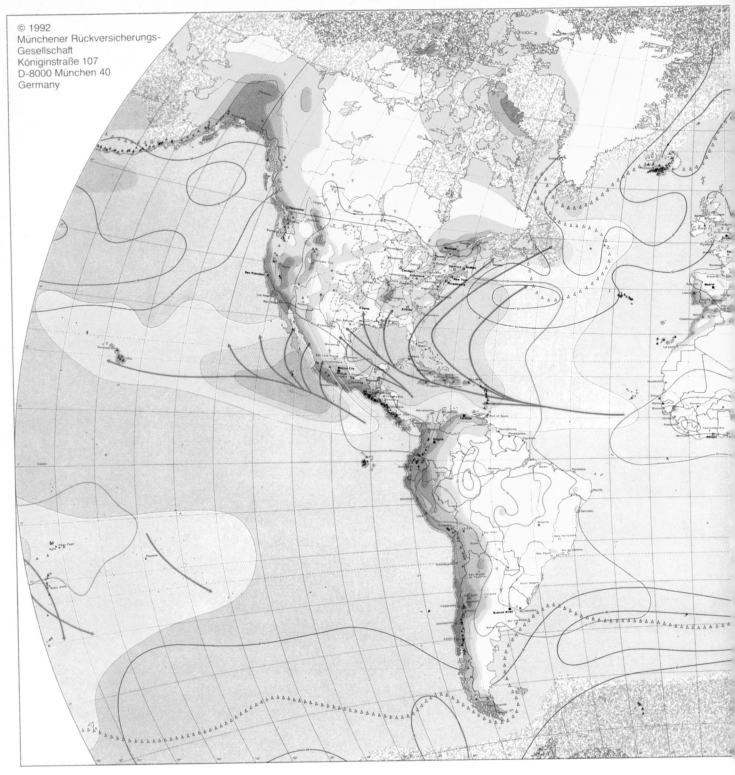

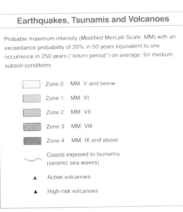

Earthquakes, Tsunamis and Volcanoes

Probable maximum intensity (Modified Mercalli Scale: MM) with an
exceedance probability of 20% in 50 years equivalent to one
occurrence in 250 years ("return period") on average, for medium
subsoil conditions:

- Zone 0   MM V and below
- Zone 1   MM VI
- Zone 2   MM VII
- Zone 3   MM VIII
- Zone 4   MM IX and above

Coasts exposed to tsunamis
(seismic sea waves)

▲ Active volcanoes

▲ High-risk volcanoes

Further Natural Hazards, Other

- △ △ △ Limit of iceberg drift
- Temporary pack ice
- Permanent pack ice
- Sea fog frequency above 30% (July)
- —100— Isoline of thunderstorm days per year

▫ Bombay	more than 1 million inhabitants
◌ Chimbote	100,000 to 1 million inhabitants
○ Townsville	less than 100,000 inhabitants
● Bonn	capital city
▫ Sydney	MR office abroad

State borders
(These should not be regarded as official.)

Rivers

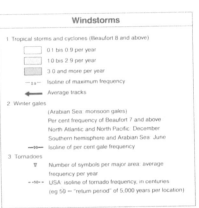

Windstorms

1 Tropical storms and cyclones (Beaufort 8 and above)

- 0.1 bis 0.9 per year
- 1.0 bis 2.9 per year
- 3.0 and more per year
- — 2.9 — Isoline of maximum frequency
- ←—— Average tracks

2 Winter gales

(Arabian Sea: monsoon gales)
Per cent frequency of Beaufort 7 and above
North Atlantic and North Pacific: December
Southern hemisphere and Arabian Sea: June
—10— Isoline of per cent gale frequency

3 Tornadoes

▽ Number of symbols per major area: average
frequency per year

-50- USA: isoline of tornado frequency, in centuries
(eg 50 = "return period" of 5,000 years per location)

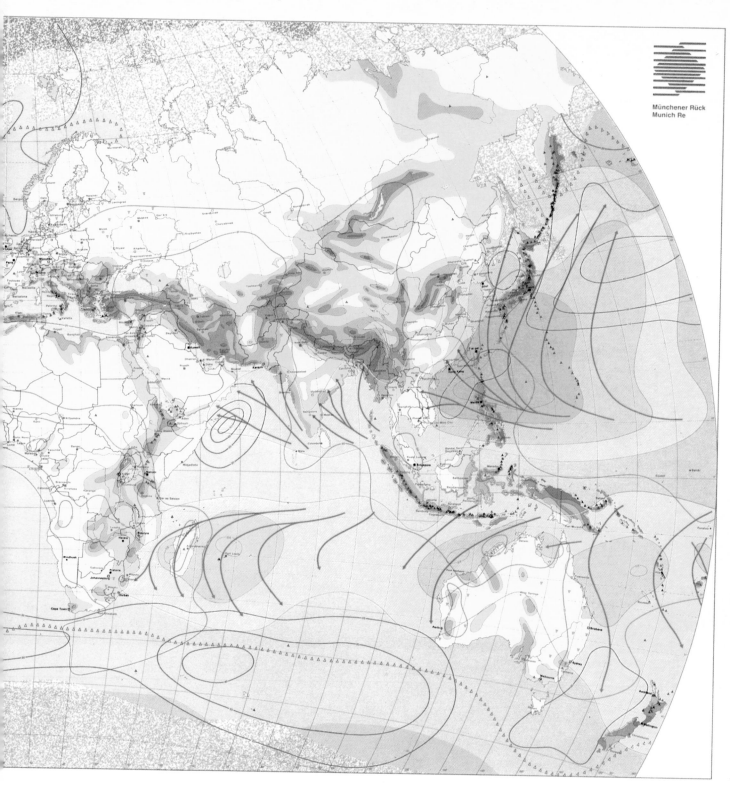

Plate 27   *World map of
natural hazards.*
*Courtesy of Münchener Rück
(Munich Re, Germany).*

Plate 28 *A pyroclastic cloud produced by the eruption of Mount Pinatubo in the Philippines in June 1991.* Courtesy of Alberto Garcia/SABA Katz Pictures.

dependence on nuclear power, and even mothball nuclear plants. Decommissioning nuclear power stations also has an environmental and cost problem.

5 In the future, hydrogen energy, a form of chemical fuel, which could be easily transported and readily stored, may provide a major part of the world's energy demand.

6 Renewable energy resources include: hydro-electric, wind power, tidal energy, wave power, solar energy, geothermal energy, and biomass energy. Technologies are being developed to increase the efficiencies of these fuels. Environmental problems, however, are associated with renewables, including the disruption of ecosystems, visual pollution, and the current high production costs. Also, established energy businesses and cartels operate to minimize the potential competition for as long as possible.

7 Cumulatively, private vehicles use large amounts of energy and contribute to poor air quality and other forms of environmental pollution. Improved public transportation systems and better urban planning could lead to a considerable reduction in pollution. Liquid biofuels, including biodiesel and bioethanol, used as substitutes for conventional petrol and diesel, could provide environmentally more friendly alternatives. Although the technology to use these substitutes is sufficiently developed, the hydrocarbon and allied chemical industries have strong vested interests in slowing the introduction of biofuels, and without a distortion of the traditional fuel markets through favourable tax incentives and subsidies it seems unlikely that biofuels will become important until the middle of the twenty-first century.

8 Attempts to reduce fuel consumption and improve air quality, including a reduction in the emissions of greenhouse gases caused by the combustion of fossil fuels, have involved proposals for the introduction of a carbon tax, increased use of energy-efficient technologies, energy conservation, and a greater reliance on renewable energy resources.

Chapter 7: Further reading

Blunden, J. and Reddish, A. (eds) 1991. *Energy, Resources and Environment*. London: Hodder & Stoughton.
A well-written and well-illustrated textbook, dealing with the nature of energy and mineral resources, their extraction, refining, and disposal, and the environmental problems associated with obtaining these resources. The book explores the possible alternatives to conventional energy resources, including substitution and recycling, energy conservation, solar energy, wind and water energy. It also considers the political implications of using such alternative energy resources. There is a whole chapter devoted to an examination of the politics associated with the disposal of radioactive waste.

Cassedy, E.S. and Grossman, P.Z. 1990. *Introduction to Energy Resources, Technology, and Society*. Cambridge: Cambridge University Press.
In this textbook, Cassedy and Grossman explore energy issues, and examine the benefits and problems associated with energy technology. The book is in three parts: Part I, dealing with energy resources and technology; Part II, examining power generation, the technology and its effects; and Part III, evaluates energy technology in the future. This book will prove useful as supplementary reading for students with a general interest in energy issues, particularly energy technology.

Johansson, T.D., Kelly, H., Reddy, A.K.N. and Williams, R.H. (eds) 1993. *Renewable Energy for Fuels and Electricity*. London: Earthscan Publications/United Nations.
A comprehensive and authoritative assessment of the technical and commercial potential for renewable forms of energy.

Stockholm Environment Institute 1993. *Energy without Oil: The Technical and Economic Feasibility of Phasing Out Global Oil Use*.

Yergin, D. 1991. *The Prize: The Quest for Oil, Money and Power*. London: Simon & Schuster.

The river went on raising and raising for ten or twelve days, till at last it was over the banks. The water was three or four foot deep on the island in the low places and on the Illinois bottom. On that side it was a good many miles wide; but on the Missouri side it was the same old distance across – a half a mile – because the Missouri shore was just a wall of high bluffs.

Mark Twain,
The Adventures of Huckleberry Finn (1894)

The landscape is fashioned by a wide variety of natural processes. These processes include volcanic eruptions, the slow inexorable drift of continents and seafloor spreading, earthquakes, and the formation of rock. Earth-surface processes include wind and ice action, ocean currents and tides, the flow of water over the Earth's surface, erosion, landsliding, and many others.

Meteorological processes such as wind and rain constitute the weather. Biological processes such as the growth of plants and animals, death and the decay of organic matter, and the spread of species are enacted within the global theatre of natural hazards as well as catastrophes caused by humans.

The energy that drives the natural processes

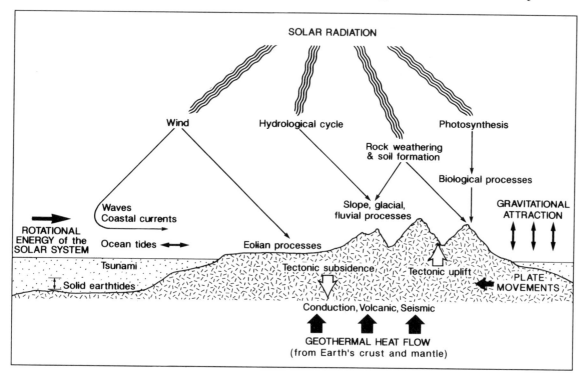

Figure 8.1 *The major energy sources and associated processes for Planet Earth. After White (1986).*

comes from three main sources (Figure 8.1). The most important source of energy is from the Sun reaching the Earth in the form of shortwave radiation. Some of this radiation is converted into longwave radiation towards the infrared end of the spectrum and beyond it, which heats the Earth's surface and atmosphere. This heat energy is responsible for global and local variations in air temperatures and pressures and ultimately controls the circulation of gases within the atmosphere and across the globe to give weather. It also controls the state of moisture (water) in the air and hence the form of precipitation. Shortwave radiation such as ultraviolet light is essential for providing the energy for life. Green plants need this radiation for photosynthesis to form carbohydrates from carbon dioxide and water that go to build plant tissues. Plants in turn provide the primary food source for animals.

The second major source of energy comes from within the Earth itself. This is predominantly heat energy produced by the radioactive decay of elements such as uranium and thorium which are present in rocks at depth. The same heat energy allows rocks to behave plastically and to flow at depth or even partially melt to produce rock melts called **magmas**. The magmas may rise toward the surface of the Earth and can pierce it to form **volcanoes**. The flow of rocks in the Earth's mantle at depths greater than 5–8 km below ocean floors and more than 35–50 km below continents provides a fundamental mechanism for the horizontal and vertical movement of the cooler surface layers of the Earth or crust. The continents also flow like extremely viscous liquids but at rates so slow that for most purposes they can be considered as solid and rigid. It is these movements that cause stresses to develop within the crust and energy to be released violently in the form of earthquakes.

The third main energy source responsible for many Earth processes exists because of the gravitational forces that attract masses towards each other. Sir Isaac Newton stated in his **Law of Gravitation** that the larger the mass of an object, the greater the gravitational attraction it will have on other masses. The mass of the Earth is considerable compared to bodies present on its surface and hence objects are attracted by the gravitational force towards the centre of the Earth – an anecdotal observation expressed in the apple falling on Newton's head! The attractive force of the Earth for an object much smaller than itself is so large that things fall to the ground. The Earth also attracts bodies such as the Sun and the Moon, which, in turn, attract the Earth. It is this force that keeps the planets in motion orbiting the Sun, and maintains the orbits of moons around their respective planets. Gravitational forces are important for Earth surface processes such as the movement of materials down slopes like water (as rivers or slope wash), rock and debris (landslides), snow avalanche, and ice (glaciers). The gravitational attraction of the Sun and Moon on the Earth is responsible for the daily rise and fall of the sea as tides, in most places often twice a day as the semi-diurnal tidal cycle.

Many of the natural processes mentioned above can constitute natural hazards capable of injuring or killing (see Table 8.1), or causing damage to buildings and agricultural land. A natural hazard becomes a catastrophe if a situation develops in which the damage to people, property, or society is sufficient to cause a long

Table 8.1 *Risk of death from involuntary hazards.*

Involuntary risk	Risk of death/person/year
Struck by automobile (USA)	1 in 20,000
Struck by automobile (UK)	1 in 16,600
Floods (USA)	1 in 455,000
Earthquake (California)	1 in 588,000
Tornadoes (Midwest)	1 in 455,000
Lightning (UK)	1 in 10 million
Falling aircraft (USA)	1 in 10 million
Falling aircraft (UK)	1 in 50 million
Explosion, pressure vessel (USA)	1 in 20 million
Release from atomic power station	
At site boundary (USA)	1 in 10 million
At 1 km (UK)	1 in 10 million
Flooding of dike (Netherlands)	1 in 10 million
Bites of venomous creature (UK)	1 in 5 million
Leukaemia	1 in 12,500
Influenza	1 in 5,000
Meteorite	1 in 100 billion

Source: After Dinman (1980), in Smith (1992).

recovery or rehabilitation process. Floods, hurricanes, tornadoes, **tsunamis**, volcanoes, earthquakes, and large fires are the most common natural hazards that produce catastrophes. Particular natural hazards are concentrated in various geographical regions (Plate 27). The reasons for this distribution are discussed in this chapter.

During the last few decades, there has been an apparent increase in damage done by natural hazards (see Table 8.2). This may just be a function of increased reportage and media coverage, with the enhanced ability to transfer news and events more efficiently and effectively. It may, however, be a function of increased population and its concentration in marginal areas, that is, regions of the world less suited to habitation because of the inherent associated risks from natural disasters. It may be due, in part, to human-induced environmental change, leading to the crossing of critical thresholds, above which natural disasters are much more likely. If climatic hazards during the last few decades are considered, for example, it seems that the fiercest storm this century was experienced in September 1988 when Hurricane Gilbert devastated the Caribbean with reported wind speeds up to 349 km/hour. In the south Pacific between 1941 and 1980, five severe hurricanes were recorded in the Fiji islands compared with six between 1981 and 1985. The North Atlantic has also experienced severe storms, with three major hurricane depressions between October 1987 and January 1990. In southern England, the storm of October 1987 was the most violent storm experienced in this area in 300 years (Thompson 1989).

There have also been severe droughts since the 1960s in the Sahel, with 1984 being the driest year this century. Recent droughts in the Midwest states of the USA were reminiscent of the **'dust bowl'** conditions of the 1930s. Extreme drought conditions have been more frequent in southern Britain in the last 15 years, with four severe drought years – 1976, 1983, 1989, and 1990 (Thompson 1992). Whether such conditions are produced by natural fluctuations in the Earth's climate or were stimulated by human activity has yet to be resolved. This chapter examines the causes and consequences of the main natural hazards.

The effect on populations of a disaster resulting from a natural hazard may be direct, involving injury, death, and damage to property, or indirect through a knock-on effect from reduced economic resources caused by the catastrophe. Sometimes, disasters occur through everyday natural processes that we take for granted. Just such a case occurred on 26th July 1987, when the Greek government declared a state of emergency because a national heatwave had resulted in the death of over 700 people.

Natural hazards are generally considered in terms of their magnitude (the intensity of the energy released) and their frequency (how often the event will recur). There is an inverse relationship between magnitude and frequency: the larger the magnitude, the less frequent the event. The impact of a natural hazard on humans and other organisms is not directly related to the magnitude of the natural event. An earthquake in mountainous and unpopulated regions clearly will not have the same effect as if the earthquake had occurred in the vicinity of a crowded city. The devastation is dependent upon human factors, such as the concentration of population within the area, and the remedial or preventative measures that a society may have made for the likelihood of the natural hazard occurring. A society that is prepared for a natural hazard is more likely to suffer less than one where totally unexpected and unpredicted events occur.

There are many examples where the magnitude of a natural hazard does not necessarily relate to the scale of the ensuing disaster. The

Table 8.2 *Federally declared disasters within the USA during the period 1965–85.*

Type of disaster	Number	Federal outlay (thousands of current $)	Federal outlay (thousands of 1982 $)
Ice and snow events	19	151,427	205,511
Hurricanes/tropical storms	39	1,173,141	1,947,939
Earthquakes	7	203,881	405,706
Dam and levee failures	7	55,764	80,806
Rains, storms, and flooding*	337	1,684,702	2,439,852
High winds and waves	2	125,313	120,536
Coastal storms and flooding	7	158,261	205,357
Tornadoes	109	441,685	648,352
Drought/water shortage	4	1,134	5,344
Totals	531	3,995,803	6,059,403

*Includes land, mud, and debris flows and slides
Source: After Smith (1992).

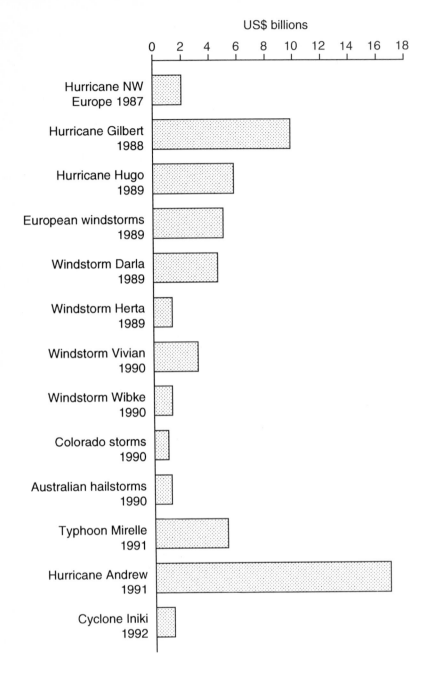

US$ billions

Figure 8.2 *Catastrophic insurance losses (in US $ billion) for selected major global natural disasters from 1987 to the present. Redrawn from* The Times *(5 April 1993).*

quake in north-east China, which killed over 650,000 people, with some estimates putting the fatalities up to well over 1 million people. Northwest China is one of the most densely populated regions of the world while southern Chile has a considerably smaller population density. So, the difference in fatalities between these two examples is a function of the size and density of the populations in an area affected by a quake.

Urbanized areas are clearly more at risk than the rural open countryside because potentially large numbers of fatalities can result during a disaster. As the population increases and we place more demands on the Earth's limited resources, natural processes become more serious as potential hazards. With increasing awareness and understanding of the Earth's natural processes, however, many remedial measures can be taken to safeguard against a likely catastrophic event. As a result, the number of internationally reported disasters has been decreasing each year. The total deaths due to natural hazards, however, is on the increase. This is simply because disasters affect areas with large and growing populations. In monetary terms, the total loss of property due to damage has also been increasing as urban and rural areas increase in size and become more sophisticated.

In order to reduce the likelihood of a major disaster, the inexact science of risk assessment and risk management has grown up. The risk of a particular hazard is related to the probability of its occurrence multiplied by the predicted consequences. It is very difficult to assess the consequences of a likely event and to estimate the probability of its occurrence. As a result, the awareness of individuals and policy-making bodies such as governments varies considerably. Although much is understood about the different types of natural hazards, frequently all too little is done to reduce the risks. Additional problems arise because of the poor communication between scientists working on aspects of particular natural hazards, the policy makers and the media.

From a financial perspective, the insurance industry has a vested interest in risk assessment for natural hazards. Within the past few years, there has been a string of natural disasters associated with severe weather, and the insurance companies have had to pay out large sums of money to meet the claims (Figure 8.2). There

largest recorded earthquake of recent time occurred in southern Chile in 1960, and caused the death of approximately 5,700 people. The energy released by this earthquake was more than 10 times that of the 1976 Tangshan earth-are people who believe that the world weather patterns are changing rapidly because of global warming influenced by human activities, while others argue that global climate changes are both natural and cyclical over a time frame of hundreds of years.

Minimizing the risks from natural hazards takes many forms. Good land planning, hazard-proof constructions, insurance, evacuation programmes, and disaster preparedness are among the most common. In less-developed countries, the populace often have no choice but to bear the brunt of a disaster with all its attendant loss of life and damage. The degree to which 'adjustments' are made depends on the type of hazard, how the population perceives the hazard and, not least, the cost of taking preventative action. Severe snow blizzards, for example in northern Canada, are not considered a hazard by the Canadians as they are a common occurrence during the winter, whereas in the UK they represent a dangerous natural hazard because they occur infrequently and the population is generally unprepared for such freak conditions. When severe snow blizzards occur in the UK, they often result in some fatalities, particularly of the infirm and elderly, and consequently they attract considerable media coverage because they are perceived as a natural hazard.

In order to cope better with natural hazards, it is necessary to understand the causes and effects of such hazards, and something of the dynamics of natural processes, that is, the way they come about and just how they occur. Natural hazards can be considered in three groups of processes: geological, meteorological, and biological.

Geological hazards are of two main types. First, there are those hazards that are driven by the Earth's internal energy. These include earthquakes, volcanoes, and tsunamis. The second type may be described as hazards resulting from land surface processes or Earth surface processes. It is these processes that are primarily responsible for shaping the landscape. Surface processes are dependent on the atmospheric, climatic, and weather conditions, the vegetation type and cover, topography, drainage patterns, bedrock type and geological structure, the tidal regime in coastal regions, and the way in which the land is used. The ability of objects to fall and do damage by virtue of gravity, also called gravitational

potential energy, is very important as it is responsible for major processes such as the movement of rock, ice, and snow masses down slopes, the flow and form of rivers, and tides. Solar energy is also important because it controls the climatic conditions which influence these processes. Hazards directly or indirectly caused by gravity and climatic conditions include landslides, rock and snow avalanches, river flooding, collapsing soils, icebergs, and **jokulhalaups**.

Meteorological hazards are driven primarily by the Sun's energy. These include tornadoes, hurricanes, floods, lightning strikes and resulting fires, droughts, cold fronts, fogs, hailstorms, blizzards and snow, windstorms, sandstorms, and frosts.

Biological hazards include epidemics, proliferation of pests, endemic parasites, and the invasion of areas by insects or plants. The spread of these hazards is influenced by the local and regional conditions, which include climate and the available food sources for pests. These conditions are essentially ultimately controlled by the Sun's energy.

Some natural hazards fall into more than one of the above categories as they may be the result of two or more causes. A corollary of this is that similar effects may be produced by different causes. Floods, for example, may be produced by meteorological factors such as heavy rainfall, sea swells resulting from typhoons, and spring snow melts. They may also be the result of geological factors, such as the bursting of landslide-dammed lakes, or they may be caused by processes originating within the Earth, such as earthquake-generated tsunamis. In coastal areas, submergence due to land surface movements associated with an earthquake can effectively be instantaneous. Examples include the famous Alaskan Good Friday Earthquake that caused widescale subsidence of several metres along the coastal regions of Anchorage.

Some natural hazards are induced or initiated by previous hazards. A good example is where landslides and fires commonly occur after many large earthquakes. Approximately 200,000 people were believed to have died as a result of an earthquake in Gansu Province, central China, in 1920. This was not the result of collapsing buildings during the quake, but because of major landslides produced by the mobilization of loess as the ground shook (Derbyshire *et al.* 1991).

Similarly, the Great Earthquake in San Francisco was followed by extensive fires which could not be extinguished due to the depleted water supply resulting from broken water mains produced by the earthquake. It was these fires that were believed to have claimed the majority of the deaths associated with the quake. Another example of the domino effect with natural hazards is where epidemics such as cholera and hepatitis often follow a flood disaster. This is because drinking water becomes contaminated, and medical help and supplies are limited due to poor physical communication, poverty, or the destruction of the normal channels of communication and transport. This has been a particularly big and recurrent problem in developing countries, such as in the flood-prone delta regions of Bangladesh.

Natural hazards that appear initially to be natural may actually owe their origin to anthropogenic causes, that is human interference with the natural environment. Landslides, for example, may be the result of badly managed land, deforestation, or construction works. Such landslides are particularly common in tropical mountain areas as in the Himalayas and the countries of South-East Asia. It is here that extensive deforestation is taking place on steep slopes that commonly experience high rainfall. Similarly, some river flooding may owe its origin to badly managed catchment areas, channelization of rivers and adjacent urban settlements. In these cases the flow of water into the main stream has been accelerated by human activities.

Human-induced earthquakes are another hazard, as experienced in Denver, Colorado during the period 1962–5. The quakes were due to the disposal of chemical waste under ground at the Rocky Mountain Arsenal. Numerous small quakes in Nevada are believed to have been triggered by underground nuclear testing. Many Earth scientists believe that human interference with the atmosphere could result in global warming and changes to the Earth's atmospheric circulation that, in turn, may lead to an increase in natural hazards.

The El Nino event during 1982–3 is frequently cited to demonstrate the adverse effects of possible global warming. This event occurred because of changes in the direction of the trade winds which resulted in a warming of equatorial Pacific water. This increased the amount of heat energy to the atmosphere supplied by the ocean and consequently caused droughts in eastern Australia, Central America, and East Africa. It also increased precipitation, resulting in the flooding of large areas of Bolivia, Peru, Ecuador, and the western coastal regions of the USA and, in addition, the warm oceans may well have caused more hurricanes, thus further adding to the occurrence of natural hazards during 1982–3. El Nino events appear to have a frequency of about every seven years.

The range of types of natural hazard are considerable. To help understand and appreciate these in more detail and to evaluate the preventative measures that can be taken to predict a likely catastrophe and prevent a disaster, the main natural hazards will be considered.

Plate 8.1 *Reconstruction of the earliest seismometer built by Chang Heng in* AD *132 and now displayed at the Institute of Seismology, Academic Sinica, in Lanzhou, China. When an earthquake occurred the urn would rock and one of the dragons would drop a small ball from its mouth into the mouth of the frog below, and therefore the frog would indicate the direction from which the earthquake wave travelled.*

Earthquakes

Since the beginning of this century, more than 1,500,000 people have died because of earthquakes. On 19th April 1906, an earthquake shook San Francisco, with up to 1,000 people being killed through falling debris and the associated firestorms. On 25th December 1972, about 10,000 people are believed to have lost their lives when tremors over a two-hour period shook the Nicaraguan capital Managua. In 1988, an earthquake rocked the region around Spitak and Leninakan, in Armenia in the former Soviet Union, and killed 25,000 people. And so the list goes on.

Earthquakes pose a particularly serious natural hazard, with all their related phenomena such as landslides, tsunamis (the 1960 Chilean earthquake tsunami wave travelled right across the Pacific to cause damage in Japan – see Figure 8.3), fires, liquefaction of the ground (a change

to a liquid-like state), and virtually instantaneous changes in land surface elevation which can cause flooding of coastal areas and the disruption of groundwater supplies and communication links (Plate 8.1). The energy released during an earthquake or its magnitude is measured on the Richter scale. The amount of energy released by the largest earthquake recorded (M8.5–8.7) was equivalent to 60,000 1-megaton TNT bombs or 60,000 hydrogen bombs. The effects of earthquakes are also measured semiquantitively using the Mercalli scale of intensity (see Table 8.3). The intensity of a particular earthquake at a location is dependent upon the ground

Table 8.3 *Mercalli scale of earthquake intensity.*

Scale	Intensity	Description of effect	Maximum acceleration (mm/sec²)	Corresponding Richter scale	Approximate energy released in explosive equivalent
I	Instrumental	detected only on seismographs	< 10		1 lb TNT
II	Feeble	some people feel it	< 25		
III	Slight	felt by people resting; like a large truck rumbling by	< 50	< 4.2	
IV	Moderate	felt by people walking; loose objects rattle on shelves	< 100		
V	Slightly strong	sleepers awake; church bells ring	< 250	< 4.8	
VI	Strong	trees sway; suspended objects swing; objects fall off shelves	< 500	< 5.4	small atom bomb, 20,000 tons TNT (20 kilotons) hydrogen bomb, 1 megaton
VII	Very strong	mild alarm; walls crack; plaster falls	< 1,000	< 6.1	
VIII	Destructive	moving cars uncontrollable; chimneys fall and masonry fractures; poorly constructed buildings damaged	< 2,500		
IX	Ruinous	some houses collapse; ground cracks; pipes break open	< 5,000	< 6.9	
X	Disastrous	ground cracks profusely; many buildings destroyed; liquefaction and landslides widespread	< 7,500	< 7.3	
XI	Very disastrous	most buildings and bridges collapse; roads, railways, pipes, and cables destroyed; general triggering of other hazards	< 9,800	< 8.1	60,000 1-megaton bombs
XII	Catastrophic	total destruction; trees driven from ground; ground rises and falls in waves	> 9,800	> 8.1	

Source: After Bryant (1990).

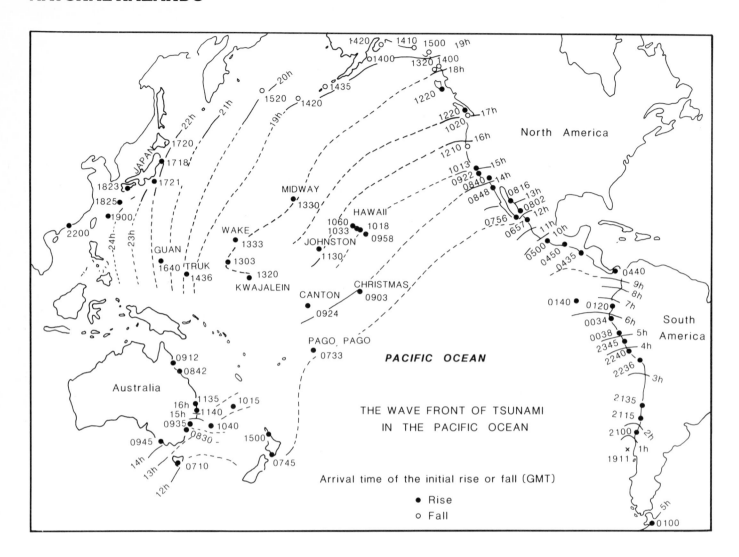

Figure 8.3 *Tsunami wave front in the Pacific Ocean following the May 1960 Chilean earthquake.*

conditions, the distance from the source of the earthquake (or focus), and its magnitude.

Earth scientists explain the cause of earthquakes using the Plate Tectonic Theory. In this theory, the outer layers of the Earth (lithosphere), comprise seven large plates. The plates 'float' on top of hot molten rock called the **asthenosphere**, which flows at very slow rates and is responsible for generating motion in the above lithospheric plates (Figure 8.4). The plates move at rates typically of a few centimetres per year. The crust of the Earth consists of the upper layers of the lithosphere made of two main types: oceanic crust which is thin and dense, and continental crust of thicker and lighter rock which form the continents. Today, Earth scientists accept a substantially modified view of plate

tectonics. No longer are the continental plates considered as rigid and independent of one another. Instead, they are regarded as extremely viscous liquids so that mountain belts can only maintain their height by virtue of a continuous 'push' of compressional forces. Take away these push forces and the roots of the mountains will literally flow away under the force of gravity (as well as being washed away through erosion) until the elevation of the continental crust is at about sea level. Of course this process operates at a slow rate measured in millimetres per year.

The relative motion of plates produces stresses (measured as force per unit area) along their boundaries. The strains or movement produced by these stresses may be gently relieved if the plates can slide reasonably easily against each

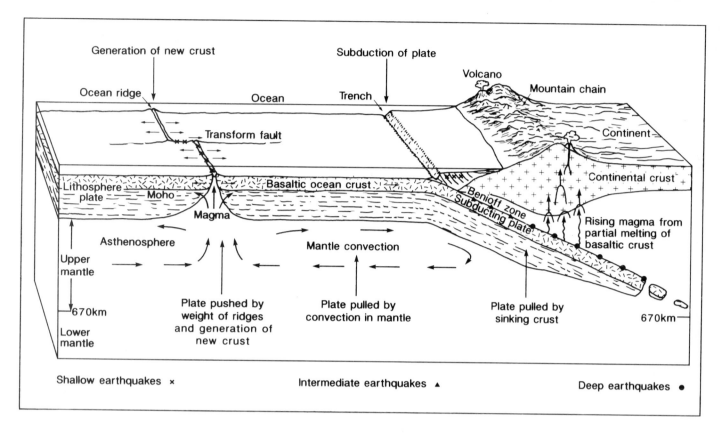

Figure 8.4 *The structure of the outer layers of the Earth, the major physiographic features, and the plate tectonic mechanisms responsible for the generation of new oceanic crust, sea-floor spreading, the consumption or subduction of oceanic crust, earthquakes and volcanicity. After Selby (1985).*

other. If, on the other hand, there is no easy and smooth slip, stress builds up because of friction forces between plate boundaries. This stress build-up may be suddenly and violently released. The result is a burst of energy which is referred to as an earthquake. The build-up of strain is not only along the actual boundary between the plates, but also along zones of highly localized strain associated with the boundaries. These zones comprise highly deformed rock which has **folds** and is cracked.

Large cracks along which the Earth's surface has been displaced are called geological faults. It is along these geological faults that the sudden release of accumulated strain occurs and where earthquakes are most frequent. This view is supported by examining the distribution of past earthquakes throughout the world. Since the globe comprises seven main plates, the plate boundaries are limited to several well-defined zones. We know that it is along these zones that the majority of earthquakes occur. The three

most important zones are the Alpine–Himalayan Belt where the continental Eurasian plate is colliding with the continental African and Indian plates; the Circum-Pacific Belt where the oceanic Pacific plates are colliding with the Eurasian, South American, Australian, and North American continental plates; and the Mid-Atlantic Belt where the oceanic crust of the Atlantic is spreading apart, to generate new ocean crust along the centre of the Atlantic Ocean.

An understanding of the mechanism of faulting is essential in earthquake hazard assessment. Clearly, the earthquake hazard is a result of the accumulated build-up of strain and its sudden release. Seismologists (geologists or physicists who study earthquakes) can study the hazards from earthquakes in a number of ways. For example, zones susceptible to earthquakes can be identified. This can be done by assembling data from past earthquakes and studying their distribution. The focus of an earthquake – the position and depth within the Earth from where

Plate 8.2 *A view looking across the remains of a small town in the Garwhal Himalaya, northern India. This was totally destroyed when an earthquake (magnitude 7.1 on the Richter scale) shook the area in October 1991. Seventy people and numerous cattle died as their poorly constructed houses and barns collapsed around them. Until this event there had been no large earthquakes recorded in the region during historical times, and the area was considered a seismic gap within an otherwise highly seismic mountain belt. At present, a major dam is being constructed at Tehri, less than 100 km down valley from the epicentre of this earthquake, causing concern for the future safety of those people living down valley.*

the earthquake may have originated – can be located using an array of seismometers (instruments used for detecting earthquakes) which are distributed around the world and within earthquake-prone areas (Plate 8.2). Active geological faults may be identified when the earthquake focus has been located. The fault can be traced across a region using geological mapping and subsurface seismic techniques and, to a certain degree since movement along the fault may disrupt drainage and create scarps or small lakes parallel to its length, topographic mapping using the surface expression of the fault. Once an active fault has been identified, movement along it can be studied using precise surveying techniques. The accumulation of strain may be measured using **strain gauges** – instruments which deform under stress, the amount of strain being converted electronically into absolute measurements.

Strain may be localized along a fault so that stress is released and movement occurs just within one zone, while stress may continue to accumulate elsewhere along the fault. Once

stress has been released as strain along one fault segment, it is unlikely that a major earthquake will occur along this stretch for some time, probably not before the stress has been released from other lengths of the fault which are still being stressed. Using this principle of **seismic gaps**, as well as other studies, seismologists are able to estimate the earthquake risk along particular stretches of faults. This has been done for the San Andreas fault in California, where seismologists have identified two major seismic gaps where stress has been accumulating for a long period compared to the other stretches of the fault which have experienced continued movement and release of stress.

Seismologists also seek evidence regarding the **recurrence intervals** of earthquakes. Archaeological evidence and historical records are used to compare the time intervals between successive earthquakes. This has been particularly successful in China where accurate and detailed records have been kept for more than 2,000 years (Plate 8.2). Geological evidence is also used such as the detailed study of the displacements of sediment layers which lie astride fault lines. One of the most famous studies was made in California at Pallet Creek which traverses the San Andreas fault. Here sediments which had been deposited by streams record a history of the last 1,700 years. The displacements of the sedimentary layers were measured and dated using radiocarbon dating techniques. The sediments showed 12 earthquake events, giving a recurrence interval of about 145 years. The last movement of the San Andreas fault at Pallet Creek occurred about 120 years ago. This suggests that an earthquake is likely in this area within the very near future. Such detailed studies provide predictions for the San Andreas fault, which suggests that there is a 25 per cent probability that an M8 earthquake will occur along the San Andreas fault east of San Diego within the next few decades, and a 75 per cent probability of an M6–7 earthquake along a stretch south of San Jose over the next few decades. Either event will cause large-scale devastation to parts of California.

The short-term prediction of earthquakes makes use of phenomena that have been observed before an earthquake. These are called precursors and include changes in ground level; emission of radon gas; a change in the electrical resistivity (the ability of the ground to carry an electric current) of the ground; a decrease in the velocity of small earthquake waves recorded in the area; a decrease in the number of small earthquakes, although often there may be many foreshocks; and changes in water levels within wells. These phenomena have been explained using what the seismologists call the 'dilatancy-diffusion model'. The model suggests that there is dilation of the ground accompanying the increased build-up of stress before a major earthquake event. As the ground dilates, cracks develop allowing gases such as radon to escape from the Earth's interior, water to enter the ground and decrease its electrical resistivity and/or produce changes in water levels within wells. A decrease of earthquake waves is explained because they have to travel though air and water, which slows them down. After a major earthquake event, several small after-shocks may occur before the region stabilizes and stress once more begins to build up.

On the basis of such evidence, scientists may be able to make accurate earthquake predictions in the future. The first successful prediction was made in 1975 in Haicheng, China. About 90,000 people were evacuated before an M7.5 earthquake totally destroyed 90 per cent of Haicheng. Research into earthquake prediction still has a long way to go, however, before accurate medium- and long-term predictions can be made. For example, the 1992 Nicaragua earthquake generated a tsunami which was disproportionately large for the earthquake surface-wave magnitude of 7 (Kanamori and Kikuchi 1993). The reason for this discrepancy was probably that slip on the plate interface filled with soft, water-rich sediments of the **accretionary prism** caused the rupture process to be slower than in most subduction zones where earthquakes are associated with thrusting, that is, the strain rate was slower due to the intrinsic weakness of the sediments. An important practical corollary of this scientific observation is that to reduce the hazards from this type of earthquake, tsunami warning systems using long-period (> 100 s) waves are necessary.

The prevention of earthquakes is also a major area of study. Scientists suggest that the stress accumulated in particular faults may be relieved slowly by underground nuclear explosions or by the pumping of water into the fault zone.

Legislation allowing experiments and such methods to be used, however, is a long way off – we simply do not know if such action would prevent the likely occurrence or inadvertently increase the likelihood and magnitude of an earthquake. In the USA, scientists from the Earthquake Engineering Research Center at the University of Berkeley, California, have released the results of laboratory experiments showing that buildings supported on specialized steel ball bearings have a significantly improved chance of withstanding the impact of severe earthquakes. The spherical ball bearings rest at the centre of concave steel 'bowls', and during an earthquake they absorb much of the energy by deforming and moving in response to the cyclic stresses of the quake. These results were presented in May 1993 at the National Earthquake Conference in Memphis, Tennessee, and suggest that the ball bearings can reduce the stresses on buildings by up to about 80 per cent. The results have proved so impressive that the US Court of Appeals building in San Francisco, which suffered damage during the 1989 earthquake, is to have 256 ball bearings installed underneath it. In the future, buildings constructed from smart materials will be earthquake-compensating in relation to passing earthquake shock waves and, therefore, much more able to withstand severe earthquakes.

Volcanoes

Most of the world's active volcanoes are situated in the less-developed and developing countries where there are high rates of population growth. Living space, particularly on the fertile volcanic soils, is at a premium. In the more industrialized, developed nations there are also increasing numbers of people living on the slopes of active volcanoes, for example Mount Vesuvius in southern Italy and Mount Etna in Sicily. Volcanic eruptions pose a serious natural hazard.

At present, there are approximately 600 volcanoes that are considered active above sea level, and several thousand extinct or dormant volcanoes which still retain their volcanic shape. On average, 5–15 of these volcanoes are active every month, emitting hot fluids and **lava**. Most of the active volcanoes are found within the Pacific region associated with the Circum-Pacific Belt, which is often called the 'Ring of Fire'.

Recent notable eruptions include Mount Pinatubo in the Philippines, about 100 km north of the capital Manila, which erupted after about 600 years of dormancy, with violent outpourings of ash and lava on 12th June 1991 (Plate 28). A column of ash and vapour was ejected about 25 km into the atmosphere. Some 3–5 km³ of magma and 20 million tonnes of sulphur were released into the atmosphere, with predictions from simple climate models indicating that this eruption would cause a global cooling of about 0.5°C within the next year. During the two-week eruption, local villages were covered by more than 30 cm of ash. A typhoon exacerbated the situation with high winds and torrential rain. At least 136 people were reported killed, with hundreds injured and more than 100,000 Filipinos fleeing their homes. At the nearby American Clark Air Base with 14,500 US servicemen and their dependants, a nuclear alert was issued because the cruise missiles, in transit from B-52 bombers in Guam to the USA following the Gulf War, were not in the deepest underground bunkers usually reserved for them: volcanic ash began to block ventilation shafts as personnel were ordered to abandon the base to nature (Plate 29). The 36 cruise missile warheads were evacuated from the base by nuclear technicians, who braved the worst of the eruption, and were then shipped out on 17th June aboard the USS *Arkansas*. The missiles themselves, together with 3,300 tonnes of munitions, remained in the bunkers at Clark Base. On 21st June, exploding ammunition dumps in the vicinity of the still-smouldering volcano caused panic in the local population. Although the eruption of Mount Pinatubo was not particularly large or devastating by historic records, this detailed look at the eruption shows that there are many facets of a volcanic hazard, including the additional inherent dangers in storing nuclear weapons in such areas.

Also on 3rd June 1991, a volcano in the Mount Unzen complex, near Nagasaki in Japan, erupted resulting in the death of 38 people. In 1792, an eruption of this volcano had killed about 15,000 people. Unlike the volcanic eruption of Mount Unzen which was successfully predicted, Mount Pinatubo's explosions came as a surprise, basically because unlike Japan, the Philippines lacks a sophisticated and well-funded volcano monitoring system and, in any event, the volcano had not erupted for 600 years. Most of the volcanoes in the Circum-Pacific region are related to **convergent plate boundaries**, either oceanic–oceanic plate collision or oceanic–continental collision. In addition, some volcanoes, such as those which form the Hawaiian islands, are believed to be the result of more localized hot magmas rising to the surface from **plumes** within the mantle. The latter geologists call 'hot spots'. Other volcanoes are associated with divergent plate boundaries where new oceanic crust is formed, such as those along the Mid-Atlantic Ridge. Iceland is an example of the surface expression of such volcanoes that have risen from the seafloor. There are two main types of volcano: those associated with divergent oceanic **spreading ridges**, oceanic–oceanic collision, and oceanic hot spots which tend to produce relatively fluid, low-viscosity lavas. The eruptions are usually gentle and the volcanoes have very gentle gradients. Second, there are volcanoes associated with continental regions which tend to have rather viscous lavas and often produce very violent eruptions, and steep volcanic cones.

Fortunately, much volcanic activity usually affects sparsely populated regions, but when major eruptions occur, the destruction may be huge and for nearby large populations it is catastrophic. The effects of volcanic activity are of several main types. These include lava flows, ejection of volcanic ash and rock material, poisonous gases, **lahars**, and fires.

Lava flows usually move relatively slowly, and tend not to cause major loss of life. They will destroy any building, construction, or farmland, however, that lies in their path. Particularly threatening have been the lava flows that have advanced from Mount Helgafell in Iceland since 1973 and which threaten to block the main harbour on the island of Heimaey, Iceland's main fishing port. Much effort has been put into containing the lava flows by the construction of walls and also by hydraulically chilling the hot lava to redirect its flow. Such measures have limited and temporary effectiveness. In the early part of 1992, Mount Etna in Sicily began to increase its volcanic activity. Lava continued to pour out of the main vent and fissures, something that began in December 1991, and flowed at such a rate that Zafferana village in the path of the advancing lava had to be abandoned. A canal formed which funnelled the lava, allowing it to maintain its high temperature and liquidity and so move forward at up to about 16 km/hour. Attempts were made to impede the path of the lava by channelling it into an artificial pool, but this filled in early April and the lava continued to flow over the top. By mid-April, soldiers were resorting to dropping huge concrete anti-terrorist barriers by helicopter on to the lava and detonating controlled explosions in a further attempt to divert it away from the village. It was possible to slow the advance of the lava temporarily but not to halt its progress completely.

The eruption of hot volcanic ash and rock creates one of the most severe volcanic hazards. Ash may be ejected over hundreds of miles and may stay in the atmosphere for many years. The immediate effect of the fallout of the ejected material is to destroy animal life and crops, and to cause structural damage to buildings and contaminate water supplies. Probably the most famous example of loss of life due to the ejection of ash occurred in AD 79 when Vesuvius in Italy erupted, burying Pompeii, killing 16,000 people, and effectively preserving the city until its rediscovery by archaeologists in 1595.

Ash that stays in the atmosphere for longer periods may affect the amount of solar radiation reaching the Earth's surface and cause darkened days. It may also affect global atmospheric circulation. The summers following the great eruption of Krakatoa in Indonesia in 1883 were recorded as being darkened, with spectacular sunsets produced by the scattering of light from the increased dust content in the air. These were artistically captured by John Turner in many of his paintings of this period.

Mudflows are produced by hot fluids, rock, and water, often from melting snow present near the summits of volcanoes. The water or melted snow and other hot fluids help mobilize rock and ash which, along with associated collapse of slopes during the eruption, results in a *mélange* of material flowing down slopes at fast speeds. One of the most catastrophic lahars occurred in 1985 in Columbia, when a mixture of ash and melted glacial ice flowed from Nevado del Ruiz down a valley killing over 22,000 people. The previous year geologists had predicted the event would occur if the volcano erupted, but no preventative action was taken to warn or evacuate the population.

Monitoring active volcanoes in order to predict eruptions is common practice in many parts of the world (see *Journal of the Geological Society of London* 1991). Large amounts of money are spent each year, particularly in Japan and in the Cascade Range in the USA, monitoring active volcanic mountains. The prediction of a volcanic eruption is based on the phenomena associated with the movement of magma towards the surface from depth. This will initiate small earthquakes and may cause a detectable warping of the ground. The upward motion will heat up groundwater and release small amounts of various gases through the soil and via small vents called fumaroles. Small eruptions of lava and ash may also occur before the large eruptive event. Monitoring of these phenomena is useful in predicting large volcanic eruptions. Much was learnt by studies of Mount St Helens in the Cascade Range, western USA, by monitoring its eruption in 1980 (Plate 30). In addition, detailed information may be gathered regarding the effects of volcanoes by studying the deposits of previous eruptions.

A relatively new technique for monitoring active volcanoes is through remote sensing from satellites using infrared radiation. The thermal energy given off by volcanoes can be picked up in the infrared wavelengths using devices such as the AVHRR (Advanced Very High Resolution Radiometer) carried by satellites run by the National Oceanic and Atmospheric Administration (NOAA).

Lake-water overturn

The sudden release of toxic gases is commonly associated with volcanic eruptions, but there are other natural processes that may be responsible. The build-up of large amounts of dissolved gas at the bottom of lake waters and their catastrophic release can be a potentially serious natural hazard.

The most catastrophic effects of toxic gas release that have been recorded occurred in western Cameroon late in the evening on 26th August 1986 as the result of carbon dioxide and hydrogen being released from the volcanic Lake

Nyos, near the town of Wun about 200 miles north of the capital Yaounde. A poisonous cloud of gas swept along the valleys north of Lake Nyos, leaving in its path about 1,700 people dead, along with numerous livestock. Although carbon dioxide is not normally poisonous, because most living organisms have evolved to cope with the atmospheric concentrations, it is a very heavy gas and when present in large quantities will form a thick layer above the ground to displace the lighter oxygen so essential for respiration. At the time of the disaster, many people believed that the cause was volcanic, but now scientists think that it was due to the sudden overturn of the stratified water column in Lake Nyos under particular weather conditions. It is believed that gases held under pressure in the bottom layers of Lake Nyos were released by the mixing of lake water due to particularly heavy rainfall which increased the head of fresh water.

The release of dissolved carbon dioxide from water occurred in an **endothermic reaction**, that is, one in which additional energy for the reaction is required, consequently the temperature of both the gas and water involved cooled down. The rising gas bubbles expanded and cooled further. It has been estimated that the escaping gas may have been up to 10°C cooler than the deep lake water from where it originated (Freeth 1992). Since the carbon dioxide was much denser than the air, the gas cloud spread across the lake and cooled the surface water, causing it to sink and promote the further rise of deep lake water. A positive feedback loop was probably created, in which cooler surface waters sank to greater depths, resulting in larger quantities of deeper lake water coming to the surface. These deep waters were rich in ferrous bicarbonate and, upon coming into contact with the oxygen-rich surface waters, oxidized to produce the characteristic reddish-brown colour, due to the formation of hydrated ferric oxide.

Landslides

The movement of material down slope under the influence of gravity is referred to as **mass movement**. Landslides are one category of mass movement involving falling, sliding, and/or flowing of rock or weathered rock material down and out of a slope. Movement is often along well-defined surfaces confined to a limited portion of a slope. There are many different types of landslides classified on the basis of the type of material involved, plus the speed and mechanism of movement. Landsliding may involve simple fall of debris (rock fall), collapse and fall of rock/snow (avalanche), plastic flow of material (debris flow), sliding of rock (rock slide) or clay and debris (mudslide), the rapid flow and spread of the ground (Earth/sensitive clay flow), and rapidly moving mixtures of debris and air and/or water (flowslides).

Flowslides are the most disastrous type of landslide due to their incredible speeds and the fact that they occur with little or no warning. Torrential rainfall is probably the main cause of mudslides. In mid-June 1991 in northern Chile, after 10 times the annual rainfall fell in under four hours near the city of Antofagasta, a mudslide caused the death of 112 people and injured 750. The mudslide then baked as hard as concrete in the sun, thereby adding to the problems of cleaning up after the catastrophe.

Earthflows are also a big problem, especially in Norway and Canada where there are large areas of land consisting of sensitive clays. Of all the types of landslides in North America each year, earthflows account for the greatest amount of damage and financial cost. **Sensitive clays** cause problems because they deform and fail under stress, or due to changes in their moisture content or the chemical composition of the groundwater. Failure may occur on very gentle slopes upon which buildings or excavations have been constructed.

Landslide disasters are often composite, that is, they are triggered by other hazards such as earthquakes, volcanic eruptions, and floods. Landsliding costs the USA more than US $2 billion each year with other countries such as Japan (> US $1.5 billion), and Italy (> US $1 billion) following close behind. It is estimated that there are over 600 deaths, not including major catastrophes, each year in the Pacific Basin. For example, in 1970, 18,000 people were killed in Peru as rock debris mixed with snow and ice travelled down the slopes of Mount Huascaran at speeds in excess of 300 km/hour, destroying the town of Ranrahirca. This failure, like many other catastrophic landslides in Peru, was initiated by an earthquake. Landsliding tends to be a recurrent problem, especially in the tropics, and it is not helped by persistent re-urbanization soon after the disaster on the still unstable slopes.

The susceptibility of a slope to failure is dependent on many factors. These include the gradient and length of the slope, the geotechnical properties of the rock (such as the rock strength), cohesion, and whether the rock has discontinuities such as joints and bedding planes. The amount of water entering a slope will be a very important factor, which is a function of vegetation, drainage, soil type, the structure of the rock (i.e., whether the rock has joints or voids), and the duration and intensity of rainfall. If there are large quantities of loose rock material or soil that are capable of being moved, then erosional processes such as undercutting of the slope by rivers, the sea, or glaciers and the possibility of earthquake or volcanic activity will determine its likelihood of failing. Finally, the use humans make of slopes or adjacent regions can drastically affect slope stability. Particularly susceptible areas are those where there are steep and long slopes with large quantities of weathered rock, high rainfall, the possibility of earthquakes, poor drainage, intense erosion, and poor vegetation. Such conditions exist in many tropical South-East Asian countries, especially where deforestation is taking place rapidly, since the vegetation previously acted as a protective cover. The number of landslides in such regions has increased because of human activities.

Human activities have also been responsible for many landslides in urban areas as slopes are exploited for urbanization and the creation of artificial slopes. A particularly famous example occurred on Hong Kong island, where large tower blocks were built on, and adjacent to, steep tropical slopes. In 1972, major slope failure occurred on one of these slopes in a residential area, the Mid Levels, completely destroying a newly constructed tower block and severely damaging an adjacent tower (Plate 31). The most famous failure of an artificial slope occurred in the small mining village of Aberfan in south Wales and was caused by the failure of a badly managed coal waste tip. The event occurred during a morning in October 1966, sending coal waste mixed with water down slopes at speeds of about 16–32 km/hour to engulf a school and adjacent buildings. A total of 144 people were

killed, most of them school children.

It was through such events as Aberfan that much attention is now given to the potential of a slope to fail. Most large constructions in developed countries will make detailed site investigations to examine the ground conditions to prevent loss of life and preserve property. There is still a reluctance among engineers, however, especially in less-developed countries, to contract detailed site investigations, even though the cost of such studies is small compared to the costs encountered should a catastrophe occur. In an attempt to produce landslide hazard maps that can be used by planners and policy makers, geologists are commissioned in some countries to produce maps showing the distribution of causal factors that may initiate landslides. The varied causes of landslides and the increased land-use pressures still make this one of the most common natural hazards.

River flooding

In any season of the year, somewhere in the world, river floods cause the loss of life and the devastation of agricultural land. In mid-September 1992, three days of torrential monsoon rains caused many rivers to swell, such as the Indus, resulting in extensive flooding in northern Pakistan and India, leaving more than 2,000 dead and hundreds of thousands of people homeless. The Ganges–Brahmaputra–Megna drainage basin is just one example of a very densely populated and low-lying coastal region subject to repeated major flooding (Figure 8.5) (Plate 32). In the USA this century, hundreds of people have lost their lives due to flooding (Figure 8.6).

River flooding may occur due to high rainfall, melting snow in spring, or the emptying of lakes when a natural or artificial dam is breached. The amount of water overspilling a river bank is a measure of the magnitude of the flood. Hydrologists who study flooding also refer to the magnitude of floods in terms of recurrence intervals. Generally the larger the flood, the less frequent its occurrence. The area of land that is naturally flooded, often annually, is called the 'floodplain' of the active river channel. The hazard of flooding is measured as a function of its destruc-

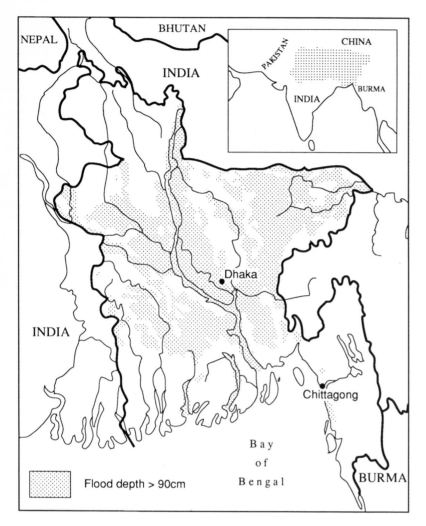

Figure 8.5 *Areas of Bangladesh prone to flooding to depths greater than 90 cm in a normal year in relation to the major rivers. Inset map shows the location area of the Ganges-Brahmaputra-Megna drainage basin. After Rogers et al. (1989), and Smith (1992).*

regions, for example, torrential rainstorms may only occur once or twice a year and when they occur the rainfall is usually very intense and lasts only a short time. In humid high-latitude regions, in contrast, rainstorms may occur intermittently throughout the year, and are usually less intense. In deserts, therefore, river channels tend to fill up quickly and **flash flooding** is common whereas in high latitudes flash floods occur less frequently.

The rate at which water can enter river channels during rainstorms particularly depends upon the vegetation cover which acts to intercept the falling rain and may stop it from reaching the ground. The ability of the soil and bedrock to seep up water, the steepness and irregularity of the slope, and the number of gullies or rills on that slope, will also control the rate of movement of water towards drainage channels. The speed at which run-off water travels through the ground to rivers will determine whether or not flooding is likely.

Regions that experience large floods occur where there is little vegetation, steep slopes, and many gullies, soils which are impervious to water, and rocks which allow water to flow quickly to rivers. This is because the water level rises more quickly in the streams and overspills the banks before the river has time to discharge further downstream to the sea. Such floods are called 'downstream floods' because the effect of the flooding progresses downstream.

Floods produced by the breaching of dams and melting snow tend to be 'upstream floods' which are confined to smaller areas, and their effects are more limited downstream. Flooding may be particularly disastrous along coastal areas if high river discharges coincide with high tides. The high tides prevent the river from releasing the large quantities of water so that coastal areas are flooded. This was the case in 1953 in the Netherlands, when 1,835 people were drowned as a result of a high sea swell in conjunction with rivers discharging large quantities of flood water into the North Sea.

In most developed countries, water height in rivers is monitored using **stream gauges**. If the water level rises to a critical height, then flooding is imminent and a flood warning may be issued. In addition, rain gauges are used to monitor the progress of a rainstorm and, from experience and study of a particular river, warnings may be

tive capacity in terms of life and property. Hydrologists advising engineers and planners in the design of structures and buildings will provide information on the size of a possible flood that may be expected to occur with a recurrence interval of 10, 25, 50, or 100 years. The magnitude of a natural flood produced by a rainstorm is related to the intensity and duration of that rainstorm, the rate at which the water flows across the land surface toward rivers, and the rate of flow of water through the ground to drainage systems.

The nature of a rainstorm depends upon the climatic conditions for that region. In desert

issued if flooding is likely. Rainfall monitoring is more useful than flood gauging because there is a time lag between the peak of the rainstorm and the rise of river level while the rainwater flows into the main river. Hence, there is more time available to issue an effective warning.

Flood hazard maps are useful in assessing the extent of likely damage and the area of land that needs to be evacuated. Such maps are produced using historical data and evidence from old river strand lines (high water marks). In addition, they are useful in providing planners and legislators with information for further development.

Unfortunately, today the risk from flooding is increasing in some areas because of conditions associated with human activities. The development of a region usually means that natural vegetation is cleared, and roads, buildings, and **storm sewers** constructed with the channelization of adjacent river courses. The clearing of vegetation reduces the ability of the ground to intercept rainwater. The rain may then flow quickly over the impervious surface of concrete and asphalted buildings and roads to the rivers along storm sewers and channelized streams. All this means that the rate at which water enters the main drainage channel or river increases as does the height of water. The likelihood of flooding is correspondingly increased. Urban planners are beginning to consider these factors in their attempt to reduce the likely hazards from floods. One method used to mitigate the effects of floods is to construct **retention ponds**.

Amongst the most recent devastating flooding in the USA was that of the Mississippi river in mid-July 1993, the worst on record in the past 20 years. Following a period of torrential rainfall, the Missouri river rose to its highest recorded level at St Louis of more than 15 m above normal, in the early hours of Monday 19th July, causing very extensive flooding of the farmlands and settlements, and leading to the death of more than 30 people. In Iowa, one of the worst-affected states, more than 50,000 people were flooded out of their homes. More than 250,000 people were left without clean drinking water. As a result, President Clinton promised large amounts of federal aid to mitigate the effects of this flooding.

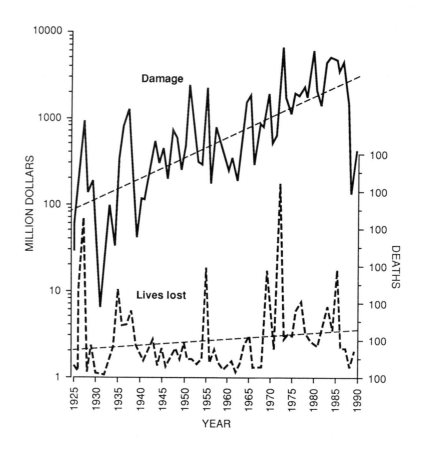

Figure 8.6 *Annual deaths and economic losses caused by flooding in the United States, for the period 1925–89. Damages are in US $ million adjusted to 1990 values. After Smith (1993).*

Glacial hazards

There are four main types of glacial hazard: (1) glacial advances and retreats, (2) glacial floods, (3) ice falls, and (4) floating/capsizing icebergs. The degree of hazard depends on the geography of the region and the dynamics of the glacier. Most glaciers are in remote, commonly mountainous areas where populations are small. However, as some of these areas become more populated due to increasing demands on land space because of growing populations, and as the areas are utilized for leisure activities, the hazards become more apparent.

Glacial advances constitute a hazard when ice approaches a settlement threatening to override it. The most famous example of this is the Mer de Glace in the Chamonix Valley, France, which threatened the villages of Les Bois and Le Chatelard in the seventeenth and eighteenth

Plate 8.3 *Abandoned village and its dry fields in the Karakoram Mountains, northern Pakistan. This settlement would have once been a fertile area, but because glaciers in this area have continued to retreat over the last century, their waters cannot be redirected into this area without the use of expensive pumps.*

centuries, but since then has retreated. Glacier movement is usually slow, being of the order of a few metres per year. However, **surging glaciers** occur where rates are measured in the order of several metres per day. These surging glaciers may pose a threat to settlements, but more commonly they act as dams blocking tributary river valleys and form lakes which have the potential to drain catastrophically.

A.T. Wilson (1964, 1969) has suggested that a glacier surge of the Antarctic ice sheet could have considerable effects on global climate. Such a surge would considerably increase the aerial extent of **ice shelf**, thereby increasing the Earth's albedo with a consequent global cooling and increased formation of ice sheets in the northern hemisphere, in turn initiating renewed glaciation. Breakup of an ice shelf would decrease the albedo, favour rapid melting of ice sheets and, therefore, the termination of a glaciation – associated with a rapid rise in global sea level. A.T. Wilson (1969) argues that if surges occurred, they would cause a rapid rise in sea level, in 100 years or less as the ice melts, with renewed ice storage being associated with much slower falls in global sea level in the order of 50,000 years. Interglacial pollen profiles should, therefore, record a rapid but temporary marine transgression beginning at the break of climate, and although some evidence exists for such profiles in the UK and USA, these cannot be unequivocally linked to surging, but may be the result of other factors such as localized tectonic subsidence.

Other glacial floods relate to the release of water from within or beneath the ice. The most spectacular examples of these floods have occurred in Iceland where volcanic activity beneath glaciers has melted large quantities of ice with the resulting floods being of extremely high magnitude, and having catastrophic effects on the landscape.

The advance of valley glaciers downslope may also threaten settlements and people working in glaciated areas, especially where the slopes over which they flow are convex. Here, the glacier extends and may tumble down slopes as ice falls, colliding with and destroying obstacles in its path.

Where settlements depend on glacial meltwaters for irrigation, a retreating glacier may constitute a hazard because the flow of meltwater into irrigation systems is reduced. In regions like the Karakoram Mountains, northern Pakistan, a number of glaciers have retreated during this century, resulting in several villages being abandoned, such as the one shown in Plate 8.3.

Icebergs are one of the most common and hazardous glacial phenomena. They are produced where individual glaciers flow into the sea or at ice shelves where extensive platforms of glacial ice form, such as in Antarctica. The glacier ice floats and may break up by a process called calving along its margin to form blocks of ice which may range in size from a few metres to several tens of kilometres. In September 1986 an enormous piece of the Antarctica Filchner ice shelf some 600 km across broke off and formed three separate icebergs which together covered an area of 13,000 km². These icebergs floated into the Weddell Sea to be grounded on the shallow seafloor, although one of them, called the A24 iceberg, floated free early in 1991 to travel into the southern Atlantic Ocean where it finally melted (Vaughan 1993). In autumn 1991, an iceberg formed by calving of the Ross ice shelf in Antarctica to create an iceberg floating towards the Falkland Isles measuring 50 × 52 km, or approximately four times the area of the Isle of Wight off England. The movement of icebergs away from glaciers is governed by ocean currents and to a lesser extent by wind. The routes are often complex and irregular and the iceberg may last a considerable time as melting of ice involves considerable energy and is, therefore, very slow. In addition, the movement of icebergs may be halted for long periods if the sea freezes around them.

These icebergs are a persistent problem to shipping in the seas in mid- and high latitudes as was dramatically illustrated on 14th April 1922 by the iceberg that collided with the Titanic, which subsequently sank. In areas where icebergs are common, continuous information regarding their position is issued to shipping on VHF radio. An area particularly prone to iceberg hazard is off the north-east coast of Canada in the Labrador Sea. Icebergs arrive in this region from calving of glaciers in the Queen Elizabeth Island (Canadian High Arctic: Plate 8) and north-east Greenland, having floated down the Davis Strait. They may even travel as far as New York. Iceberg monitoring and management is taken very seriously in the north-west Atlantic off eastern Canada and has involved such schemes as firing jets of water on to the icebergs to redirect their movement, although the enormous size of some of these icebergs makes controlling their direction impossible.

Tornadoes

Tornadoes are violent wind storms that take the form of a rotating column of air moving with speeds in excess of 116 km/hour. Tornadoes may occasionally reach speeds of 480 km/hour, but their average speed across land is about 45 km/hour, and the average path distance is approximately 26 km. The typical diameter of

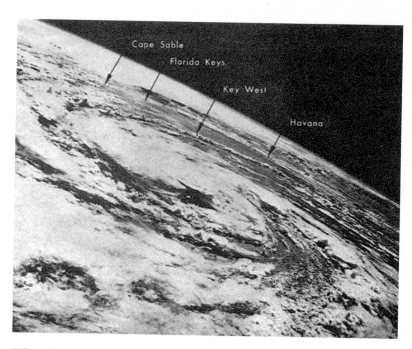

Plate 8.4 *Tornado.*

Courtesy of NASA/Lunar and Planetary Institute.

the spiralling air column is 150–600 m, though some have been recorded up to 1.5 km.

As they travel across country, tornadoes cause great devastation, with loose objects, including people, being carried skywards. Buildings and crops are severely damaged and destroyed. Houses with closed windows may explode because of the sudden creation of low pressures outside the building induced by the passage of the tornado (Plate 8.4). In the USA alone, about 750 tornadoes are reported every year. Occasionally, several hundred people may be killed in one day, for example in a region between Canada and Georgia, USA, on the 3rd April 1974, 300 people died due to tornadoes.

More recently, in Kansas and Oklahoma in the central USA on Friday 26th April 1991, 27 people were killed (mainly in a caravan park near Wichita where approximately 400 mobile homes were destroyed with 22 people left dead) as up to 50 separate tornadoes ripped through a region that is euphemistically called Tornado Alley, because of the frequency of its storms. These tornadoes started to occur because of the development of a complex depression near the eastern Rockies, formed as cool air from the north-west interacted with very warm and moist air masses advected north from the Gulf of Mexico.

Tornadoes are compared using the **Fujita Intensity Scale**. They are usually associated with severe thunderstorms when cold air from continental areas converges with warm wet air from oceanic areas. The convergence of these two types of air mass produces a cold front. This is a particularly common phenomenon in mid-latitude continental areas during the summer, for example in the mid-USA which is particularly prone to tornadoes between the months of April and September. The convergence of these cold and warm air masses produces high-pressure differences which create a tornado. The pressure at the centre of a tornado may be only 10 per cent of the surrounding air. This pressure difference sucks air from all directions towards the low-pressure centre, resulting in winds spiralling inwards and upwards and often forming thick clouds. The detailed mechanism by which tornadoes form is still not fully understood, something that makes their prediction difficult.

Since exact predictions are impossible at present, and tornadoes are short-lived, they are among one of the highest-ranking natural hazards. In the USA, a tornado watch is issued in areas where cold fronts develop and tornadoes are likely. When a tornado has been sighted, a warning is issued but, because the path of the tornado is often unpredictable, it is difficult to determine the likely path of damage. In some instances, tornadoes can be located using conventional radar. However, the new technology of doppler radar will allow quicker and more accurate identifications of tornadoes to be made in the near future.

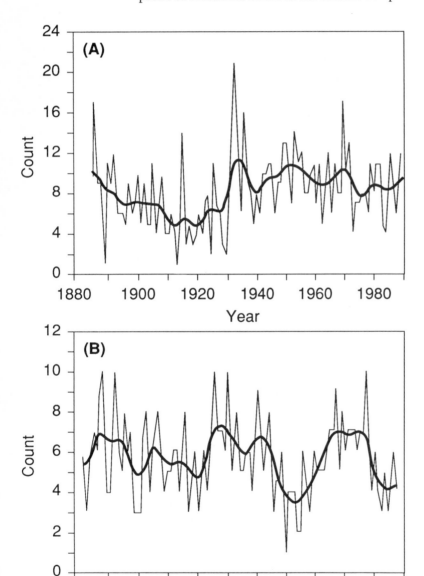

Figure 8.7 *Estimated number of tropical cyclones in (A) the Atlantic, and (B) the north Indian Ocean over the past century. Data in (B) are less reliable before 1950. After IPCC (1990).*

Tropical cyclones

Every year tropical cyclones, which vary considerably in frequency (see Figure 8.7), cause loss of life and damage to property (see Figure 8.8). Tropical cyclones are also known by various names. They are commonly called hurricanes in the Caribbean, typhoons in the north-west Pacific, and cyclones in the Indian Ocean. Cyclones are zones of low pressure, in which the low-pressure region causes the inflow of winds from surrounding regions. In the tropics, between the latitudes of 5 and 20°, these cyclones may produce winds that can reach 300 km/hour, in which there is an intense asymmetric vortex of moving air with average diameters of about 600 km and heights of about 12 km. The vertical development of thick clouds and heavy rain is an integral feature of a tropical cyclone. The eye of the hurricane is relatively tranquil, but away from this centre winds spiral round and upwards.

Tropical cyclones constitute a hazard in three main ways. First, the strong winds can produce

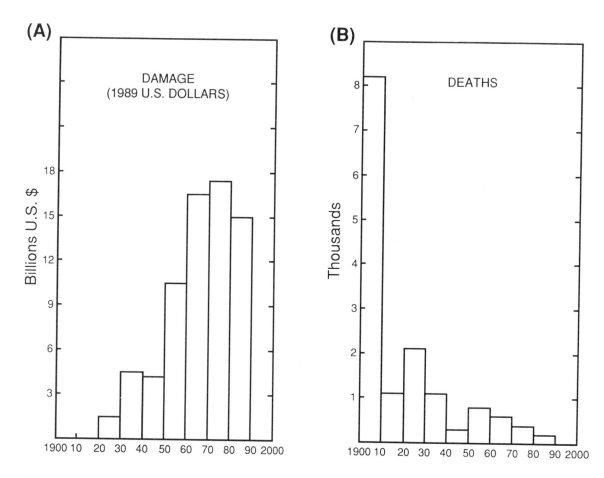

Figure 8.8 *Losses of (A) property, and (B) life in the continental United States due to tropical cyclones for the periods 1915–89 and 1900–89, respectively. After Gross (1991).*

considerable damage (Plate 33). In 1900 in the southern coastal regions of Texas, for example, over 6,000 people died in Galveston and more than 200 people died in smaller towns, as hurricane winds devastated the region. This was the worst natural disaster associated with a hurricane in US history.

The second major problem is the storm surge produced by a tropical cyclone where its eye comes onshore from the open sea. The low barometric pressure caused by the hurricane leads to the development of a dome of water several metres high and 65–80 km long. Such **'coastal set-up'** is a particular problem in low-lying delta regions such as Bangladesh. In 1970, for example, a surge of just 2 m was responsible for the deaths of as many as 500,000 people in Bangladesh. Again, in Bangladesh on 29th/30th April 1991, following nine hours of very strong typhoon winds gusting up to 225 km/hour, and

seas with tidal waves up to 6–7 m high, the low-lying coastal areas and offshore islands at about sea level suffered severe flooding and destruction as an estimated 125,000 people lost their lives, thousands of homes were destroyed, about 10 million were left homeless and hungry, and the harvest was wiped out. Because of the poor sanitary conditions and heat, a cholera epidemic followed the cyclone and many more people lost their lives. The worst flooding and loss of life occurred south of Chittagong, the second city and main port. Poverty drives the Bangladeshi people to eke out a precarious living on these flat-lying fertile islands and coastal plains covering about 140,000 km^2, generally less than 3 m above sea level, and which are prone to repeated flooding. Following the last really severe tropical cyclone in this region about 20 years ago, 500 shelters were planned, with about 50 completed at the time of the cyclone, but despite warnings to the population and even allowing for the inadequacy of the shelters, it is believed that many people simply did not leave their homes in time. From those who did, there were reports of arriving at shelters to find them full and so being turned away. The scale of these disasters is quite simply awesome.

The third problem is the heavy rainfall, commonly associated with tropical cyclones, which causes inland freshwater flooding. Hurricane Agnes, which was moderate in magnitude, produced extensive floods that were responsible for 122 deaths and US $2 billion of damage to property. On Friday 11th September 1992, on the Hawaiian island of Kauai, Hurricane Iniki severely damaged the homes of more than 50,000 inhabitants, but fortunately without loss of life, when winds gusting to 256 km/hour and torrential rain swept the island. Waves 7 m high crashed over coastal highways and the island lost all power and telephone services. It was the most powerful hurricane to hit the Hawaiian islands this century. One month earlier, more than 50 people lost their lives and thousands of homes were damaged when Hurricane Andrew roared through the Bahamas, south Florida, and Louisiana.

Meteorologists explain the formation of hurricanes using a model they call the '**heat engine**'. This heat engine is initiated by the evaporation of warm water (> 27°C) from tropical seas. Condensation of this water releases energy to warm the air and make it buoyant so that it rises. The result is to reduce the pressure above the sea and cause an inflow of air, that is, wind. As a result of the Earth's rotation, the wind spirals in an anticlockwise direction in the northern hemisphere and a clockwise direction in the southern hemisphere. The path of a tropical cyclone is also influenced by the Earth's rotation, directing tropical cyclones in a clockwise direction in the northern hemisphere, and in an anticlockwise direction in the southern hemisphere. If the wind speed does not exceed 61 km/hour, the cyclone is described as a tropical depression and if less than 119 km/hour, a tropical storm. It is only when the air speed exceeds 119 km/hour that the cyclone is described as a tropical cyclone, hurricane, or typhoon. The continuous supply of atmospheric moisture is essential to the continued existence of a tropical cyclone, once the tropical cyclone begins to enter a continental area, its energy is quickly dissipated because of the drop in air moisture from sea to land.

The prediction of tropical cyclones is very accurate and their formation and development are continuously observed and monitored using weather satellites. Warnings are issued to populations when a tropical cyclone approaches coastal regions so that preparatory measures can be taken. Unfortunately, many of the areas which are very prone to tropical cyclones are in less-developed countries where resources for preventative measures are inadequate so that large numbers of people still die as a result of tropical cyclones each year.

In the 1960s Project Stormfury, a co-operative venture between the US Navy and the National Oceanic and Atmospheric Administration in the US, began to investigate the possibility of modifying or destroying a hurricane. A process called 'cloud seeding' was tested in which silver iodide crystals were dropped into the air so that water vapour condensed on the particles to produce rainfall, which in turn would reduce the heat energy produced by a hurricane, and hence slow down the process of hurricane formation and motion. Project Stormfury actually seeded two hurricanes during the duration of the project, and the scientists involved in the project believed the seeding significantly reduced wind speeds. Many authorities, however, were sceptical about the results from the project, so it was terminated in 1972. Since then little attention has been given to modifying hurricanes, although their development has been simulated using elaborate numerical and computer modelling.

Thunderstorms

There are three main hazards associated with thunderstorms: torrential rain, hailstorms, and lightning. Thunderstorms are produced as warm air rises rapidly from near the surface of the Earth. As the air rises, more air from the surrounding regions is dragged inwards which, in turn, is undraughted. The rising air cools with height, and water vapour in the air begins to condense, forming thick dark billowing clouds which eventually spread out at heights of about 6–12 km to form the characteristic anvil-shaped

Plate 29 *Destruction resulting from the eruption of Mount Pinatubo in the Philippines in June 1991. The burial and collapse of a Filipino village (A) and Clark Airbase (B), due to thick ash being deposited during the eruption.* Courtesy of US Geological Survey.

(A)

(B)

Plate 32 *Severe flooding in Bangladesh, showing a submerged village after 2 months of monsoon rain.*
Courtesy of Steve McCurry/ Magnum.

Plate 31 *Human-induced landslide on the steep slopes of the Mid-Levels in Hong Kong, caused by the construction of a large tower block, the remains of which can be seen scattered in the landslide debris.*
Courtesy of the Geotechnical Control Office, Hong Kong.

Plate 30 *The eruption of Mount St Helens in May 1980. The lower ash clouds were produced by pyroclastic flows while the upper ash clouds spread out in the upper parts of the tropopause, causing coarse particles to fall towards the ground.*
Courtesy of Comstock.

Plate 33 *Before and after Hurricane Andrew at Biscayne Florida in 1992.*
Courtesy of Comstock.

Plate 34 *Lightning storm in Arizona, USA.*
Courtesy of Comstock.

Plate 35 *Swirls of dust kicked up by hot winds near Timbuktu in drought-stricken Sahel, Africa. This degradation of once-fertile land may be the result of human-induced climatic change.*
Courtesy of Steve McCurry/ Magnum.

Plate 37 *Curua Forestry Station in central Amazonia, established to undertake experimental studies on methods for efficient farming and reforestation.*
(A) Selective logging.
(B) Clearance of forest.
Courtesy of Mike Eden.

Plate 36 *Centre pivot irrigators making the desert fertile at Yuma, Arizona, USA. These produce some of the USA's highest crop yields. Each wheel irrigates 130 acres and may produce as many as 10 harvests of grain per year. Technologies like this provide hope for regions experiencing the effects of desertification, and a means of feeding the growing populations of the world.*
Courtesy of Comstock.

(A)

(B)

Plate 38 *Ships stranded in the dried-up Aral Sea, former Soviet Union, Central Asia.* Courtesy of Fred Mayer/ Magnum.

tops to the clouds. Torrential rainfall results from the large amount of condensation. Associated with the rapid draught of air, the lower portion of clouds become negatively charged. When sufficient electrical potential has built up compared with the positively charged ground, an arc of electrically charged atomic particles, called electrons, passes from the cloud towards the ground, carrying a current of as much as 60,000–100,000 amperes. This arc is the lightning we observe during a storm (Plate 34). The air heats up due to this electrical charge, expands, and finally explodes, producing the noise of a thunder clap. The details of this process are still not fully understood. Hail may also be produced during a thunderstorm as raindrops which formed within the storm are carried up under the updraught into the thunderstorm cell, become frozen, and form ice pellets. These pellets increase in size by the attachment of freezing droplets of water. When the size of the hailstones increases such that the updraught can no longer support their weight, the hailstones fall to Earth under the influence of gravity. Some of these hailstones may be larger than hens' eggs, which may cause great damage to crops and property.

There are between 1,500 and 2,000 thunderstorms throughout the world at any one moment, with more than 5,000 lightning strikes every minute. On average lightning strikes kill five people in the UK and about 95 in the USA every year. Those at greatest risk are people who work outdoors and those who participate in outdoor sports. People out of doors may form **positive point sources** to which electrical charges are attracted. If lightning is discharged through a person, the large currents are likely to result in the person being severely burnt or they may induce heart failure and subsequent death. However, people indoors have also been known to experience electric shocks produced by lightning, usually if the person is earthed to the ground while holding a water mains or talking into a telephone.

The most destructive product of a thunderstorm is the hail. The annual damage from hail each year in the USA totals over US $300 million. Most of this loss results from the destruction of crops. Hail-suppression programmes have been in progress since the 1960s in the former Soviet Union, where hail is particularly hazardous to large areas of farmlands. This has involved **cloud seeding** to initiate heavy rainfall and to dissipate cloud production, and therefore the heavy updraughts produced in a thunderstorm that are responsible for elevating rain droplets and their subsequent freezing to form hailstones. However, many scientists are not happy with these experiments and currently cloud seeding of thunderstorms is neither practised nor permitted by governments.

Droughts

Droughts are periods of increased dryness due to precipitation falling far short of that expected for a region. They may occur in every type of climate, arid to humid, tropical to tundra, and have profound economic implications for many regions as crops fail and cattle die. Individual droughts vary considerably in nature, effects, and duration. Their effects are most evident in semi-arid lands, where increased aridity leads to evaporation of water from the soil and consequently leads to poor plant growth, increased soil erosion, dust storms, and deficient and polluted salty water. This is particularly detrimental in developing countries which are strongly dependent on subsistence agriculture and are capable of doing little to mitigate the problems. Particularly hard hit in recent years have been the countries of sub-Saharan Africa (Plate 35). In 1983–5 hundreds of thousands of people died as a result of famine due to poor plant growth and reduced agricultural yields in Ethiopia and the Sudan. Some of the largest droughts have occurred in the Indian subcontinent, for example an estimated 3–10 million people died in the 1769–70 drought and in 1876–8 3.5 million people died near Madras. The problem of drought in India has been greatly eased during recent decades with the development of water schemes such as the Rajasthan Canal and the Tabela Dam and associated irrigation schemes using the waters from the Indus river. Many aid programmes have concentrated on providing reliable clean water to drought and poverty stricken areas in order to help reduce the problems of water shortages and the spread of disease (Plates 5.3 and 8.5).

In addition droughts are also common in developed countries, for example, the countries

of western Europe, and the USA. Here the associated problems are generally less, resulting in decreased water levels in lakes and reservoirs, decreased agricultural yields, possible forest fires, the decrease in the production of hydroelectric power and the use of water for recreational purposes.

Droughts owe their origin to climatic variations that result in decreased precipitation, although some people believe the effects of droughts are largely caused by bad land management. Specific mechanisms for this are unclear, but climatologists consider these variations as quasi-cyclical events. These are natural cycles of weather changes, but the duration and onset of droughts cannot be predicted at present. The effects of a drought also depend on other factors, such as the land's productivity capacity, the increased demands on soils and surface waters, and the disaster preparedness and awareness of a society.

Relieving the effects of drought in vulnerable countries depends on short- and long-term planning, with a consideration of the water demand in terms of land-use change, increased urbanization, industrial development, and other water-related activities. Planning should be supported by monitoring the drought, and developing technologies and projects such as irrigation systems (Plate 36).

Pests

Pests are usually regarded as any organisms that are judged to be a threat to human health, comfort, and endeavours. Commonly, pests multiply and spread rapidly and, therefore, compete with humans for available food sources such as crops. Pests include: (1) micro-organisms such as fungi, bacteria, and viruses; (2) invertebrates such as protozoa, flatworms, nematodes, snails, slugs, insects, and mites; and (3) vertebrates such as rabbits, and many rodents.

Of particular concern are pests which transmit disease, and these include: mosquitoes transmit-

Plate 8.5 *Providing clean water at Communidad Santa Martha, El Salvador.*

Courtesy of Rhodri Jones/Oxfam.

ting malaria and yellow fever; tsetse fly transmitting sleeping sickness; human louse, fleas, mites, and ticks transmitting typhus; blood-sucking bugs carrying the flagellate protozoan Trypanosoma cruzi, which is responsible for Chaga's disease (American Trypanosomiasis) causing swelling and lymph node enlargement, fever, and heart failure; sandflies transmit *leishmaniasis*; and from the roundworm, *Onchocerca volvulus*, causing river blindness (Onchoceriasis). The potentially most serious diseases and disease-carrying pests are discussed below.

Most species that have become pests appear to have adapted to take advantage of human activities associated with increased demands for food production. However, in some cases, the spread of pests has occurred as a result of entirely natural causes, such as locust plagues. Locusts (part of the Orthopteran family Acrididae) have plagued humans for centuries: the Biblical accounts of the plague in Egypt during the time of Moses are well known, whilst in more recent times there has been the devastation in the prairie farms of Canada and the US during the 1870s by the Rocky Mountain locusts and migratory grasshoppers (*Melanoplus spretus* and *M. sanguinipes*), and the plagues of locusts in West Africa and the Mediterranean. Locusts can travel vast distances, for example, in 1869 desert locusts (*Schistocerca gregaria*) reached England probably from West Africa. *S. gregaria* is known to be able to fly in swarms up to at least 1.5 km high. The size of swarms can be large, for example in 1889 a flight across the Red Sea was estimated to be approximately 5,000 km^2 in size. Once formed, a plague is virtually impossible to stop.

Methods of control include: destroying the eggs laid by locusts; digging trenches to trap the nymphs (the young stage of the locust, not yet able to fly); using hopperdozers, which are wheeled screens that cause the locusts to fall into troughs containing water and kerosene; using poisoned bait; and spraying swarms and breeding grounds with insecticides. Eradication of the nymphs before they mature and begin to swarm is amongst the most effective methods. A study of the life cycle of the locust is important for the prevention of plagues. In 1945 the Anti-Locust Research Centre was established in London to record and project migrations of swarms, and to study the biology of the locust in order to try to reduce its effects. One of the main problems with the control is that locust swarms tend to appear sporadically and unpredictably. This is attributed to the 'Phase Theory' which suggests that there are two phases of locust: a normal state or solitary phase in which it adjusts its colour to match the immediate surroundings, has a low metabolic and oxygen-intake rate and is sluggish, and the gregarious phase, in which it has an affinity to group, is black and yellow in colour, has a high metabolic and oxygen rate, and is very active. If one or other of these two phases develops within a group dominated by the other phase, it will mature into the other phase. The gregarious phase is considered to be a physiological response to violent fluctuations in the environment. Migratory swarms form in marginal areas where suitable habitats are scarce. A succession of favourable seasons causes restricted populations to expand and forces them beyond these marginal areas. When less favourable environmental conditions return, the enlarged populations in the marginal areas must go back to the relatively small permanently habitable areas, and this causes overcrowding leading to swarming.

Today, the transmission of pests is a global problem because of the ease of world travel and global trade. Strict regulations on importation and quarantine of hazardous foods substances and crops, and travel of people suffering from diseases or non-vaccinated people with their associated pests, are enforced by many countries. The risks from pests have increased over recent decades with the change from natural vegetation and traditional small-farming techniques, to monocultural farming systems which are intensive and large. This has led to greater hazards as it provides a more uniform food source for plant-eating species to increase rapidly in number and the uniform plant cover is easily invaded. The introduction of new crops to an area may also transfer previously harmless species from their natural habitat to the new abundant food source. Cultural practices such as fertilization, irrigation, and the use of modern harvesting techniques, which leave large amounts of plant litter in the field, also enhance the ability for pests to increase in numbers. The elimination of species that prey on pests and the evolution of insects which become insecticide-resistant exacerbate the problem.

Box 8.1

Common infectious diseases in developing countries

Malaria

Malaria is a serious and chronic relapsing infection in humans, apes, rats, birds, and reptiles, and is characterized by periodic paroxysms of chills and fever, anaemia, and enlargement of the spleen which may lead to fatal complications. Malaria is caused by the blood sporozoa of the genus *Plasmodium* and is normally transmitted by anopheline mosquitoes. It can also be transmitted by hypodermic needles and blood transfusions. Hyperendemic areas include Central and South America, North and Central Africa, the countries bordering the Mediterranean, the Middle East, and East Asia. In many parts of Africa and South-East Asia, entire populations are infected almost continuously by the disease. It is most common in the tropics because conditions are most favourable to the mosquito.

There are several types of malaria: the most widespread is Vivax (tertian) malaria because it is able to withstand therapy and remains chronic. Falciparum (subtertian or malignant tertian) malaria has the most severe symptoms, tends to cause fatality, and is confined to the hottest tropical regions, particularly in West Africa. Quartan malaria is present throughout the Mediterranean. Infection from more than one of these forms of malaria can occur in any individual. Effective treatment for malaria has been known since the 1700s, when the bark of the chona tree was used with its most active ingredient being quinine, but it was not until much later that the mosquito was recognized to be the transmitter. Anti-malarial drugs may be used to prevent the disease or as a suppressive to reduce or eradicate it. Some individuals have a natural resistance to malaria, for example those who carry the sickle-cell trait. An acquired immunity decreases the susceptibility to the disease. Control on mosquito populations is the best prevention, but the use of insecticides such as DDT in the 1950s and 1960s caused other widespread ecological problems. Global eradication programmes have been under way since 1955, when major concern was expressed by the WHO. Unfortunately, many of the countries still threatened by malaria epidemics are poor, and it is difficult to implement control and eradication schemes.

The control of pests has a long history dating from the ancient Chinese who used predator ants to control foliage-feeding insects, to the introduction of the Indian Mynah bird to Mauritius by Westerners in 1762 to control the red locust, and to sophisticated modern physical, chemical, and biological controls. Pest control really began to be studied in the eighteenth and nineteenth centuries with increasing population growth, industrialization, and agricultural expansion. Methods included physical or chemical control. Physical methods involved the use of sticky barriers, flooding, and burning, but these were only effective for short periods of time. Chemical pesticides began with the use of poisonous plants, such as ground tobacco which was used in France in 1763 to kill aphids. Other natural products which were widely used included nicotine, petroleum, kerosene, creosote, and turpentine. Inorganic products included Paris green, lime sulphur, Bordeaux mixture, hydrogen cyanide, and lead arsenate, the use of which started in the 1800s. Biological means were also used, and particularly successful was the introduction of the vedalia beetle (*Rodolia cardinalis*) from Australia to control the spread of cottony-cushion (*Icerya purchasi*) in California in 1888, saving the citrus fruit industry. Plants resistant to pests were also introduced, for example the *Phylloxera*, an aphid-like insect that attacked the European vine, was controlled by grafting onto the vine a more resistant American strain.

The discovery of the insecticidal properties of synthetic organic compounds during the Second World War gave hope for the possibility of pest-free crops. These included DDT (dichloro-di-phenyl-trichloroethane) and BHC (benzene hexachloride), and herbicides such as 2,4-D (2,4-dichlorophenoxyacetic acid). But the serious ecological problems associated with the use of these has led to their ban in many countries.

Current pest controls try to minimize the use of pesticides and combine them with the use of biological methods. These integrated methods include: the breeding of pest-resistant crops; crop culture methods that inhibit pest proliferation; the release of predators or parasites; the use of traps which are baited with the pests' **pheromones**; the disruption of reproduction by the release of sterilized pests; and the application of chemical insecticides.

Yellow-fever

Yellow-fever infects humans, all species of monkeys, and other small mammals. The virus is transmitted by mosquitoes, either between individuals (urban/classical yellow-fever) or from a forest mammal to a person (jungle yellow-fever). Yellow-fever causes headaches, backache, rapidly rising fever, nausea and vomiting, severe haemorrhages in the mucous membranes, destruction of liver cells resulting in jaundice, and it may result in death (dependent on the virus strain). Typically, the course of the fever is rapid and convalescence is long, but it produces a lifelong immunity in the victim. The disease has plagued the tropics and subtropics over the past two centuries. The USA was subjected to devastating epidemics and outbreaks, as far north as Boston, but the last serious outbreak occurred in 1905 in New Orleans and other parts of the southern states. During the nineteenth century, similar plagues paralysed industry and trade throughout the West Indies, Central America, Spain, Italy, and England.

The disease is completely preventable by the use of live-virus vaccines or by eradication of the mosquito. One of the most famous eradication campaigns was undertaken to control the *Aedes aegypti* mosquitoes during the construction of the Panama Canal, a course of action which saved many thousands of lives.

Typhoid (typhoid fever)

The bacterium *Salmonella typhi* enters the body via contaminated food or water and is absorbed into the bloodstream, causing blood poisoning. Following a 10–14-day incubation period, headaches, lassitude, aching, fever, restlessness resulting in loss of sleep, loss of appetite, nosebleeds, coughs, and diarrhoea or constipation result. Fevers with very high temperatures occur after a further 7–10 days and continue for about 10–14 days, at which time the fever begins to subside, but fluctuates diurnally. Complications include gall bladder inflammation, pneumonia, encephalitis, meningitis, and even heart failure.

Major epidemics have been caused by the pollution of water supplies, contaminated food

Box 8.1

Common infectious diseases in developing countries

Environmental diseases

Biotic agents are among one of the largest natural hazards threatening both humans and other organisms. These include infectious diseases spread by parasites, bacteria, and viruses, which include hepatitis B, schistosoma (bilharziasis), cholera, typhoid, poliomyelitis, and malaria. Additionally, environmental diseases include **neoplastic diseases** such as environmentally induced cancers initiated from drinking-water supplies or caused by areas where radioactive rocks produce high background radiations. Little information is available on environmentally induced cancers, and the connection between the environment and the prevalence of cancers remains uncertain.

Estimates of mortality are difficult to determine for many diseases, commonly because a disease contributes to death but may not appear on any death certificate or statistics, and inadequate records are kept in many poor developing nations. Morbidity statistics are published annually by the World Heath Organization, and provide one of the most reliable estimates. For example, from these tables, it appears that annually there are more than 10 million cases of malaria, over 50,000 cases of polio, some 200 million people infected by schistosomiasis, and approximately 50,000 cases of cholera. Young children, the infirm, and aged are amongst the most susceptible to infection, and there is commonly a clear correlation between poverty and disease, hence the associated high infant mortality rates in developing regions.

The transmission and prevalence of infectious diseases are a function of climate, water supply, sanitation, and socio-economic factors such as nutritional status, hygiene, and population density. Many diseases are transmitted by hosts, such as malaria by the mosquito, schistosomiasis by fresh-water snails, and sleeping sickness by the tsetse fly. Migration and increased populations of the host help to spread disease. Eradication of

Box 8.1

Common infectious diseases in developing countries

and milk, and by flies and persons transmitting the disease. Shellfish grown in polluted waters are a particular cause of the disease. Many cured victims continue to pass on the disease for life, transmitting the bacteria in their faeces where it may continue to live for months and infect others.

Prevention involves proper sewage treatment, filtration, and chlorination of waters, and even the exclusion of carriers from working in the food industry and restaurants. Prophylactic vaccinations are given to people travelling in affected regions, but their effectiveness is limited. Typhoid was prevalent throughout the world until the beginning of this century, but improved waste disposal management and water treatment have reduced its incidence. It is in the developing countries where sewage treatment plants are inadequate or non-existent, and where clean water supplies are scarce, that this disease remains a serious problem

Sleeping sickness (African Trypanosomiasis)

Sleeping sickness is transmitted by the tsetse fly which infects humans and cattle with the flagellate protozoans *Trypanosoma gambiense* and *T. rhodesiense*. Sleeping sickness causes fever, inflammation of the lymph nodes, and affects the brain and spinal cord, leading to lethargy and frequently death. It is hyperendemic in Central Africa causing mass mortalities of cattle and people. Treatment of an infection by *T. gambiense* includes the use of the synthetic arsenical (tryparsamide), but this may result in optic neuritis and loss of vision. Suramin sodium can be used for both infections, but once an infection by *T. rhodesiense* has developed to the toxemic stage, treatment is ineffective.

Great efforts have been made to control sleeping sickness, including: the isolation and treatment of victims; protection from bites; eradicating the tsetse fly by clearing its habitat around villages and using insecticides; the use of prophylactic doses of suramin and daimidine compounds for persons entering affected areas; relocation of entire villages from endemic zones to disease-free areas; and in extreme cases, the extermination of the wild game reservoir for the disease. In some areas, these controls have been very effective, particularly in the Cameroons where the locals have persisted with the use of tryparsamide to reduce the problem to very low levels.

the host or the host's environment helps to contain the spread of disease. Control methods include the spraying of pesticides in lakes and irrigation dykes to kill mosquito larvae and snails carrying schistosomiasis. Most of these infectious diseases can be cured with medical help, especially if they are diagnosed early enough, but they usually have long and debilitating effects. Prevention is considered the best method of control, but unfortunately this can be expensive and difficult to manage in poor developing countries. Therefore infectious diseases still rank high as a natural hazard. Box 8.1 deals with the most widespread and devastating diseases.

Conclusion

Natural hazards are among one of the major causes for concern on Earth. Natural processes that become a hazard have highly variable causes and effects. In order to help mitigate the effects,

it is necessary to understand the dynamics of the processes involved and to suggest ways that preventative measures can be executed within the socio-economic framework of the area threatened.

Natural hazards are not always detrimental to humankind. Some may provide a service, such as the renewal of mineral nutrients to the soil on floodplains during flooding and adjacent to volcanoes during volcanic eruption, or the creation of new land by Earth movements and **volcanism**.

Finally, the extent to which humankind is altering the environment is not fully appreciated. If the result is a change in the frequency and magnitude of these natural processes, this, in turn, may affect the occurrence of natural hazards. The following chapter examines the effects on the land, important in helping to understand the consequences of human actions, which may have possible implications in terms of natural hazard assessment.

Schistosomiasis (bilharziasis)

Schistosomiasis is caused by parasitic flatworms of the family Schistosomatidae (commonly called blood flukes) which, as part of their life cycle, live in the blood vessels of humans and other mammals. There are three main types of fluke (*Schistosoma japonicum*, *S. mansoni*, and *S. haematotobium*) which are geographically defined and are endemic in South-East Asia, Africa and South America, and the Middle East and southern Europe. The female fluke is 10–25 mm in length, and may release 300–3,500 eggs daily into the blood of a mammal. The release of eggs within the body produces tissue damage, causing allergic reactions, inflammation, coughs, skin eruptions and swelling, tenderness of the liver, and faeces and urine may contain blood. Chronic stages of the disease affect the body's vital organs, leading to fibrous thickening and loss of elasticity which may result in serious liver damage, stone formation in the bladder, and bacterial infections of the urinary tract. In extreme cases, eggs may be lodged in the brain and lungs to cause death. The disease is spread by the victim releasing eggs from the intestine or bladder into their faeces and urine. The eggs hatch when they reach water, and the

larvae swim to a snail host where they develop within the skin of the snail. Once developed, the fork-tailed larvae (cercariae) leave the snail and swim in water until they find a mammal host where upon they penetrate the skin and feed within the blood system, thereby continuing their life cycle.

Early diagnosis and treatment usually ensures recovery from this disease, but since this disease is common in poor countries, where many people cannot afford medication, prevention provides the best mechanism of control. Preventative measures require a break in the life cycle of *Schistosoma*, by reducing the possibility of contact between the larvae and potential host during bathing or working in standing bodies of water, or irrigation canals, places where the host snail abounds and cercariae are released. Attempts have been made to reduce the snail population with molluscicides, which include sodium pentachlorophenate, dinitro-o-cyclo-hexylphenol, and copper sulphate. Commonly, irrigation schemes and reservoir construction increase the potential habitats for the host snails and, therefore, have increased the environmental health risks in such regions.

Box 8.1

Common infectious diseases in developing countries

CHAPTER 8: KEY POINTS

1 Natural hazards are typically unpredictable, and may cause death, injury, and the destruction and damage of agricultural land, buildings, and communities. The effects of a natural disaster on a community depend upon factors such as the magnitude and extent of the disaster, how prepared the affected population is, and its economic resources to mitigate a potential disaster and/or clean up afterwards. The magnitude and frequency of various natural disasters form part of any risk assessment, and corresponding insurance.

2 Geological hazards result from the Earth's internal (tectonic) processes and Earth surface (geomorphic) processes. Earthquakes and volcanic eruptions are a consequence of tectonic processes and are generally located along plate boundaries. Damage may be local and related to building collapse, fires, landsliding and subsidence, tsunamis, flooding, the release of poisonous gases, and associated hazards such as contaminated or depleted water sources, disease, famine, injury, and death. Volcanic activity may result in regional and global climate changes. Earth scientists are developing more reliable means of predicting earthquakes and volcanic eruptions.

3 Geomorphological hazards, commonly included with geological hazards, include landsliding, river flooding, glacial hazards, and soil erosion. The environmental effects depend upon their magnitude and frequency, which are a function of climate, geology, vegetation, and human activity.

Mitigating the effects depends upon a knowledge of the dynamics of Earth surface processes, the accurate identification of high-risk zones, improved land-use practices, and the implementation of protective measures to reduce the effects.

4 Meteorological hazards are driven by the Sun's energy and are controlled by atmosphere–ocean systems. The effects are determined by the magnitude and patterns of weather systems, and the disaster preparedness of the threatened population. Included amongst these hazards are tornadoes, tropical cyclones, flooding by heavy rainfall and storm surges, disease associated with contaminated water sources, thunder and lightning damage, hail storms, and droughts.

5 Biological hazards include pests and environmental diseases. Pests cause great destruction to argriculture and lead to the spread of disease. Such hazards can be controlled by physical, chemical, and biological means, but most effectively by integrating all these approaches. Environmental diseases are one of the greatest hazards to health, and can lead to millions of deaths each year. Amongst the most serious diseases to affect humans are malaria, yellow-fever, typhoid, sleeping sickness, and schistosomiasis. Most diseases can be successfully tackled with good preventative education, early inoculation against disease, improved hygiene, and various artificial controls on the spread of the transmitting agent, for example spraying crops. Prompt medical attention can help to cure individuals of many diseases.

Chapter 8: Further reading

Bryant, E.A. 1991. *Natural Hazards*. Cambridge: Cambridge University Press.
An interdisciplinary treatment of a variety of natural hazards including oceanographical, climatological, geological, and geomorphologic hazards. This book is suitable as an introductory undergraduate text.

Chester, D. 1993. *Volcanoes and Society*. London: Edward Arnold.
An informative text on volcanic activity examined from both a geological, and socio-economic and political perspective. It is well illustrated with many interesting examples of the processes and effects of

volcanicity worldwide. It is an ideal text for university and college students with Earth science and social science backgrounds, wishing to pursue the issues associated with environmental risk assessment and natural hazards.

Smith, K. 1992. *Environmental Hazards: Assessing Risk and Reducing Disaster*. London: Routledge.
A comprehensive book covering most of the major environmental hazards, including seismic, mass movement, atmospheric, hydrological, and technological hazards. It integrates both the Earth and social sciences and is suitable for undergraduates from a variety of backgrounds, as well as being an important reference source for teachers and researchers.

The parched eviscerate soil
Gapes at the vanity of toil,
Laughs without mirth
  This is the death of earth.

T.S. Eliot,
'Little Gidding', *Four Quartets*

CHAPTER 9
Human impact on the Earth's surface

With the rapid growth of the world's population, many societies have been demanding more from the Earth's resources and affecting its land surface at ever increasing rates. Prehistoric evidence shows that in Palaeolithic times the early hunter-gatherers used fire and, accidentally or intentionally, burnt extensive areas of forest. The early agronomists burnt large areas of land to create farmland or pasture, modified the soil by ploughing, altered the drainage by irrigation, introduced or bred new animals and crops, and altered the natural vegetational structure of many regions. In more recent times, humans have destroyed enormous tracts of natural vegetation, excavated large areas of land, greatly modified the landscape, and even created new land.

Unfortunately some renewable resources are being used at rates that exceed the speed at which they can be regenerated. Nowhere is this more apparent than the destruction and deforestation of the rainforests. A hectare of forest can be destroyed within an hour, but it may take several decades for the forest to regenerate itself. A report published in 1991 by the UN Food and Agricultural Organization estimates that the current destruction of the tropical rainforests is occurring at a rate of 40 million acres per year, mainly as a result of human activities. Secondary effects complicate the problem. For example, rapid degradation of the forest soil accompanies deforestation, the nutrients being washed out by rain. In addition, the organic compounds are no longer replaced in the soil. It may take decades of slow regeneration before the soil can support a forest again. Other effects may lead to changes in slope stability, the amount of soil erosion, increased sediment washed into rivers, changes of climate within a small region, and the increased occurrence of floods.

There are many examples of how uncontrolled or excessive exploitation of the land's natural resources (including vegetation, fossil fuels, minerals, water, and land) can have a profound effect on the natural environment, both in terms of ecosystems and the aesthetic beauty of landscapes. This chapter will consider the main effects caused by the exploitation of resources on vegetation, soils, oceans, and the landscape.

Human impact on vegetation

Vegetation is important to humans as a primary source of food, as a building material, in manufacturing industries, as a fuel and as medicine. Early in human history, people gathered plants and began to cultivate selective types. With agricultural activity came the associated changes in the shape of the landscape.

The first human impact on vegetation, which is still prevalent, is the use and misuse of fire. Even though over half of the fires that occur are natural, resulting from lightning strikes or spontaneous combustion of decaying organic material, the rest can be attributed to accidental or deliberate burning by humans. Accidental fires may result from agricultural uses, cigarettes, rubbish tips, children playing with fires, camp fires, trains, motor vehicles, and arson. Deliberate burning is used to clear land, though it can be used to help improve the quality of the soil in arid regions through adding fresh organic material, or as an aid to reduce widespread fires. Fires cause a reduction in the natural vegetation; they threaten wildlife, humans, and property. Fire produces secondary problems associated with the clearance of vegetation, such as soil erosion, flooding, and wind erosion. However, some plants, such as the Jack Pine, depend on fire to initiate the dispersal of seeds, their pods bursting upon excessive heating.

Fires are more abundant in arid hot areas of the world, such as along the coastal regions of the Mediterranean, Australia (particularly near Adelaide), the north-western states of the USA, and the grasslands of central Africa. In these areas, fires devastate hundreds of square kilometres of land. Smaller fires are frequent throughout the world, causing only limited damage. Continued and frequent burning, however, often reduces the capability of an area to regenerate itself and to replenish its natural vegetation. Particularly problematic is the frequent misuse of slash and burn techniques in the rainforests. Here, extensive areas of forest are cut down and/or burnt to provide pasture land for cattle ranching, the area degenerates rapidly and the forest is prevented from regeneration.

The domestication of animals also has a major impact on the land surface. Heavy grazing of cattle leads to trampling and compaction of the soil, reducing its capacity to hold water and altering its structure. Ultimately this leads to soil erosion, both by wind and water. Selective grazing of particular plants may lead to changes in the nature of the vegetation cover. In the UK, for example, heavily grazed pastures in Scotland are dominated by bracken, a successful plant which survives because it is particularly distasteful and prickly to sheep and cattle. The growth of trees may be inhibited as young saplings are favoured by grazers. Grazing, however, may have positive

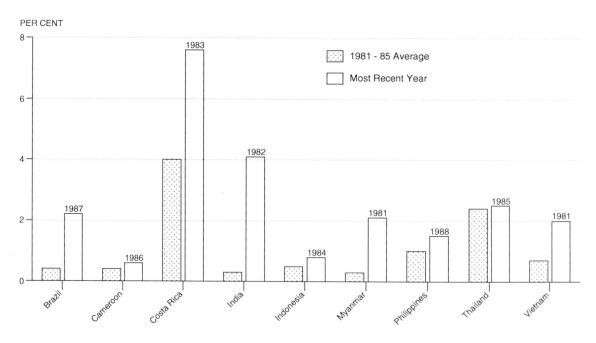

Figure 9.1   *The present extent of tropical and equatorial rainforests, affected and unaffected by deforestation, together with the calculated extent of rainforests by the year 2000, if current levels of deforestation continue. After Mannion (1991).*

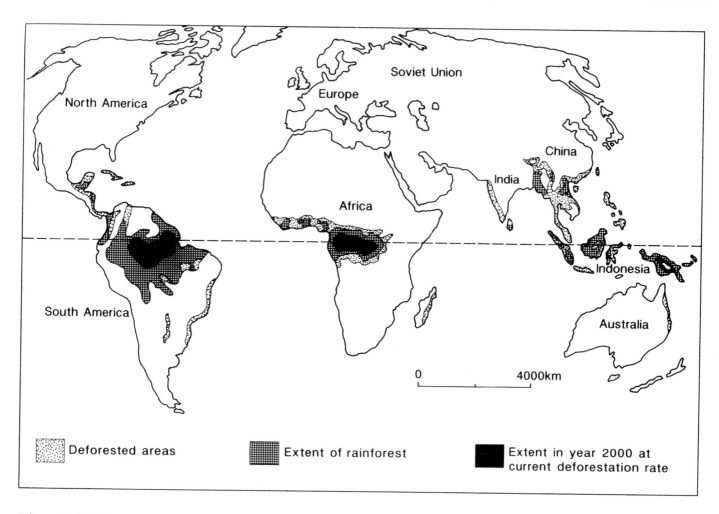

Figure 9.2 *The average percentage of closed forest (i.e., without open spaces) cleared in selected tropical countries in 1981–5, with data supplied for the most recent years. After World Resources Institute 1990.*

effects on the land because the animals provide faeces, a natural fertilizer rich in nitrates and other nutrients. Animals also help propagate seeds, and grazing may increase species diversity by opening up new communities and creating new niches. Nibbling of plants may also encourage vigorous growth. Many scientists believe that grazing in grassland areas in fact has little effect on grass, which is well adapted to withstand grazing.

Major problems are created when humans try to rear domestic animals in regions not suited to their lifestyle, especially when they alter the natural vegetation to grassland. This is particularly true of large areas of Brazil where the rainforest has been cut down to produce pasture land. The rainforest soils are poor in nutrients and are quickly deprived as grass provides little

litter to replace the much-needed nutrients. As a result these areas can only be used for pastures for a few years. Then the area becomes sterile and soil erosion quickly devastates the region, removing the soil and the deeply weathered bedrock.

Deforestation: the rainforests

Deforestation involves the deliberate removal of forest to create new agricultural or urban land, to provide wood for building and manufacturing industries, for the exploitation of minerals and fossil fuels, to create reservoirs for water supplies and hydroelectric energy, to build highways, for fuel, or as a result of defoliants used to help locate enemies during wars. The tropical and equatorial rainforests are shrinking at alarming rates because of deforestation (Figures 9.1 and 9.2), and there is little sign of a real slowdown in this destruction.

Humans have cleared forests throughout history. Major forests once covered most of central Europe, but during the eleventh century a major phase of deforestation began which within 200 years almost cleared most of the forests in Europe. Similarly, in North America before the first colonialists arrived, forest occupied some 170 million hectares between the Mississippi and the Atlantic seaboard. Today only about 10 million hectares remain. Presently the greatest rates of deforestation are occurring in the humid tropical and equatorial regions of the world, where the last big forests exist. How long these will last can only be guessed.

The rate at which these tropical and equatorial rainforests are being cleared is frightening. A few thousand years ago rainforests covered about 14 per cent of the land's surface, whereas today they cover only 7 per cent. Much of this has been lost over the last 200 years, most after the Second World War. In a study by the UN Food and Agricultural Organization published in October 1992, the most thorough to date and involving satellite and aerial photograph reconnaissance in 88 countries, it was estimated that the rainforests are disappearing at the rate of one acre per second, equivalent to the combined size of England and Wales being lost annually. This annual rate of destruction is running at 50 per cent more than a decade ago, half of which is taking place in Latin America, where 20 million acres of rainforest are devastated annually. As long ago as 1981, the United Nations estimated that one-fifth of the rainforests then existing would be destroyed by the end of this century.

These rich and complex ecosystems will be gone before long, depriving us of a wealth of diversity and the potential use of many of their unique biological compounds, often of great medical value. Reduction of this vast area of vegetation may upset the nutrient cycles, especially the oxygen and carbon dioxide cycles, of which trees form an important component. Figure 9.3 represents the recycling of nutrients in selected ecosystems. Notice the greater rate of flow associated with the tropical rainforests. This in turn may have long-term effects on climate. In the shorter term, water flow may increase over the land's surface, as rain will fall directly to the ground, no longer being impeded by the vegetational cover, which will then flow quickly across the surface. This increased rate of water flow over the land's surface may lead to an increase in the magnitude and frequency of flooding, soil erosion, increased sediment loads in rivers, slope instability, and degradation of adjacent land. Much of the flooding in regions such as Bangladesh has been attributed to the deforestation of the Himalayas, from where the main rivers originate which now drain the mountains more rapidly.

There are many interwoven complex local issues involved in the deforestation of the tropical rainforests. Landowners and others involved in exploiting the rainforests for a quick profit have little concern for the long-term survival of their environment. In South America, particularly in Colombia, large amounts of tropical rainforest are being cleared by drug barons to grow coca to produce cocaine and crack, and to harvest poppies for heroin. The profit motive is simply enormous, with a profit mark-up of around 4,000 per cent between Colombia and London (*Panorama*, BBC, 16th November 1992). But there are more prescient and brave people who have fought and are fighting for the long-term interests of the rainforest. One such individual in Amazonia was Chico Mendes, a Brazilian rubber tapper and self-taught environmentalist (see Revkin 1990). He was murdered in December 1988 by ranchers whose only concern was a short-term profit.

The destruction of the rainforests of Amazonia is at least now well publicized. The rainforests of Madagascar, with their unique plant and animal species, have suffered such devastation through forest clearance by the land-hungry population that the rich biological heritage is in real danger of mass extinctions. The natural desire of the Madagascans to better their lives today may be mortgaged at an intolerably high cost to the future generations. Fortunately, concerned international agencies are now working in concert with many Madagascans to preserve what remains of the rainforests with their precious inhabitants.

One of the problems in trying to slow the rate of deforestation in the rainforests is their geographical and political location. Most of the rainforests are in less-developed countries, many of which are in conflict or at war with adjacent countries. Unfortunately, the economy of most of these countries is dependent to a large degree on the exploitation of the rainforests. At present, for example, one of the most destructive projects will decimate over 2,000 km² of Brazilian rainforest. This is the construction of a reservoir and hydroelectric plant at Tucurui in the Amazon. Brazil believes this will help reduce its expenditure and dependence on fossil fuels needed to produce electricity, which constitutes a major part of Brazil's budget. The Brazilian government plans larger projects for the future, which will devastate even more of the rainforest.

Unfortunately, there are many problems associated with such constructions. Clearance of such huge areas of forest usually takes the form of large-scale flooding. The trees are just left to rot in the flood waters. Their decay leads to acidification of the water which produces poisonous hydrogen sulphide and explosive methane gases. These acid waters often corrode the turbines that produce the electricity in the hydroelectric power plant. Replacement is costly and difficult. The turbines may also become clogged with decaying logs or rafts of weeds that proliferate soon after flooding. Other problems include the spread of diseases such as schistosomiasis and malaria which are associated with large water bodies. Epidemics and fatalities are also associated with many of these developments in the less-developed countries. The indigenous population becomes affected by diseases introduced by the new workers and die in large numbers. This was a particularly alarming problem and a major issue during the construction of the Trans-Amazon Highway across Brazil. All these problems are difficult and expensive to remedy. Many people believe that as a result such projects as Tucurui will never pay for themselves, and should not be attempted.

Internationally little is being done to control deforestation, although on a national scale many countries are beginning to enforce legislation controlling the degree to which loggers can exploit the forest. Unfortunately, many of these restrictions are difficult to enforce or are unrealistic. For example, in the Philippines, the Bureau of Forestry Development has enforced logging standards so that the maximum acceptable damage to a logged-over forest does not exceed 40

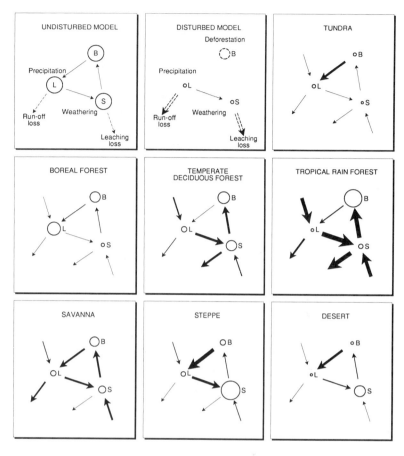

Figure 9.3 *Nutrient cycling in selected ecosystems. In the so-called 'disturbed model' after deforestation, litter decomposes relatively rapidly because the ground is warmer and less water is lost by evaporation and plant transpiration, so more nutrients are lost by surface-water run-off and leaching. The circle size is proportional to the amount in the organic pool. Arrow widths indicate the quantity of nutrient flow which is expressed as a proportion of that stored in the sources pool. B = biomass, L = litter, S = soil. After Tivy (1982).*

per cent. However, loggers select the best trees, clearly degrading the strength of the forest and depleting it of its best gene stock. In addition, loggers often return to the forest long before it has been regenerated to its original glory, destroying yet another 40 per cent of the forest and making it even more degraded and their heavy machinery may cause damage to the adjacent vegetation and the soil (Plate 9.1). Experimental studies, such as those being undertaken at the Curua Forestry Station in the Amazon, are providing important information on methods of sustainable exploitation and regeneration of the rainforest. These studies, however, are only beginning to aid in checking

Plate 9.1 *The use of heavy machinery in deforestation in the Brazilian tropical rainforest.*
Courtesy of Dr M. Eden.

the rate of deforestation in developing regions.

In the insatiable desire for timber, there are many extravagant and unsubstantiated claims made by retail outlets as to the source of the solid wood sold. The timber trade is concerned about profit and has a vested interest in convincing consumers that its trade does not damage the rainforests. Most logging still destroys forests whether or not they are tropical. There has to be a real will on the part of the trader and consumer alike to use timber only from genuinely sustainable sources. International legislation may prove to be the only satisfactory way forward on this issue.

It's not just the rainforests

Perhaps less well advertised than the destruction of the rainforests is the human threat to Brazil's unique **savannas**. Burman (1991) has pointed out that about 3 million square km of Brazil – an area larger than the Mediterranean basin – consists of non-forest habitats. Numbered amongst these are the wonderfully named *cerrado*, the *caatinga*, and *campo rupestre*, all of which can be grouped under the term savanna. Because these areas do not have the obvious luxuriant vegetation and diversity of species seen in the rainforests, they have tended to be regarded as trash vegetation and hence open to thoughtless exploitation. They are home, however, to a unique and diverse flora and fauna, including **bromeliads** and orchids. Open-cast mining occurs on a large scale. There is surface mining for gold in Bahia on the *campo rupestre*, and manganese extraction on the Serra do Cipo. The careless exploitation of the savannas through the planting of inappropriate crops such as soya bean and sugar cane, and the development of pine plantations, or eucalyptus, all serve to decimate habitats where, because of the extremes in climatic conditions, plants are already under natural stress. Plant collectors and the local population take many of the beautiful, rare, and valuable flowers and plants at rates much greater than they can be replenished naturally. As the flora diminishes, or is wiped out, the animals and insects lose their habitats and disappear from the savanna.

Deforestation not only affects the immediate rainforest and its ecosystem, but it may also have drastic consequences on adjacent regions. This is well illustrated in Jamaica where deforestation on steep slopes has led to intense soil erosion and landsliding. This in turn has increased the sediment loads in rivers, which eventually drain into seas rich in coral life. Corals forming the main component of reefs are particularly susceptible to sediment concentrations in water, and sediment reduces the amount of light reaching them and also poisons their tissues. As a result, many of the corals around Jamaica have not been able to tolerate this sediment and the reefs have become degraded. This in turn results in the death of other animals that depend on the reef for food, and are part of the coral reef ecosystem.

Desertification

Deforestation and the degradation of other vegetation, particularly near the margins of deserts, have caused once fertile/vegetated land to become barren in a process called '**desertification**'. Factors that contribute to the expansion of desert regions also include bad land management and poor farming techniques. Desertification and its associated problems are devastating many parts of the world, especially developing countries. These include the desert margins of the Sudan, the Sahel region of the southern Sahara desert (Plate 35), the Gobi desert in China, and the Kalahari desert in southern Africa. The margins of these deserts have advanced as much as 100 km in the last couple of decades. The United Nations Environmental Programme (UNEP) has calculated that about 60 per cent of the 3.3 billion hectares of agricultural land outside humid areas are affected to some degree by desertification.

Although the United Nations claims that desertification, caused by human activities, is continuing to intensify, many argue that this may not be true (Pearce 1992). The United Nations world map of desertification identifies key areas of desertification but states that the margins of error fall within ± 10 per cent (UNEP 1992). Any change since the last survey in 1977 falls within this error margin. It is not possible, therefore, from these data to draw any firm conclusions about the degree of desertification. Furthermore, mapping desertification is difficult because remote sensing often gives a false impression of the amount of vegetation cover. Scrublands, for example, are difficult to distinguish from true desert. However, dryland areas of Africa have continued to report that desertification has increased since 1977. Scoging (1993) believes that reclamation and irrigation programmes mitigate the problems of desertification which are greatly exaggerated.

Many scientists also argue that desertification has not occurred as a result of human activities (Hulme 1989), and that land degradation attributed to cattle herding and overgrazing, particularly around watering holes, does not lead to desertification in the Sahel region (Pearce 1992). They believe that cattle herders possess an innate knowledge and sensitivity towards the land, and therefore contribute little to its degradation. Instead, they argue that the apparent effects of desertification may be the result of natural fluctuations in global and regional climate, such as

droughts which are inherent to drylands. Additionally, it is well known that desert margins oscillate by tens of kilometres on a scale of years, depending on variations in rainfall (Hulme 1989). Droughts, however, may be the result of human-induced global warming, but this assertion remains unproven. Street-Perrott and Perrott (1990) argue that fluctuations in aridity are due, at least in part, to the development of the North Atlantic Deep Water (NADW), which is an important control on precipitation in North Africa. By studying lake levels and sediments, they have demonstrated longer-term changes in aridity in North Africa.

During the Last Glacial Maximum, the Sahara and Sahel regions of North Africa were extremely dry, although there were periods when precipitation in North Africa was much higher than at present, for example 9,000 years BP. Gasse et al. (1990) believe that precipitation may have been 125–130 per cent more than today. The evidence for this comes from palaeohydrological indicators, including oxygen and carbon isotope ratios measured in inorganically precipitated carbonates, and also from the mineralogy of sediments from Sebkha Mellala in the Sahara, Algeria, and Bougdouma in the Sahel, Niger. These data show that during deglaciation the transition from arid to humid conditions occurred synchronously. Comparisons with other palaeoclimatic continental records in Europe, and deep-sea sediments from the North Atlantic Ocean, show the synchroneity of changes in the coupled ocean–atmosphere system (ibid.), controlled by changes in the dynamics of the North Atlantic and the meltwater history of the Laurentide ice sheet (Street-Perrott and Perrott 1990).

There is much debate as to whether desertification is reversible. Though many attempts have been made, such as irrigating and revegetating desert margins, and though many have proven successful, it is not known how long an area can sustain itself when maintenance is discontinued.

There are natural '**badlands**', semi-arid landscapes severely eroded and dissected into fantastic geometrical shapes of pinnacles and deep gullies, for example in the USA along the Greybull river in the Bighorn Basin of Wyoming or the Big Badlands of south-western Dakota, along the White and Cheyenne rivers on the eastern and southern edges of the Black Hills.

But even in these semi-arid areas, the bad management of the land can turn the fertile soil that is available into a 'dust bowl', a story so well told in John Steinbeck's novel, *The Grapes of Wrath*, with its images of tattered refugees trudging westwards through Oklahoma to the promised land of California. In the 1920s, wheat farming was very successful in this region. In the early 1930s, however, several years of drought and bad maintenance of the land resulted in millions of acres of wheat shrivelling and thousands of cattle starving to death. In addition, a succession of large wind storms driven by cold air originating from the polar regions in northern Canada blew away the unprotected topsoil, producing devastating dust storms carrying soil many thousands of miles. This dust also caused many respiratory diseases such as '**dust pneumonia**'. By the late 1930s the climate had become wetter and government intervention, aided by the advice of soil scientists, returned much of the wheat land to pasture, and introduced new farming practices such as crop rotation, tree planting to act as wind breaks, and the addition of fertilizers to the soil.

One of the Central Asian tragedies of desertification has taken place in the inland Aral Sea which was once the fourth-largest inland water (Plate 38). It was drained by irrigation to support a cotton monoculture, to the extent that it has now virtually disappeared to be replaced by a desert of sand, salt, and obsolete fishing boats.

Other regions have experienced, and still suffer, similar problems. Drastic changes in farming practices, the trend towards monoculture, increased mechanization, and other bad farming practices are usually to blame. Effects are often seen in developing areas where conservative traditional methods of subsistence farming are frequently replaced in favour of crops that can be sold quickly for ready money. These crops are often not suited to the region and result in environmental degradation.

Human introduction of exotic plants and animals

The introduction of new plant species, accidentally or intensively, into a new area can have both positive and negative effects. Plants that invade an area may be more successful than the indigenous flora. They can begin to dominate the region, which may be advantageous if they provide a new food supply or they may help reduce the problems of erosion. The introduction of the Himalayan shrub, rhododendron, to many areas of the UK and other regions of Europe is an example of how successful some of these plants can be. Though rhododendrons are helpful in stabilizing slopes, it is very difficult to eradicate them and many park keepers are plagued with them. The accidental introduction of plants in some areas can also have devastating ecological consequences. For example, the accidental introduction of the **Dutch Elm disease** fungus imported in timber to the UK caused the death of many thousands of elm trees during the 1970s.

Human introduction of animals to a region may have devastating effects on its vegetation. This is well illustrated by the introduction of grazing animals such as goats, donkeys, cattle, rabbits, and hares to tropical islands. In Laysan island, Hawaii, for example, much of the vegetation was cleared in a short time and the number of plant species was considerably reduced by the introduction of goats.

Human pollution and vegetation

The human impact on the natural vegetation may also result from air and water pollution. Acid rain, for example, is considered to be a major culprit in the destruction of million of trees in high- and mid-latitude regions (see Chapter 4). Pollution will reduce plant growth and can lead to other forms of environmental degradation, such as soil erosion. This loss of vegetation and a reduction in floral diversity will deprive us of the world's most essential and beautiful renewable resources.

Human impact on soils

Soil is another great natural resource. It is a combination of mineral and organic matter, structurally arranged in layers, and capable of supporting plant and animal life. Soils cannot exist without plants, and plants are dependent on soils for support, air, water, and nutrients.

Soils are highly variable in nature. This variation includes their structure, layering, colour, range of particle sizes, chemistry, nutrients, acidity, temperature, water content, thickness, organic content, and its associated biota. These properties vary because of differences in the parent material, climate, topography, organic content, and the amount of time it has had to develop. Changes in one or more of these factors may drastically alter the soil properties, changing its nature and ability to support particular plant species. These changes can happen very easily, having profound effects on the soil and the landscape such as vegetation reduction, soil erosion, slope instability, increased flooding, and more sediment in rivers. The major changes induced by human activities include chemical changes (salinization and **laterization**), structural changes (compaction), hydrological changes, and soil erosion.

There are many chemical changes within a soil which can be initiated by humans. The most widespread and problematic are salinization and laterization. Salinization involves the accumulation of salts such as sodium chloride, potassium chloride, calcium sulphate, and sodium carbonate within a soil. This makes the soil alkaline, caustic, and generally restricts or inhibits plant growth. Salinization may also lead to secondary problems such as soil erosion resulting from poor plant growth.

Salinization may occur naturally in semi-arid and arid areas where **evapotranspiration** or direct evaporation from the soil exceeds precipitation. It may also occur in coastal regions which have saline groundwater. In areas where the evaporation of water from the soil is high, water is drawn upwards and evaporated from the soil surface. Hence salts are left behind and are concentrated near the surface of the soil. This results in a hard salty layer within the soil called a salt pan. Salinization can be induced by irrigation and water abstraction. The abstraction of water leads to a rise in the groundwater table, driving salts towards the surface. In coastal regions, withdrawal of underground freshwater, which floats on top of underground saline water (originating from the sea), pulls the saline water beneath it nearer to the ground surface, contaminating the fresh water in the soil. In addi-

tion, water in wells may become saline and therefore useless. This is a particularly big problem in coastal regions such as California and Israel, and islands such as Bahrain and Long Island, New York. This was also a problem in cities such as London and Liverpool during the middle of the last century, until effective management of groundwater resources was introduced.

Irrigation also enhances salinization by increasing the height of the water table in the immediate and adjacent areas over which irrigated water is spread. This leads to the evaporation of water from within the soil, providing a process by which soil salts can be concentrated and drawn towards the ground surface. With the rapid expansion of irrigation schemes in the last 20 years, the UN estimates that as much as 25 per cent of irrigated areas have become affected by salinization, making it a major land management problem. For example, the percentage of soils affected and water-logged amounts to 50 per cent of the irrigation areas in Iraq, 23 per cent in Pakistan, 30 per cent in Egypt, and 15 per cent in Iran.

Laterization of the soil is a major problem in the tropics where soils are enriched in aluminium and iron oxides. These metal oxides accumulate due to strong tropical weathering. Minerals in rocks are decomposed releasing metal ions into the soil water. These are transported, precipitated, and concentrated by seasonal wetting and drying of the soil as layers of metal oxides. Problems start to occur when these lateritic layers become exposed to air. They become hard and inhibit plant growth leading to soil erosion and its associated problems. Exposure of these layers may also be due to soil erosion, often as a result of deforestation. In addition, deforestation may lead to increased evaporation of water from the soil, enhancing the process of laterization. The extent to which laterization is a problem has not been fully assessed, but particular problem areas include northern India, the Cameroons, and central Africa. Unfortunately, populations in these countries are heavily dependent on soil for subsistence agriculture and often degradation is difficult to reduce due to increasing pressures of population growth on those lands.

Changes to a soil structure – its mutual arrangement of grains – can also have a profound effect on its properties. These include the soil's ability to retain water, to allow water to enter and flow through it, the strength of the soil, the degree to which a plant can penetrate it as well as the withdrawal of water from it, and its resistance to erosion. The main way soil structure can be altered is through compaction, the pushing together of soil grains. This may be done by vehicles driving over its surface, overgrazing, trampling along public footpaths, or by ploughing, which compresses the soil immediately below the ploughed surface.

Compaction reduces the ability of a soil to retain water, to hold air, and to allow water to enter the soil and it increases its hardness. This retards or inhibits plant growth and it may enhance processes such as soil erosion by wind and water. Soil compaction is a worldwide problem, but it is often greatest in developed countries where vehicles are common. Soils are commonly ruined adjacent to building sites, and on old battle fields or military training areas where heavy vehicles are particularly common. Unfortunately, compaction of soil is one of the

hardest soil problems to remedy and it may take many decades before a soil can regain its original structure.

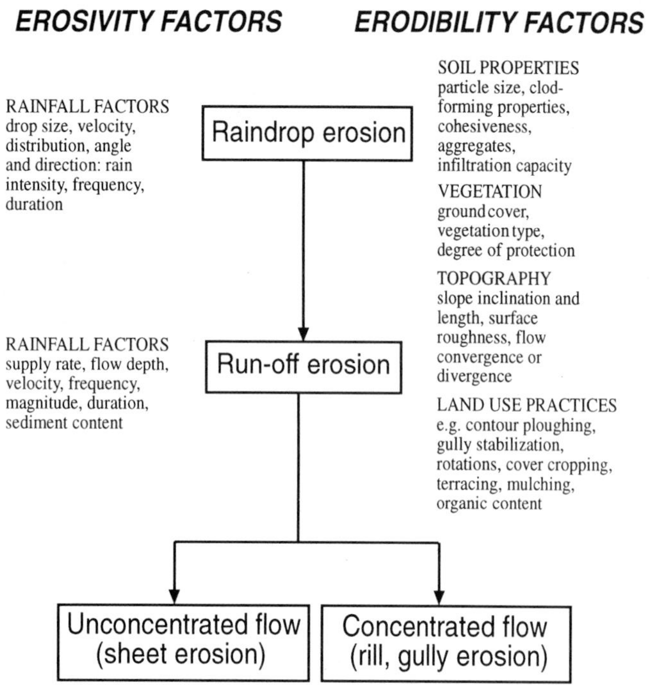

Many soils are drained to allow the increased growth of crops, the planting of trees or the construction of settlements. In Great Britain, for example, over 60 per cent of the land surface is artificially drained. Draining soil may produce several undesirable effects. Previously wet organic rich soil may begin to decay, producing unpleasant methane and hydrogen sulphide gases. This in turn increases the acidity of the soil and lowers its productivity. Then the soil volume may decrease as water is extruded, leading to subsidence and structural problems for buildings. The water table may be lowered, affecting adjacent regions and their soils. Some soils expand and contract due to changes in water content, which will affect engineering structures. In addition, the drainage of soils may lead to the rapid flow of water over the land's surface into rivers, increasing the likelihood of flooding.

Humans also alter the chemistry of soils by the addition of organic or artificial fertilizers. This may help increase the agricultural productivity, but it may also be detrimental to the soil, especially if the fertilizers are incorrectly applied. This may lead to the deterioration of the soil, a reduction in vegetation, soil erosion and other associated phenomena. Factors leading to increased soil erosion are shown in Figure 9.4, and Figure 9.5 shows processes of soil degradation.

By far the greatest impact on the soil is caused by soil erosion. This includes the abrasion of water running over the surface, the breakup of soil due to the impact of rain drops, and the deflation of soil particles by the wind. The various causal factors that may initiate soil erosion have already been discussed, such as deforestation, grazing, salinization, laterization, and compaction. Many of these factors are interlinked and should not be considered in isolation. They may also be accelerated by bad farming techniques, urbanization, constructions, mining, wars, and fires. Some of the worst-affected areas are the result of a combination of these. These areas include many of the coastal regions of the Mediterranean Sea, where fires and wars have led to vast areas of badlands. Particularly problematic are the consequences of soil erosion. These include the increased likelihood of flood-

Figure 9.4 *Important factors contributing to soil erosion. After Cooke and Doornkamp (1990).*

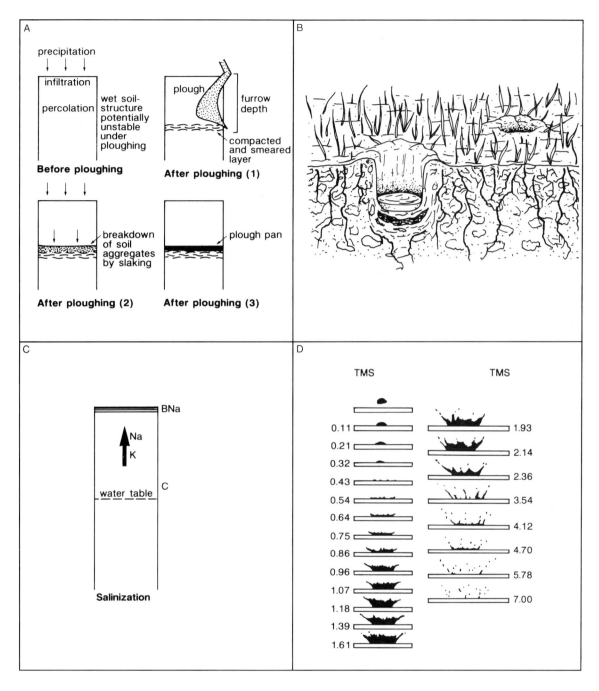

Figure 9.5 *Processes of soil degradation. (A) Formation of a plough pan. Redrawn after White et al. (1984). (B) Effects of animal hoof imprints on soil structure. Redrawn after Batey (1988). (C) Salinization of soil. Redrawn after Knapp (1979). (D) Rain splash (time-scale in milliseconds). After Ghadiri and Payne (1980).*

ing, increased sediment loads often leading to silting up of reservoirs, and landsliding. Plate 9.2 illustrates the problem of careless exploitation of tropical soils and the regolith. Destruction of vegetation and the exposure of weathered gran-

ites leads to deep gullying and slope instability. The highest erosion rates in the world produced these deep gullies in the thick wind-blown (or loess) of the Loess Plateau, Gansu Province, central China (Plate 9.3). Loess is easily eroded

HUMAN IMPACT ON THE EARTH'S SURFACE

Plate 9.2 *Deeply weathered granite in Thailand which has become exposed during the construction of a new road. The slopes are beginning to become gullied and if erosion continues this will lead to slope stability problems which will threaten the highway.*

particularly during heavy rainfall events, and sediment is washed into the Yellow river and redeposited on the floodplains. Plate 9.4 shows a damaged irrigation dyke on the Loess Plateau, central China. This is common in the fields of the Loess Plateau, produced by the collapse of a soil pipe which formed by seepage of concentrated flows of water from fields. The dyke has also

Plate 9.3 *The eroded landscape of the Loess Plateau, central China.*

become partially infilled with sediments that have been eroded from the fields, blocking the drainage. Careful monitoring and management of erosional and depositional processes have to be undertaken to care for this environment, because the region is intensively farmed to support very high population densities.

Much can be done to try to retard soil erosion and conserve soil resources. These include revegetation, crop management, slope run-off control, construction of gabions and retaining walls, and the dissemination of information regarding good land-use practices. There is still, however, much to be achieved with regard to the correct use of one of the most valuable natural resources.

The Himalayan dilemma – a lesson in caution

Since the mid-1980s, the media have highlighted the large-scale environmental degradation in the mountains of central Asia, principally the Himalayas (Plate 9.5). The perspective put forward has been that increased health care, medicine, and the reduction of malaria in the post-war years from the 1950s have led to a rapid growth of population in the Indo-Gangetic Plains south of the Himalayas. Within 27 years the population doubled. At the same time, the indigenous populations in the mountains grew, but at less severe rates as health care was more difficult to provide in the more remote mountain regions. The population growth led to rapid deforestation of the Himalayan foothills but also in some higher-altitude areas. This was due to the increased demand for timber for fuel and building, together with the need to clear more land for crops to feed both the population and animals. Some observers (e.g. World Bank 1979) suggested that the rate of deforestation was so rapid that by the year 2000 little forest would be left in the Himalayan regions.

Where forest has been cleared on steep slopes, there are cut terraces to form a stepped topography to cultivate wheat, barley, maize, and – at lower altitudes – rice. It was argued that this deforestation and change in land use would lead to increased slope instability and an accelerated rate of soil erosion once the protective tree cover

Plate 9.4 *A damaged irrigation dyke on the Loess Plateau, central China.*

travel further for timber. This meant that animal dung was substituted for wood as a major local fuel. The use of dung for fuel meant that it was not used to fertilize the soil. This further decreased the soil quality. In turn, the poorer-quality soils became more easily eroded, since the organics helped to bind the soil particles together. A vicious cycle of enhanced landsliding and flooding ensued. The prospect appeared bleak for the Himalayas.

These arguments superficially appear very convincing and are still widely used by environmentalists. However, from their work with UNESCO and the World Bank, as well as many other research projects, Ives and Messerli (1989) suggested that many of these arguments were naive. For example, there is no widespread evidence for deforestation on the scale suggested. There were certainly areas that were actively being deforested at alarming rates, such as the Hengduan Mountains in Sichuan (China) and the lower hills of Nanga Parbat in Pakistan. Some areas, however, such as the Garhwal Himalaya, although they were degraded early in this century, have had a long history of forest management with active reforestation. Areas such as the Khumbu Himalaya (Nepal) have a substantial part designated as a national park, the Sagarmatha National Park (Everest). The future

was removed that had acted to stabilize any local soils. Furthermore, the deforestation caused enhanced surface run-off of rainwater, so further increasing the soil erosion and leading to flooding. Some of the devastating floods in Bangladesh in the early and mid-1980s were attributed to high river discharges due to increased run-off in conjunction with the occurrence of tropical cyclones.

Landslides were believed to be more common as a form of hill-slope failure and erosion once the number of trees had been severely depleted. The sediment loads in rivers increased because of the accelerated soil erosion and greater frequency of landslides; the rapid sedimentation up or siltation of reservoirs and hydroelectric power schemes were also attributed to this environmental degradation. Furthermore, the growth of islands in the Bay of Bengal was regarded as a consequence of the increased sediment discharge.

As the forests became smaller, villagers had to

Plate 9.5 *Mountain people finishing a tree to provide supports for a new house in the Nanga Parbat Himalaya, northern Pakistan. Deforestation for domestic use such as this is essential for these peoples, but it is difficult to reconcile their needs for development with broader environmental issues.*

for the Himalayas is not so bleak after all.

Another major assumption in the arguments for alarming deforestation was that terrace-style farming degrades the valley slopes. Most of the terraces in fact increased slope stability as overall gentler slopes were created, and in some cases the best terraced slopes were on ancient landslides because they actually provided lower gradients and therefore increased slope stability. There are a few examples of farmers actually initiating landslides to produce more stable farmable slopes. The terraces are designed so that surface water run-off can be controlled, and the water efficiently utilized in irrigation. In fact, farmers have shown an acute awareness of the need for natural fertilizers and seldom use dung as a fuel.

The assumption that the increased siltation of rivers in the Himalayas and drainage basins is linked directly to the degradation of soils brought about by human activities requires closer scrutiny. Most of the rivers receive much of their water from high mountain glaciers which also supply enormous quantities of silt and mud. Indeed, silt and mud may be supplied from accelerated soil erosion caused by human activities, but how does this compare with the amounts of sediment supplied by natural erosional processes? To date, few studies have addressed this question. The silting up of reservoirs is occurring at a disconcerting rate but it is difficult to determine accurately the expected lifespan of such constructions because of the problems in measuring sediment budgets in big rivers. Most estimates are based on suspended sediment loads, where little consideration is given to the sediment transported along the river bed (the traction load), which is substantial in the fast-flowing Himalayan rivers. In addition, the theory suggests that landslides provide vast quantities of sediments to the rivers, but many studies have shown that a landslide rarely reaches the river and most of the landslide sediment is dumped on mid-slopes and benches along the river valleys. Little consideration has been given to whether the growth of islands in the Bay of Bengal have formed by entirely natural sedimentation processes or by human activities.

The paradigm of massive environmental degradation in the Himalayas has proved inaccurate. This case study epitomizes the dangers of environmentalists only seeing what they want to see, and so drawing conclusions without looking closely at the evidence, which may indicate the exact opposite.

Attributing increased flooding to environmental degradation is also dangerous. There is little information on the climatology of the Himalayas. This is partially a function of the difficulty of gathering long-term scientific data in such a politically sensitive region. Climatological data are considered to have military importance and are not freely available. Where data are available, they appear to suggest that there are 12-year cycles in the intensity of the monsoon, similar to the El Nino Southern Oscillations in the southern Pacific. A coincidence of heavy rainfalls during a strong monsoon and a tropical cyclone would certainly produce increased flooding. Climatic cycles on a longer time scale are more difficult to reconstruct, and it is not possible at present to assess the relative importance of the human impact from that of the natural system.

What is a fact in the Himalayan region is that the human population has considerably increased, and this has undoubtedly produced many socio-economic problems and in some areas caused environmental degradation. What Ives and Messerli emphasize, however, is that many projected environmental scenarios are not necessarily as simple and straightforward as one might like to think. There is a need to examine the physical, social, economic, and political aspects of environmental issues. Their book provides a counterbalance to the more alarmist statements and books on the human impact on the environment. Of course, such a book can be used by individuals and groups with a vested interest in doing nothing about conserving the natural environment or indeed those actively engaged in environmental degradation. It is very much a double-edged sword.

Human impact on the oceans and seas

The oceans and seas cover more than two-thirds of the Earth's surface. They contain submarine trenches that are deeper than the highest mountains. Life almost certainly evolved from the sea, and there is still more species diversity in the sea

Plate 39 *Beside causing the all too familiar personal anguish, wars result in dramatic changes to the landscape and vegetation. (A) Bomb craters produced during the Vietnam war.* Courtesy of Rex Features. *(B) The destruction of Delnice Forest in Yugoslavia by the explosion of an ammunition store.* Courtesy of Greenpeace.

(A)

(B)

Plate 40 *Mining for coal in a West Virginia strip mine, USA.*
Courtesy of Comstock.

Plate 41 *Opencast mining for lignite in Germany.*
Courtesy of Greenpeace/ Vennemann.

Plate 42 *Strip mining for diamonds in South Africa. Beach sediments 200 m thick are removed in order to expose gravel which contains diamonds. The activities are very extensive, creating large-scale disruption to coastal ecosystems.* Courtesy of Fred Mayer/ Magnum.

Plate 43 *Intensive mining of emeralds in Brazil. The miners face considerable danger from slope failure and flash flooding.*
Courtesy of Magnum.

Plate 44 *Dioxin contamination of the Love Canal, near Niagara Falls, New York State. This incident threatened a nearby community, and the clean-up operation utilized financial aid from the US Superfund.*
Courtesy of Greenpeace/Visser.

Plate 45
Urbanization on a grand
scale. Manhattan Island,
New York. Central Park
in the middle ground
provides an important
amenity within this
concrete jungle.

Plate 46   Venice, which
is threatened by subsidence
caused by the withdrawal
of groundwater.

Plate 49 *Dense urban life in Kowloon, Hong Kong. In some areas of Hong Kong, population is so dense that it is common for families of six or more to live in crowded flats which are <500 square ft in area. Sleeping and working schedules have to be organized to allow people the convenience of space.*

Plate 50 *Heavy traffic congestion in Washington DC.*
Courtesy of Greenpeace/Visser.

than anywhere else on Earth. Many of the food chains or food webs start with organisms inhabiting the seas and oceans. The ocean–atmosphere system regulates global climate. It is a sensitive thermostat. The seas and oceans are a rich food and mineral resource, but over-exploitation and pollution threaten this vast wilderness. Humans still tend to feel that the vastness of the sea makes it an ideal dumping ground for virtually every type of waste, including toxic chemicals and nuclear waste.

Since the introduction into the North Sea of steam trawlers in the 1880s, vast numbers of fish have been caught more easily and with less cost. The increased fish catches have brought a greater demand for fish, both for domestic consumption, and for fish meal and fertilizers. Between the 1950s and 1980s, the global catch of marine fish rose from 20 million to 70–80 million and then levelled out (Smith and Warr 1991), leading to a significant depletion of fish stocks, with the impact on the marine ecosystems not yet determined. Figure 9.6 shows the dates of the beginning of overfishing for different fish pop-

ulations in the North Atlantic. In order to mitigate the environmental effects caused by overfishing, catches should be regulated, for example on the basis of the number of fish that can be caught. This is a global problem necessitating international conventions and treaties, because it involves political disputes over fishing grounds, linked to individual nations' diets, and employment for fishermen and the allied industries. A more local solution to overfishing is the development of fish-farming practices. Intensive fish farming, however, can create environmental problems, because the introduction of fish food and nutrients into coastal waters, to increase fish yields, may pollute adjacent waters and seriously disrupt the existing aquatic ecosystems.

One of the areas of growing concern is the future exploitation of the seafloor. Breuer (1991) makes a good case for the international adoption of a strategy for the seabed. There are large mineral reserves, for example in the form of **manganese nodules**. Some 30 years ago, these nodules were thought to be worth mining from the seafloor, but the fall in metal prices over the

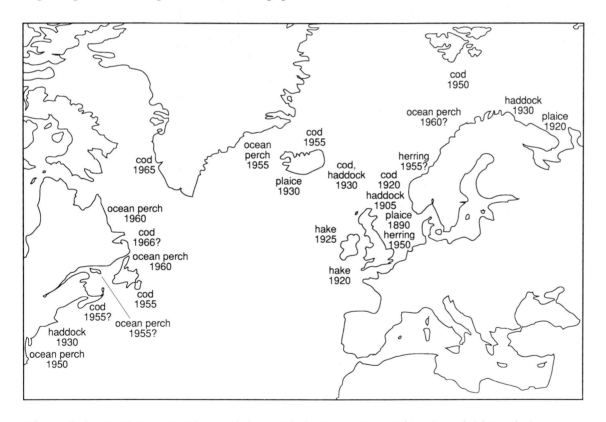

Figure 9.6 *North Atlantic fishery with dates of the beginning of overfishing for each fish population. Redrawn after Smith and Warr (1991).*

intervening years has made this seem increasingly unlikely in the near future. When the situation changes, as it may well, then there will need to be international laws and conventions to control the extent and manner in which any exploitation takes place. There are also '**black smokers**' in the deep sea, especially where new oceanic crust is forming and there is a lot of hydrothermal activity from the submarine volcanoes. Black smokers are tall chimneys built of metal sulphides that have been precipitated from the metal-rich fluids escaping from igneous rocks below the seafloor. Their name derives from the fact that the hot escaping (hydrothermal) fluids produce a very fine suspension of metal sulphide particles in the seawater that look like black smoke when illuminated by submersibles and camera flash. Besides a lot of iron, the black smokers also discharge high concentrations of copper, zinc, lead, silver, and gold. While the technology to mine these deposits is not yet developed, it must only be a matter of time. Indeed, the Japanese government has provided US $24 million over five years to evaluate the economic potential in manganese nodules and polymetallic sulphide resources on the seabed. International discussions and agreements are needed now, not when the developers move in.

The United Nations Convention on the Law of the Sea (UNCLOS) states that the oceans and seas are a 'common heritage of humankind' which: cannot be appropriated by any individual, institution, or country; must be managed by and for the benefit of humankind as a whole; must be conserved for future generations; and must be reserved for exclusively peaceful purposes. To date, this convention has been ratified by 45 nations, but requires ratification by a further 15 before it can take effect. Human chemical pollution of the oceans and seas is dealt with elsewhere (Chapter 5).

Human impact on the landscape

The land's surface is an important resource, allowing us to build settlements, produce communication links, and have farmland and recreational areas. It is also beneath this land that mineral resources and fossil fuels are to be found. But in order to exploit these we have had to cut away at the land surface and dump the rock waste, which in turn creates new landforms.

Human activities are continually modifying the landscape, creating pits, ponds, spoil heaps, terraces, cuttings, embankments, dykes, canals, reservoirs, and areas of subsidence. War also causes dramatic and significant landscape modifications as is illustrated in Plate 39. Geomorphologists consider humans as an important landforming agent and refer to us as geomorphological anthropogenic agents. Many consider that for large regions of the world humans are an important factor in contributing to the landscapes we see today.

The rapid increase in population has placed great demands on the available living space. The trend towards urbanization has led to an increase in the size of settlements at an incredible rate and the exploitation of marginal lands. In the latter, natural processes constitute a hazard to people settling in these regions. Advances into these areas are often accompanied by rapid deforestation and devegetation, resulting in soil erosion, flooding, and other associated problems.

Subsidence is a problem in areas where the ground is permanently frozen. The upper layers of this frozen ground may melt in spring and refreeze in the winter. Building on these permafrost areas may cause the ground to thaw to a greater extent and deeper than normal. The ground may shrink and subside, and water is extruded, resulting in structural damage to buildings. This is a particular problem in the settlements in Arctic regions such as Alaska, where special engineering structures have to be designed to combat the freezing and thawing ground.

The pressure for land has become so great in some regions that new land has been created by coastal reclamation. This is particularly well illustrated in settlements such as Hong Kong which has one of the highest population densities in the world. Several million people live on a coastal strip in an area of about 30 square kilometres, bounded to one side by the sea and on the other by steep unstable mountains. As a result, vast areas of land have had to be reclaimed from the sea in recent years and much more reclamation is planned in the near future. This will probably be at the expense of rich fishing grounds and ecologically valuable wetlands.

Coastal construction may also upset the balance of erosion and deposition of sediment along coastal stretches. Sediment in coastal regions is continuously being moved by the activity of sea waves. Often it moves parallel to the beach's length, a process known as **longshore sediment drift**. A construction, such as a sea wall or harbour, may inhibit the erosion and hence the supply of new sediment moving by longshore drift to an adjacent region. As a result the region that should have received sediment may then become depleted of its own sediment as it is continuously being moved away. The original sediment stockpile may have acted as a natural coastal defence, protecting the coastline from erosion and associated coastal processes such as landsliding. This is particularly common along the coasts of southern Britain, where several areas of coastal retreat are attributed to anthropogenic processes produced by coastal constructions. The resultant erosion and concentrated deposition cost large sums of money in damage and dredging each year.

A coastal construction, such as a pier, may also cause sediment to accumulate around it, inhibiting sediment drift. Such accumulation may hinder shipping and may even lead to the infilling of harbours. In many cases this can be used to help reclaim land, such as Miami Beach in Florida. It is important to remember that modifications made to a coastal region in one area may have profound effects on adjacent regions, as sediment movement is a continuous cycle of deposition and erosion.

Mineral extraction

Opencast and subterranean mining activities provide humankind with the wealth that has sustained the growth of civilization and the quality of life that we enjoy. It is hard to find many items in constant daily use, let alone the luxury goods including works of art, that have not come from the ground. Mineral extraction is an integral part of civilization. Nothing is gained without some cost to individuals and the environment. The debate will always be focused on an acceptable level of pollution, inconvenience and cost in return for a product from mining. Different individuals and communities will have different perceptions of the degree of acceptable social

and economic cost, especially if they live near the mines.

There are many problems associated with mineral extraction which include chemical pollution and disfigurement of the landscape, as well as disturbances to the natural rates of Earth surface processes. Mineral extraction results in the construction of tips, increased sediment loads in rivers, soil erosion, and the pollution of water sources and adjacent land, which often lead to vegetation and soil degradation. Even as early as the Neolithic period of prehistory, humans were modifying the Earth's surface in search of its mineral wealth. Many excavations in the chalk hills of southern England represent Neolithic pits dug to extract good-quality flint for tool making. With time the extraction of metallic ores became important to make metal tools and jewellery. There are many examples of Roman mines still visible in the landscape throughout Europe and north Africa. Fossil fuels have also been extracted from early times. Excavations for peat were widespread, and it is believed that the lakes and waterways of the Norfolk Broads in eastern England owe their origin to the removal of more than 25 million cubic metres of peat prior to the fourteenth century.

The coming of the Industrial Revolution saw a major growth in the extraction of iron ore and

Plate 9.6 *Quarrying for stone on Portland Bill, southern England.*

coal. As a consequence, large opencast and shaft mines were constructed, some to great depths (Plates 40 and 41). In some regions, for example in south Wales, so much coal was removed from under ground that subsidence of large stretches of valleys has occurred, causing structural damage to buildings and, in places, landsliding.

Disfigurement of the landscape is particularly well illustrated in large strip and opencast mines, one of the largest being Bingham Canyon Copper Mine in Utah which covers an area of over 7 km^2 and extends to a depth of more than 700 m. Goudie (1993) pointed out that estimates for the annual amounts of movement of soil and rock resulting from mineral extraction each year may be as much as 3,000 billion tonnes, whereas the amount of sediment carried by natural processes into the sea is far less, 24 billion tonnes each year. Other examples include the china clay workings in Cornwall, south-west England, and the quarrying for limestone on Portland Bill, southern England (Plate 9.6), where excavations are seen as an eyesore. Others would argue that these pits give the region a very important source of employment and income. These examples emphasize just how great a role humans play as geomorphological agents.

Plates 42 and 43 show contrasting styles of environmental degradation resulting from the large-scale mining activities in South Africa and Brazil. Even small-scale mining can have devastating effects on the environment.

Much of the rock waste produced by mineral extraction is piled up to form tips or is used to infill pits and produce new landforms. However, some of the waste may be carried into rivers, choking up their courses and ultimately altering the river patterns. Unfortunately, mineral waste often contains poisonous substances such as arsenic, and cadmium. These may pollute rivers and groundwater sources, poisoning fish, plants, and drinking supplies. This is particularly environmentally devastating to regions where lead is mined, such as in mid-Wales, and the very big copper mines, such as Ox Tedi in Papua New Guinea, where vast numbers of fish and plant life have died due to the pollutants.

Mining can cause great problems, particularly when the extracted waste materials or overburden are rich in toxic minerals. This problem was highlighted in February 1992 when waters flooding Cornwall's last tin mine, Wheal Jane, flowed out of mine workings into streams and to the coast. The silty sludge in these waters was rich in toxic metals, especially iron, arsenic and cadmium. After much local public concern and media attention, in March 1992 the mining company was forced to pump the mine waters into a tailings-dammed pond to filter out the pollutants.

Mining of radioactive metals such as uranium also causes concern. This was the case at Elliot Lake, Ontario during the 1950s and 1960s. Here, radium was the waste product of uranium mining. The radium was deposited along with rock waste as tailings. When it rained, radium was flushed through the tailings into streams or the groundwater system. Radium can cause bone cancer if ingested in drinking water or consumed, for example by eating fish. Attempts were made to reduce the removal of radium through flushing from the tailings by chemical processing and by using the tailings as a backfill within the mine. It seems that these techniques, however, were not totally successful and there is still considerable debate over their effectiveness.

Underground mining also causes considerable environmental problems. For example, mining may result in subsidence, which may be related to the extraction of rock material and water. This has been a major problem in the USA where the Bureau of Mines has estimated that over 32,000 km^2 has been affected and may well rise by another 10,000 km^2 by the end of the century. Particularly affected areas are in the Illinois Basin, western Pennsylvania and northern West Virginia where coal mining activities have caused subsidence. Below the Cheshire plain in central England, the extraction of salt has resulted in severe local subsidence and damage to buildings. Of course, subsidence can be fatal for those actually engaged in the mining operation. British history is littered with many infamous mining disasters, to the extent that they form a major part of prose, poetry, and music.

Dust is yet another unpleasant and potentially harmful by-product associated with mining activities. Ill health, the destruction of vegetation, and pollution of the atmosphere and water supplies are not uncommon. This is particularly well demonstrated in Sudbury, Ontario in Canada, where metal smelters release vast quantities of gases into the atmosphere together with particulate mineral matter. Few detailed and widely publicized studies of the effects of this latter type of pollution are documented, yet it is the everyday by-product of mineral extraction.

Hazardous/toxic substances and waste

In the USA, the major Federal statute on solid waste is the Resources, Conservation, and Recovery Act (RCRA), passed in 1976. Under the Act, solid waste is defined in Section 1004(27) as:

> any garbage, refuse, sludge from a waste treatment plant, water supply treatment plant, or air pollution control facility and other discarded material, including solid, liquid, semisolid, or contained gaseous material resulting from industrial, commercial, mining, and agricultural operations, and from community activities, but does not include solid or dissolved materials in irrigation return flows or industrial discharges which are point sources subject to permits under section 1342 of Title 33, or source, special nuclear, or byproduct material as defined by the Atomic Energy Act of 1954, as amended (68 Stat. 923).

In solid waste management, Federal efforts have focused on the disposal of so-called 'hazardous' substances, although they represent a small part of the total solid wastes. Hazardous waste is defined under RCRA 1976 Section 1004(5) as:

> a solid waste, or combination of solid wastes which because of its quantity, concentration, or physical, chemical, or infectious characteristics may [a] cause, or significantly contribute to an increase in mortality or an increase in serious irreversible, or incapacitating reversible illness; or [b] pose a substantial present or potential hazard to human health or the environment when improperly treated, stored, transported, or disposed of, or otherwise managed.

Contaminated land

Most industrialized countries have large areas of land which have become contaminated by various substances, although local very high levels of trace metals in toxic doses occur naturally. Issues surrounding the environmental damage and health risks associated with such land have really only surfaced in the last two decades. Redevelopment of land for housing has opened up many of these issues, for example in London, the large housing development of Thamesmead to the east of the City was constructed on a former munitions complex of the 1,000-acre Woolwich Arsenal, closed in the 1950s.

All countries that have initiated national action plans for cleaning-up contaminated land have begun with lists or registers of potentially contaminated sites. For example, in the USA, the Environmental Protection Agency has a hazard-ranking system, Canada has a national classification system, and Denmark also ranks sites potentially affected by oil pollution. Such registers enable a systematic protocol for action priorities, and resource allocation, including making funds available for any clean-up.

Dealing with contaminated land, both in terms of liability and financing clean-ups, and ensuring that those responsible actually pay for any costs incurred, remains a problem for governments around the world. In the USA and the EC, there are major new initiatives incorporating the polluter-pays principle (see Chapter 10). Many countries already have limited systems of strict liability (joint and several liability) in statutory and common law, but there is a need for more far-reaching, international legislation in order to address issues of equity, economic efficiency, and liability across nations' borders.

In the USA, legislation was enacted by Congress in 1980 in the Comprehensive Response, Compensation, and Liability Act (CERCLA), under which a Federal trust fund (Superfund) was established to permit a swift response to any immediate and future health hazards caused by contaminated land. This Act was partly in response to the Lowe Canal disaster in which a leaking toxic waste dump led to the evacuation of hundreds of families and the contaminated area being declared as a federal disaster area (Plate 44). The original land owners had sold on the land with the condition that they had no future liability. In 1986, CERCLA was amended by the Superfunds Amendments and Reauthorization Act (SARA), and included the authorization for collecting further revenues. In a 1989 report from the US auditor general's office, it was estimated that 135,000–425,000 sites may come within the US Environmental Protection Agency's (US EPA) criteria for the treatment of contaminated land – originally, the scheme was assumed to involve only 1,000–2,000 sites.

Superfund is a Federal instrument, and many states have their own systems running in partnership with this. It is essentially a mechanism for raising money with the aim of spreading liability costs across a broad spectrum of producers, consumers, and tax payers, and much of which is provided by the US Congress (under SARA). Congress voted an extra US $8.6 billion to the fund for the period 1986–91. Additional monies include a US $2.75 billion petroleum tax, a US $2.5 billion corporate environment tax from companies with an annual turnover greater than US $2 million, US $1.5 billion from general revenues, and US $1.4 billion from a tax on chemical feedstocks. Also, US $300 million is budgeted from interest on the money in the fund, together with a further US $300 million recovered from liable parties in various clean-ups. The US EPA pursues potentially responsible parties (PRPs), of which there are an estimated 14,000 notified on only 250 sites, and the PRPs naturally then pursue others and so the spiral of litigation continues. It is hard to obtain reliable figures for the enormous transaction costs of Superfund, partly since those involved are reluctant to provide information.

Under Superfund, there are four categories of PRPs who are both jointly and severally liable for any clean-up costs: the owner or operator; the prior land owner; those responsible for the disposal of hazardous substances; and those responsible for transporting the hazardous substances. Furthermore, those responsible for any contamination and pollution remain responsible at any time after the event, that is, there is no statute of limitation. Such retrospectivity raises many issues, such as the extent to which it is fair to seek liability for actions that at the time of their implementation were legal but have subsequently become unlawful. Lenders' liability has proven to be amongst the most contentious issues associated with Superfund, mainly because

there is no discrimination between lenders with a purely legal rather than operational relationship, even to the extent that the US EPA has proposed a means of mitigating its effects. Joint and several liability for pollution has also caused considerable problems because of the difficulties in apportioning blame. So far, although Superfund has only used its financial resources for on-site costs, the US Environmental Protection Act contains a clause permitting liability to extend to claiming 'natural resources' damage off-site.

On balance, Superfund represents a commitment by Americans to recognize and define the burden of liability for environmental damage, and to monitor pollution and clean-up. However, in operation, Superfund is associated with substantial transaction costs, estimated at 40–60 per cent, and a plethora of involved parties seeking to re-apportion liability and costs to the extent that the legal machinations are in danger of causing Superfund to grind to a halt.

In the EC, the US experience of the Superfund has helped to mould opinion and guide the legislators in making proposals on civil liability as a result of environmental damage caused by waste disposal ('waste' being defined as anything that is unwanted by the producer). Like its US counterpart, the proposed EC legislation is focused on the polluter-pays principle, although there is discussion over 'green taxation' where potential polluters pay a pollution tax to the government which could be used to defray the costs of any clean-up. A potential stumbling block for the EC is the avoidance of any legislation that could permit polluters to operate with relative impunity in one member state rather than another, simply because of any differences in the strictness of laws within the EC. On 1st September 1989, an EC draft Directive was submitted to the Commission on Civil Liability for Damage caused by Waste, which was an attempt to instigate the polluter-pays principle throughout the Community. Two years later, on 28th June 1992, major amendments were made to the original draft Directive, and at the time of going to press the EC has still not adopted the draft Directive. The proposed Directive, whilst not actually defining waste, specifically excludes oil and nuclear waste. If adopted, albeit with modifications, the producers and handlers of waste would be liable for damage caused to individuals, property, and the environment. In contrast to the USA, it remains unclear whether banks and other financial institutions that provide support may be liable. Another significant feature of the draft Directive is that common interest groups, such as Greenpeace and Friends of the Earth, can initiate lawsuits, or become involved in existing lawsuits.

Furthermore, under the EC Freedom of Information Directive, environmental and other interest groups now have more access to sensitive environmental data. It is often claimed that one of the major differences between the US Superfund and the EC draft Directive is that the latter does not have a retroactive effect (i.e., cover 'old' pollution), but the issue of retroactive liability remains contentious, with the legal profession casting doubt on this assertion.

Within the EC, legislation dealing with contaminated land varies considerably between Member States. In the UK, for example, the Environmental Protection Act 1990 provides for waste regulatory authorities (WRAs) to inspect and monitor landfill sites which are no longer licensed, and to take any remedial action if the site is likely to pose a risk to human health and/ or cause environmental damage. Any incurred costs can be sought from the land owner, who may not have been the operator, but essentially the EC draft Directive focuses liability primarily on the producers and handlers of waste. The liability is strict in that the producer of waste has a civil liability irrespective of the fault on its part, and where the producer cannot be identified then the holder of the waste becomes the deemed producer with the liability being 'joint and several'. The disposer of waste has a legitimate defence if they were deceived by the producer of waste as to the actual nature of the waste, a third party contributed to negligence through an act or omission, and *force majeure*. An important feature of any liability is that it does not impose an inequitable burden on a party to contamination for the full clean-up costs where its contribution has been minor. For example, a producer may manufacture a relatively innocuous waste which, when mixed with a substantially more toxic substance, gives rise to serious injury or damage. In these circumstances, liability should be limited proportionately to the actual contribution to the harm. In this context, the UK House of Lords Select Committee on the European Communities (1990) crystallizes the issue:

> In general, it is the role of the criminal law to enforce these norms which society requires to be respected – for instance the appropriate standards for emissions to air or water from industrial enterprises – and that of the civil law to give redress to those who may have suffered as a result of the environmental damage – for instance when a farmer's fields have been covered in oil from a crashed tanker. If the criminal law is thus primarily a means of promoting the precautionary approach, the making good of damage or compensation for it is largely left to the civil law.

A facet of the UK Environmental Protection Act 1990 was the provision for the establishment, under Section 143 of the Act, of 'public registers of land', whereby each waste regulation authority would maintain a register recording land which is being or has been put to contaminative use, but excluding, for example, substances such as radon gas or naturally occurring arsenic. Information could be excluded from a register if it were to be regarded as commercially sensitive by the authority maintaining the register, or on appeal, by the Secretary of State, and if the information is regarded as being contrary to the interests of national security. Under the Act, each waste regulation and waste collection authority would ensure open inspection of the register by the public, at no cost. The Section 143 registers, contrary to popular belief, were not conceived to identify actually contaminated land, but rather to provide a database of information on sites where land-use has had the potential to leave contamination. Commercial confidentiality is an important and controversial issue in the management of waste and environmental injury or damage, and private companies are frequently reluctant to disclose information that can be seen as potentially harming their interests.

Probably because the government was strongly influenced by various commercial pressure groups, on 24th March 1993 the Secretary of State for the Environment withdrew the proposals for statutory registers of contaminative uses of land under the 1990 Act on the grounds that the proposals would have led to the inclusion of sites not actually contaminated, while omitting others that are actually contaminated, that once registered a site could never be removed even if any contamination had been satisfactorily dealt with, and as many uncertainties remained about the liability and action that would be taken to clean up sites. The UK government plans, instead, to instigate a new wide-ranging review of the problems in this area. Prior to this withdrawal, it appears that under the Department of Environment's revised proposals published in July 1992, the area of contaminated land would have been reduced to 10–15 per cent of that originally envisaged and estimated at figures between about 100,000 ha (Fowler 1993) and 50,000 ha, affecting 50,000–100,000 sites, although only a small proportion of these pose an immediate threat to public health or the environment (reported in the trade magazine *Surveyor*, 29 January 1992). UK scientists at the government's Warren Spring Laboratory have completed a review of appropriate technology to clean up contaminated land, yet other civil servants are still debating a national policy for cleaning up contaminated land.

Any legislation that imposes a civil liability on polluters to pay in full the costs of any clean-up will meet strong opposition from the myriad groups with an economic vested interest in profit with minimum responsibility. Land registers are also unpopular with entrepreneurs since they permit both rapid and easy access by the media, environmental monitor groups, and other interested parties, to detailed information on 'land quality'. Inevitably, the introduction of land registers would make certain land sites harder to sell and develop, but this should be of secondary

Box 9.1
Contaminants

There are various definitions of contaminants and contaminated land, but contamination is not synonymous with pollution, although clearly it is a necessary condition which can lead to pollution. The types of activities which could be considered as contaminative uses of land are listed below.

Chemical industries – the production, refining, and bulk storage of organic and/or inorganic chemicals, such as fertilizers, pesticides, paints, dyestuffs, inks, soaps, detergents, pharmaceutical products, cosmetics, toiletries, pyrotechnic materials, fireworks, recovered chemicals, etc.

Energy industries – the use of natural and/or synthetic substances in the production of energy, for example fossil fuel or nuclear fuel in power stations.

Extractive industries – extraction, handling, and storage of substances from mines, for example metal ores, tailings, coal, petroleum products, etc.

Metals – processing of metals (recovery, refining, production, finishing treatments, etc.) by any means, and their use to manufacture products in processes from heavy engineering to metal working, even in small businesses in the marketplace. Scrap metal handling and processing are included.

Non-metals – production and/or refining of non-metals, for example production of cements, bricks and associated products, lime, gypsum, asbestos, fibres, ceramics, glass, vitreous enamels, etc.

Rubber industries – manufacture of synthetic and/or natural rubber products, such as vehicle tyres, etc.

Engineering and manufacturing – manufacture of motor vehicles, aircraft, aerospace equipment, ships, railway/tramway equipment, electrical/electronic products, mechanical engineering/industrial plant, etc.

Infrastructure – maintenance, repairing, dismantling of industrial equipment, for example railways/tramways and rolling stock, roads, fuel stations and road-haulage vehicles, aerospace facilities and equipment, including aircraft, docks and marine vessels, etc.

Waste disposal – treatment, storage and/or disposal of sewage, radioactive materials, landfill, scrap, other effluents or substances, including the cleaning of tanks and drums, etc.

Foods – manufacture/processing of animal by-products (excluding slaughter-house butchering), animal feedstuffs, pet foods, etc.

Agriculture – burial of diseased livestock.

Textile industries – manufacture of various textiles and leathers, for example production of carpets or other floor coverings, including linoleum, fabrics, and leather products.

Timber and timber products – treatment, including coating, of timber and timber products, such as wood preservatives, etc.

Paper, pulp, and printing industries – paper and pulp manufacture, including processes associated with printing works, etc.

Miscellaneous – various processes such as building and industrial plant demolition, some high street processes such as dry-cleaning, the running of laboratories for research and/or educational purposes, etc.

concern to conserving the natural environment. The bottom line is that the business world faces 'Hobson's choice' of either cleaning up its act by becoming environmentally responsible, or facing financial ruin by the regulators. Indeed, within the EC, the Commission believes that the time is fast approaching for compulsory environmental insurance, partly influenced by Continental European insurers. In France, Assurpol is an insurance pool of 75 companies, which offers insurance coverage on a global basis (excepting the USA and Canada), mainly for French companies, with members having a responsibility to reinsure the risks ceded to the pool in proportion to their share of the total capacity of the pool. Assurpol, underwriting up to Fr 330 million per claim (with retrocession or re-reinsurance to the Italian pool), covers liability for both accidental and non-accidental (gradual) environmental damage. The Italian counterpart, Inquinianento, is an insurance pool with a capacity up to £20 million per claim. In the Netherlands, there is a pool called MAS, but this only deals with relatively small claims of up to DFL 16 million per claim, and in 1992 Denmark initiated a pool.

Finally, a relatively new issue concerns the adoption of compulsory environmental audits. In Europe, the EC is working on a proposal for a voluntary environmental audit. In the long term, environmental audits may become compulsory, something that environmental pressure groups are actively campaigning for. It is important to appreciate that unless there is international comparability in environmental legislation and any 'green tax', the net result of over-zealous, market-leading legislation is potentially to disadvantage seriously those producers who are in the countries with the strictest legislation. Such arguments must be balanced by the need for urgent government action on a whole range of environmental issues.

Urbanization

Since ancient times, cities have played a central part in the economic, political, and cultural development of societies. Cities serve as the commercial and administrative focus for nations, and generally provide the main places for both the production and consumption of goods and services. Presently, 45 per cent of the world's population live in urban areas: 37 per cent in less-developed countries or LDCs and 73 per cent in the more-developed countries or MDCs (ASCEND 1992). In other words, contrary to popular belief, most of the metropolitan areas where much of the world's population dwells are situated in the developed countries. Currently, about 20 per cent of the world's population inhabits metropolitan areas, and 33 per cent lives

Soil clean-up involves the treatment to remove, stabilize, or destroy contaminants using physical, solidification, chemical, biological, and thermal methods. Essentially, contaminants can be disposed of by landfills or containment. Low disposal costs have made landfill an attractive option, using either containment or attenuation of the waste, but these costs are expected to rise steeply as landfill sites become scarce, and controls on environmental pollution become more stringent. These changes are leading to the development of alternative, innovative technologies.

Containment involves isolating contaminants from the environment to minimize any liquid or gaseous interchange, whereas attenuation attempts to minimize the movement (e.g., by adsorption) and/or reduce the toxicity (e.g., by degradation). Currently, landfill is cost-effective and commonly practised, but design/construction problems and long-term uncertainties about the persistence of toxic substances can present problems for landfill. The appropriateness of various innovative technologies for the disposal of contaminants will depend on many factors, including the type of toxic substances, their concentration and distribution, the soil type and hydrological cycle in relation to the contaminants, and a risk assessment from natural hazards (e.g., earthquakes).

Physical processes do not destroy contaminants and, therefore, are often seen as the first stage of their multi-stage destruction and/or stabilization. *Ex situ* processes (those away from the site of contamination) include particle separation techniques, exploiting differences in physical properties such as density, weight, grain size, magnetic susceptibility, and surface chemical properties, or physical extraction (e.g., washing, steam stripping). *In situ* processes include soil washing (using aqueous solutions, acids, and/or

Box 9.2
Contaminated soil clean-up technologies

HUMAN IMPACT ON THE EARTH'S SURFACE

Table 9.1 *Urban and rural populations, settlements, and labour.*

Region	Urban population as % of total		Average annual population change 1960–90 (%)		Cities with at least 1 million			Total labour force 1990 (000s)	Women as % of labour force 1990	% of 1980 labour force in		
					% total of pop.		No. of cities 1990			Agri-culture	Industry	Services
	1960	1990	Urban	Rural	1960	1990						
World	34.2	45.2	2.8	1.3	12.2	14.8	276	2,363,547	36.1	51	21	28
Africa	18.3	33.9	4.9	2.1	5.7	9.2	24	242,784	34.4	69	12	19
North and Central America	63.2	71.4	2.0	0.7	28.7	31.8	44	189,258	37.4	12	29	58
South America	51.7	75.1	3.6	0.1	23.4	32.8	29	104,465	26.4	29	26	45
Asia	21.5	34.4	3.7	1.5	8.3	11.3	115	1,436,522	35.3	66	15	19
Europe	61.1	73.4	1.2	(0.7)	16.5	17.0	36	231,702	38.6	14	39	47
Former USSR	48.8	65.8	2.0	(0.4)	12.4	15.3	24	146,634	48.0	20	39	41
Oceania	66.3	70.6	2.0	1.3	31.8	32.2	4	12,181	37.0	20	28	52

Source: World Resources Institute (1992).

in cities with populations greater than 100,000 (Angotti 1993). By the year 2000, it is estimated that 51 per cent of the world population will inhabit urban areas, rising to 65 per cent by the year 2025 (United Nations Population Fund 1991). In 1960, seven out of the world's 10 largest urban agglomerations were in North America, Europe, and Japan, with New York, London, and Tokyo at the top, whereas at present seven out of the top 10 are in less-developed countries, Mexico City being the largest with a population of more than 20 million (ASCEND 1992) (Plate 45). Table 9.1 shows some of the trends in urbanization (including settlement and labour force data) for various regions between 1960 and 1990, from which the trend towards increasing global urbanization is evident.

The future of cities in the context of sustainable development (see Chapter 10) is explored in the book *The Living City: Towards a Sustainable Future*, edited by Cadman and Payne (1990), and contains a useful set of essays on these topics. In this book, Robertson examines four alternative scenarios for the future of cities: (1) 'Decline and Disaster', envisaging a future in which severe breakdown follows from the failure of cities to adapt either to changes in economic function or, in the case of the developing world, to uncontrolled expansion of the urban population; (2) 'Business-as-Usual', offering a 'top-down' approach in which cities rely upon the benefits of

Box 9.2

Contaminated soil clean-up technologies

surfactants), soil vapour extraction techniques (using forced air, induced air, or steam), and electroremediation (in which a direct current passes through an array of electrodes placed in the soil to induce the contaminant to flow in the pore water to the electrodes, followed by removal of the toxic solution to a water treatment plant).

Solidification processes involve commercially viable and well-proven technologies, although some techniques are at an R&D stage. Solidification technologies are grouped according to the way in which the contaminant is bound, being divided into organic and inorganic techniques: (1) organic, as thermosetting, thermoplastic microencapsulation, or macroencapsulation; (2) inorganic, as cement-based, lime-based, vitrified, liquid silicate, or pozzolan-based (a pozzolan material acts like cement, and contains silicates of

alumino-silicates which react with lime and water to form stable insoluble compounds). The organic binding systems are expensive, and at bench-scale trial stage, whereas the inorganic treatments are seen as having the better potential. Generally, it is extremely difficult to fix organic contaminants using inorganic binding processes. Cements have been routinely used to treat radioactive and other toxic wastes. Lime-based systems have also been widely used to stabilize or solidify toxic waste. Silicate-binding processes involve using powdered aluminium silicate oxides and alkaline solutions of alkali metal polysilicates, which condense to form a high compressive-strength solid in a reaction which gives out heat energy (exothermic reaction). Vitrification, either *in situ* or *ex situ*, involves the melting of contaminated soil to produce a glass-like material which is inert.

conventional economic development to trickle down to their more disadvantaged citizens; (3) 'Hyper-Expansion', in which new technology replaces conventional employment 'releasing' an increasing proportion of the workforce to extended periods of leisure; and (4) the 'Sane, Humane and Ecological' alternative, where changes in work and a continued trend towards decentralization lead to the development of greater self-reliance at both a personal and urban level. In the latter scenario – if it were to materialize – greater emphasis would be placed on enabling urban dwellers to take control of their own future development, with community-led ('bottom-up') urban revival. Also, such a scenario provides more emphasis on the resourceful, self-reliant city, in which there is a minimum waste of resources, by energy conservation and recycling.

Urbanization can cause many problems associated with changes in the hydrology of an area. New canals may have to be constructed and rivers canalized. These, together with buildings, paths, and roads, produce an impermeable surface over which water will flow. The use of storm sewers increases the rate and amount of water entering rivers. All these result in an increase in the magnitude and frequency of flooding events. The extraction of groundwater for domestic use in large settlements may also cause major alterations to the land's surface, such as salinization,

Plate 9.7 *Traditional houses in Bangkok, constructed along one of its many waterways. Rain water is collected in pots (middle left) for domestic use. With the rapid urbanization of Bangkok, water is now extracted from the ground, which has led to large-scale subsidence, thereby threatening both traditional dwellings such as this and the modern city.*

and other effects include ground subsidence. This is well illustrated in Venice where the pumping of groundwater for industrial purposes has caused the gradual subsidence of buildings and increased flooding during the winter. Similar effects are occurring in Bangkok and in the Fenland in the UK.

Box 9.2
Contaminated soil clean-up technologies

Chemical processes are applied either to destroy or to convert the contaminants to potentially less harmful substances. Techniques include oxidation, reduction, extraction, neutralization, hydrolysis, mobilization, electrochemical processes, polymerization, and chemical dechlorination. For most of these processes, the contaminated soil needs to be treated as a slurry or at least with the contaminants in an aqueous solution (e.g., groundwater). As with physical processes, chemical techniques can be applied either *in situ* or *ex situ*.

Biological processes (or 'biotreatment') generally use microbial biodegradation to make any contaminants benign or to remove them. Biodegradation involves the breakdown of harmful substances, which can then be left at the treatment site or removed. Biotreatments tend to be time-consuming and frequently unsuccessful in removing all of the contaminants. Also, biotreatment technologies require more research before the full implications of their use can be evaluated.

Thermal processes involve either incineration of the contaminants within the soil, or a two-stage process of volatilization and pyrolysis followed by the removal or destruction of the toxic substances from a gaseous phase (by condensation or further combustion). Thermal processes are most appropriate for soils contaminated with organic substances, although some of these chemicals are difficult to combust, and often the process is incomplete. Some inorganic chemicals (e.g., mercury and cyanides) can be combusted, but may leave toxic residues in the ash, thereby posing further disposal problems. Sandy, silty, loamy, and peaty soils respond to thermal processes better than clay soils, due to handling problems with the latter.

Pirazzoli (1973) showed that there had been a mean sea level rise of 6 mm per year in Venice between 1950 and 1970. This rise was primarily the result of the withdrawal of groundwater for use within the city, and the dredging and embanking of the bay which enhanced the tidal amplitude to make Venice more susceptible to flooding (Plate 46). Bangkok has rates of subsidence estimated at between 3–4 cm per year, with a maximum rate of 10 cm per year. As in the case of Venice, these problems are related to groundwater withdrawal from sandy aquifers, but they are also due to the compaction of clays because of the overlying weight of major building structures (Rau and Nutalaya 1982). Rau and Nutalaya estimate that Bangkok may subside below sea level within the next 20 years (Plate 9.7).

In his book, *Metropolis 2000* (1993), Angotti advances the view that there are essentially three categories of metropolis which reflect general historical tendencies and planning models – although in reality every urban system contains elements of all three: (1) the US metropolis; (2) the Soviet metropolis; and (3) the dependent metropolis.

In the US metropolis, land use is segregated with the fragmentation of social groups and political institutions. The centre is densely developed and the suburbs are an urban sprawl. The population is highly mobile and dependent upon the motor vehicle, and it consumes relatively large amounts of non-renewable resources, planning is determined mainly by the interplay between the automobile and petroleum monopolies, together with local real estate interests. It represents the quintessential twentieth-century capitalist urban development.

In the Soviet metropolis (perhaps more aptly renamed as the Russian metropolis), the archetypal twentieth-century socialist metropolis, the urban population is more integrated *vis-à-vis* its social and political structure, there is limited social mobility, and little consumer choice. There is an administrative/residential centre, and relatively high density suburbs, with transport by mass transit systems (e.g., metro and bus systems in St Petersburg and Moscow). Planning follows a highly centralized administrative/commercial structure.

The dependent metropolis reflects the particular history and dynamics of urbanization and planning in the developing countries of Africa, Asia, and Latin America. Although there are important differences between many of these metropolises, they all share a strong dependency upon the developed, capitalist world and typically are associated with considerable social deprivation and poverty. Planning strategies show a wide range of patterns, including large amounts of *ad hoc* development. Angotti (1993) argues that metropolitan planning has failed to adapt to present needs, because it was rooted in the premetropolitan era of the industrial city, and is based upon 'simplistic notions of master planning and master building that ignore the complex division of labor and functions characteristic of the metropolis. It is strongly influenced by Utopian thinking and philosophical idealism'. The remedy advocated by Angotti is for there to be a much greater emphasis upon neighbourhood planning, which has the function of integrating the family or household with the metropolis, residence, and workplace.

Conclusions

Clearly, humankind has radically altered the Earth's surface, with accelerated impact in recent times. There is a need to understand the natural systems and the interaction between various Earth-surface processes and the impact of human activities in order to be able to predict the consequences of human actions, and to manage resources in sympathy with the natural environments. In doing so, people should take cognizance of the Himalayan dilemma (above). Urbanization tends to be an *ad hoc* process of colonization of the natural environment. A means to reducing the actual and potential environmental damage caused by urbanization is to plan urban centres that provide a real sense of community, with good health care and educational provision, have adequate recreational open spaces, are clean and are not dominated by private and commercial haulage vehicle traffic, are developed on a human scale, maintain co-existing structures in scale, and minimize the discrepancies between housing for the rich and the poor.

CHAPTER 9: KEY POINTS

1 Rapid population growth has resulted in increased demands upon the Earth's resources, which have led to accelerated environmental degradation, and have precipitated potentially serious global climate change.

2 The human impact on land has been enormous, as land-use has changed, natural vegetation is cleared for agricultural use, settlements and urbanization increase, reservoirs are created, minerals are extracted, and more land is developed for recreation purposes. Acute concern is now widely expressed over the deforestation of boreal and tropical forests, the degradation of grasslands and wetlands, and desertification. Such destruction of natural ecosystems has led to a reduction in biodiversity, and impoverishment of soils. In attempts to counter the deleterious effects of land misuse in some areas, exotic plants and animals are being introduced, and indigenous fauna and flora are being carefully monitored and encouraged.

3 Human impact on soils has caused considerable damage, commonly because of poor agricultural practices, excessive water extraction, poor irrigation methods (e.g., leading to salinization), defoliation (particularly resulting in laterization), and compaction by heavy vehicles and animals. The cumulative effects of these can be disastrous to countries whose economies are heavily dependent on agriculture. The amelioration of these poor practices and the improvement of soil quality require an understanding of the chemistry of soils and nutrient supply cycles.

4 Human impact on the oceans and seas results from pollution by dumping and accidents, overfishing, mineral extraction (e.g., phosphates) and the removal of rare and important marine life such as corals. The seas are an available resource which require more careful research in order to avoid irreversible damage being caused to their ecosystems, which could have a knock-on effect on the atmosphere and, ultimately, terrestrial life.

5 The exploitation of Earth resources inevitably produces waste, some of which may be hazardous/toxic (contaminants). Until the past few decades, much of this waste has been disposed of without any real concern for the damage to ecosystems, and frequently under the auspices of 'not in my backyard'. Today, as environmental issues are becoming more focused, there is much greater awareness of contaminants and contaminated land. Clean-up technologies are more readily available and preventative measures are being instigated in many countries. Many nations and international organizations are adopting a philosophy of the 'polluter-pays' principle. Responsibility for cleaning up contaminated land has led to the introduction of legislation, for example in the USA and throughout Europe. New and forthcoming legislation aims to identify the polluters and arrange for appropriate levels of compensation to injured parties, but because such laws are in their infancy, there are many teething problems, exemplified by the US Superfund.

Chapter 9: Further reading

Angotti, T. 1993. *Metropolis 2000: Planning, Poverty and Politics.* London: Routledge.
This very readable book on urbanization offers an analysis of metropolitan development and planning in all parts of the world, and under different economic and environmental conditions. The first four chapters are devoted to an examination of metropolitan development in the United States, the former Soviet Union (FSU), and the 'dependent metropolis' of the developing world. The last three chapters consider the problems of urban planning theory and practice in the metropolis and its communities. Throughout, Angotti advances the principle of 'integrated diversity' and emphasizes linked neighbourhood planning with a broader vision of a planned metropolis.

Cooke, R.U. and Doornkamp, J.C. 1990. *Geomorphology in Environmental Management*, 2nd edn. Oxford: Oxford University Press.
A comprehensive text which highlights the importance of geomorphology in environmental management and risk assessment. Suitable for college and university undergraduate students at all levels, and provides a useful reference source for teachers and researchers. Topics include mass movement, catchment studies, erosion and weathering problems, neotectonics, aeolian environments, and glacial systems.

Eden, M.J. 1989. *Land Management in Amazonia.* London: Belhaven.
This book considers the tropical rainforest as a global resource and its vital importance in sustaining the life

on Earth. The competing needs of conservation and development in Amazonia are assessed in terms of climate, geomorphology, hydrology, soils, ecology, diverse histories, and current impact of human intervention. Case studies are presented where, for example, there are attempts to adapt resource-use systems of native peoples to encourage the more effective and less harmful exploitation of the rainforests. Conservation issues are addressed, including the role of national parks and interpretive land management. A good supplementary book for students concerned with environmental risk assessment, particularly relating to the rainforests.

Goudie, A.S. 1993. *The Human Impact on the Natural Environment*, 4th edn. Oxford: Blackwell.
A very useful undergraduate textbook which addresses the ways that human activity has changed, and is changing, the Earth's surface. The book is well illustrated, and includes a comprehensive bibliography. Topics which are dealt with include desertification, deforestation, plant and animal invasions, marine pollution, climatic change, and environmental uncertainty.

Gradwohl, J. and Greenberg, R. 1988. *Saving Tropical Forests*. London: Earthscan Publications.
A text which provides case studies from throughout the world to show how the destruction of the tropical forests might be slowed or even stopped, and how sustainable management could be achieved. A thought-provoking book for students, teachers, and policy makers.

Grainger, A. 1990. *The Threatening Desert: Controlling Desertification*. London: Earthscan Publications.
An interesting book which describes the distribution and processes of desertification, and the successes and failures which have accompanied the various attempts to combat desertification as set out by the Plan of Action resulting from the 1977 Nairobi United Nations Conference on Desertification. The book argues for a new International Plan of Action to control the increasing threat to the natural environment posed by desertification.

Ives, J.D. and Messerli, B. 1989. *The Himalayan Dilemma*. London: Routledge.
This book addresses the complex dynamics and environmental systems in the Himalayas, and considers the problems of reconciling development and conservation. It includes a look at the interaction between human activities and the natural environment, and is a useful text for advanced undergraduate courses, teachers, and policy makers.

Morgan, R.P.C. 1986. *Soil Erosion and Conservation*. London: Longman.
A useful textbook aimed at undergraduate and postgraduate students who are studying soil erosion and conservation as part of any Earth science or environmental science course. The book provides an introduction to the subject including the magnitude, frequency, rates and mechanics of wind and water erosion, erosion hazard assessment, methods of measurements, modelling and monitoring, and strategies for erosion control and conservation practices.

Nebel, B.J. and Wright, R.T. 1993. *Environmental Science: The Way the World Works*, 4th edn. Englewood Cliffs: Prentice-Hall.
An undergraduate environmental textbook with a central theme of sustainability. There are four sections in this book: Part I, What ecosystems are and how they work; Part II, Finding a balance between population, soil, water, and agriculture; Part III, Pollution; Part IV, Resources: biota, refuse, energy, and land. The text has various elements that provide teaching aids, e.g., learning objectives, review questions, etc. While we found this book useful, it has the somewhat irritating presentation style of very well-drawn and sophisticated diagrams alongside over-simplistic, naive artwork. The book is aimed at college students taking environmental courses.

Poore, D. 1989. *No Timber Without Trees: Sustainability in the Tropical Forest*. London: Earthscan Publications.
Based on a study for the International Tropical Timber Organization, this book reviews the extent to which natural forests are being sustainably managed for timber production and how these practices could be improved. The book places timber production in the wider context of tropical rainforest conservation. Examples are drawn from Queensland, Africa, South America, the Caribbean, and Asia. Interesting and easy reading for students of environmental science.

Revkin, A. 1990. *The Burning Season: The Murder of Chico Mendes and the Fight for the Amazon Rain Forest*. London: Collins.
An inspiring, but also sad book which emphasizes the beauty of the tropical rainforests and the need for conservation in Amazonia. The book provides an ecological, historical, and industrial outline of life in the forest, focusing on the life and environmental work of Chico Mendes, a rubber planter who strived for sustainable development in the forest in which he lived. Mendes' success is reducing the exploitation of the forest by cattle ranchers cost him his life when he was murdered in 1988.

I am not yet born; O hear me.
Let not the bloodsucking bat or the rat or the stoat or the
club-footed ghoul come near me.

I am not yet born, console me
I fear that the human race may with tall walls wall me,
with strong drugs dope me, with wise lies lure me,
on black racks rack me, in blood-baths roll me.

I am not yet born; provide me
With water to dandle me, grass to grow for me, trees to talk
to me, sky to sing to me, birds and a white light
in the back of my mind to guide me.

I am not yet born; forgive me
For the sins that in me the world shall commit, my words
when they speak me, my thoughts when they think me,
my treason engendered by traitors beyond me,
my life when they murder by means of my
hands, my death when they live me.

I am not yet born; rehearse me
In the parts I must play and the cues I must take when
old men lecture me, bureaucrats hector me, mountains
frown at me, lovers laugh at me, the white
waves call me to folly and the desert calls
me to doom and the beggar refuses
my gift and my children curse me.

I am not yet born; O hear me,
Let not the man who is beast or who thinks he is God come near me.

I am not yet born; O fill me
With strength against those who would freeze my
humanity, would dragoon me into a lethal automaton,
would make me a cog in a machine, a thing with
one face, a thing, and against all those who
would dissipate my entirety, would
blow me like thistledown hither
and thither or hither and thither
like water held in the
hands would spill me.

Let them not make me a stone and let them not spill me.
Otherwise kill me.

Louis Macneice, 'Prayer before Birth'

The natural ecosphere under threat from human activities

The greatest challenge that confronts society and governments today, and the generations to come, is the sustained development and intelligent management of this (some would argue, over-populated) planet. In essence, societies must be able to supply sufficient food, raw materials, and energy to all nations without compromising future generations, and without leaving a wasteland of environmental degradation.

Humans, unlike other animals, have the ability, if not always matched by the foresight, to appreciate the responsibility for managing this planet wisely, and the capacity to control human actions and monitor their impact on the environment. We can look into Earth from Space, communicate rapidly across the world and even from Space to Earth, prevent and remedy many diseases, manufacture many items to make life more comfortable and enjoyable, and construct complex cities. Humans can inhabit nearly every environment on Earth. Many of these environments, together with issues surrounding environmental stress, were discussed in previous chapters of this book.

Broadly, there are four main components of the ecosphere which are identified as potentially under threat from human behaviour. First is the climatic system, which includes the destruction of the ozone layer, the emission of gases into the atmosphere such as the greenhouse gas carbon dioxide, which cumulatively could lead to significant global warming, and the pollution of the atmosphere resulting in phenomena such as acid rain. Second, there is the interaction between the organic and inorganic components of the ecosphere, that is, the global circulation of nutrients, or the nutrient cycles. This includes the mobilization of carbon, nitrogen, and phosphorous, so that an imbalance results, with excessive depletion in some parts and enrichment in other parts of these natural cycles. Third, humans threaten the hydrological cycle, for example by the withdrawal and pollution of water, droughts, and floods, and activities which contribute to processes of erosion and deposition of sediment to silt up major rivers and estuaries. Lastly there is the direct or indirect human influence on the extinction of endangered species, the reduction of biological diversity and the changes in the vegetational character of various regions of the world. Here, the main threat posed to other species is a consequence of the rapid growth of the human population.

The survival and evolution of life on Earth is, in essence, about being adaptable to changing circumstances; the alternative is extinction. This appears to be a truism both for species and individuals. The conditions leading to mass mortalities in a species may be different from those which cause the elimination of an entire species of organism. Mass mortalities generally do not cause the extinction of an entire species, but rather act as a catastrophe that leaves enough of a population for recovery to some equilibrium level, perhaps similar to the pre-catastrophe value. Major earthquakes, volcanic eruptions, and the impact of small meteorites are examples of natural disasters which can wipe out geographically restricted populations, but which allow recovery of the species numbers. While these arguments are true for the natural way in which life has evolved on Earth, any rational person would hardly countenance a nuclear holocaust or any other anthropogenically precipitated disaster as an acceptable catastrophe. At least, as compassionate and caring people we should value the life of other fellow humans in the same way as we might wish to be treated.

Scientists could take a dispassionate long-term perspective, for example on a geological time scale, and say that the human species is bound to become extinct like so many species before. It is inevitable, so why worry? The Earth will survive, the human species will not. The same philosophy could be applied equally to all the Earth's fauna and flora. With or without human intervention, various species have reached near-extinction levels. An extreme viewpoint would be to accept that given the inevitability of extinction, what does it really matter whether it occurs today or tomorrow? You might think that nobody could be quite so *laissez-faire* about the environmental impact of human activities. But this is exactly how many human activities and attitudes could be construed. The scant regard often shown for the environment is symptomatic of the prevalent attitude that somebody else can clean up after us. Humankind can no longer bequeath such a legacy to future generations. Concern and interest in the environmental impact of human activities are needed now. The will to translate that concern and interest into remedial and preventive action is also required in these decades. These issues involve all of us. Individuals may 'come for the environmental ride as mindless passengers', but societies invariably endeavour to help with the driving, and steer away from irrevocable environmental damage.

Life on Earth may be robust for many species, even as far as withstanding the impact of global nuclear carnage. Insects, for example, would survive, but it seems unlikely that higher species such as *Homo sapiens* would.

The lifespan of humans is short relative to that of Earth or geological time, which is measured in tens to hundreds of thousands and millions to hundreds of millions of years. The extinction of humans as a species may also be an inevitable natural process, but such a conclusion is no reason for apathy and complacency over the consequences of polluting the environment. Nor should scientists demand absolute proof of cause-and-effect before ceasing to use or dump certain 'suspect' chemicals. Where a reasonable degree of doubt exists about the consequences of human actions, then there are, perhaps, sound reasons for being cautious and conservative. Polluters are only too eager to employ scientists who are willing to bury their heads in the sand over environmental pollution. If human activities destroy the habitability of Planet Earth, there will be no second chance. No opportunity will exist for those same scientists and technologists to undo the damage with a contrite heart. Apologies to future generations for our inept management of the environment are unacceptable. Humankind must avoid the sins of commission and omission, but instead seek to be guilty of only one thing – being accused of acting overcautiously.

Before discussing ways in which humans can manage the planet intelligently, plan for and implement sustained development, there is a need to consider the social, economic, cultural, and political aspects of global environmental issues. Prominent amongst the things humans must come to terms with is the size of the world's population, and the levels at which sustainable development for the whole world is both achievable and acceptable.

World population

The population of the world (Figure 10.1) doubled from around 2.5 billion in the middle of this century to about 5 billion by 1987. By the middle of the 1990s, world population will be about 5,292,200,000, an increase by 75 per cent since 1960, and projected by the United Nations Population Division to increase by 60 per cent over the 1990 figures to 8,488,600,000 by the year 2025, triple the 1960 level, and reach about 11.3 billion by the year 2100 (Figure 10.2).

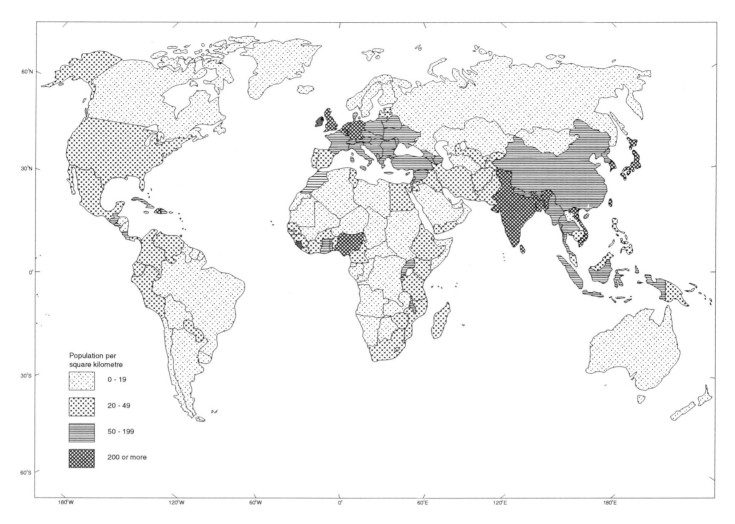

Figure 10.1 *Population density, calculated by dividing a country's population by its total surface area (square km of land and inland water areas). After World Bank (1992).*

Since the eighteenth century, world population has increased eight-fold and the average life expectancy has doubled. Two thousand years ago, the population was a mere 200 million: it took 1,500 years to double, the most recent doubling taking only 27 years. The joint US National Academy of Sciences and Royal Society of London document published in February 1992, using data from the UN Population Fund's 1991 report which noted an acceleration of population growth since 1984, and assuming a sustained decrease in fertility towards the replacement level of 2.1 offspring per woman per lifetime, stated that world population may reach 10.5 billion in 2050, with around 90 per cent of it in the developing countries. The United Nations Population Division also predict that 95

per cent of this expansion will take place in the less-developed countries such as India, China, Bangladesh, Pakistan, the Philippines, Indonesia, Vietnam, Iran, Mexico, Brazil, Egypt, Kenya, Tanzania, Zaire, Nigeria, and Ethiopia. Although the rate of increase in world population is slowing, the absolute numbers continue to increase and, therefore, exert social, environmental, and economic pressure on the available global resources.

The developing countries are growing most rapidly, while the developed countries have more or less stabilized (Figure 10.3) (Plate 47). World fertility and mortality data are summarized in Figure 10.4. In some developed countries, such as Germany, there has been a decrease in the birth rate over the past decade and the

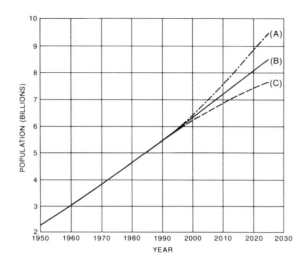

Figure 10.2 *Projected growth of world population at the current rate of 1.74 per cent per annum until the year 2000, and a decline thereafter of (A) 0.98 per cent per annum. (B) 0.59 per cent per annum. (C) if the growth rate were to increase to 1.9 per cent per annum. After Keyfitz (1989).*

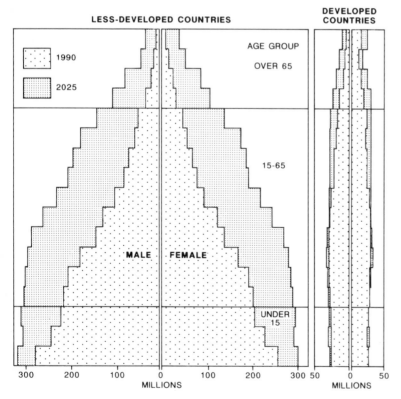

Figure 10.3 *Age distribution of populations of the less-developed and the developed countries in 1990 compared with that projected for 2025. In the less-developed countries, the population will continue to grow rapidly with an expanding labour force. The percentage of old people will also increase with respect to the young, requiring greater care. After Keyfitz (1989).*

population is declining steadily. The developing countries are least able and least likely to cope with the problems of feeding, and suppling housing and clothing to their people, whose numbers are expected to double in the next 100 years.

It is instructive to compare the estimated population growth and energy consumption for developed and less-developed countries as shown in Table 10.1. Clearly, if population growth is considered in terms of energy consumption, the more-developed countries are growing at a faster rate than the less-developed countries. Statistics similar to these could be calculated for the growth of pollution and waste disposal. Population growth, therefore, is not just a problem for less-developed countries, but will obviously have knock-on effects in developed countries.

Poor people have expectations on time scales different from those in the developed world. They have to consider their immediate needs, whereas planning for future generations and global considerations are unrealistic for them. Many live on the edge of subsistence, with their immediate concern being where their next meal is coming from or the yield and condition of the next harvest (Plates 10.1 and 10.2). In many developing countries, life expectancy is often much lower than in the developed world. According to United Nations estimates, in Africa and parts of Asia, child mortality rates still remain very high: in Afghanistan, 30 per cent of children die before the age of five, 19 countries throughout Africa will suffer a 20 per cent death rate of their children before reaching five years, and globally, more than 14 million children under five years die each year (World Resources Institute 1991). For those who survive, poor people also tend to be less educated, and are unable to afford the luxury of good medical care or schooling. All these factors hinder good management and sensible development, whatever these actually happen to mean. The transition to a more capitalist, market-led economy within China and the former Soviet Union has caused major political unrest within these very large populations. Tiananmen Square (1989; Plate 48) and the unrest and bloodshed in Moscow (1993) epitomize the potential scale of violence which could erupt.

Throughout much of the world, the change

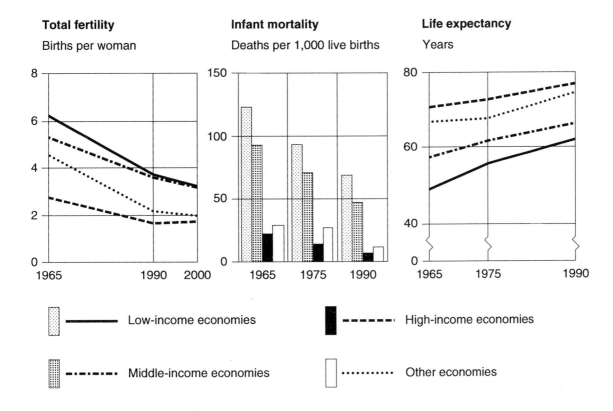

Figure 10.4 *World fertility and mortality. After World Bank (1992).*

to mechanized corporate cultivation techniques, replacing small family-run farms, has meant that large-scale intensive production methods are now commonplace. As a consequence, more of the agricultural land has degraded into desert, and where land is still farmed the land prices have soared so that people have been driven from the countryside to large urban centres. It is here that the displaced people suffer alienation and anomy. Rather than being producers, many more become dependent on others who produce. Urbanization is a phenomenon of the twentieth century. The epitome of the urban sprawl is Mexico City, which is the largest city in the

Table 10.1 *Per capita energy consumption and projected growth in per capita energy consumption at present rates for the developing and developed countries.*

	Total energy consumption (petajoules)	Total population in 1990 (millions)	Per capita energy consumption (petajoules)	Net population growth per 1000, 1995–2000	Growth of energy consumption per person at present energy consumption rates
Less developed	76,396	4,087.0	1.8×10^{-5}	19.2	3.5×10^{-4}
More developed (including centrally planned countries)	261,641	1,205.2	2.1×10^{-4}	4.0	8.7×10^{-4}

Source: World Resources Institute (1990).

Plate 10.1   *Scavenging is an everyday part of life for families, Communidad Veuda De Alas, municipal rubbish dumps, El Salvador.*

Courtesy of Rhodri Jones/Oxfam.

Plate 10.2   *Self-sufficiency practised at Communidad Santa Martha, El Salvador. Programmes such as these, which are self-initiated rather than imposed by external aid bodies, provide amongst the most sustainable means of development and, importantly, can empower communities to independence.*

Courtesy of Rhodri Jones/Oxfam.

world, with a population of about 19.4 million, and predicted to reach 24.4 million by the turn of the twenty-first century. The extreme crowding in Mexico City means that millions of people exist in squalid conditions of dire poverty, without an adequate supply of clean water and with poor sanitation. Crime and disease abound. Despite the overcrowding and bad living conditions, people still flock to urban centres such as Mexico City because there is the possibility of a job, better standards of health care and educational opportunities. For many, however, these opportunities remain unrealized. This scenario is repeated the world over in many lesser-developed countries, especially, but not uniquely, where populations are large and concentrated (Plates 10.3 and 10.4). Some cities, for example Hong Kong, have also attracted large

numbers of political refugees from politically unstable areas and regions under conflict. In Hong Kong, the influx of Chinese from the mainland and the Vietnamese boat people have caused many urban problems, associated with the rapid growth of population in the 1960s and 1970s (Plate 49). There is growing concern today that environmental refugees will be a continuing problem in the future, as sensitive areas undergo environmental changes related to population pressures.

Until the twentieth century, with its enormous strides in medical science that have allowed people to live much longer and survive illnesses that would previously have proved fatal, population levels did not pose a threat to the environment. Large families were common because of the high rates of infant mortality, the shorter

lifespan of individuals, and the desire for economic and family security in old age. Death, birth, and longevity conspired to maintain population levels at a fairly steady state, albeit with modest increases. Seventeenth-century Europe, for example, had sufficient available land to support a larger population than actually existed at that time.

What will limit population growth? Will it be a lack of food? Can societies feed these numbers and what will be the costs to the environment as more land is utilized to produce food? The world is crowded with starving people, yet there are many countries with a food surplus. The World Commission on Environmental Development (WCED) estimates that a five- to ten-fold increase in the world's economic activity is needed during the next 50 years to support this population explosion. Increased economy activity can only mean increased exploitation of raw materials, land, energy, and agriculture.

One of the exacerbating factors in some nations with high levels of malnutrition and starvation is civil war and politics, as in the Sudan. Urbanization takes people away from the land and, therefore, removes some pressure from the land as a resource. Urbanization, however, while removing some pressures from the land resource, tends to concentrate atmospheric pollution, with exhaust from motor vehicles, domestic and industrial processes, and causes greater stress-related problems. Ghettoes and crime become prevalent.

Urbanization allows easier access to medical care and by concentrating health and welfare resources supports larger populations. City environments, because of the greater degree of medical care and access to other social provision, can allow the population more easily to reach levels that cause serious environmental damage and pollution.

It seems that food alone may not be the brake on the growth of the world population. The world's current food production is actually

Plate 10.3   *Large urban areas flourish in both the developed countries (as in New York (A)), and in the developing countries (e.g. make-shift dwellings in Caracus, Venezuela (B)).*

Plate B courtesy of Rob Potter.

greater than the rate at which the world population is increasing. If food production can continue at present rates, it is estimated that there will be enough food for a stable world population of 10 billion within 100 years. However, it is estimated that the world population may reach the 10 billion mark within the next 50 years (see Figure 10.2). Instead, it may be that human activities stress the environment to a point where it cannot recover quickly enough to support the population levels. Certain natural resources may become so depleted that a lack of suitable substitutes could precipitate crises. The exhaustion of fossil fuels such as oil provides one such example of a scenario in which the world economy could quickly reach a crisis point and go out of control.

Economic security seems to be the main reason for a reduction in the birth rate. In general, the least-developed nations have the highest birth rates. Ironically, large families actually do offer a cushioning from the vagaries of poverty. Despite this, many families in developing countries remain unable to support themselves adequately. Where both families and nations desire lower birth rates, family planning programmes have been instigated. Nations have to be economically and socially receptive to the use of family planning. For these reasons, family planning programmes, with the widespread distribution of contraceptives, have met with little success in Pakistan, Kenya, and Nepal, whereas they have proved successful in China, Indonesia, Thailand, and South Korea.

The underlying debates over birth rates, world population, and a need for more efficient, less harmful use of the environment, are never-ending. There are those who claim that the actual population levels are not the real problem, but rather the use of inefficient and bad farming practices which create unnecessary food shortages. A contrary perspective is that inefficient and bad practices exist and need rectifying, but should not be cited as an excuse for inaction in controlling population levels. There are religious groups who argue that it is morally wrong to control the growth of populations artificially through the use of contraceptives. Individuals and policy makers have to take these arguments on board and, naturally, implement action, if any, which is appropriate and acceptable within their cultural, social, and economic framework. There are no easy answers here. Perhaps the most important aspect in the control of population is to educate individuals so that they become more aware of how their child-bearing and child-rearing affects their country's socio-economic fabric and environment, as well as the larger-scale global environment – the latter being of little interest to humans in dire poverty.

William C. Clark (1989) has likened the growth of the world's population to the growth of bacteria in a laboratory culture. Bacteria in cultures grow rapidly from distinct nuclei, expanding and encroaching on other colonies. Eventually, the bacteria die through competition and inadequate food supplies. Likewise, human populations increase from centres, and settle-

Plate 10.4 *The daily commuter crush on trains in Tokyo. Urbanization and overcrowding, causing environmental stress and associated problems, are not restricted to the poor, developing countries, but affect even the most affluent nations.*

ments expand in size from villages to large metropolises, creeping along communication links, continuously expanding and growing as pressures increase for living space and land to produce food. Will the global human population eventually die as it increases further? Will this expansion result in conflicts such as more wars and greater food shortages? Will pollution also limit economic growth, and reduce life expectancy by as much as 50 per cent in the twenty-first century (Meadows *et al.* 1972)?

The view that population growth will limit itself is not new. As early as 1798 the English economist, the Reverend Thomas Malthus argued that population growth would limit itself because of the finite food supplies. He argued that if population growth exceeded this critical level, then the scarcity of food would cause famine and war, thereby reducing the population to some sort of equilibrium level. Other cata-strophes, such as plagues and disease, would further limit population size.

Modern adherents to **Malthusian views**, the so-called neo-Malthusians, believe that accelerating population growth will lead to an increase in pollution and environmental degradation (e.g., Ottaway 1990). Contemporary neo-Malthusian organizations, which include Population Concern, the United Nations Population Fund (UNFPA), the World Commission for Environment and Development, and the Worldwatch Institute, believe that the lowest common denominator causing environmental degradation is too rapid population growth. Opposed to these views are the anti-Malthusians who believe that it is the distribution and organization of people, not the absolute numbers, that cause the problems. The latter group further argue that this then leads to an inappropriate use of technology, overconsumption and inequalities in wealth and life chances, which ultimately induce environmental degradation.

Neo-Malthusian views were revived in the late 1960s and early 1970s with the publication of Paul Ehrlich's book *The Population Bomb* (1968) and the report to the Club of Rome *Limits to Growth* (Meadows *et al.* 1972). The latter book was based on a computer model which predicted that the world population would deplete the present world resources within the next 100 years. Ehrlich's book suggested that fast-growing population meant that the Earth's capacity to feed humans had almost been reached and a minor change in agricultural productivity, such as the failure of a monsoon in Asia, would plunge the world into political chaos which would result in famines, food shortages, and wars – a doom-and-gloom scenario of cataclysmic proportions.

A contrary perspective was put forward in 1981 by J. Simon in his book *The Ultimate Resource*. Simon suggested that population growth does not in itself present an intractable problem for the future of the Earth, but rather that an increase in population would lead directly to higher living standards and economic development. This improved lifestyle and life chances would come to fruition, because humans would be forced into ever-greater innovation and technological advances to meet the additional need to sustain the increased population. These arguments represent a very optimistic view of human survival and embody the belief that humans will always be able to adapt to new and changing global problems.

Geographers such as Moore Lappe and Schurman, in their book *Taking Population Seriously* (1989), were also opposed to the neo-Malthusian views and arguments on the basis that they are oversimplistic and deterministic. Instead, Moore Lappe and Schurman considered what they perceive as the causes of rapid population growth. Their analysis led them to the conclusion that population problems are predominantly the result of imbalances between people's reproductive choices, or rather the lack of them. They emphasized the fact that many women have poor and frequently subservient economic, social, and cultural positions which often leave them with little choice other than to reproduce. Furthermore, they suggest that one of the main ways to reduce rapid population growth is to improve the socio-economic and political conditions and rights of women, for example, through improved education, wealth creation, and the sharing of that wealth. Moore Lappe and Schurman believe that this would then release women from their social and economic deprivation, which in turn would tend to favour smaller families.

In an ideal world, individuals and family units should be allowed to make choices about birth control in the light of as much information as possible. In reality, such an argument is untenable and impracticable. Despite the fact that it is impossible to make individuals fully aware of the consequences of their actions, this is no justification for under-education.

AIDS

Not only is AIDS a global issue of great concern, because of the personal risks to health and as a threat to the health of communities and nations, but it is also a global environmental issue since its spread may affect the sustainable development of some developing countries, for example in parts of Africa.

It was only in 1981 that scientists first recognized AIDS, the acquired immune deficiency syndrome. Already 10–12 million people have been infected with the human immunodefi-

ciency virus (HIV) which leads to AIDS. The World Health Organization predicts that this figure will treble within the next eight years, with up to 40 million people being infected by the year 2000. Unfortunately for many people in the developed countries, it takes the death of international celebrities such as the film star Rock Hudson in 1990, or the rock star Freddie Mercury in 1992, to highlight one of the greatest problems facing humankind today. The spread of HIV has still not been fully assessed, mainly because such studies involve gathering data on people's personal sexual activities and preferences, together with an in-depth understanding of complex behavioural and socio-economic factors.

HIV causes the collapse of the body's immune system, which then makes it fatally susceptible to any infection that a healthy person would fight with ease. AIDS is extremely complex and is not a single virus. In Africa there are at least 100 different strains. The virus attacks T4 helper **lymphocytes** and other cells of the immune system which are present within the bloodstream. These cells play a key role in fighting diseases and are required as a fundamental part of the human immune system. HIV can only replicate itself inside human cells. In order to do this, the HIV inserts its genetic material into the healthy cells as a genetic parasite. HIV enters the bloodstream in blood from an infected person. This infection can happen in a number of ways: during intravenous drug use, including from affected needles; during sexual activity, including heterosexual acts; or from an infected mother to her baby before the child is born.

Once in the bloodstream, HIV may lay dormant for a latent period of 10–15 years. The virus is then activated by secondary co-factors, which are not yet fully understood or even properly identified, but some have suggested may include conditions such as gastroenteritis, tuberculosis, thrush, or malnutrition. Once activated, the T4 helper cells become depleted in number. When their count falls below a certain threshold the person is said to have developed full-blown AIDS. At this stage they are susceptible to every sort of infection and death will occur as a result of the patient contracting any complication, even a cold. HIV is very variable, developing into many strains in different people within the same city. So, a vaccine against one strain may be ineffective against another.

HIV is known as a **retrovirus**. All organisms, apart from certain viruses, contain their genetic material in the form of **DNA** (deoxyribonucleic acid). Retroviruses are different because their genetic material is in the form of **RNA** (ribonucleic acid), so they have a special enzyme to convert the RNA into DNA which is called reverse transcriptase. HIV enters a cell by attaching itself to a so-called receptor molecule on the surface of a cell. Current research suggests that this receptor is something named CD4, present on the surface of T helper lymphocytes, **macrophages**, **microglial cells** in the brain and **dendritic cells**. Inside a cell, the active HIV manufactures more viral proteins to reproduce more viruses. The new viruses, called virions, bud off the cell's outer membrane and go on to infect further healthy cells.

There are two different HIVs, HIV-1 and HIV-2. HIV-1 has a worldwide distribution, whereas HIV-2 occurs mainly in West Africa. AIDS results from both strains of HIV, but HIV-2 appears to take longer to do so. The origin of HIV is not known. A whole new science has developed to answer this question of provenance called **seroarchaeology**. To date, the oldest recorded case is from a man in Zaire in the 1950s. The main spread of the virus was in Africa in the 1970s. Table 10.2 shows the global HIV positivity rates and the number of AIDS cases reported from several African countries. The HIV positivity rates are frightening as the potential number of AIDS cases in the next 10–15 years is projected to a staggering level. The World Health Organization (WHO) estimates that the spread of the virus in Africa is doubling every 34 weeks, and the transmission of the virus to babies is approximately 30 per cent efficient. The WHO worst-case scenario predicts that 40 million people will be HIV positive by the year 2000, with 6.5 million cases in Africa. Measuring the positivity of the AIDS virus is difficult because of sampling bias, including screening of volunteers, army personnel, prostitutes, blood donors, and mothers tested soon after they have given birth. The latter group may prove most reliable for estimating future trends since they are usually the most representative of the global population.

These frightening statistics predict large mortality rates in the next few decades. This will have profound psychological, social, and economic effects. It will be particularly devastating in Africa where the rates are among the highest and economic and social welfare are least. The control of the spread of the AIDS virus is also more difficult in these countries. The propagation of knowledge about the AIDS virus and its control is difficult in Africa because of the overall poor level of education and communication. The use of condoms to reduce the spread of the virus by the transmission of bodily fluids during sexual activity is still not widely accepted in many

Table 10.2 *(A) Global HIV positivity rates. (B) AIDS cases reported and case rate in Africa. (C) Incidence of AIDS in developing countries in 1992.*

(A)

	Males	Females
Sub-Saharan Africa	1 in 40	1 in 40
North America	1 in 75	1 in 700
Caribbean/Latin America	1 in 125	1 in 500
Western Europe	1 in 200	1 in 1,400
Oceania	1 in 200	1 in 1,400
Asia and South-East Asia	1 in 2,500	1 in 3,500
Eastern Europe/former USSR	1 in 4,000	1 in 20,000

Excludes AIDS sufferers. South African data not included

(B)

	No. cases reported	Case rate
Namibia	311 (1985–90)	6.5/100,000
Malawi	12,074 (1990)	37.1/100,000
Botswana	216 (1991)	2.3/100,000
South Africa	1,011 (1991)	0.9/100,000

Source: WHO (1991).

(C)

Country	AIDS cases per million head of population in 1992
United States	1,200
Spain	441
Switzerland	417
France	403
Italy	272
United Kingdom	120
Sweden	89

Source: R. Anderson (1993).

African countries. To compound the situation, condoms are often unavailable. Locally, the social mores mean that people are often unwilling to use them, as condoms have been traditionally associated with birth control and are considered to be the white person's means of suppressing the black African. In addition, sexual behavioural patterns and fertility rites differ from those in many western developed and industrialized societies, something that further accelerates the spread of the AIDS virus.

Although there is much research into HIV and AIDS, it seems unlikely that a comprehensive vaccine will be developed within the next 10 years. AIDS, therefore, is one of the biggest threats to humankind in the near future. It may yet prove to be another sinister Malthusian check on population, but to regard AIDS as an acceptable and natural way of controlling population would be tantamount to adopting a neo-Nazi philosophy. Probably the main reasons for the

lack of action on AIDS in Africa is a lack of money, where many basic health problems compete for a relatively small pot of money, such as vaccination programmes, and improving water quality for human use.

There has been much discussion recently over whether epidemiologists have greatly exaggerated the potential spread of AIDS into the heterosexual community in developed countries, and even a denial that there is an epidemic in sub-Saharan Africa. In developed countries, there are relatively sophisticated statistics that reveal the temporal evolution of AIDS within different risk groups. Since HIV has an incubation period of up to about 10 years, the current reported incidences of HIV infection have a large margin of error associated with any figures, making it difficult to model the spread of HIV. In developed countries, there has been a steady increase in reported incidences, but there are large differences between countries (Table

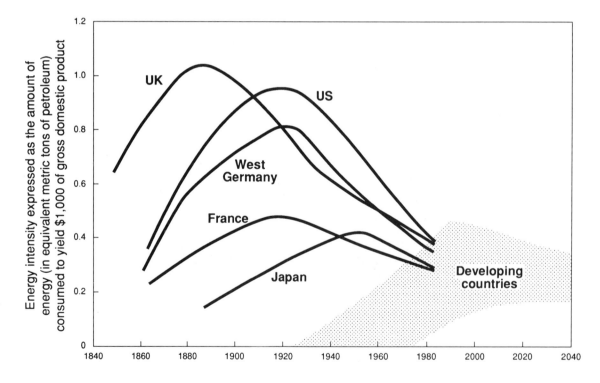

Figure 10.5 *Energy intensity versus time in industrialized and developing countries. In industrialized countries, the energy intensity ratio (ratio of energy consumption to gross domestic product) rose, then fell. Because of improvements in materials science and energy efficiency, the maxima reached by countries during industrialization have progressively decreased with time. Developing nations can avoid repeating the history of the industrialized world by using greater energy efficiency. It is unrealistic, however, to expect developing countries to reach an energy-efficient development path very quickly, given the capital constraints and industrial weakness these countries confront. After ASCEND (1992).*

10.2C). R. Anderson (1993) highlights this, and argues that over 1992 there has been a significant increase in incidences (24.8 per million) in the heterosexual group, suggesting that the epidemic is still in its early stages. Estimates of future incidences are based on mathematical models which attempt to mimic the transmission of the disease, and such models are revised as more data become available. Back-calculating methods such as this show that the rate of transmission has been slowing down in two of the major risk groups, homosexual men and intravenous drug users, whereas the incidence in heterosexuals has been increasing since 1984. Reported incidences of HIV infection within the developing countries, particularly throughout Africa, are very difficult to calculate because of the poor databases.

The rise of the consumer society

The rate of increase in the consumption of natural resources is a cause of concern to environmentalists. Since the beginning of this century, energy consumption has increased by 80-fold, manufacturing by 100-fold, and over 9 million square km of land (much of this was forest) has been converted to agricultural land. Patterns of energy use are very different in the developed, industrialized, and developing countries (Figure 10.5). Water resources and water quality also show significant differences between the developed and developing countries (Figure 10.6).

Withdrawal of water by human activities has increased annually by more than 100 square km and the amount of sediment loads in rivers has

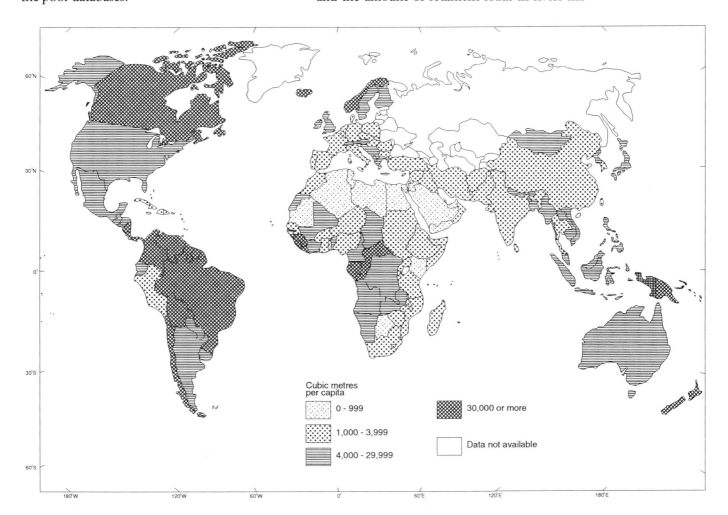

Cubic metres per capita

0 - 999	30,000 or more
1,000 - 3,999	Data not available
4,000 - 29,999	

Figure 10.6 *World map showing annual renewable water resources. The average amount of water available per person per year is calculated by dividing a country's annual internal renewable water resources by its population. After World Bank (1992).*

risen by 300 per cent since the eighteenth century. In addition, since then, industry has polluted the environment by doubling the amount of methane in the atmosphere, adding an extra 25 per cent of carbon dioxide, increasing metals such as lead, cadmium, and zinc 18-fold, and metals such as arsenic, mercury, and nickel by two-fold. This is particularly disconcerting as the polluters are currently only a small proportion of the world's population; these are mainly the people of the rich industrialized countries. Gross domestic product (GDP), expressed as either an absolute figure or per capita, shows a very large difference between the rich and poor nations (Figure 10.7), and clearly without sufficient economic power it is extremely difficult for the developing world to improve overall life chances and economic prosperity, and deal adequately with pollution. Virtually any environmental indicator at different national income levels reveals the gross imbalance between rich and poor, developed and developing countries (see Figure 10.8).

Prior to the 1970s, natural resources were seen both as plentiful and essentially limitless, but with the publication in 1972 of the report *Limits to Growth* presented to the Club of Rome (Meadows *et al.* 1972), this widely held belief began to change rapidly. The report claimed to show that existing patterns of global resource use would lead to a collapse of the world's socio-economic and political systems within the next century. The report altered many people's perceptions about sustainable growth, and focused attention on the scarcity and finite nature of many natural resources. In turn, the changed economic mood led to the stockpiling of commodities throughout the 1970s and early 1980s. As a result of this siege mentality, commodity prices rose sharply, in part caused by the large increase in oil prices in the early 1970s, amid fears and beliefs amongst oil producers that there would be cartels to rival the Organization of Petroleum Exporting Countries (OPEC). The 1975 second Club of Rome report, *Mankind at the Turning Point* was much less gloomy over these issues.

Today, opinion and mood have shifted about the reserves of non-renewable natural resources. Improved methods of assessing the amount of

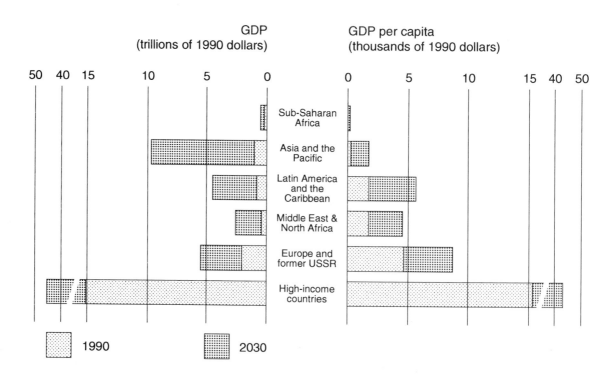

Figure 10.7 *GDP and GDP per capita in developing regions and high-income countries, 1990–2030. Data for 2030 are projections. After World Bank (1992).*

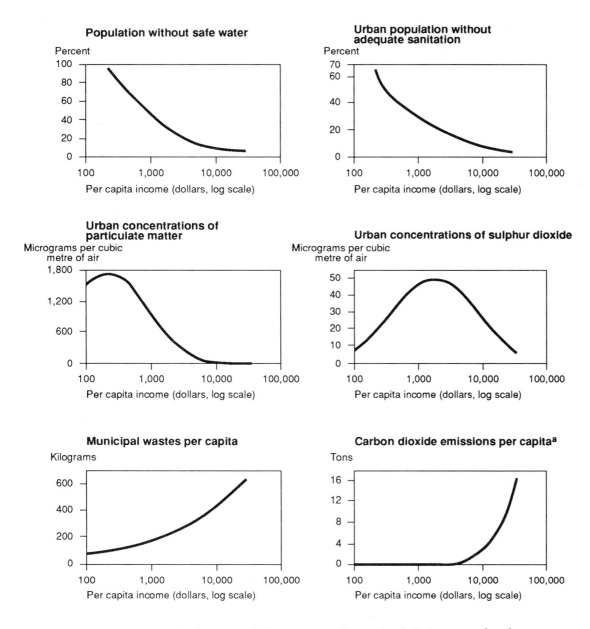

Figure 10.8 *Environmental indicators at different country income levels. Estimates are based on cross-country regression analysis of data from the late 1980s. a = emission levels from fossil fuels. After World Bank (1992).*

various resources, particularly non-renewable energy resources, together with a prolonged economic recession in the West, especially in the late 1980s up to the present day, have resulted in a subtle change of attitude towards a perception that although resources may be finite, many will last longer than was projected in the 1970s. This changed global view, and the economic recession in the West, produced a glut of many minerals on the world market and commodity prices correspondingly collapsed in the early 1980s. This was particularly devastating during the large fall in crude-oil prices in 1986. It is now perceived that the low demand and low prices of exports are a major world economic problem. This has led to a decline in government revenues, widespread unemployment, and little prospect of economic growth. Particularly badly hit are the developing countries, especially much of Africa and South America, which desperately need foreign currency to help pay off their national debts. The current world surplus in many commodities has caused a shift in international debate from the finite nature of many resources to an emphasis on the distribution of these resources, and the concept of sustainable development.

As consumers, human activities waste vast amounts of energy, emit large amounts of pollutants into the atmosphere and hydrosphere as by-products from manufacturing and agriculture, and produce mountains of refuse from overconsumption. It is estimated that the average person in developed countries produces 2–3 kg of refuse each day. In developing countries, it is less than 1 kg, with much of the refuse being recycled or utilized in other ways. Ironically, the greatest dissemination of knowledge concerning the consequences of human actions occurs in the developed countries, yet it is these developed countries which already have the technology, capability and capital to control and manage these activities correctly. All too often, profit motives outweigh environmental concerns. Unfortunately, as pressure grows on developing countries to mimic the developed countries' path of progress, to catch up and become consumers, the consequences for the natural environment may become irreversible and catastrophic.

It is heartening, however, that not all the adverse trends are still increasing. Over the past few decades, for example, the rate of increase in human-induced extermination of vertebrates, especially marine mammals, has been slowing down as a result of increased pressures by organizations such as the World Wide Fund for Nature (WWF). The WWF, however, estimates that humans are still exterminating approximately 1 per cent of the world's species annually. Many species, for example of whales and dolphins, are threatened by thoughtless human activities.

The release of sulphur, lead, and radioactive fallout has also declined due to government and inter-governmental legislation, with the instigation of monitoring and control of the release of pollutants from factories. The use of lead-free petrol is encouraged, through favourable pricing and information. The testing of nuclear weapons has been outlawed.

Agro-economics

For many developed, industrialized countries,

the share of agriculture in the GDP is generally less than 10 per cent (see Figure 10.9) – calculated by taking the value of an economy's agriculture sector and dividing it by gross domestic product, a figure that says nothing about absolute values of production. Even today, humans are unable to manage present global population levels. One-quarter of the world's population (1,250 billion people) goes hungry during at least one season of the year. At the same time, food mountains grow in the European Community, just to control the food prices in wealthy countries. More than one-third of the world's population lives below the poverty level. The world's population has been divided into two classes: the affluent peoples of developed countries who dominate the northern regions of the globe, and the poor peoples of the developing countries who are mainly in the southern regions of the globe. This North–South division is getting wider; while the North gets richer the South gets poorer. Agro-economics seem to skew towards feeding the northern hemisphere at the expense of much of the southern hemisphere.

Data from the United Nations Environment Program (UNEP) suggest that approximately 60 per cent of the estimated 3.3 billion hectares of agricultural land not in the humid regions are affected to some degree by desertification. The validity of these figures can be questioned. The definition of desertification and the criteria used to assess it are negotiable. Nevertheless, this large figure suggests that desertification is one of the major problems facing us today (Crosson and Rosenberg 1989). If societies are to feed the population of the world, then soil productivity has to be maintained. The destruction of once fertile land through poor management can lead to accelerated erosion, waterlogging, and salinization of irrigated land, and degradation of farmlands or range lands in the arid, semi-arid, and sub-humid regions, as well as reduction of the genetic diversity of its plants and animals. These deleterious effects of bad management degrade the land, ultimately converting it to a desert in a process now referred to as desertification.

Agricultural problems have increased since the Second World War. Trends in farming practices have meant a shift towards mono-cropping, further resulting in a reduction of genetic diversity in any region. Vast areas of forest have been cleared resulting in such adverse effects as soil erosion, flooding, and landsliding. The use of pesticides and fertilizers has begun to poison water supplies, as well as many animals which form vital links in food chains and food webs. With environmental degradation come the problems of social instability, local and regional conflict, further depletion and degradation of resources, hunger and death, and environmental refugees seeking new lands. These are the grapes of wrath that the environment gives back in return for careless management of land resources.

Civil wars and politics conspire to leave agricultural land undeveloped, destroyed, and in some regions overintensively farmed. In addition, urbanization accounts for increased migration from rural to urban areas, attracting peasants from their farms to overcrowded cities. The continued growth of economic blocks with protectionist policies and associated trade barriers has created a climate of distrust between the USA and the European Community, something that has been the subject of considerable debate and diplomacy, for example in the November 1992 Uruguay Round of General Agreement on Tariffs and Trade (GATT) discussions. In Europe amongst the member states of the EC, there is also a growing consensus that the EC Common Agricultural Policy (CAP) is in need of change. In the EC, the 1992 CAP budget was 36 billion ecu, or US $46.3 billion, out of a total

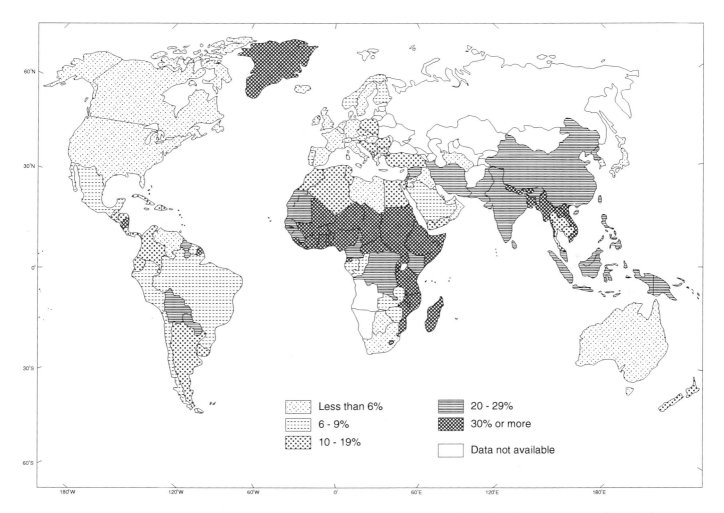

Figure 10.9 *World map showing the share of agriculture in GDP, calculated by taking the value of an economy's agriculture sector and dividing it by gross domestic product. The shares say nothing about absolute values of production. For economies with high levels of subsistence farming, the share of agriculture in GDP is difficult to measure because of problems in assigning subsistence farming its appropriate value. After World Bank (1992).*

62.5 billion ecu, and since 6.8 per cent of the European population are engaged in farming they receive 55 per cent of the entire EC budget. It is for reasons such as this excessively high financial burden that the CAP needs reform.

It is the responsibility of the industrialized countries to help their developing counterparts. Ironically, the richest 15 per cent of the world's population consume half the world's energy and use more than one-third of its fertilizers. The vast majority of people in the developing countries use little or no fertilizer, rely heavily on human and cattle power, and use little energy for lighting, cooking, and heating. One-quarter of the world's population consumes 80 per cent of its goods and owns three-quarters of its wealth.

On a more optimistic note, a new US-based centre intends to help developing countries take advantage of new advances in agricultural biotechnology, with support from foundations and private companies. In April 1992, the International Service for the Acquisition of Agribiotech Applications (ISAAA) announced that its base for the next five years is to be in the USA in New York State at Cornell University, Ithaca. Amongst the aims of the ISAAA are the intent to help farmers increase crop yields but, at the same time, reduce their dependence on pesticides. Figure 10.10 shows the recent percentage increase in global production of cereal crops, measured against the growing world population (also expressed as a percentage), the improved yields coming about through the greater use of fertilizers, pesticides, the introduction of new strains of high-yield crop varieties, and better land management.

Many previous attempts to transfer technology to developing countries have failed through a lack of money, technical skills, and infrastructure in those countries. The ISAAA hopes to remedy this by tapping the extensive knowledge held in many private companies and transferring it to developing countries where there is a politically acceptable climate and without disrupting the traditional agricultural practices. The organization has already raised millions of US dollars, and will concentrate on 10 target countries within Latin America and the Middle East/ Africa, and Asia. Collaborative projects on plant biotechnology in Mexico, Taiwan, and Costa Rica have been instigated, with many others under negotiation.

Agenda for action: sustainable development

If societies are going to manage this planet more thoughtfully, there are two main questions that need to be asked: 'What kind of planet do we want?' and 'What kind of planet can we get?'

These questions are central because they ask what size of population the planet can support, and what size of global population can realistically be sustained. How much poverty is acceptable and what living standards are achievable? How much climatic change is tolerable and how much climatic change is unavoidable? How

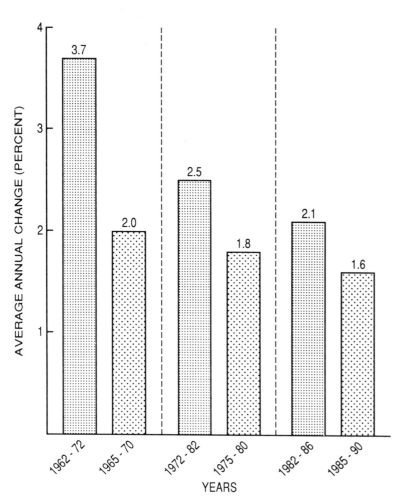

Figure 10.10 *The increase in total production of cereals (dark shading) and the world's population (light shading). After Crosson and Rosenberg (1989).*

much pollution can be tolerated, and can the oceans or the Earth's interior store anthropogenically created pollutants, and will the atmosphere disperse them? How will human actions help support maximum global biodiversity? Can we conserve many of the endangered species under the threat of climatic change, pollution, and land degradation, and is it desirable to do so? Easy as it is to ask these questions, any responses represent the outcome of very many complex issues relating to individuals' and societies' perceived and actual vested interests.

These questions need to be answered soon, as the rate of change is fast. Time appears to be against us. What used to be local incidences of pollution or degradation now involve several nations and may have global consequences. If individuals and societies are going to answer the panoply of questions raised above, then there is a vital need to understand the complex interactions within each of the global systems such as atmospheric or oceanic-current circulation. Science needs to provide us with information that permits policy makers to make sensible decisions about the environment. Technologies need to be developed and implemented to help conserve resources, prevent unnecessary pollution and help restore the environment wherever appropriate. Also, international co-operation must continue and develop so that there can be effective global environmental management.

The idea of **sustainable development** was first presented in an international forum in the World Conservation Strategy, by the International Union for the Conservation of Nature (IUCN 1980). In essence the concept of sustainable development invokes present development of available resources without compromising the ability of future generations to meet their needs. There are many people who would argue that this is an abstract ideal which is impossible to meet fully. How can this generation understand the needs of future generations, even before they have been born, yet alone formulate their needs! The practical application of the notion of sustainable development, however, must involve a greater environmental awareness, both from governments and individuals. The IUCN argued that three priorities should be incorporated into all development programmes. These included maintenance of ecological processes, the sustain-

able use of resources, and the maintenance of genetic diversity. Emphasis was placed on maintaining the present physical environment and the report was therefore criticized as being anti-development. Critics also argued that the report concentrated too much on attacking the symptoms of environmental degradation rather than analysing the causes. In addition, it was also criticized as having an anti-poor bias, as the report suggested that poverty was one of the major inhibiting factors in achieving sustainable development. These criticisms led to the reformation of the IUCN in 1984 and the creation of the World Commission on Environment and Development (WCED), which became known as the Brundtland Commission after its Chair, Gro Brundtland of Norway. The reports of the Brundtland Commission aimed at maximizing growth without jeopardizing people and future resources. Emphasis was placed on economic quality being just as important as quantity.

The aims of the WCED, as stated when it was established in 1983 as an independent commission by the United Nations, were to make recommendations which can be used for sustained global development. The WCED included 23 commissioners from 22 countries. Their global enquiry into the state of the world involved analysis of available information, the commissioning of reports by specialists, debates with world leaders, panels of experts, public inquiries, and across-the-board interviews. This provided a broad spectrum of data and opinions. In October 1987 the Commission published its report, entitled *Our Common Future*. It emphasized that the basic needs of all people must be met to secure their common survival; that the poor should be given economic priority, not just through any altruistic reasoning, but because they have the potential to help world economic growth by providing additional markets which would ultimately improve the world economy.

The report presented strategies for sustained development. It recognized three major obstacles that would hinder sustained growth. These were social, institutional, and political in nature. It emphasized that an integration of both economic and ecological systems is paramount to the success of sustained development. The Commission further suggested that ministries of finance and ministries of the environment should not be isolated, but combined, sharing the responsibilities of development programmes. It argued that financial gains must offset the environmental damage that economic activities inflict on the environment. Also, the exploitation and depletion of natural resources should be at a rate that is not greater than the rate at which they can be replenished naturally, allowing the environment to recover so that future industries will have raw materials. It emphasized that industry must play its part in replacing resources and restoring the environment, in cleaning up its proverbial act. The report, therefore, argued that poverty, resource depletion, and environmental stress arise from disparities in economic and political power. Sustainable development at a global level can only be achieved by major changes in management of the planet, something that must involve a new global psychology, that is, a fresh way of thinking about economic change and society's relationship with the environment.

To achieve these goals, the report suggested that changes within the present political, economic, and technological systems must be made. This would allow effective citizen participation in decision making, and provide the capacity for self-correction. It would also allow economic systems to generate surplus commodities, and provide freely exchangeable technological know-how, in order to encourage self-reliance and sustained development. Additionally, organizations should be instigated that encourage and permit the common, international discourse aimed at solving global problems that arise out of the present form of disharmonious, nationally oriented economic development.

Our Common Future presented an ambitious and broad agenda for trying to achieve global sustainable development. Unfortunately, the report failed to identify the specific barriers that inhibit the changes that are needed. In other words, the idealistic goals are well defined, but the mechanisms by which these may be achieved were left as vague statements. Perhaps the most significant aspect that arose from the report was the realization by governments and international agencies that sustainable economic growth cannot be considered seriously without addressing the environmental consequences of any action.

In a major study by the London Environmental Economic Centre (Pearce *et al.* 1989), three major themes were identified as lying at the heart of sustainable development. The first was a consideration of the proper value of the environment. Problems in evaluating the value of the environment and assessing the seriousness of resource degradation are immense, and no present framework for this exists. Clearly, there is a need for just such an institution, with well-defined guidelines for undertaking such a task.

Box 10.1

Recommendations of the Stockholm Initiative on Energy, Environment and Sustainable Development (SEED), November 1991

Developing countries

1. Developing-country governments should develop and implement programmes for improving power sector efficiency, both in supply and demand. These programmes should focus on greatly improved performance compatible with an integrated energy strategy and environmental sustainability.

2. Developing-country governments should support efficient alternatives to capacity expansion for utilities through better utilization of existing capabilities, and the development of independent private power facilities. Tariff reforms that make the sector creditworthy should be an integral part of such measures. However, the political, economic, and social conditions in individual countries underscore the need for a country-specific approach in addressing these issues.

3. Developing-country governments should strengthen financial mechanisms, institutions, and associated policies and regulations to provide innovative lending in supply and demand side power sector efficiency, including direct lending for private sector initiatives. Sector financing entities, including development financing institutions with portfolios in industrial modernization, agriculture, the environment, and housing, are targets for such institutional reforms.

Bilateral and multilateral institutions

1. Bilateral and multilateral institutions should dramatically alter their investment priorities to support end-use efficiency, sustainable and reliable operations and maintenance programs, and private sector initiatives, in addition to traditional investments in supply.

2. Bilateral and multilateral institutions should provide financial and technical support to

The second theme is the need to set realistic short- and medium-term temporal frameworks for implementing change. A part of this issue involves the accurate prediction of future needs, especially as resource values change. Finally, there is the need to provide for the least advantaged in society, in order to reduce the gap between rich and poor. These are difficult challenges in a competitive world where, all too often, personal and national greed exist.

One of the most popular definitions of sustainable development was submitted to the United Nations General Assembly in the WCED report (1987, p. 43) *Our Common Future* (see above), as 'development which meets the needs of the present without compromising the ability of future generations to meet their own needs'. A definition like this cannot be the basis of a theory and, indeed, the multiplicity of possible interpretations allows anyone to construe it to mean virtually anything they like. In other words, it is a slogan liberally laced with rhetoric for politicians to use but has little tangible meaning.

Sustainable development, for many, carries the connotation of economic growth without deleterious effects on the environment and human beings. Development without destruction is another concept of sustainable development, as espoused in a report by the UK Department of the Environment (1988).

The actual meaning of development is highly emotive. Seers (1977) has argued that if poverty, inequality, and unemployment are decreasing without a loss of self-reliance, then development is taking place. Of course, this concept is a quagmire of meanings and ambiguity. What is one nation's perception of development may not be another's. To some degree or other, most nations in Europe, and North America, would subscribe to a view of development as focused upon industrialization, urbanization, and democratization within a capitalist economy. Many nations in the developing countries might take a contrary view. The definition is not purely one of semantics. For international dialogue to take place at a level that can foster good will and peaceful coexistence between nations, it is imperative that each understands the others' aspirations and meaning of development. These issues are beyond the scope of this book, but they are tackled in some depth in W.M. Adams' book, *Green Development: Environment and Sustainability in the Third World* (1990). In this book, Adams argues that sustainable development appears to be acceptable to many governments, such as the British government, precisely because in their interpretation it does not demand a radical change of policy direction. Of course, this is the crux of the matter. If sustain-

---

improve the legal and regulatory framework as well as the management and institutional performance of power utilities.

3. Bilateral and multilateral institutions should expand their financing to cover joint ventures in environmentally sound electric power-related technology co-operation.

4. Bilateral and multilateral institutions should provide insurance for private sector power projects to enable capital mobilization from commercial and other markets.

5. Bilateral and multilateral institutions should commission a study investigating the lack of progress of private sector involvement in developing country power sectors.

6. Bilateral and multilateral institutions should create a fund in specific countries to support the availability and delivery of critical spare parts to ensure high system availability.

Institutional linkages

1. Bilateral and multilateral institutions should, together with developing countries, perform long-term power and environmental sector appraisals to formulate policy reform packages and investment priorities for public and private entities.

2. Existing bilateral and multilateral networks in energy and environment should be strengthened and expanded to link with developing country financing institutions and recognized centres of excellence.

3. The SEED recommendations should be widely disseminated to relevant agencies, developing country governments, and the private sector, they should also be presented to the refocused World Bank/UNDP Energy Sector Management Assistance Program (ESMAP) programme and its consultative group of donors and developing country representatives.

Box 10.1
Recommendations of the Stockholm Initiative on Energy, Environment and Sustainable Development (SEED), November 1991

able development can only be achieved by governments radically changing their policy direction at some considerable political cost to themselves, then how many will actually do so? It seems more likely that in such a scenario they will argue for minimal change and mortgage the problems requiring radical reform to some future and unspecified date.

Allowing for the enormous disparity in meaning over the term development between various individuals, organizations, agencies, and governments, there are perhaps some predictable mores. Most governments in office would probably argue that sustainable development or sustainable growth is realistic and that it forms an integral part of their policy direction. On the other hand, many environmental pressure groups argue that sustained growth, while achievable in the near future for some countries, is not possible for many without deleterious effects upon the environment and for other humans in different parts of the world. Not only are these different perspectives semantic but also they reflect fundamentally different sets of values. In many ways, the future survival of many ecosystems may well depend upon the coming together of such diverse perceptions and goals. Any mature judgement of such adopted positions must accept that the way forward lies less in openly hostile criticism of one group by another, environmentalist against government, and more in the mutual exploration of areas of common concern and co-operation. Environmentalists and governments can and must work together to conserve the natural environment and provide an acceptable level of sustainable development, something that may not be synonymous with sustainable growth.

Development should not be restricted to the relatively affluent countries, but seen as a global concern and target. The United Nations is the obvious international forum to encourage these aims and turn an ideal into reality (Plate 10.5). The widely perceived view of a North–South divide, portrayed by Willy Brandt in reports published in 1980 and 1983, and which laid the foundations for much of the WCED report (1987), needs to be eliminated. The South is frequently depicted as in a state of economic, political, and social crisis, and subject to extreme environmental risk. It is also portrayed as subject to problems of debt, and periodic hyper-inflation, growing poverty, repeated failed crop production, and falling commodity prices. The North is often shown as the antithesis of the South, with problems that are surmountable over much shorter time spans. These views, while not always accurate, have a large measure of truth in them.

The fundamental causes of the present North–South dichotomy lie in a number of reasons, including the widely held belief that colonization and imperialism over many centuries has been the major influence. In the early 1980s, the world lurched back into a major recession following the previous recession of 1974–5, and with it the amount of foreign aid and commercial loans made available to the developing countries decreased substantially after 1981. The indebtedness of the developing countries correspondingly rocketed. Against this economic panoply, the early 1980s saw a major failure of agriculture to keep pace with the production of foodgrain. The crop failure and food shortages in the Sahel are one example of this. Desertification and a shortage of fuelwood, together with the logging of tropical rainforests, were the attendant environmental degradation.

In order to ameliorate the North–South divide, there is a need for much greater economic co-operation and a focus on the possibility of global sustainable development, whatever this actually means for different countries. Any analysis must link environmental resources and their use to socio-economic and political considerations. Any proposed solutions to environmental issues must be set within an international political and ideological context. Many environmentalists have failed to do this and, therefore, found that their messages have sounded like a bell in a vacuum. Indeed, the WCED report stressed the concept of basic needs and environmental limits set not by the environment itself but, rather, through technology and social organization. As pointed out by Adams (1990, p. 59), the WCED report represented 'a subtle but extremely important transformation of the ecologically-based concept of sustainable development, by leading beyond concepts of physical sustainability to the socio-economic context of development'. In the same book, Adams goes on to argue that green development is perceived not so much as the way the environment is managed, but about who has the power to decide how it is managed. At its core, green development is seen as the attempt to redirect change to maintain or enhance the power of the poor to survive without hindrance and to direct their own lives.

The following section is an idealistic and provocative shopping list of the kinds of issues that we feel should at least form the basis of international discussion in an attempt to provide a global framework for sustained development, with or without sustained growth in any single country.

An attempt at sustained development must, initially, be stimulated with the aid of grants and incentives, with emergency aid programmes to fight poverty. The Worldwatch Institute estimated in 1988 that the annual expenditure to subsidize sustained development and help bring the developing world above the poverty level would be approximately US $45 billion, increasing to approximately US $150 billion by the year 2000. Sadly, the present expenditure of the UNEP's environment fund is less than US $100 million, falling far below the level of financial help needed to get out of the present quagmire. The cost would, therefore, have to be offset by other means. One such method of offsetting the costs of development for the developing world is to provide relief for the national debts.

In 1988, the World Bank (which has a policy of lending funds raised by floating bonds in the wealthier countries), and the International Monetary Fund (IMF) met in Berlin to develop such programmes which would incorporate money lending and would be concerned with promoting environmentally sound practices. The new Environmental Department of the World Bank, for example, will conserve land in lieu of other natural land destroyed by development projects such as a major reservoir or irrigation scheme. One such policy known as 'Wildlands' has been established to help conserve endangered species and habitats. So far over 40 projects in 26 countries protect 60,000 square km, the largest of which covers 19,000 square km of rainforest in the Rondoma Province of Amazonian Brazil. The World Bank has also established the Consultant Group of International Agricultural Research (CGIAR). This organization is particularly concerned with increasing food production by improved agricultural techniques and new hybrid strains of crop with greater yields. The World Bank has also been working with the World Resource Institute to facilitate financial arrangements for the protection and sustained development of habitats. Sources of revenue for such programmes could include taxes imposed on international trade.

Another way of addressing the problem of poverty is to look at the world's military expenditure. Taken as a whole, military expenditure totals US $1 trillion annually, that is, US $2.7 billion per day. Developing countries have increased their military budgets in the past 20 years by five times. Developing countries spend large percentages of their gross national product (GNP) on the military, for example, Nicaragua currently spends 12.4 per cent of its GNP, Mongolia 10.4 per cent, Somalia 9.6 per cent, and Ethiopia 9.3 per cent. When compared with the expenditure on health provision in the same countries, the contrasts are revealing: for example, Nicaragua spends 4.6 per cent of its GNP, Mongolia less than 0.8 per cent, Somalia less than 0.8 per cent, and Ethiopia less than 0.8 per cent. The WCED recommended a reduction in the expenditure on military resources, so that the money freed from this source would then become available for more constructive use. Achieving such a reduction in arms is not as easy. In a worst-case scenario, as environmental degradation and climatic change heighten global tensions, and as people are forced to abandon their degraded land and encroach on other already-populated lands, armed conflict might become more commonplace.

Poverty

The WCED emphasized that poverty was one of the major causes of the accelerated depletion of the Earth's resources, and the degradation of its forests, soils, species, fisheries, water, and atmosphere. One of the major goals must be to eradicate poverty. They estimated that an annual average national growth of 3.2–4.7 per cent would be necessary to keep pace with the growing global population. Given the disparities between population growth rates throughout the world, an average growth in national income of 5 per cent would be needed within the developing countries of Asia, 5.5 per cent in Latin America, and 6 per cent in Africa. These countries experienced growth of this magnitude during the 1960s and 1970s, but during the 1980s rates dropped to below this level. The decline is blamed on population growth, deteriorating trade (often the result of protected markets in the industrialized countries), reduced resources, and long-term national debts.

Long-term national debt and its non-repayment are probably the biggest destabilizing factors in reducing the growth in national income. The accumulated debt of developing countries is approximately US $1 trillion, which attracts an annual interest of US $60 billion. In 1984, the flow of money from industrial to developing countries was reversed with more than US $43 billion exchanging hands each year. In 1988, the 17 most debt-ridden countries paid out US $31.1 billion more than they received in aid. If such debts did not exist, the developing world would be a long way towards eradicating poverty.

The WCED believes that increased living standards would bring about a significant reduction in population growth, something that has been true for industrialized countries this century. For the immediate future, the Commission proposes improved education and empowering women, particularly in order to help improve the dissemination of knowledge concerning birth control and family planning. This will further help reduce population growth, which in turn would help to reduce poverty.

Agriculture

The WCED supports the use of reforms and public policy making, with the intelligent formulation of policies and grants aimed at reducing poverty. The European Economic Community (EC) provides grants and subsidies to farmers to encourage them to overproduce, to exploit marginally productive areas, to waste ground- and surface-waters on irrigation, and to use pesticides and fertilizers excessively. These policies place undue pressure on the land, encourage pollution and environmental damage, and are clearly not sensible ways of utilizing finite, exhaustible resources.

The Organization for Economic Co-operation and Development (OECD) subsidizes farms in Western countries with US $300 billion each year, far in excess of the money available for soil and water improvement. The WCED suggested that considerably more money should be used to improve the quality of the soil, water, and vegetation on farmland which would be beneficial in the long term to the farmers instead of providing subsidies for food production. This would then offset the cost of remedial measures resulting from the degradation of land, or the need to acquire new land due to bad farming practices. Agricultural production should be shifted from the industrialized countries to the developing countries so that the latter may become self-sufficient. Unfortunately, this may be undermined by governments in industrialized countries offloading surplus food, thus reducing the political pressure within developing countries to reform their own agricultural policies and improve their output. Governments should also be encouraged to increase the productivity and efficiency of their farming while at the same time improving the environment. Encouraging multiple crops and intercropping of nitrogen-fixing plants should reduce the need for fertilizers, and so decrease demands on irrigation. Improved irrigation techniques and the reduction of the use of pesticides, substituted by biologically engineered pest-resistant strains of crops, can all contribute to the prudent management of agricultural land.

In developed countries, industrialization and mechanization are causing many modern agricultural systems to become more energy-intensive. Government subsidies are frequently paid to farmers not to grow certain crops. Global food surpluses are a facet of the international commodity markets, and provide a stark contrast to the dire poverty, malnutrition, and starvation in many poor countries. These global inequalities need to be addressed much more effectively.

Energy conservation

With increased growth, the demand for energy will increase. This increase can be offset by increased energy production and efficiency, but most important of all conservation of energy. In many cases, energy efficiency and energy conservation are used as interchangeable terms, although strictly this is incorrect – efficiency involves maximizing the energy output for a given input, whereas conservation may simply mean using only the least necessary amount of energy (even with inefficient sources of power and/or machinery). However, in this book where these terms are not differentiated, the reader can assume that energy efficiency and conservation are linked.

Increased energy production would have drastic environmental effects, probably with enhanced global warming as more carbon dioxide and other greenhouse gases are added to the atmosphere. The WCED favoured conservation measures such as the recycling of aluminium, steel, paper, and glass, and advanced technologies to reduce the consumption of energy drastically.

The disparities between the energy used per capita in the developed and developing world are a source of some concern. Poor people use very little energy, usually only sufficient for subsistence, whereas in developed countries energy is liberally used to power a plethora of domestic appliances such as dishwashers and air-conditioning units. Poor countries need cheap and reliable energy. In rural areas, electrification improves the lifestyle of those in poverty. Additionally, it can reduce environmental stress, because forests suffer less rapid degradation as a result of the gathering of fuelwood, and cattle dung which is often burnt as a fuel can be used as a fertilizer for crops. A key consideration in promoting a change in patterns of energy consumption in the developing countries is to encourage energy efficiency. Box 10.1 summarizes the recommendations of the Stockholm Initiative on Energy, Environment and Sustainable Development (SEED), November 1991.

Energy-saving technologies should be encouraged. Many of the newer industrialized countries, such as Taiwan, South Korea, and Brazil, have already implemented some of the new technologies which provide them with large cost savings. Governments should also be encouraged to promote conservation policies. The USA, for example, reduced its domestic production by 23 per cent between 1973 and 1985 due to increased efficiency, and thereby demonstrated cost effectiveness of conservation, and the ease with which such measures can be introduced. It has been estimated that in the UK, energy use could be reduced by 20 per cent by using cost-effective measures.

Not only should there be government encouragement for industrial/commercial energy conservation, but also domestic energy savings should be encouraged. More energy-efficient domestic habitations need to be constructed. Sensible town planning should reduce the distance a person has to travel to and from the workplace, in tandem with better and efficient, reliable, cheap public transport. Private vehicle use should be discouraged through offering acceptable alternatives, not simply through punitive measures within inner cities and town centres. This latter issue is a major concern because private cars account for a considerable amount of energy consumption and pollution. Pedestrianizing city centres needs to be undertaken sensitively, because the complete prohibition of private vehicles from shopping areas in towns and cities can discourage people from using the facilities, and encourage the development of large, out-of-town shopping complexes. Such shopping complexes tend to destroy the heart and character of towns, reduce commercial competition and employment, and are difficult to reach, particularly for the elderly and disabled.

A study by the Stockholm Environment Institute (1993), commissioned by Greenpeace, using conventional assumptions concluded that it is technically and economically feasible to reduce current global oil use by 50 per cent within 40 years, and the use of oil and other fossil fuels could be entirely phased out during the twenty-first century. In more detail, the study developed a fossil free energy scenario (FFES), and concluded that global oil consumption could fall from 120 exajoules (10^{18}J) today to 59 exajoules in the year 2030, and related global carbon dioxide emissions from oil could fall by 50 per cent by 2030, and 75 per cent by 2075. The study's recommendations included: (1) the introduction of tough new fuel-efficiency standards for all vehicles; (2) government support for public transport, and the discouragement of private vehicle use within urban areas; (3) a doubling of energy R&D budgets within 10 years, the bulk of which should be used for energy efficiency and renewables; (4) the removal of massive subsidies on oil, such as oil exploration tax breaks; (5) the introduction of pollution taxes for oil and other fossil fuels to

MANAGING THE EARTH

Plate 10.5   (A and B) United Nations General Assembly and conference room at the UN headquarters, New York. World peace, improved life chances for all, and sustainable development are dependent upon international co-operation and the implementation of internationally binding treaties and agreements forged through organizations such as the United Nations.

reflect the true costs of major oil spills and pollution damage, the effect of which would be to more than double current oil prices; (6) the establishment of a UN agency for Technologies for Renewables and Energy Efficiency (TREEs) to promote the development of these technologies through training, financial support, and information; and (7) Multilateral Development Bank (MDB) lending for energy projects to be re-oriented towards energy and smaller-scale renewable projects – less than 1 per cent of the World Bank power sector lending is currently used for energy-efficiency projects.

The introduction of alternative and renewable energy sources, instead of conventional fossil fuels and fuelwood, would help reduce pollution and the amounts of carbon dioxide, sulphur dioxide, nitrogen oxides, and other harmful gases being released into the atmosphere. Such technologies include mini-hydroturbines, solar power, wind and tide power, and biogas. The introduction of environmentally friendly policies, alongside shifts in accepted practices or conventions, could be eased into place and stimulated with government subsidies, the cost of which could be calculated against that of having to take remedial action to restore the environment. The WCED report suggested that, in many cases, many of the smaller technological schemes may actually be more appropriate to an area than the conventional, harmful practices. Mini-hydroelectric schemes, for example supplying energy in a mountainous area, may be more

appropriate in terms of management, maintenance, and cost than a large nuclear or oil-fired power station in the same region.

One of the most important considerations for sustained development is the merger of environmental and economic decision-making bodies. The inequalities between environment and economic agencies are far too large to resolve, and the WCED believes that both should work together, symbiotically, so that one would benefit the other. The environmental agencies have been hindered by mandates which are too narrow, small budgets, and little or no political muscle. While the various economic agencies have become very powerful over the years, this has not necessarily been as true for environmental agencies, which are often cast in the role of watch-dog with power to criticize but no legal instruments to enforce environmental legislation. Penalties for damaging the environment are commonly wholly inadequate and derisory. The environmental agencies must have much more power in order to make the trade and sectoral agencies responsible for the pollution and other damage which they may inflict on the environment.

Working together, market-place incentives could be developed which are both economically viable and beneficial to the environment. Taxes could be increased on energy, other easily exhaustible resources, and polluters, while at the same time being offset by a decrease in taxes (corporate and value added) on the labour force.

This would encourage organizations to conserve energy and other resources. As well as using energy more sensibly, societies must develop both 'carrots' and 'sticks' to encourage and enforce a greater respect for the environment.

Energy conservation must run hand-in-hand with the polluter-pays principle. Incentives should be offered to companies that develop and manufacture energy-efficient, environmentally-friendly industrial and domestic goods and devices, for example through tax rebates or reduced taxation.

The polluter-pays principle

Pollution could be considerably reduced through a taxation system which offers incentives to organizations with a good track record of being environmentally responsible. Ideally, such measures could ultimately lead to optimum, and possibly sustained, development. The introduction of the 'polluter-pays principle' is a good example of the type of legislation which is necessary. This was introduced by member countries of the OECD in 1972 and it makes industries responsible for conserving or protecting the environment, with costs offset by the consumers. Unfortunately, many governments are still too slow to apply this principle, mainly because it is seen as too onerous to implement as it can be very difficult to make an absolute assessment of the environmental degradation produced by a specific pollution incident, let alone the long-term pollution. Atmospheric pollution, for example, may have global as well as regional effects. Who pays, and in what proportion? The issues surrounding market-based economic instruments (MBIs), for example a pollution charge or tax, versus direct legislation to control pollution, are a source of long-standing debate. The main advantage of MBIs is that they can help industries to allocate resources efficiently through the most economic methods, but those polluters with the largest abatement costs may decide to pay a 'pollution tax' as the cheapest option, that is, they may see an MBI system as a 'licence to pollute'. If pollution remained at environmentally unacceptable levels under an MBI system, then the pollution taxes, which would have to be empirical in any event,

would require adjustment in order to achieve prescribed environmental standards. Ideally, any pollution tax would be set at a level at least equal to the environmental damage done, something that is no easy matter.

An alternative approach to paying for cleaning up the environment, and/or preventing pollution, through a direct polluter-pays principle is for governments to levy some form of pollution tax which is assessed by an independent panel of scientific/technical experts and lawyers, the revenue being devoted solely to environmental pollution control. The tax could be a blanket tax to industry with concessions for companies and organizations that meet certain, specified targets, and additional penalty taxes for pollution incidents. Any pollution tax would have to be effective whilst not proving so punitive that many industries could not survive.

In the USA, by the mid-1970s the issue of waste and waste disposal was regarded as one of the major socio-economic problems to confront. As a result of the political pressure, the Comprehensive Environmental Response, Compensation and Liability Act 1980 (CERCLA) was enacted, otherwise known as 'Superfund'. This Act provides for the creation of a trust fund, mainly financed from special industrial taxes, from which the monies raised can be used for clean-up operations in both derelict and uncontrolled sites. If it is possible to identify the owner or operator of a site at the time of waste disposal, even extending to the waste carriers, then the Federal authorities look to these parties to pay for the clean-up. In some instances, the banks have been liable as 'deemed polluters'. Ideal as the US example of the polluter-pays principle may seem, there are many who regard it as a failure, because the parties involved in environmental litigation have sought reimbursement from their insurers which, in turn, has led to very high litigation costs between both the insurers and reinsurers, to the extent that only about 12 cents per dollar spent is actually used in the physical clean-up. In the USA, the current average cost of a Superfund clean-up is estimated as US $30 million (US RAND's Institute for Civil Justice survey). Other countries, and the EC, are looking critically at the issues surrounding the US Superfund, prior to introducing their own legislation (see Chapter 9, and POST 1992b).

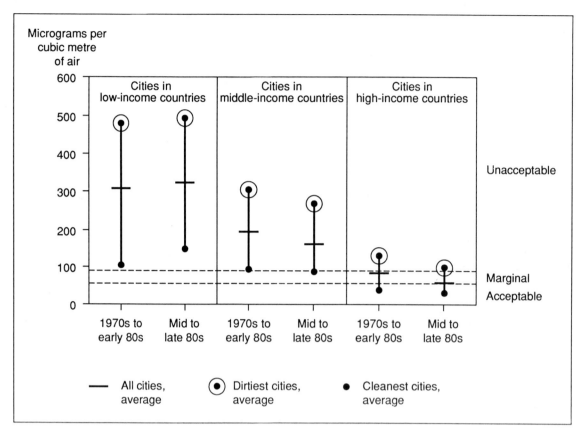

Figure 10.11 *Urban air pollution levels and trends: concentrations of suspended particulate matter across country income groups. Data are for 20 urban sites in low-income countries, 15 urban sites in middle-income countries, and 32 urban sites in high-income countries. 'Cleanest cities' and 'dirtiest cities' are the first and last quartiles of sites when ranked by air quality. Periods of time series differ somewhat by site. World Health Organization guidelines for air quality are used as the criteria for acceptability. After World Bank (1992).*

Atmospheric pollution control

Worldwide, urban air pollution (expressed as concentrations of suspended particulate matter across country income groups) has shown a downward trend from the 1970s to the late 1980s within cities in the high- and middle-income countries, in contrast to a converse pattern for cities in low-income countries (Figure 10.11). Figure 10.12 shows air pollution in developing countries for selected pollutants. The condition of the atmosphere represents one of the major current global concerns. Even though several conventions on global atmospheric change and protection have taken place in recent years, the agreements and treaties are not universally accepted, adhered to, or enforced. In 1984, for example, 19 countries signed an agreement to reduce sulphur emissions by 1993, but the

four biggest culprits, the USA, UK, Poland, and Spain, did not sign the agreement.

Today, the global implications of pollution are beginning to be appreciated by both scientists and the wider community, along with a recognition that countries should be responsible for more than just their own backyards. In 1988, the first world conference on the atmosphere was sponsored by the World Meteorological Organization and UNEP, with 37 countries participating in the theme of 'The Changing Atmosphere'. They called upon governments to develop global and national plans for the protection of the environment and to initiate development of an International Convention for the Protection of the Atmosphere. They asked for the introduction of taxes on the use of fossil fuels to provide money for the newly established 'World Atmosphere Fund' which would go to

Figure 10.12 *Selected air pollutants in developing countries for three scenarios, 1990–2030. (A) Emissions of particulates from electric power generation. (B) Lead emissions from motor vehicles in urban areas in developing countries. The calculations are based on the following data and assumptions. Growth rates for per capita income and population are as given by the World Bank (1992). Per capita income elasticity of demand for vehicle fuels equals 1.2, and fuel price and congestion price elasticities equal −0.5 and −0.6 respectively. The average life of vehicles is 15 years. Gasoline and diesel fuels each account for about half the total consumption. Efficiency reforms include congestion charges (based on data from the Singapore Area Licensing Scheme) and higher fuel taxes (assumed to rise over a 25-year period to levels now found in Europe). Pollution abatement measures include emission controls and the gradual introduction of cleaner fuels over a 25-year period. Under this scenario, lead emissions gradually drop to the bottom of the shaded band; emission levels of particulate matter, hydrocarbons, and sulphuric oxides fall within the band, and nitrogen oxides are at the top. After World Bank (1992).*

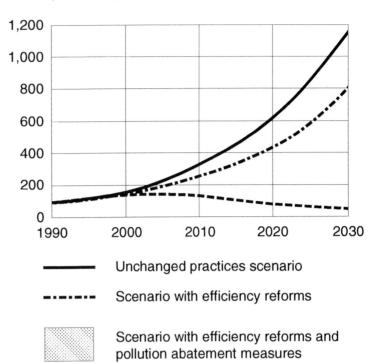

(A) Index (1990=100)

——— Unchanged practices scenario

–·–·–· Scenario with efficiency reforms

▒▒▒ Scenario with efficiency reforms and pollution abatement measures

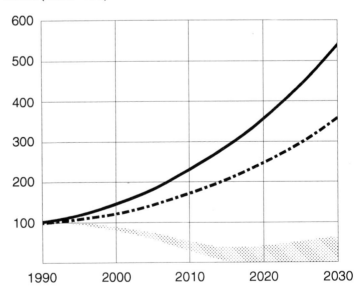

(B) Index (1990=100)

developing countries to offset the consequences of any future global warming and associated sea level rise. They requested a reduction of carbon dioxide emissions and other atmospheric greenhouse gases, and aimed to promote public awareness, research, and the use of technologies to reduce the input of pollution into the atmosphere. They also pushed to strengthen the Montreal Protocol which, in 1987, called for a reduction of CFCs by 50 per cent by the year 2000, and a complete ban on CFCs after that date. The latter recommendation was pursued and addressed at an international forum in Helsinki in May 1989. It was agreed by the 86 countries that attended the forum that CFCs should be totally banned by the year 2000. Norway pledged 0.1 per cent of its GNP to help developing countries follow suit and offset the likely expenditure incurred while it changed over to the use of alternative chemicals.

Traffic congestion in city centres is a major modern problem (Plate 50). Traffic may even grind to a complete standstill, termed 'gridlock'. In one of the most publicized incidents of gridlock, on 29th October 1986, a traffic accident on the San Diego freeway in Los Angeles caused an eight-hour traffic jam involving tens of thousands of motorists. Road pricing, together with improved public transport, is needed to tackle the problem. This system began in Singapore in 1975, and is simply the imposition of an extra government tax on vehicle use and parking within cities. People wishing to use private cars in a major city centre must purchase a licence on a pay-and-display basis, and calculated at a daily rate. Whatever the technology and future developments in producing cleaner vehicles, society will have to be encouraged to use cars less often in urban areas. Such a change in attitudes can only come about through the introduction of cheap and improved public mass-transit systems in tandem with new technologies for cleaner vehicles.

Unlike New York, San Francisco, Tokyo, Paris, or London, Los Angeles never had an

Table 10.3 *Contribution of road transport to air pollution in selected cities. These data emphasize the role played by road transport emissions in contributing to poor air quality in many large urbanized areas throughout the world.*

Region	Year	Total pollutants from all sources 10^3 tonnes	Percentage attributable to road transport					
			CO	HC	NO_x	SO_x	Particulates	Total
Mexico City	1987	5,027	99	89	64	2	9	80
Sao Paulo	1981	3,150	96	83	89[a]	26	24	86
	1987	2,110	94	76	89[a]	59	22	86
Ankara	1980	690	77	73	44	3	2	57
Manila	1987	500	93	82	73	12	60	71
Kuala Lumpur	1987	435	97	95	46	1	46	79
Seoul	1983	x	15	40	60	7	35	35
Hong Kong	1987	219	x	x	75	x	44	x
Athens	1976	394	97	81	51	6	18	59
Gothenburg[b]	1980	124	96	89	70	2	50	78
London	1978	1,200	97	94	65	5	46	86
Los Angeles[b]	1976	4,698[c]	99	61	71	12	x	88
	1982	3,391[c]	99	50	64	21	x	87
Munich	1974/5	213	82	96	69	12	56	73
Osaka	1982	141	100	17	60	43	24	59
Phoenix	1986	1,240[d]	87	64	77	91	1	28

a   Includes evaporation losses from storage and refuelling
b   Percentage shares apply to all transport. Motor vehicles account for 75–95% of the transport share
c   Excluding particulate matter
d   Includes 490,000 tonnes of dust from unpaved roads
x   = data not available
Source: World Resources Institute (1992).

Box 10.2

Reducing motor vehicle pollution

Many of the world's major cities are suffering the ever-increasing problems of serious traffic congestion and its attendant atmospheric pollution. For the motor car, California's emission laws are amongst the toughest in the world. They ensure that cars in the state emit about one-tenth of the main pollutants – carbon monoxide, nitrogen oxides, and hydrocarbons – compared to the 1960s. It is these stringent standards that have led to catalytic converters and electronic fuel injection systems being commonplace in California. Necessity has proved to be the mother of invention. Catalytic converters remove pollutants from the exhaust gases before they leave the car, and electronic fuel injection systems make combustion more efficient by reducing the amount of pollutants. Despite these changes, California's Air Resources Board, which monitors the atmospheric pollution, found that on more than 200 days in 1990 Los Angeles failed to meet the air quality guidelines. Although cars are now manufactured with cleaner emission standards, the growth in population and increased car ownership has actually led to a deterioration in air quality in the state. The current emphasis is now focused on encouraging the use of cleaner petrol, known as 'reformulated gasoline' because it evaporates more slowly. In California, from November 1992

underground railway, and the public transport which has existed has been inadequate. Not surprisingly, therefore, motor vehicles provide the principal means of transport, with all of the associated atmospheric pollution and the infamous smogs. An estimated 8.8 million people live in Los Angeles County, owning (at 1993 figures) about 6 million motor vehicles; by the year 2010, the population is predicted to exceed 10.2 million. Atmospheric pollution caused by road transport, both through individual and corporate activities, is now recognized as a major problem in many urban centres. Table 10.3 shows the contribution of road transport to air pollution in selected cities. In order to combat the problems of motor vehicle congestion and pollution, Los Angeles has embarked upon a programme to construct a metro system linked to a more integrated surface public transport system. The first part of this 30-year programme came into effect on 30th January 1993, when a 7 km long (Red Line) underground route with five stations was opened to connect downtown Los Angeles with Hollywood. Completion of this 46 km long Red Line is projected for the year 2000. The Los Angeles 'Metro' scheme is an example of the ways in which large urban centres are attempting to combat traffic congestion and the associated poor air quality, but such projects can only become viable if people are prepared to accept a changed lifestyle. In California, tough new legislation has been introduced to combat and reduce motor vehicle pollution (see Box 10.2).

Changing attitudes

If change is to come about in the ways discussed here, and societies are to aim for sustained development, then the attitudes of individuals and governments will have to change. Communication and understanding between all concerned and involved groups will need to be improved. This is particularly so between scientists and policy makers, who so often appear to speak different languages. The scientists need to deliver their arguments on a level and with language easily understood by policy makers and the general public. Jargon should be abandoned in favour of simplicity and clarity.

When examining problems and processes, scientists often present several different scenarios, with the results and conclusions being presented in a form that is too complex for policy makers to appreciate readily and, therefore, to make decisions upon. Scientists also endeavour to be objective and non-committal. These may seem to be laudable aims, but all too frequently they mask the real subjectivity of science, and serve only to send out ambiguous messages. But, there is a real underlying problem, because most environmental issues are associated with varying degrees of uncertainty, therefore policy makers and scientists must develop ways of working with these uncertainties.

How is change to come about? Ruckelshaus (1989) suggests that change is a three-phase process. First, the world leaders need to transmit environmental values to both the public and the

Box 10.2
Reducing motor vehicle pollution

petrol companies were forced to add oxidizing agents to petrol in order to reduce carbon monoxide emissions by converting it to carbon dioxide before emission. In 1993, this was followed by tougher controls on diesel fuel, and in 1996 vehicles will be encouraged to use 'phase 2 reformulated gasoline' which will produce less benzene in the gas emissions.

In California four new categories of low-emission cars are being demanded of the motor industry. By 1994, transitional low-emission vehicles (TLEVs) must be on the market which emit half the reactive organic gases of a conventional car, and by 1997 low-emission vehicles

(LEVs) emitting one-quarter, and ultra-low-emission vehicles (ULEVs) producing one-eighth of the reactive organic gases, must also be available. In 1998, zero-emission vehicles (ZEVs) must be available, a category currently met only by electronic vehicles. Electric cars, whilst much cleaner than conventional cars, are more limited in range and slower, and recharging a battery is much slower than filling up at a petrol station. In order to tackle these potentially less appealing features associated with electric cars, companies such as Chloride in the UK are developing high energy-density batteries, for example the sodium-sulphur battery.

private sectors. Second, motivation is needed to initiate and drive those changes, and finally, and most difficult of all, institutions are needed which can translate the agreed policies into action.

The first of these phases is well under way at present. During the past year, leaders of the USA, the former USSR, UK, France, and Brazil have all made environmental statements. During the Economic Summit in Paris in July 1989, the leaders of the group of seven major industrial countries (G7) discussed environmental issues together for the first time. They called for worldwide policies to be developed to pursue the goal of sustained development. The meeting was nicknamed the 'First Green Summit' and a seven-page document with 19 clauses was produced, the Paris *Communique*. Interestingly, the discussions may have been stimulated by the growing interest of politicians in environmental issues with a real awakening of latent environmental sympathy, rather than vote-catching. Hopefully, the latter interpretation is too cynical and unfair on professional politicians. This change in public consciousness was demonstrated by the large number of parliamentary seats won by environmental parties during the European election earlier that year, together with the realization by politicians that if they want to remain in power they need to continue to attract the voters to their party. There is a possible cloud on the horizon to darken this optimism, however, because at the last G7 summit, very little was said about environmental issues, and in the last UK election agenda such issues were apparently put on the proverbial back-burner. The interest and concern at government level in environmental issues may be on the wane.

Incentives to protect the natural environment which should be introduced by governments should include pricing policies to cover the cost of environmental degradation. Some of the methods by which governments can motivate organizations to conserve energy, restrict pollution, and recycle materials have already been discussed. Other incentives might include offsetting the emission of carbon dioxide in a factory by planting trees. Citicorp, for example, one of the major money-lending institutions, has a policy of writing off debts in exchange for areas of land in South America which will be designated as conservation areas and made into national parks.

The development of institutions which can motivate and enforce policies has only recently begun. Unfortunately, organizations such as the UNEP, the Human Dimensions of Global Change Programme (HDGCP), the WWF, the International Geosphere–Biosphere Programme (IGBP), and many others, are given very small budgets and have far too little political and economic power. This situation contrasts sharply with such institutions as NATO, the World Bank, and the multinational corporations. There is clearly a need for an international environmental organization comparable to these in size, budget, and power, with the aim of sustained development of Planet Earth. This organization should integrate all the environmental facets we have discussed throughout this book, and have a truly global, not just North–South, perspective.

An Agenda of Science for Environment and Development into the Twenty-first Century

The International Conference on an Agenda of Science for Environment and Development into the 21st Century (ASCEND 21) was convened by the International Council of Scientific Unions (ICSU) in Vienna during the last week of November 1991. ICSU is an international organization, primarily of scientific unions and scientific committees principally concerned with the natural sciences, and ASCEND 21 was produced, at least in part, as a preparation for the June 1992 Rio conference, which has become well known as the 'Earth Summit'. A summary of the influential ASCEND recommendations is given here because they provide a good example of the ways in which the professional scientific/technological community is responding to global environmental issues.

Members of the international scientific community participating in ASCEND came to a consensus on the major problems which affect the environment and hinder sustainable development. The specific areas identified as of the highest scientific priority through which the scientific community could begin to attempt to find solutions were listed as: population and per capita resource consumption; depletion of agricultural/land resources; inequality and poverty; climate change; loss of biological diversity; industrialization and waste; water scarcity; and energy consumption (ICSU 1992). ASCEND recommended the following:

1   intensified research into natural and anthropogenic forces and their interrelationships, including the carrying capacity of the Earth and ways to slow population growth and reduce overconsumption;
2   strengthened support for international global environmental research and observation of the total Earth system;
3   research and studies at the local and regional scale on: the hydrologic cycle; impacts of climate change; coastal zones; loss of biodiversity; vulnerability of fragile ecosystems; impacts of changing land use, of waste, and of human attitudes and behaviour;
4   research on transition to a more efficient energy supply and use of materials and natural resources;
5   special efforts in education and in building up of scientific institutions as well as involvement of a wide segment of the population in environment and development problem-solving;
6   regular appraisals of the most urgent problems of environment and development and communication with policy makers, the media, and the public;
7   establishment of a forum to link scientists and development agencies along with a strengthened partnership with organizations charged with addressing problems of environment and development;
8   a wide review of environmental ethics.

These recommendations pose a challenge, not just for the international scientific community, but for the whole of humankind. More specifically, if these aspirations are to be realized, then it is of paramount importance that they are adopted by those with the political and economic power to support the scientists and technologists in their endeavours to find appropriate solutions. With potential solutions, the scientific/technological community must then convince the rest of society that their solutions are morally and ethically acceptable, something that involves them in clearly defining the problem(s) and translating any complex solutions into intelligible and easily understood arguments.

Earth Summit, Rio de Janeiro

On 3rd June 1992, more than 100 world leaders and 30,000 other participants met at an extraordinary meeting in Rio de Janeiro for the beginning of the United Nations Conference on Environment and Development, popularly known as the Earth Summit. The summit was the brainchild of the Canadian millionaire, Maurice Strong, who had been a member of the WCED and secretary general of the first UN environmental conference in Stockholm in 1972. Amongst these leaders were US President George Bush, Chancellor Helmut Kohl of Germany, Prime Minister John Major from Britain, and Prime Minister Kiichi Miyazawa of Japan. Even the Dalai Lama attended to join the delegation of clerics, artists, and green-minded parliamentarians. The central focus of the agenda to discuss the future of the Earth was treaties on biodiversity, climate change, and the so-called Agenda 21 to address the problem of the twenty-first century (see Appendix 3). Amidst much media hype and world attention, delegates from the rich and poor nations met, having come to Rio with differing expectations and perceptions of the major problems facing humankind and the ways of tackling the issues. History may say that too much was expected of this summit at the time but that it marked the beginning of a continuing dialogue between the rich and poor nations over the management of the planet.

Over twenty years ago in June 1972, the first Earth summit took place in Stockholm as the United Nations Conference on the Human Environment. After two weeks of intense negotiations, a declaration of principles and an action plan emerged. At that time, in 1972, a number of present key concerns had not surfaced. For example, the hole in the ozone layer over Antarctica, let alone the hole above the Arctic or Europe, had not developed. But, at that time, much of the destruction of the rainforests and threats to biodiversity were well under way. In 1972, about one-third of the Earth's tropical rainforests had been destroyed and around 0.5

per cent of the remainder was being lost each year, equivalent to 100,000 km² annually. In 1972, the world's population was 3.84 billion (72 per cent living in developing countries), whereas in 1992 it was 5.47 billion (77 per cent living in developing countries). As an example of the increasing threats to biodiversity, 1972 saw just under 2 million African elephants, reduced to about 600,000 in 1992, mainly because of ivory poaching. In 1972, about 3 million people were refugees fleeing war, a number that has risen to an estimated 15 million in 1992. For 1992, the global military spending on arms and armed forces is projected to be just under US $800 billion (at 1988 prices), compared with US $600 billion (also at 1988 prices; figures from UN agencies, World Resources Institute, and the World Wide Fund for Nature, reported in *The Independent*, 3 June 1992). Since 1972, some new issues have come to the fore whilst others remain just as poignant 20 years on. The 1972 summit took place in the shadow of the Cold War with the planet divided into rival East and West blocs, and obsessed with the nuclear arms race. In 1992, the world political stage had altered dramatically. This legacy formed part of the road to Rio.

The rationale behind the 1992 Earth Summit was that with the relaxation of Cold War tensions, combined with the increased awareness of the growing ecological crisis, the conference offered a rare opportunity to persuade nations to look beyond their national interests and come to some kind of agreement over the management of the planet. Of course, countries went to Rio with very different perceptions and goals. The rich developed countries of the North have become accustomed to a lifestyle and share of the world's resources that they are not willing to sacrifice. The poor developing countries of the South, for their part, are consuming irreplaceable global resources at a rate which is causing concern about the ecology of the planet and threatening sustainable development. So, at face value it seems fairly clear that the nations of the world must abandon the self-destructive practices in favour of sustainable development. This is where the simplistic arguments break down. What is sustainable development? Who will compensate the developing countries for not destroying the tropical rainforests or for the over-exploitation of other natural, non-renewable resources? What is

a fair and just level of aid to developing countries? Will the affluent countries of the North sacrifice some of their lifestyle in order to help the developing nations? And so the strands of argument run on. There are no easy answers, and it is no wonder that many environmentalists and some nations expected too much from Rio and were disappointed by the lack of international agreement.

As the Earth Summit came to a close on Sunday 14th June, many people around the world were asking if it had all been worthwhile. The answer has to be that Rio was a qualified success. The Earth Summit ended with more than 100 world leaders, led by the Brazilian President Collor de Mello, adopting a charter for sustainable development together with a new United Nations body to supervise its implementation. When the summit ended, 152 countries had signed the Biodiversity convention and 150 the Climate Change convention. The richest nation on Earth, the USA under President George Bush, refused to sign the Biodiversity convention, claiming in a presidential election year that it would cost the American economy precious jobs and financial resources it did not have available. It was a week in which George Bush claimed that 'America's record on environmental protection is second to none, so I did not come here to apologize.' Despite this major setback, for most of the delegate nations and independent environmental groups, Rio did mark an important opportunity for the developing and developed countries to express ideas together.

At Rio plans were drawn up for a declaration of principles for the pursuit of sustainable development, known as 'Agenda 21' (for the twenty-first century), which included plans for a Desertification treaty, a Forestry convention and the establishment of a United Nations Sustainable Development Commission to oversee its implementation. The Climate Change convention, although rather dilute in substance, committed signatories from the developed countries to set their own targets for greenhouse gas emissions within six months whereas developing countries were given up to three years. The world's first agreement on forests was instigated, with a set of principles on forest management and conservation, but the agreement was neither legally binding nor technically a part of Agenda

21. At the Earth Summit, the poorer nations had hoped for commitments from the richer nations for considerable additional financial aid, but received only a lukewarm response. Many developed nations still fall a long way short of the UN target aid figure of 0.7 per cent of their GNP. Rio must be viewed as the start of an ongoing international dialogue between the developed and developing nations about global environmental issues. The arguments will continue over just how successful Rio was and it will be many years before its true historical significance can be evaluated.

A manifesto for living

There are ways of living more sensitively with the natural environment. The following guidelines represent our aspirations for the optimal use of the Earth's resources. Many are controversial and should form the basis for discussion/debate.

Feeding the world and eliminating national poverty. IMF figures show that for the 15 most indebted countries, 1988 actually witnessed a reverse flow of funds from the poorer to the richer countries, from South to North of US $24.5 billion, bringing the net outflow since 1982 to US $164 billion. At the same time, developing country debt levels, from 1981 to 1987, rose from US $748 billion to US $1,195 billion (Huhne 1989, Ekins 1992).

The first priority is the goal of eliminating the poverty that leads to the major famines. Through both self-help, including freely given expertise and advice, to emergency aid where appropriate, compassion and concern for fellow human beings should ensure that starvation and malnutrition do not exist alongside extravagant affluence in the developed nations. Much more international debt incurred by the poorest developing countries could be written off. This action would have the immediate effect of providing a realistic opportunity for at least some of the poorest nations to shake off the intolerably heavy burden of debt which impedes both the will and the means to recovery.

Control of population growth. World population is too large, and it is growing too rapidly, for a reasonable share in the available natural and artificial resources. The main cause of stress on the environment is over-population. There are quite simply too many people wanting more than is available, at least for those seeing a 'North' lifestyle as desirable. Even allowing for the inequitable distribution of the world's resources, the human species cannot grow at the predicted rates without creating even more environmental problems.

Over-population is the most difficult global problem to tackle. It is hard to advise a couple in a very poor part of the world not to produce as many children, especially where infant mortality rates are high and they may rely on their children to provide family support later in life. The practical aspects of birth control, particularly through artificial methods, are often actively discouraged by some of the world's major religions. Perhaps the optimum way of encouraging slower population growth is through increasing the overall quality of life and standard of living for many people in the poorer nations. It is only through increased personal and national security that populations seem to stabilize at sensible levels. Over-population is an intensely emotive issue, but it cannot be ignored. The problem can be solved only through concerted international efforts to feed the world, provide acceptable standards of hygiene, health care and educational provision, and give people the security to see that it is not possible, at least for nations, to breed their way out of poverty – something which may be possible on a family level.

Improving basic medical care. Longevity and good health correlate extremely closely with wealth. The richer nations enjoy a standard of health provision which is far above that available to all but a few in the poorer nations. This imbalance of the most fundamental human provision, after food, should be rectified through greater international direct aid to countries where health care is limited. Richer nations should provide a larger proportion of their GNP as grants to train more medical personnel from the poorer nations, and supplement such bursaries with the provision of basic medical supplies to those countries where easily treatable diseases are still prevalent.

Expanding educational provision at all levels. Education should be provided free of charge at the point of discharge and demand, at all age levels and, wherever possible, in order to raise

Plate 10.6 *Education programme at Communidad Santa Martha, El Salvador.*

Courtesy of Rhodri Jones/Oxfam.

the public awareness of the role of the individual in society and his or her relationship with the natural environment. By increasing the overall educational standards in society, the potential for a more concerted effort by more individuals to take an active part in the solving of common environmental problems can be increased. Without a rudimentary education, people are more likely to act out of fear and ignorance and cannot be expected to take an active interest in national, let alone international, issues. Education is the greatest legacy that can be given to future generations of children and society in general (Plate 10.6).

Energy conservation. The natural energy resources available to us are finite, that is, they are limited. They will not last for ever. Undue waste should be avoided, and an attempt made to reduce excessive energy consumption. Renew-

able energy resources with minimal environmental pollution should be developed and used in preference to the non-renewable, finite resources such as fossil fuels. Technologies for energy efficiency and energy conservation should be encouraged through favourable tax regimes and other fiscal incentives. Governments should discourage the wasteful 'disposable-commodity ethos' that is now so prevalent, and which is very wasteful of energy resources.

An estimated one-half of the world's pollution and one-fifth of the greenhouse effect result from the motor car. In order to conserve energy and reduce the global impact of this pollution, there is a very urgent need to develop and market cleaner and more energy-efficient motor vehicles. But most important of all is the need to encourage the increased use of public mass-transit systems through both highly subsidized public transport and heavy financial burdens on the use of private transport to and from large cities and other places of work. Although cities with severe pollution problems, such as Los Angeles, are beginning to attempt to reverse the car culture, far too few large urban centres have tackled the problem. This can be partially achieved by sensible town planning and the provision of efficient and cheap public transport systems, subsidized by government if necessary. Thus, part of the solution to energy conservation can be found in resource sharing. Finally, governments should have an energy policy that includes support for a diverse range of energy resources, and is sensitive to the need to maintain acceptable levels of employment.

Resource sharing. People waste energy and materials, as well as contribute unnecessary pollution to the ecosphere through the selfish use of resources. Sharing transport and other resources more efficiently not only provides for a general improvement in the quality of people's lives, but also frees much more potential resources for others who may be less fortunate. To encourage resource sharing, countries need to introduce government and industrial incentives.

Recycling resources and materials. Enormous quantities of materials are wasted. The present throw-away, disposable culture creates unnecessary pollution, and squanders energy resources and precious natural materials. This is particularly the case in the developed countries whereas in developing countries recycling is a necessity of life (Plate 10.7). With more recycling should come the opportunity to dissipate commodities more widely, and therefore increase the quality of life for many more people and nations. As with resource sharing, government- and industry-funded incentives are necessary to stimulate more recycling of resources and materials.

International co-operation on global issues. An international, rather than solely national, perspective on global issues and especially environmental issues, must be encouraged. It must be ensured that the common fate of humanity is fully appreciated, which is intimately and inextricably bound up in the collective actions as families, neighbourhoods, villages, towns, large urban centres, counties, cantons, countries, or groups of nations, and leads to a more equitable distribution of resources and wealth, equality of

Plate 10.7 *Ethiopian refugees on the Tihama Plain, North Yemen, making use of metal waste from old cars to make simple farming tools to sell in the market for food. This type of alternative technology illustrates the ingenuity of refugees and people in developing countries, their fight for survival and the potential for recycling.*

opportunity and a harmonious co-existence with the natural environment. Such lofty ideals may appear unachievable, but they are nevertheless goals worth striving for. Through increased internationalism, individual societies and nations are more likely to understand the aspirations of others and to provide for the easy interchange of ideas and resources.

There is a real need for more effective international policing of incidents of pollution, particularly where they occur in less-industrialized, poorer countries and involve multinational companies. The United Nations and World Bank are obvious instruments which could be used to bring pressure to bear on offenders and non-compliant multinational enterprises. It took until the first week in October 1991 for the Supreme Court in India to rule that the US chemical giant Union Carbide is no longer immune from criminal prosecution for the Bhopal incident in 1984. The poisonous cloud of methyl isocyanate which was released on that fateful day killed about 3,000 people and injured another 200,000. So far, the Bhopal victims have waited 10 years for justice. Such delays as this are unsatisfactory for everyone concerned.

On a promising note, in early October 1991, the western nations and developing countries

agreed to examine the ways in which the General Agreement on Tariffs and Trade (GATT) affects the environment. Countries might refuse to import certain goods if they are considered to have been produced in a manner harmful to the environment. At present, these sorts of import restrictions are unlawful under the terms of GATT. A new international environmental code, backed up by the world's leading financial institutions and groups, could give real muscle to global concerns over the environment.

Reducing military expenditure. The futile arms race for ever-refined means of destroying each other is sapping valuable intellectual, financial, and material resources that should be devoted to other ends, such as feeding the world, health care, education, and generally improving the overall quality of life. Even those military strategists and politicians most committed to the arguments of keeping the peace through arms cannot fail to see the sheer waste that the arms race has created. But military expenditure will only be reduced where a nation or country has confidence in its future survival and a feeling of security against attack from others (Plate 10.8). Greater global co-operation across a broad front of economic, political, and environmental issues has to come before any country will seriously countenance reducing its military expenditure, including nuclear weapons, and thereby free money for any pressing social and environmental needs.

Non-nuclear future. This issue is probably the most contentious item of this agenda, and it is easier to write such a slogan than it is to adopt and effect. The use of nuclear power and the manufacture of nuclear weapons, however, bequeaths an unacceptable legacy and burden on future generations. Its polluting capacity is now well demonstrated and the concept of atoms for peace has been shown to be a chimera as the frantic arms race seems to gather pace from day to day, with the proliferation of ICBMs, SLBMs, ALBMs, MIRVs, ERWs, and the Star Wars technology (see Chapter 6).

Plate 10.8 *While large amounts of badly needed money are invested in arms and ammunition, inevitably less is available for social programmes to improve life in the poorer, developing nations. Outside municipal rubbish dumps, El Salvador.*

Courtesy of Rhodri Jones/Oxfam.

Nuclear weapons are simply too terrifying to use again. Robert McNamara, a US Defence Secretary, put the issue well when he stated that 'You cannot make a credible deterrent out of an incredible action.' The nuclear arms race has to be halted, with a reversal in the build-up of nuclear arms arsenals. Environmentally more friendly energy programmes need to be developed so that future generations are not the custodians of the radioactive waste, including contaminated processing plants, that are left behind.

The scaling down of nuclear arms arsenals, halting the nuclear arms race, and the decommissioning of nuclear power stations, cannot be achieved overnight. There is a need for multilateral arms reductions, with acceptable and effective verification procedures, to remove nuclear weapons. This can only come about in a climate of international trust, goodwill, and co-operation. Alternative energy technology may have to be improved before existing nuclear power stations are decommissioned. Finally, many countries without easily obtainable alternative energy resources may well argue for a nuclear future. It will be incumbent on those nations who prefer a non-nuclear stance to set out realistic alternatives and to offer economic incentives to potential nuclear nations, for example other fuels and energy at competitive prices, or the technology to develop other viable energy resources. The risks of reactor meltdowns are extremely low, but we would point out that the potential long-term consequences of a major nuclear accident, albeit very unlikely, pose an unacceptable level of danger for the environment, including to humans.

Another major problem for nuclear power concerns what to do with radioactive waste. How can society ensure its safe custody and care into the future? Beside the accidents that occur in countries that, at the time of the incident, have political stability which makes any clean-up operations easier, as nuclear power plants proliferate, so too does the likelihood of such plants becoming part of politically unstable parts of the world, simply because it is often impossible to predict where these areas may develop. The demise, and break-up, of the former Soviet Union means that safety standards in the nuclear industry may decline, thereby increasing the risk of serious nuclear accidents. Tomsk-7 is, perhaps, just such an example. Political instability within and between nations with a nuclear weapons capability, nations that previously may have been stable, will inevitably make the world a less safe place. Humans have, time and again, demonstrated an almost fatalistic inability to avoid wars with all the human misery that accompanies them, and with such a record that we really should not entrust either ourselves or others with radioactive materials for ostensibly peaceful or military purposes.

Ethical investments. Those with the financial capability and power to invest money should take moral responsibility for the ways in which their money may grow. Investors should avoid lending their money to companies and organizations involved in polluting and destroying the natural environment. Depending upon one's own ethics and morals, the list of acceptable investments will vary greatly, but at least individuals should recognize that it is not sufficient to make money in ignorance of the means by which this is done. Making ethical investments is not always easy, not least because initial investment portfolios may change, through buy-outs, mergers, reinvestment, etc., to include 'undesirable' activities.

Practising efficient and environmentally sound farming. Farming practices should be efficient but not to the detriment of the environment. Fertilizers should not pollute water resources or harm other aspects of the natural environment, for example by nitrates being leached in dangerous quantities into local water resources. The economics of farming need rationalization, for example in the EC where the common practice of 'set aside' is pursued, in which, in order to maintain a relatively high market price, farmers are paid for not growing specified crops. On a global scale this seems irrational because of the intensive use of fertilizers to increase crop yield elsewhere, and as parts of the world suffer drought and food shortages alongside food mountains. Economic blocks are an inevitable consequence of any capitalist cash nexus, but organizations such as the United Nations should work to reduce the agro-economic divide between the North and South, developed and developing countries.

When there is so much starvation and undernourishment throughout large parts of the world, precious agricultural land should not be wasted in rearing and grazing excessive numbers of livestock, which then require additional land for grain crops to feed the cattle, just so that people in the developed world can overeat. By consuming less meat, diets may be healthier, and the land freed by this change will allow the production of more grain crops to feed a larger proportion of the world's population.

Leaving designated natural wildernesses undeveloped and unexploited. The ecology of many as yet unspoilt and unplundered wildernesses, the Arctic and Antarctica, large tracts of the tropical rainforests and deserts, is in a precarious and fragile balance. These regions of the world often play a vital part in regulating global climate, both through positive and negative feedback mechanisms. It is through their preservation and maintenance that individuals can ensure the continued survival of life on Earth as it now exists. International treaties must be negotiated to protect these last remaining bastions of much of the planet's rare species of fauna and flora.

One of the recent encouraging news stories about leaving some of the natural wildernesses unexploited and unpolluted is over Antarctica. After two years of battles and campaigns, the anti-mining lobby seems to have won a stay of execution. In 1990, Australia, France, Belgium, and Italy proposed that Antarctica be designated a World Park, a global conservation area free from exploitation. The USA and Britain opposed this suggestion, but in May 1991 Japan and Germany, both countries that the USA and Britain counted on to support their case, changed sides and undermined the pro-mining lobby. This volte-face by Japan and Germany paved the way for the United States and Britain to follow suit in July, leading to the ratification of a new Antarctic Treaty in the autumn of 1991, on the 30th anniversary of the existing Antarctic Treaty. Indeed, on 3rd July 1991, the USA was the last nation to sign the protocol after unsuccessfully holding out for an exclusion clause if a minority group wished to mine as a joint venture.

The comprehensive new treaty, ratified in Bonn on Friday 11th October 1991, prohibits the mining and exploitation of Antarctica for the next 50 years. After that, any nation wishing to exploit any mineral wealth in Antarctica will require the agreement of at least 75 per cent of the signatories to the treaty. The new treaty also

Plate 10.9   *Large quantities of human rubbish are still being dumped from (A) Antarctica (as shown by this photograph taken in December 1992 of an Antarctic base in the South Shetland Islands), to (B) the Arctic (abandoned vehicles in the tundra landscape of northernmost Russia, photographed in August 1993).*
Plate A courtesy of Gary Nichols.

includes safeguards to stop Antarctica being spoiled through tourism and waste disposal. There are proposals to keep tourism offshore, and to discourage the construction of hotels and encampments, which would bring their own pollution, sewage, and waste problems to Antarctica. The treaty, initialled by 23 of the 26 member nations, only comes into force after it has been formally accepted by the respective governments.

Amongst the articles of the treaty, some of the basic provisions were: Article I, Antarctica shall be used for peaceful purposes only. All military measures, including weapons testing, are prohibited. Article II, freedom of scientific investigation and co-operation shall continue. Article IV, the treaty does not recognize, dispute, or establish territorial claims. Article V, nuclear explosions and disposal of radioactive wastes are prohibited. Article VII, treaty-state observers have free access to any area and may inspect all stations, installations, and equipment.

In 1993, anticipating a higher profile for research in Antarctica, the US $240-million a year Antarctica Program, which is funded through the National Science Foundation (NSF), was upgraded from a division within the directorate to a programme within the office of the NSF director. This programme strives for a balance between scientific exploration and environmental protection. Within the USA, concerns over the sensitive regulation and mon-

itoring of Antarctica, and allied research, have led to an ongoing debate as to who should control the US interests in this continent, and whether this is best done through the NSF or another body such as the National Oceanic and Atmospheric Administration, and the Environmental Protection Agency.

Sadly, Greenpeace have closed their independent monitoring base on Antarctica because of the economic costs – about US $1 million per year. Without an organization such as Greenpeace to monitor activity on Antarctica, there is a danger that aspects of the treaty could be violated without world opinion being alerted to any potential dangers. Plate 10.9(A) illustrates one such violation on the Antarctic base in the South Shetland Islands, while Plate 10.9(B) shows that the Arctic wildernesses are also being polluted by human waste. The problem remains truly global in extent.

Conclusions

In order to optimize the international effort on environmental issues, there is a need to rationalize resources by integrating the effort of the many small organizations and pressure groups that currently exist, each with their own overheads and expensive experts, often duplicating work done elsewhere or discovering what other experts in other organizations have expensively discovered before. An international organization with participants of ministerial rank is desperately needed to steer towards sustainable development. This is achievable if societies and individuals act quickly, efficiently, and intelligently. Environmental groups should endeavour to collaborate and, perhaps, pool some of their hard-won power and influence, together with expertise and other resources. This is not an easy thing to do where organizations have established a power base that they may jealously guard.

But, amongst all of this suggested international collaboration at the highest and most expert levels, independent environmental pressure groups must continue to operate free from any bureaucratic structures, not least because this offers them a chance to suggest and lobby for radical solutions. Such radicalism is necessary since, even if seen as extreme, it provides a climate of debate in which nations are more likely to find sensible solutions to local and global environmental issues. Ministerial participation, as a prerequisite to all international environmental commissions and other organizations, will always have the inbuilt propensity for getting bogged down in side issues and looking

for solutions that tend to preserve the *status quo*. We would advocate greater international co-operation on global environmental issues, and at the same time support for the role of independent pressure groups, not least as watch-dog bodies.

Finally, it is very hard for individuals and societies to move away from an anthropocentric or exclusively human-centred, to a more biocentric or life-centred perspective of Planet Earth. Indeed, such an approach may be impossible, simply because of our humanity. But what is certain is that without a drastic change in present societal values, norms, and mores, this generation may well be counted amongst the last of humans to inhabit the Earth. There are finite resources, yet an apparently infinite number of ways to squander them. The history of life on Earth, with its record of past climate change, together with the environmental impact that humans have made over a very short time span, show that there are lessons to be learned. People can only learn if minds are open, if individuals and governments are willing to discover the fragility and sensitivity of the natural environment before it is too late.

Key points

CHAPTER 10: KEY POINTS

1 The four principal components of the ecosphere under threat are the climatic system, the nutrient cycles, the hydrological cycle, and biodiversity.

2 Population growth is a cause of major concern because of the stress that it imposes on the environment, although some argue that current world resources are capable of adequately sustaining an even larger global population.

3 Enormous differences in wealth, life chances, health, education, and social provision exist between the developed and less-developed nations. In various parts of the world, such differences in access to the means of life have initiated wars and political instability, thereby contributing to environmental stress and degradation. Urbanization and population pressures have concentrated pollution, poor housing, disease, and poverty into large megalopolises. In many cases, these social and

environmental issues can only be tackled by international co-operation, and the defrayment of much of the so-called 'Third World debt'.

4 Agro-economic problems include overintensive land-use (e.g., associated with the cultivation of industrial monocultures), inappropriate land-use, the clearing of important natural vegetation, salinization, laterization, and pollution by fertilizers and pesticides such as nitrates. Methods to mitigate these effects include farming practices which concentrate upon efficient but not overintensive crop cultivation, less emphasis on the use of environmentally harmful fertilizers and pesticides, the improvement of soil productivity, the reduction of soil erosion, and stopping salinization and desertification.

5 The rise of the consumer society has led to an increased requirement for energy and natural resources. Without careful resource allocation and planning, there is a real danger that many resources may become severely depleted, something that

could act as a limit to growth. As alternatives to conventional fossil fuels, renewable energy resources (e.g., solar, wind, wave, tide, and biomass energy) should be encouraged, together with more research into technologies such as hydrogen energy. Without a concerted global effort to develop substantial energy supplies from renewables, there may be no alternative than to place greater reliance on nuclear energy with its associated problems of radioactive waste disposal and the risk of major and long-term environmental pollution.

6   The concept of 'sustainable development' was introduced in 1980 by the World Conservation Strategy, and the arguments were developed by the International Union for the Conservation of Nature. In 1984, these international groups were absorbed into the World Commission on the Environment and Development which produced the 1987 report *Our Common Future*, a document which ostensibly provided strategies for sustainable development. These strategies included reducing world poverty, improving agricultural practices, energy conservation, reducing anthropogenic greenhouse gas emissions, recycling waste, improving technologies, and reducing the disparities between rich and poor nations. The underlying arguments and strategies for sustainable development remain controversial.

7   Atmospheric pollution has become both a regional and global issue. International agreements and conventions on atmospheric pollution control resulted in: the 1984 agreement to reduce sulphur emission by 1993; the 1987 Montreal Protocol to reduce CFCs by 50 per cent by the year 2000 followed by a total ban on CFCs; the 1988 First World Conference on 'The Changing Atmosphere'; the 1989 Helsinki agreement on a total ban on CFCs by 86 countries by the year 2000; and the June 1992 United Nations Earth Summit in Rio de Janeiro where agreements and conventions were presented to preserve global biodiversity, and mitigate any possible global climate change precipitated by human activities. It was at Rio that Agenda 21 was signed by many nations. Appendix 4 summarizes the main fluxes and geochemical cycles that control both global climate and global climate change.

8   In this chapter, we present a manifesto for the management of the Earth, aimed at maximizing the chances of achieving global sustainable development, reducing global pollution, eliminating poverty, and increasing the life chances of individuals wherever they are born. This manifesto includes feeding the world and eliminating poverty; controlling population growth; improving basic medical care; expanding educational provisions at all levels; energy conservation; resource sharing; recycling materials and waste; international co-operation on global issues; reducing military expenditure; a non-nuclear future; efficient and environmentally sound farming practices; and preserving natural wildernesses.

Chapter 10: Further reading

Adams, W.M. 1990. *Green Development: Environment and Sustainability in the Third World*. London: Routledge.
A book on the problems of development and its environmental impact in the developing (Third) world. This book addresses the problems of striving for sustainable development and constitutes an important text for students and teachers of development and environmental studies.

Brandt, W. 1980. *North–South: A Programme for Survival. The Report of the Independent Commission on International Development Issues under the Chairmanship of Willy Brandt*. London: Pan Books.

Brandt, W. 1983. *Common Crisis North-South: Co-operation for World Recovery. The Brandt Commission 1983*. London: Pan Books.

These books are important historical documents, written as a result of investigations by a group of international statesmen and leaders into the problems of inequality in the world and the failure of economic systems to tackle the issues. A spectrum of bold recommendations and reforms are proposed in order to avoid the perceived imminent world economic crisis. The authors describe different elements of the global crisis in trade, energy, and food supply, and concentrate on the overriding problem of how to provide the finance for help, and ways of compensating for the decline in financial liquidity to reverse the decline in trade and to raise the overall world economy. The WCED followed these reports with its publication of *Our Common Future*. These books should be read by all students, teachers, and policy makers concerned with global environmental issues.

Ekins, P. 1992. *A New World Order: Grassroots Movements for Global Change*. London: Routledge.
A thought-provoking book on the problems associated with and resulting from war, insecurity and militarization, poverty, the denial of human rights, and environmental destruction. Attention is given to possible solutions to these problems at a grassroots level.

Miller, G.T. Jr. 1993. *Living in the Environment: Principles, Connections, and Solutions*, 8th edn. California: Wadsworth, Inc.
An introductory text for courses on environmental science, utilizing basic scientific laws, principles, and concepts. The textbook is in seven parts: humans and nature: an overview; scientific principles and concepts; the human population; ultimate global problems; resources and resource management; pollution; environment and society. This is a well-structured, well-illustrated book that will appeal to many students involved in elementary courses, and includes exercises and a comprehensive glossary.

Moore Lappe, F. and Schurman, R. 1989. *Taking Population Seriously*. London: Earthscan Publications.
This book provides a useful analysis of the reasons for population growth. The authors discuss the need to understand the underlying social and economic causes of population growth in order to implement effective population control.

Nebel, B.J. and Wright, R.T. 1993. *Environmental Science: The Way the World Works*, 4th edn. Englewood Cliffs: Prentice-Hall.
A well-written and illustrated textbook, containing review questions and other exercise sections at the end of each chapter. For more information, please refer to Chapter 1: Further Reading.

Redclift, M. 1987. *Sustainable Development: Exploring the Contradictions*. London: Methuen.
This book argues that the development recommendations of the WCED report (1987) need to be redirected to give greater emphasis to local (indigenous) knowledge and experience if effective political action is to be taken to minimize any environmental damage. A book which is easily read, and which contains many interesting examples and recommendations.

World Commission on Environment and Development (WCED) 1987. *Our Common Future*. Oxford: Oxford University Press.
This report examines the critical environmental and developmental problems, and contains many very useful tables and figures. Emphasis is placed on economic and ecological factors which may lead to sustainable development. An essential reference source for all college and university students, teachers, and policy makers concerned with environmental and development issues.

Sarre, P. (ed.) 1991. *Environment, Population and Development*. London: Hodder & Stoughton.
A British Open University text which examines environmental issues with reference to population growth and economic and technological development. It is well illustrated and provides a good introduction for students concerned with environmental issues. Topics dealt with include population dynamics; agriculture, productivity and sustainability; urbanization, behaviour and social problems.

Chapter 1: Introducing Earth

1 Assess the various opinions regarding the role of organisms in controlling the evolution and maintenance of the Earth's atmosphere–ocean system as habitable for life.

2 How do the Earth's lithospheric, hydrospheric, and atmospheric processes sustain biological systems.

3 In your opinion, what is the most important global environmental issue that should be addressed by the international scientific community. Justify your choice.

4 The 'systems approach' is a convenient way of studying natural processes, the nature and rates of change. Describe what is meant by a systems approach, and illustrate your answer by describing specific natural systems.

5 'Chaos and unpredictability are inherent attributes in all natural systems, therefore there is no point in trying to understand the complexity of ecosystems with a view to predicting the potential impacts that human activities may have on them.' Discuss.

Chapter 2: Climate change and past climates

1 Describe the natural processes that may result in global climate change.

2 Describe the different types of proxy data that can be used to reconstruct past climates, and discuss any limitations associated with such data.

3 Why is the study of Quaternary climatic change important for an understanding of human-induced climate change?

4 Describe examples of global climate change throughout the geological record, and explain how such changes may have come about.

5 Discuss the proposition that just as politicians should study history, so too should those who wish to understand present and future climatic change study the geological record.

Chapter 3: Global atmospheric change

1 Discuss the potential roles played by anthropogenic emissions of greenhouse gases in contributing to global warming.

2 Discuss the arguments against any evidence for global warming.

3 Describe a general circulation model (GCM) and discuss the potential limitations of such models in accurately predicting future global climate change.

4 Describe the effects of volcanic activity on influencing global climate change.

ISSUES FOR DISCUSSION

5 Do you believe that international agreements will successfully tackle the anthropogenic emissions of greenhouse gases and stratospheric ozone-depleting gases? Illustrate your answer with reference to past international treaties and agreements.

Chapter 4: Acid rain

1 What is acidic deposition, and what are its consequences for the environment?

2 Describe the factors which determine the susceptibility of a region to the effects of acidic deposition.

3 Discuss what is meant by the terms: (i) acid-neutralizing capacity (ANC), and (ii) acid susceptibility.

4 Discuss the reversibility of the regional effects of acidic deposition.

5 Discuss the various international conventions and agreements aimed at reducing atmospheric pollution which leads to poor air quality and acidic deposition.

Chapter 5: Water sources and pollution

1 Describe the causes and effects of nitrate pollution in rivers and seas.

2 Discuss, with reference to specific examples, how international regional conflicts have developed because of the need to share water resources.

3 Describe the effects of oil pollution in coastal waters, and the available clean-up technologies.

4 Describe the ways in which toxic metals may become concentrated in water resources, and their possible effects on humans.

5 Discuss the pros and cons of using water meters to control the consumption of water in developed countries.

Chapter 6: Nuclear issues

1 Nuclear energy provides a clean alternative to fossil fuels such as coal, oil, and gas. Discuss.

2 Nuclear weapons and the threat they pose have kept humankind from a global war. Discuss.

3 Describe the environmental effects produced by natural and artificial radioactivity.

4 Outline the treaties and conventions that have been signed to control nuclear tests and the proliferation of nuclear arms.

5 What considerations should be taken into account for the safe disposal of low- and intermediate-level radioactive waste?

Chapter 7: Energy

1 Hydroelectric and tidal power stations provide cheap renewable energy, but the construction of such plants and their continued use results in serious environmental problems in estuaries and associated coastal waters. Discuss.

2 Discuss the viewpoint that the production of electricity from nuclear power plants is more environmentally friendly than power produced by conventional fossil fuel power plants.

3 The reserve lifetime for global oil reserves is estimated to be about 56 years. How reliable are such estimates, and what is the nature of the assumptions that go into such figures? If you were in charge of a nation's energy policy, what would you do about this prediction?

4 Explain the reasons why alternative renewable energy resources have not been utilized to a greater extent than at present.

5 What is a carbon and energy tax? Do you think it could have a positive impact on environmental pollution and help in energy conservation? What alternatives, if any, exist?

Chapter 8: Natural hazards

1 The magnitude and frequency of meteorological natural hazards have increased during the last few decades. Discuss.

2 Do you believe that the developed nations provide sufficient support to meet the effects of natural hazards in the poorer developing nations?

3 Assess the regional and global environmental effects of tectonic geological hazards caused by earthquakes and volcanic activity.

4 What constitutes a natural disaster?

5 Discuss the various means available for combating the spread of diseases.

Chapter 9: Human impact on the Earth's surface

1 Describe the environmental consequences of the exploitation of mineral resources and the measures that could be taken to reduce any damaging environmental impact.

2 Discuss the possible regional and global implications of the continued destruction of the rainforests.

3 It is difficult to retard development and enforce conservation in less-developed countries when people are merely subsisting or dying from starvation. How may this dilemma be reconciled?

4 Describe the causes of soil degradation and the environmental consequences.

5 With reference to land contamination, discuss the 'polluter-pays' principle.

Chapter 10: Managing the Earth

1 Is sustainable global economic growth feasible, and even desirable?

2 Consider the assertion that the single biggest problem facing the survival of humankind is over-population.

3 There is a North–South divide in the world between the rich and industrialized nations and the poorer developing nations. The only way to rectify this inequality of wealth and life chances is to make the rich nations so well off and comfortable that they will freely provide the aid necessary to raise the global standard of living. Discuss the arguments for and against these statements.

4 To what extent should international policy on environmental issues take cognizance of various cultural and religious groups?

5 Does society seriously undervalue education as a means of eliminating poverty and reducing disease?

Prefix	SI symbol	Multiplication factor
exa	E	10^{18} (1,000,000,000,000,000,000)
peta	P	10^{15} (1,000,000,000,000,000)
tera	T	10^{12} (1,000,000,000,000)
giga	G	10^{9} (1,000,000,000)
mega	M	10^{6} (1,000,000)
kilo	k	10^{3} (1,000)
hecto	h	10^{2} (100)
deca	da	10
deci	d	10^{-1} (0.1)
centi	c	10^{-2} (0.01)
mili	m	10^{-3} (0.001)
micro	μ	10^{-6} (0.000001)
nano	n	10^{-9} (0.000000001)
pico	p	10^{-12} (0.000000000001)
femto	f	10^{-15} (0.000000000000001)

Some conversions

1° Celsius (1°C) = 273.15 Kelvin

1 Angstrom (Å) = 0.1 nm, or 10^{-10} m, or 10^{-8} cm

1 Micron (1 μm) = 10^{-6} m, or 10^{-4} cm

1 Tonne = 1000 kg = 10^{6} g

1 Year = 3.1557×10^{7} s (approx.)

1 Calorie = 4.186 Joules (J)

1 Watt (W) = 1 J s^{-1}

Atomic number	Name	Symbol	Relative atomic mass
1	hydrogen	H	1.008
2	helium	He	4.0026
3	lithium	Li	6.941
4	beryllium	Be	9.01218
5	boron	B	10.81
6	carbon	C	12.011
7	nitrogen	N	14.0067
8	oxygen	O	15.9994
9	fluorine	F	18.99840
10	neon	Ne	20.179
11	sodium	Na	22.9898
12	magnesium	Mg	24.305
13	aluminium	Al	26.98154
14	silicon	Si	28.086
15	phosphorus	P	30.97376
16	sulphur	S	32.06
17	chlorine	Cl	35.453
18	argon	Ar	39.948
19	potassium	K	39.09
20	calcium	Ca	40.08
21	scandium	Sc	44.9559
22	titanium	Ti	47.90
23	vanadium	V	50.9414
24	chromium	Cr	51.996
25	manganese	Mn	24.305
26	iron	Fe	55.847
27	cobalt	Co	58.9332
28	nickel	Ni	58.71
29	copper	Cu	63.546
30	zinc	Zn	65.38
31	gallium	Ga	69.72
32	germanium	Ge	72.59
33	arsenic	As	74.9126

APPENDIX 2

Atomic number	Name	Symbol	Relative atomic mass
34	selenium	Se	78.96
35	bromine	Br	79.904
36	krypton	Kr	83.80
37	rubidium	Rb	85.4678
38	strontium	Sr	87.62
39	yttrium	Y	88.9059
40	zirconium	Zr	91.22
41	niobium	Nb	92.9064
42	molybdenum	Mo	95.94
43	technetium	Tc	98.9062
44	ruthenium	Ru	101.07
45	rhodium	Rh	102.9055
46	palladium	Pd	106.4
47	silver	Ag	107.868
48	cadmium	Cd	112.40
49	indium	In	114.82
50	tin	Sn	118.69
51	antimony	Sb	121.75
52	tellurium	Te	127.60
53	iodine	I	126.9045
54	xenon	Xe	131.30
55	caesium	Cs	132.9054
56	barium	Ba	137.34
57	lanthanum	La	138.9055
58	cerium	Ce	140.12
59	praeseodymium	Pr	140.9077
60	neodymium	Nd	144.24
61	promethium	Pm	145
62	samarium	Sm	150.4
63	europium	Eu	151.96
64	gadolinium	Gd	157.25
65	terbium	Tb	158.9254
66	dysprosium	Dy	162.50
67	holmium	Ho	164.9304
68	erbium	Er	167.26
69	thulium	Tm	168.9342
70	ytterbium	Yb	173.04
71	lutetium	Lu	174.97
72	hafnium	Hf	178.49
73	tantalum	Ta	180.9479
74	tungsten	W	183.85
75	rhenium	Re	186.2
76	osmium	Os	190.2
77	iridium	Ir	192.22
78	platinum	Pt	195.09
79	gold	Au	196.9665
80	mercury	Hg	200.59
81	thallium	Tl	204.37
82	lead	Pb	207.2
83	bismuth	Bi	208.9808
84	polonium	Po	210
85	astatine	At	210
86	radon	Rn	222

Atomic number	Name	Symbol	Relative atomic mass
87	francium	Fr	223
88	radium	Ra	226.00254
89	actinium	Ac	227
90	thorium	Th	232.0381
91	protoactinium	Pa	231.0359
92	uranium	U	238.029
93	neptunium	Np	237.0482
94	plutonium	Pu	244
95	americium	Am	243
96	curium	Cm	247
97	berkelium	Bk	247
98	californium	Cf	251
99	einsteinium	Es	254
100	fermium	Fm	257
101	mendelevium	Md	256
102	nobelium	No	254
103	lawrencium	Lr	257

Group number

Period	1	2	3	4	5	6	7	8	9	10	11	12	13	14	15	16	17	18
1	H																	He
2	Li	Be											B	C	N	O	F	Ne
3	Na	Mg											Al	Si	P	S	Cl	Ar
4	K	Ca	Sc	Ti	V	Cr	Mn	Fe	Co	Ni	Cu	Zn	Ga	Ge	As	Se	Br	Kr
5	Rb	Sr	Y	Zr	Nb	Mo	Tc	Ru	Rh	Pd	Ag	Cd	In	Sn	Sb	Te	I	Xe
6	Cs	Ba	La*	Hf	Ta	W	Re	Os	Ir	Pt	Au	Hg	Tl	Pb	Bi	Po	At	Rn
7	Fr	Ra	Ac†	104	105	106	107	108	109	110?								

*Lanthanides (Group 3)

(6)	Ce	Pr	Nd	Pm	Sm	Eu	Gd	Tb	Dy	Ho	Er	Tm	Yb	Lu

†Actinides (Group 3)

(7)	Th	Pa	U	Np	Pu	Am	Cm	Bk	Cf	Es	Fm	Md	No	Lr

The periodic table of elements, displayed in the modern 'long' form. Each element is denoted by its atomic number or symbol. The groups have been numbered according to the recent IUPAC recommendation. Relative atomic masses taken from synthesis by Cox (1989).

Isotopic abundance and variation

The table shows the isotopic composition of the naturally occurring elements. When there is a well-established isotopic variation in natural terrestrial sources of an element (note v), the extent of variability is normally at the level of the last significant figure quoted. In this case, the percentage abundances given are those likely to be found in laboratory samples, and not an overall average. Where no normal terrestrial variability has been established, the number of significant figures given reflect the accuracy of measurement. The table is adapted from IUPAC (1983), after Cox (1989).

Notes

v Elements for which isotopic *variability* has been established in normal samples of terrestrial origin.

e Elements for which natural samples of *exceptional* provenance (for example meteorites or the Oklo mine) may have a composition outside the normal range.

m Elements where some commercially available samples may have a *modified* composition, as a result of deliberate or inadvertent fractionation.

r Long-lived *radioactive* isotope. (The decay mode is given if known and the half-life in years.)

d *Daughter* isotope or decay product of a naturally occurring long-lived radioactive isotope.

s *Short-lived* radioactive element: only the longest-lived isotope is given. (Half-life in years except when shown otherwise.)

Isotopic compositions of the elements

Element	Mass number	Percentage abundance	Note
H			v,e,m
	1	99.985	
	2	0.015	
He			v,e
	3	<0.005	
	4	100.00	d
Li			v,e,m
	6	7.5	
	7	92.5	

Isotopic compositions of the elements

Element	Mass number	Percentage abundance	Note
Be			
	9	100	
B			v,m
	10	19.9	
	11	80.1	
C			v,e
	12	98.90	
	13	1.10	
N			v
	14	99.634	
	15	0.366	
O			v
	16	99.76	
	17	0.04	
	18	0.20	
F			
	19	100	
Ne			v,e,m
	20	90.51	
	21	0.27	
	22	9.22	
Na			
	23	100	
Mg			e
	24	78.99	
	25	10.00	
	26	11.01	
Al			
	27	100	
Si			
	28	92.2	
	29	4.7	
	30	3.1	
P			
	31	100	
Cl			
	35	75.77	
	37	24.3	
Ar			e
	36	0.336	
	38	0.063	
	40	99.600	d
K			
	39	93.258	
	40	0.012	r (β^-, EC, 1.28×10^9)
	41	6.730	
Ca			v,e
	40	96.94	d
	42	0.65	
	43	0.14	
	44	2.09	
	46	0.004	
	48	0.19	
Sc			
	45	100	

Isotopic compositions of the elements

Element	Mass number	Percentage abundance	Note
Ti			
	46	8.0	
	47	7.3	
	48	73.8	
	49	5.5	
	50	5.4	d
V			e
	50	0.250	r (6×10^{15})
	51	99.750	
Cr			
	50	4.35	d
	52	83.79	
	53	9.50	
	54	2.36	
Mn			
	55	100	
Fe			
	54	5.8	
	56	91.72	
	57	2.2	
	58	0.28	
Co			
	59	100	
Ni			
	58	68.27	
	60	26.10	
	61	1.13	
	62	3.59	
	63	0.91	
Cu			v
	63	69.17	
	65	30.83	
Zn			
	64	48.6	
	66	27.9	
	67	4.1	
	68	18.8	
	70	0.6	
Ga			
	69	60.1	
	71	39.9	
Ge			
	70	20.5	
	72	27.4	
	73	7.8	
	74	36.5	
	76	7.8	
As			
	75	100	
Se			v
	74	0.9	
	76	9.0	
	77	7.6	

Isotopic compositions of the elements

Element	Mass number	Percentage abundance	Note
Se (*contd*)			
	78	23.5	
	80	49.6	
	82	9.4	
Br			
	79	50.69	
	81	49.31	
Kr			e,m
	78	0.35	
	80	0.25	
	82	11.6	
	83	11.5	
	84	57.9	
	86	17.3	
Rb			e
	85	72.17	
	87	27.83	$r\,(\beta^-, 5 \times 10^{11})$
Sr			e
	84	0.56	
	86	9.86	
	87	7.00	d
	88	82.58	
Y			
	89	100	
Zr			e
	90	51.45	
	91	11.27	
	92	17.17	
	94	17.33	
	96	2.78	r (?)
Nb			
	93	100	
Mo			e
	92	14.84	
	94	9.25	
	95	15.92	
	96	16.68	
	97	9.55	
	98	24.13	
	100	9.63	
Tc			s
	97	–	$(EC, 2.6 \times 10^6)$
Ru			e
	96	5.52	
	98	1.88	
	99	12.7	
	100	12.6	
	101	17.0	
	102	31.6	
	104	18.7	
Rh			
	103	100	
Pd			v,e
	102	1.02	

APPENDIX 2

Isotopic compositions of the elements

Element	Mass number	Percentage abundance	Note
Pd (*contd*)			
	104	11.14	
	105	22.33	
	106	27.33	
	108	26.46	
	110	11.72	
Ag			e
	107	51.839	
	108	48.161	
Cd			e
	106	1.25	
	108	0.89	
	110	12.49	
	111	12.80	
	112	24.13	
	113	12.22	
	114	28.73	
	116	7.49	
In			e
	113	4.3	
	115	95.7	
Sn			e
	112	1.0	
	114	0.7	
	115	0.4	
	116	14.7	
	117	7.7	
	118	24.3	
	119	8.6	
	120	32.4	
	122	4.6	
	124	5.6	
Sb			
	121	57.3	
	123	42.7	d
Te			e
	120	0.096	
	122	2.60	
	123	0.908	r (EC, 1.2×10^{13})
	124	4.816	
	125	7.14	
	126	18.95	
	128	31.69	
	130	33.80	
I			
	127	100	
Xe			e,m
	124	0.10	
	126	0.09	
	128	1.91	
	129	26.4	
	130	4.1	
	131	21.2	
	132	26.9	
	134	10.4	

Isotopic compositions of the elements

Element	Mass number	Percentage abundance	Note
Xe (*contd*)			
	136	8.9	
Cs			
	133	100	
Ba			e
	130	0.106	
	132	0.101	
	134	2.42	
	135	6.59	
	136	7.85	
	137	11.23	
	138	71.70	
La			e
	138	0.09	r (?)
	139	99.91	
Ce			e
	136	0.19	
	138	0.25	
	140	88.48	d
	142	11.08	
Pr			
	141	100	
Nd			e
	142	27.13	
	143	12.18	d
	144	23.80	d, r $(\alpha, 5 \times 10^{15})$
	145	8.30	d
	146	17.19	
	148	5.76	
	150	5.64	
Pm			s
	145	–	(EC, 17.7)
Sm			e
	144	3.1	
	147	15.0	r $(\alpha, 1.06 \times 10^{11})$
	148	11.3	d,r $(\alpha, 1.2 \times 10^{13})$
	149	13.8	r $(\alpha, 4 \times 10^{14})$
	150	7.4	
	152	26.7	
	154	22.7	
Eu			e
	151	47.8	
	153	52.2	
Gd			e
	152	0.20	r $(\alpha, 1.1 \times 10^{14})$
	154	2.18	
	155	14.80	
	156	20.47	
	157	15.65	
	158	24.84	
	160	21.86	
Tb			
	159	100	
Dy			e
	156	0.06	

Isotopic compositions of the elements

Element	Mass number	Percentage abundance	Note
Dy (*contd*)			
	158	0.10	
	160	2.34	
	161	18.9	
	162	25.5	
	163	24.9	
	164	28.2	
Ho			
	165	100	
Er			e
	162	0.14	
	164	1.61	
	166	33.6	
	167	22.95	
	168	26.8	
	170	14.9	
Tm			
	169	100	
Yb			e
	168	0.13	
	170	3.05	d
	171	14.3	
	172	21.9	
	173	16.1	
	174	31.8	
	176	12.7	
Lu			e
	175	97.40	
	176	2.60	r (β^-, 3×10^{10})
Hf			
	174	0.16	r (α, 2×10^{15})
	176	5.2	d
	177	18.6	
	178	27.1	
	179	13.7	
	180	35.2	
Ta			
	180	0.012	r (?)
	181	99.988	
W			
	180	0.13	
	182	26.3	
	183	14.3	
	184	30.7	
	186	28.6	
Re			
	185	37.40	
	187	62.60	r (β^-, 7×10^{10})
Os			e
	184	0.02	
	186	1.58	d
	187	1.6	d
	188	13.3	d
	189	16.1	

Isotopic compositions of the elements

Element	Mass number	Percentage abundance	Note
Os (*contd*)			
	190	26.4	
	182	41.0	
Ir			
	191	37.3	
	193	62.7	
Pt			
	190	0.01	r (α, 6×10^{11})
	192	0.79	r (α, 10^{15})
	194	32.9	
	195	33.8	
	196	25.2	
	198	7.2	
Au			
	197	100	
Hg			
	196	0.15	
	198	10.1	
	199	17.0	
	200	23.1	
	201	13.2	
	202	26.7	
	204	6.8	
Tl			
	203	29.52	
	204	70.48	
Pb			v,e
	204	1.4	
	206	24.1	d
	207	22.1	d
	208	52.4	d
Bi			
	209	100	
Po			s
	209	–	d (α, 103)
At			s
	206	–	d (α, EC, 32m)
Rn			s
	222	–	d (α, 3.82d)
Fr			s
	223	–	d (β^-, 22m)
Ra			s
	226	–	d (α, 1,600)
Ac			s
	227	–	d (β^-, 21.6)
Th			e
	232	100	r (α, 1.41×10^{10})
Pa			s
	231	–	d (α, 3.25×10^4)
U			v,e,m
	234	0.0055	d,r (α, 2.47×10^5)
	235	0.720	r (α, 7.1×10^8)
	238	99.275	r (α, 4.51×10^9)

Appendix 3 reproduces the Rio Declaration on Environment and Development, associated with Agenda 21 – a programme of action for sustainable development worldwide – adopted by more than 178 governments at the United Nations, Conference on Environment and Development or Earth Summit, held in Rio de Janeiro, Brazil, 3–14 June 1992. While the agreements, which were negotiated over two and a half years leading up to the Earth Summit and finalized in Rio, lack the force of international law, the adoption of the text carries with it a strong moral obligation to ensure their full implementation.

Agenda 21 represents a comprehensive blueprint for global action into the twenty-first century by governments, the United Nations organizations, development agencies, non-governmental organizations and independent sector groups, in all spheres of human activity which impacts on the environment. The United Nations Commission on Sustainable Development, set up under the aegis of the UN General Assembly in response to a request of the Rio conference and comprising government representatives, will examine the progress made in implementing Agenda 21 worldwide.

The central tenet of Agenda 21 is that humanity stands at a defining moment in history, where we face the perpetuation of disparities between and within nations, increasing poverty, ill health, and illiteracy, and the continuing deterioration of the ecosystems on which we depend for our well-being and survival. Basic human needs and improved living standards for all can be met, at least to a far greater extent than at present, by integrating environmental and developmental concerns. A global partnership is the only means of achieving such aspirations, and Agenda 21 provides a framework within which this can be attempted.

The Agenda 21 programme is set out in terms of the basis for action, objectives, activities, and means of implementation. It is therefore a dynamic programme. Here, only the principles of the Rio Declaration on Environment and Development are printed in order to provide the reader with the spirit and aims of Agenda 21.

Rio Declaration on Environment and Development, Agenda 21

Having met at Rio de Janeiro from 3 to 14 June 1992,

Reaffirming the Declaration of the United Nations Conference on the Human Environment, adopted at Stockholm on 16 June 1972, and seeking to build upon it,

With the goal of establishing a new and equitable global partnership through the creation of new levels of cooperation among States, key sectors of societies and people,

Working towards international agreements which respect the interests of all and protect the integrity of the global environmental and developmental system,

Recognizing the integral and interdependent nature of the Earth, our home,

Proclaims that:

Principle 1

Human beings are at the centre of concerns for sustainable development. They are entitled to a healthy and productive life in harmony with nature.

Principle 2

States have, in accordance with the Charter of the United Nations and the principles of international law, the sovereign right to exploit their own resources pursuant to their own environmental and developmental policies, and the responsibility to ensure that activities within their jurisdiction or control do not cause damage to the environment of other States or of areas beyond the limits of national jurisdiction.

Principle 3

The right to development must be fulfilled so as to equitably meet developmental and environmental needs of present and future generations.

Principle 4

In order to achieve sustainable development, environmental protection shall constitute an integral part of the development process and cannot be considered in isolation from it.

Principle 5

All States and all people shall cooperate in the essential task of eradicating poverty as an indispensable requirement for sustainable development, in order to decrease the disparities in standards of living and better meet the needs of the majority of the people of the world.

Principle 6

The special situation and needs of developing countries, particularly the least developed and those most environmentally vulnerable, shall be given special priority. International actions in the field of environment and development should also address the interests and needs of all countries.

Principle 7

States shall cooperate in a spirit of global partnership to conserve, protect and restore the health and integrity of the Earth's ecosystem. In view of the different contributions to global environmental degradation, States have common but differentiated responsibilities. The developed countries acknowledge the responsibility that they bear in the international pursuit of sustainable development in view of the pressures their societies place on the global environment and of the technologies and financial resources they command.

Principle 8

To achieve sustainable development and a higher quality of life for all people, States should reduce and eliminate unsustainable patterns of production and consumption and promote appropriate demographic policies.

Principle 9

States should cooperate to strengthen endogenous capacity-building for sustainable development by improving scientific understanding through exchanges of scientific and technological knowledge, and by enhancing the development, adaptation, diffusion and transfer of technologies, including new and innovative technologies.

Principle 10

Environmental issues are best handled with the participation of all concerned citizens, at the relevant level. At the national level, each individual shall have appropriate access to information concerning the environment that is held by public authorities, including information on hazardous materials and activities in their communities, and the opportunity to participate in decision-making processes. States shall facilitate and encourage public awareness and participation by making information widely available. Effective access to judicial and administrative proceedings, including redress and remedy, shall be provided.

Principle 11

States shall enact effective environmental legislation. Environmental standards, management objectives and priorities should reflect the environmental and developmental context to which they apply. Standards applied by some countries may be inappropriate and of unwarranted economic and social cost to other countries, in particular developing countries.

Principle 12

States should cooperate to promote a supportive and open international economic system that would lead to economic growth and sustainable development in all countries, to better address the problems of environmental degradation. Trade policy measures for environmental purposes should not constitute a means of arbitrary or unjustifiable discrimination or a disguised restriction on international trade. Unilateral actions to deal with environmental challenges outside the jurisdiction of the importing country should be avoided. Environmental measures addressing transboundary or global environmental problems should, as far as possible, be based on an international consensus.

Principle 13

States shall develop national law regarding liability and compensation for the victims of pollution and other environmental damage. States shall also cooperate in an expeditious and more determined manner to develop further international law regarding liability and compensation for adverse effects of environmental damage caused by activities within their jurisdiction or control to areas beyond their jurisdiction.

Principle 14

States should effectively cooperate to discourage or prevent the relocation and transfer to other States of any activities and substances that cause severe environmental degradation or are found to be harmful to human health.

Principle 15

In order to protect the environment, the precautionary approach shall be widely applied by States according to their capabilities. Where there are threats of serious or irreversible damage, lack of full scientific certainty shall not be used as a reason for postponing cost-effective measures to prevent environmental degradation.

Principle 16

National authorities should endeavour to promote the internalization of environmental costs and the use of economic instruments, taking into account the approach that the polluter should, in principle, bear the cost of pollution, with due regard to the public interest and without distorting international trade and investment.

Principle 17

Environmental impact assessment, as a national instrument, shall be undertaken for proposed activities that are likely to have a significant adverse impact on the environment and are subject to a decision of a competent national authority.

Principle 18

States shall immediately notify other States of any natural disasters or other emergencies that are likely to produce sudden harmful effects on the environment of those States. Every effort shall be made by the international community to help States so afflicted.

Principle 19

States shall provide prior and timely notification and relevant information to potentially affected States on activities that may have a significant adverse transboundary environmental effect and shall consult with those States at an early stage and in good faith.

Principle 20

Women have a vital role in environmental management and development. Their full participation is therefore essential to achieve sustainable development.

Principle 21

The creativity, ideals and courage of the youth of the

world should be mobilized to forge a global partnership in order to achieve sustainable development and ensure a better future for all.

Principle 22

Indigenous people and their communities and other local communities have a vital role in environmental management and development because of their knowledge and traditional practices. States should recognize and duly support their identity, culture and interests and enable their effective participation in the achievement of sustainable development.

Principle 23

The environment and natural resources of people under oppression, domination and occupation shall be protected.

Principle 24

Warfare is inherently destructive of sustainable development. States shall therefore respect international law providing protection for the environment in times of armed conflict and cooperate in its further development, as necessary.

Principle 25

Peace, development and environmental protection are interdependent and indivisible.

Principle 26

States shall resolve all their environmental disputes peacefully and by appropriate means in accordance with the Charter of the United Nations.

Principle 27

States and people shall cooperate in good faith and in a spirit of partnership in the fulfilment of the principles embodied in this Declaration and in the further development of international law in the field of sustainable development.

Source: *Earth Summit Agenda 21: The United Nations Programme of Action from Rio.* United Nations Publication E.93.1.11 (April 1993).

APPENDIX 4
Summary of principal fluxes and bio-geochemical cycles on Earth

Note: Diagrams in Appendix 4 have been redrawn and modified from the UK Natural Environment Research Council publication *Our Future World: Global Environmental Research* (1989).

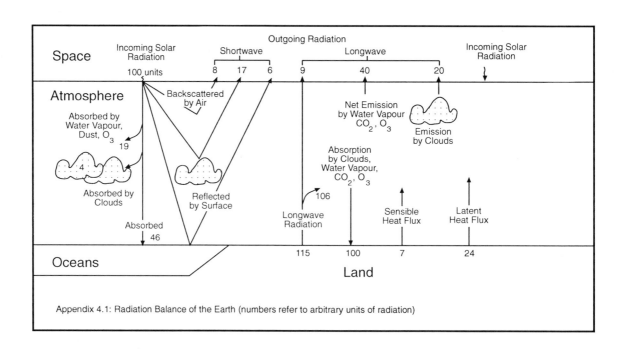

Appendix 4.1: Radiation Balance of the Earth (numbers refer to arbitrary units of radiation)

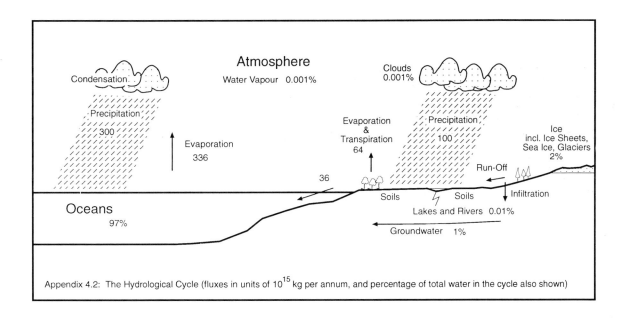

Appendix 4.2: The Hydrological Cycle (fluxes in units of 10^{15} kg per annum, and percentage of total water in the cycle also shown)

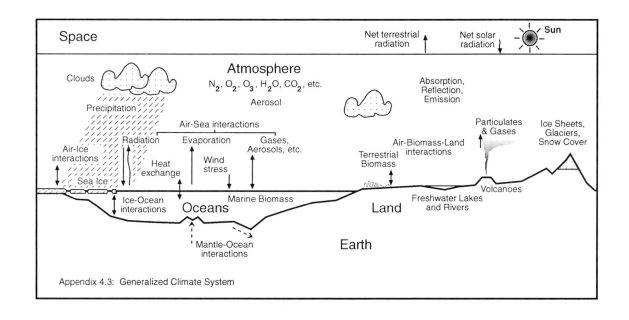

Appendix 4.3: Generalized Climate System

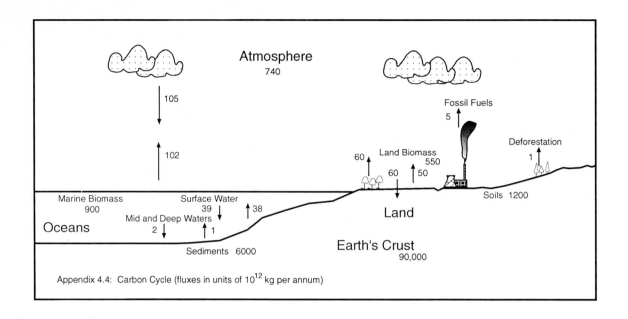

Appendix 4.4: Carbon Cycle (fluxes in units of 10^{12} kg per annum)

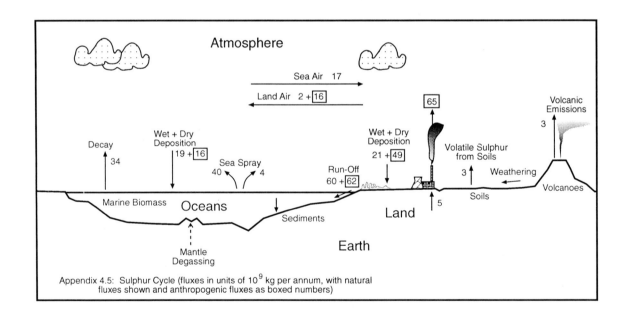

Appendix 4.5: Sulphur Cycle (fluxes in units of 10^9 kg per annum, with natural fluxes shown and anthropogenic fluxes as boxed numbers)

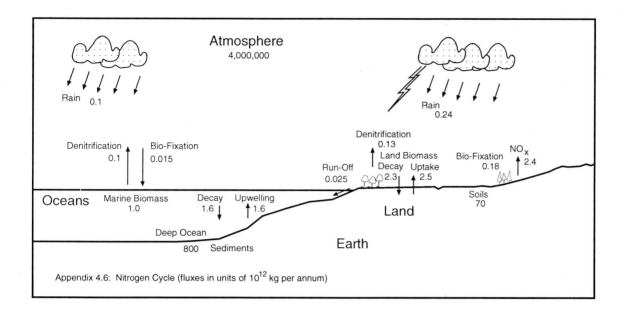

Appendix 4.6: Nitrogen Cycle (fluxes in units of 10^{12} kg per annum)

Bibliography

Adams, W.M. 1990. *Green Development: Environment and Sustainability in the Third World.* London: Routledge.

AFEAS (Alternative Fluorocarbons Environmental Acceptability Study) and PAFT (Programme for Alternative Fluorocarbon Toxicity Testing) Member Companies 1992. *Information pack.*

Allard, P., Carbonnelle, J., Dajlevic, D., Le Bronec, J., Morel, P., Robe, M.C., Maurenas, J.M., Faivre-Pierret, R., Martin, D., Sabrouk, J.C. and Zettwoag, P. 1991. Eruptive and diffuse emissions of CO_2 from Mount Etna. *Nature*, 351, pp. 387–91.

Allen, J.C., Schaffer, W.M. and Rosko, D. 1993. Chaos reduces species extinction by amplifying local population noise. *Nature*, 364, pp. 229–32.

Alvarez, L.W., Alvarez, W., Asaro, F. and Michel, H.V. 1980. Extraterrestrial cause for the Cretaceous-Tertiary extinction. *Science*, 208, pp. 1095–108.

Alvarez, W. and Asaro, F. 1990. An extraterrestrial impact. *Scientific American*, October, pp. 44–60.

American Geophysical Union 1992a. *Volcanism and Climatic Change.* UGU Special Report.

American Geophysical Union 1992b. Impact of Mid-Atlantic sewage sludge probed. *EOS*, 73, pp. 27–8.

Anderson, I. 1993. Diggers at Dinosaur Cove. *New Scientist*, 13 February, pp. 28–32.

Anderson, R. 1993. AIDS: trends, predictions, controversy. *Nature*, 363, pp. 393–4.

Anderson, T.F. and Arthur, M.A. 1983. Stable isotopes of oxygen and carbon and their application to sedimentologic and environmental problems. In: *Stable Isotopes in Sedimentary Geology*, SEPM Short Course Notes 10. Tulsa, Oklahoma: Society of Economic Paleontologists and Mineralogists, pp. 1–151.

Anderson, V. 1991. *Alternative Economic Indicators.* London: Routledge.

Angotti, T. 1993. *Metropolis 2000: Planning, Poverty and Politics.* London: Routledge.

Archibald, J.D. 1993. Were dinosaurs born losers? *New Scientist*, 13 February, pp. 24–7.

Austin, J., Butchart, N. and Shine, K.P. 1992. Possibility of an Arctic ozone hole in a doubled-CO_2 climate. *Nature*, 360, pp. 221–5.

Ayers, G.P., Penkett, S.A., Gillett, R.W., Bandy, B., Galbally, I.E., Meyer, C.P., Elsworth, C.M., Bentley, S.T. and Forgan, B.W. 1992. Evidence for photochemical control of ozone concentrations in unpolluted marine air. *Nature*, 360, pp. 446–9.

Badr, O. and Probert, S.D. 1992. Sources of atmospheric nitrous oxide. *Applied Energy*, 42, pp. 129–76.

Badr, O., Probert, S.D. and O'Callaghan, P.W. 1992a. Methane: a greenhouse gas in the Earth's atmosphere. *Applied Energy*, 41, pp. 95–113.

Badr, O., Probert, S.D. and O'Callaghan, P.W. 1992b. Sinks for atmospheric methane. *Applied Energy*, 41, pp. 137–47.

Bailey, E. 1962. *Charles Lyell. British Men of Science.* London: Nelson.

Baillie, M.G.L. and Munro, M.A.R. 1988. Irish tree-rings, Santorini and volcanic dust veils. *Nature*, 332, pp. 344–6.

Bakan, S., Chlond, A., Cubasch, U., Feichter, J., Graf, H., Grassl, H., Hasselman, K., Kirchner, I., Latif, M., Roekner, E., Sausen, R., Schlese, U., Schriever, D., Schult, I., Schumann, U., Seilmann, F. and Welke, W. 1991. Climate response to smoke from the burning oil wells in Kuwait. *Nature*, 351, pp. 367–71.

Ball, T.K., Cameron, D.G., Colman, T.B. and Roberts, P.D. 1991. Behaviour of radon in the geological environment: a review. *Quarterly Journal of Engineering Geology*, 24, pp. 169–82.

Barnola, J.M., Raynaud, D., Korotkevich, Y.S. and Lorius, C. 1987. Vostok ice core provides 160,000-year record of atmospheric CO_2. *Nature*, 329, pp. 408–14.

Barrett, P.J., Adams, C.J., McIntosh, W.C., Swisher, C.C. and Wilson, G.S. 1992. Geochronological evidence supporting Antarctic deglaciation 3 million years ago. *Nature*, 359, pp. 816–18.

Bassett, M.G. 1985. Towards a 'common language' in stratigraphy. *Episodes*, 8, pp. 87–92.

Batifol, F., Boutron, C. and de Angelis, M. 1989. Changes in copper, zinc and cadmium concentration in Antarctic ice during the past 40,000 years. *Nature*, 337, pp. 544–6.

Battarbee, R.W. 1984. Diatom analysis and the acidification of lakes. *Philosophical Transactions of the Royal Society, London*, B 305, pp. 451–77.

Battarbee, R.W. 1986. Diatom analysis. In Berglund, B.E. (ed.) *Handbook of Holocene Paleoecology and Palaeohydrology*. Chichester: John Wiley & Sons, pp. 527–70.

Battarbee, R.W. 1992. Holocene lake sediments, surface water acidification and air pollution. In Gray, J.M. (ed.), *Applications of Quaternary Research*. Quaternary Proceedings, 2, Cambridge, pp. 101–10.

Batey, T. 1988. *Soil husbandry*. Aberdeen: Soil and Landuse Consultants.

Becker, B., Kromer, B. and Trimborn, P. 1991. A stable-isotope tree-ring timescale of the Late Glacial/Holocene boundary. *Nature*, 353, pp. 647–9.

Bedding, J. 1989. Money down the drains. *New Scientist*, 122 (1663), pp. 37–41.

Beder, S. 1990. Sun, surf and sewage. *New Scientist*, 14 July, pp. 40–5.

Bekki, S., Toumi, R. and Pyle, J.A. 1993. Role of sulphur photochemistry in tropical ozone changes after the eruption of Mount Pinatubo. *Nature*, 362, pp. 331–3.

Bell, M. and Walker, M.J.C. 1992. *Late Quaternary Environmental Change: Physical and Human Perspectives*. New York: Longman Scientific.

Berkhout, F. 1991. *Radioactive Waste: Politics and Technology*. London: Routledge.

Blunden, J. and Reddish, A. (eds) 1991. *Energy, Resources and Environment*. London: Hodder & Stoughton.

Bond, G., Heinrich, H., Broecker, W., Labeyrie, L., McManus, J., Andrews, J., Huon, S., Jantschik, R., Clasen, S., Simet, C., Tedesco, K., Klas, M., Bonani, G. and Ivy, S. 1992. Evidence for massive discharges of icebergs into the North Atlantic ocean during the last glacial period. *Nature*, 360, pp. 245–9.

Bonney, T.G. 1895. *Charles Lyell and Modern Geology*. London: Cassell & Co.

Boulton, G.S., Smith, G.D., Jones, A.S. and Newsome, J. 1985. Glacial geology and glaciology of the last mid-latitude ice sheets. *Journal of the Geological Society, London*, 142, pp. 447–74.

Bowen, D.Q., Rose, J., McCabe, A.M. and Sutherland, D.G. 1986. Correlation of Quaternary glaciations in England, Ireland, Scotland and Wales. *Quaternary Science Reviews*, 5, pp. 299–340.

Bowler, S. 1993. Where the power lies. *New Scientist*, 23 January, pp. 32–6.

Boyle, E.A. 1988. *Paleoceanography*, 3, pp. 471–89.

Boyle, S. 1989. More work for less energy. *New Scientist*, 5 August, pp. 37–40.

Boyle, S. and Ardill, J. 1989. *The Greenhouse Effect*. London: Hodder & Stoughton.

Bradley, R.S. 1985. *Quaternary Paleoclimatology – Methods of Paleoclimatic Reconstruction*. London: Unwin Hyman.

Bradley, R.S. and Jones, P.D. 1992. Records of explosive volcanic eruptions over the last 500 years. In Bradley, R.S. and Jones, P.D. (eds), *Climate Since AD 1500*. London: Routledge, pp. 606–22.

Bradshaw, M. and Weaver, R. 1993. *Physical Geography: An Introduction to Earth Environments*. London: Mosby.

Brandt, W. 1980. *North–South: A Programme for Survival*. London: Pan Books.

Brandt, W. 1983. *Common Crisis North–South: Co-operation for World Recovery*. London: Pan Books.

Brasseur, G. 1992. Ozone depletion, volcanic aerosols implicated. *Nature*, 359, pp. 275–6.

Breuer, G. 1991. A strategy for the sea floor. *New Scientist*, 12 October, pp. 34–7.

Broecker, W.S. 1987. *How to Build a Habitable Planet*. New York: Eldigio Press.

Broecker, W.S. and Denton, G.H. 1990. What drives glacial cycles? *Scientific American*, 262, pp. 42–50.

Brown, L.R. and Wolf, E.C. 1984. *Soil Erosion: Quiet Crisis in the World Economy*. Washington: Worldwatch Institute.

Brown, P. 1992. AIDS: the challenge of the future. *New Scientist*, 18 April, *Inside Science*, 54.

Brown, W. 1991. Europe's lost ozone. *New Scientist*, 131, p. 1779.

Browning, K.A., Allaqm, R.J., Ballard, S.P., Barnes, R.T.H., Bennetts, D.A., Maryor, R.H., Mason, P.J., McKenna, D., Mitchell, J.F.B., Senior, C.A., Slingo, A. and Smith, F.B. 1991. Environmental effects from burning oil wells in Kuwait. *Nature*, 351, pp. 363–7.

Brundtland, H. 1987. *Our Common Future*. Report for the World Commission on Environment and Development. Oxford: Oxford University Press.

Bryson, R.A. 1968. All other factors being constant: a reconciliation of several theories of climatic change. *Weatherwise*, 21, pp. 56–61.

BSVRP (British Seismic Verification Research Project)

BIBLIOGRAPHY

1989. *Quarterly Journal of the Royal Astronomical Society*, 30, pp. 311–24.

Buffin, D. 1992. Calls to phase-out methyl bromide: a major ozone depleter. *Pesticides News*, 18, pp. 5–11.

Burman, A. 1991. Saving Brazil's savannas. *New Scientist*, 2 March, pp. 30–4.

Burton, I., Kates, W.R. and Gilbert, F.W. 1978. *The Environment as Hazard*. Oxford: Oxford University Press.

Butler, J.H., Elkins, J.W., Hall, B.D., Cummings, S.O. and Montzka, S.A. 1992. A decrease in the growth rates of atmospheric halon concentrations. *Nature*, 359, pp. 403–5.

Byrant, E.A. 1991. *Natural Hazards*. Cambridge: Cambridge University Press.

Cadman, D. and Payne, G. (eds) 1990. *The Living City: Towards a Sustainable Future*. London: Routledge.

Caldeira, K. and Kasting, J.F. 1992. Susceptibility of the early Earth to irreversible glaciation caused by carbon dioxide clouds. *Nature*, 359, pp. 226–8.

Calvert, S.E., Nielsen, B. and Fontugne, M.R. 1992. Evidence from nitrogen isotope ratios for enhanced productivity during formation of eastern Mediterranean sapropels. *Nature*, 359, pp. 223–5.

Carlisle, D.B. and Braman, D.R. 1991. Nanometre-size diamonds in the Cretaceous/Tertiary boundary clay of Alberta. *Nature*, 352, pp. 708–9.

Carson, R. 1962. *Silent Spring*, London: Paladin.

Cassedy, E.S. and Grossman, P.Z. 1990. *Introduction to Energy: Resources, Technology, and Society*. Cambridge: Cambridge University Press.

Caufield, C. 1986. *In the Rainforest*. London: Picador.

Chadwick, M.J. and Kuylenstierna, J.C.I. 1990. *The Relative Sensitivity of Ecosystems in Europe to Acidic Deposition: A Preliminary Assessment of the Sensitivity of Aquatic and Terrestrial Ecosystems*. Stockholm: Stockholm Environment Institute.

Chahine, M.T. 1992. The hydrological cycle and its influence on climate. *Nature*, 359, pp. 373–80.

Chappell, J. and Shackleton, N.J. 1986. Oxygen isotopes and sea level. *Nature*, 324, pp. 137–40.

Charles, D. 1993. In search of a better burn. *New Scientist*, 23 January, pp. 20–5.

Charles, D.F., Battarbee, R.W., Renberg, I., Van Dam, H. and Smol, J.P. 1989. Paleoecological analysis of lake acidification trends in North America and Europe using diatoms and chrysophytes. In Nortan, S.A., Lindberg, S.E. and Page, A.L. (eds), *Acid Precipitation*, vol. 4. *Soils, Aquatic Processes, and Lake Acidification*. New York: Springer-Verlag, pp. 208–76.

Chesner, C.A., Rose, W.I., Deino, A., Drake, R. and Westgate, J. 1991. Eruptive history of Earth's largest Quaternary caldera (Toba, Indonesia) clarified. *Geology*, 19, pp. 200–3.

Chester, D. 1993. *Volcanoes and Society*. London: Edward Arnold.

Clapman Jr, W.B. 1973. *Natural Ecosystems*. London: Collier-Macmillan.

Clark, R.B. 1989. *Marine Pollution*. Oxford: Clarendon Press.

Clark, R.H. and Southwood, T.R.E. 1989. Risks from ionizing radiation. *Nature*, 338, pp. 197–8.

Clark, W.C. 1989. Managing Planet Earth. *Scientific American*, 261, pp. 47–54.

Clarke, A. 1989. How green is the wind? *New Scientist*, 27 May, pp. 62–5.

Coffin, M.F. and Eldholm, O. 1993. Large igneous provinces. *Scientific American*, 269, pp. 26–33.

Colhoun, E.A., Mabin, M.C.G., Adamson, D.A. and Kirk, R.M. 1992. Antarctic ice volume and contribution to sea-level fall at 20,000 yr BP from raised beaches. *Nature*, 358, pp. 316–19.

Colodner, D.C., Boyle, E.A., Edmond, J.M. and Thomson, J. 1992. Post-depositional mobility of platinum, iridium and rhenium in marine sediments. *Nature*, 358, pp. 402–4.

Committee on Monitoring and Assessment of Trends in Acid Deposition 1986. *Acid Deposition: Long-Term Trends*.Washington, DC: National Academy Press.

CONCAWE 1992. *Motor Vehicle Emission Regulations and Fuel Specifications, Update*. Report no. 2/92 by AE/STF-3. Brussels.

Cook, E., Bird, T., Peterson, M., Barbetti, M., Buckley, B., D'Arrigo, R., Francey, R. and Tans, P. 1991. Climatic change in Tasmania inferred from a 1089-year tree-ring chronology of Huon Pine. *Science*, 253, pp. 1266–8.

Cook, J. 1989. *Dirty Water*. London: Union Papersachs.

Cooke, R.U. and Doornkamp, J.C. 1990. *Geomorphology in Environmental Management*. 2nd edn. Oxford: Oxford University Press.

Coope, G.R. 1986. Coleoptera analysis. In Berglund, B.E. (ed.), *Handbook of Holocene Palaeoecology and Palaeohydrology*. Chichester: John Wiley & Sons, pp. 703–13.

Cox, P.A. 1989. *The Elements: Their Origin, Abundance and Distribution*. Oxford: Oxford Scientific Publications.

Craig, H. 1965. The measurement of oxygen isotopes in oceanographic studies and palaeotemperatures. In *Consiglio Nazionale della Richerche Laboratoriodi Geologia Nucleare, Pisa*, pp. 3–24.

Cragg, C. 1993. Demanding plans for power cuts. *New Scientist*, 27 March, pp. 13–14.

Cross, M. 1993. A very dirty business. *New Scientist*, 23 January, pp. 28–31.

Crosson, P.R. and Rosenberg, N.J. 1989. Strategies for agriculture. *Scientific American*, 261, pp. 128–35.

Culshaw, F. and Butler, C. 1993. *A Review of the Potential of Biodiesel as a Transport Fuel*. Energy Technology Support Unit, ETSU-R-71. London: HMSO.

Daffern, T. 1983. *Avalanche Safety*. London: Diadem Books.

Dansgaard, W.S. 1984. Selected climates from the past

and their relevance to possible future climate. In Flohn, H. and Fantechi, R. (eds), *The Climate of Europe: Past, Present and Future.* Dordrecht: D. Reidel, pp. 208–13.

Dansgaard, W.S., Clausen, H.B., Gundestrup, N., Hammer, C.U., Johnson, S.F., Kristinsdottir, P.M. and Reeh, N. 1982. A new Greenland deep ice core. *Science*, 218, pp. 1273–7.

Dansgaard, W.S., Johnsen, S.J., Clausen, H.B., Dahl-Jensen, D., Gundestrup, N.S., Hammer, C.U., Hvidberg, C.S., Steffesnsen, J.P., Sveinbjornsdottir, A.E., Jouzel, J. and Bond, G. 1993. Evidence for general instability of past climate from a 250-kyr ice-core record. *Nature*, 364, pp. 218–20.

Dansgaard, W.S. and Tauber, H. 1969. Glacier oxygen 18 content and Pleistocene ocean temperatures. *Science*, 166, pp. 499–502.

Davis, G.R. 1990. Energy for Planet Earth. *Scientific American*, 263, pp. 21–7.

Dawkins, R. 1986. *The Blind Watchmaker.* London: Longman Scientific & Technical.

Dawson, A.G. 1992. *Ice Age Earth: Late Quaternary Geology and Climate.* London: Routledge.

Decker, R. and Decker, B. 1989. *Volcanoes.* New York: W.H. Freeman & Co.

Denton, G.H. and Hughes, T.J. 1981. *The Last Great Ice Sheets.* New York: John Wiley & Sons.

Denton, G.H., Wilson, S.C. and Stuiver, M. 1989. Late Weichselian and Early Holocene glacial history, Inner Ross Embayment, Antarctic. *Quaternary Research*, 31, pp. 319–82.

Department of the Environment 1988. *Our Common Future: A Perspective by the UK on the Report of the World Commission on Environment and Development.* London: HMSO.

Department of the Environment 1990. *The Householders' Guide to Radon.* 2nd edn. London: HMSO.

Department of the Environment 1992a. *Climate Change: Our National Programme for CO₂ Emissions.* London: HMSO.

Department of the Environment 1992b. *This Common Inheritance: Britain's Environmental Strategy.* London: HMSO.

Derbyshire, E., Wang Jingtai, Jin Zexian, Billard, A., Egels, Y., Kasser, M., Jones D.K.C., Muxart, T. and Owen, L. 1991. Landslides in the Gansu loess of China. *Catena Supplement*, 20, pp. 119–45.

Des Morais, D.J., Strauss, H., Summons, R.E. and Hayes, J.M. 1992. Carbon isotope evidence for the stepwise oxidation of the Proterozoic environment. *Nature*, 359, pp. 605–9.

Dinman, B.D. 1980. The reality and acceptance of risk. *Journal of the American Medical Association*, 244, pp. 1226–8.

Dostrovsky, I. 1991. Chemical fuels from the Sun. *Scientific American*, 265, pp. 50–66.

Eden, M.J. 1989. *Land Management in Amazonia.* London: Belhaven.

Ehrlich, P.R. 1968. *The Population Bomb.* New York: Ballentine Books.

Ekins, P. 1992. *A New World Order: Grassroots Movements for Global Change.* London: Routledge.

El-Sabh, M.J. and Murty, T.S. (eds) 1988. *Natural and Man-Made Hazards.* Dordrecht: D. Reidel.

Eni 1992. Biofuels in the EEC. The Commission Proposal Effects on Eventual Development in Europe Experience of United States and Brazil.

Environmental Protection Act 1990. London: HMSO.

Evernden, N. 1985. *The Natural Alien: Humankind and Environment.* 2nd edn. Toronto: University of Toronto Press.

Fahey, D.W., Kawa, S.R., Woodbridge, E.L., Tin, P., Wilson, J.C., Jonsson, H.H., Dye, J.E., Baumgardner, D., Borrmann, S., Toohey, D.W., Avallone, L.M., Proffitt, M.H., Margitan, J., Loewenstein, M., Podolske, J.R., Salawitch, R.J., Wofsy, S.C., Ko, M.K.W., Anderson, D.E., Shoeberl, M.R. and Chan, K.R. 1993. *In situ* measurements constraining the role of sulphate aerosols in mid-latitude ozone depletion. *Nature*, 363, pp. 509–14.

Fairbanks, R.G. 1989. A 17,000-year glacioeustatic sea level record: influence of glacial melting rates on the Younger Dryas event and deep-ocean circulation. *Nature*, 343, pp. 637–42.

Falkowski, P.G. and Wilson, C. 1992. Phytoplankton productivity in the North Pacific ocean since 1900 and implications for absorption of anthropogenic CO₂. *Nature*, 358, pp. 741–3.

Fan, S.-M. and Jacob, D.J. 1992. Surface ozone depletion in Arctic spring sustained by bromine reactions on aerosols. *Nature*, 359, pp. 522–4.

Fanning, K.A. 1989. Influence of atmospheric pollution on nutrient limitation in the ocean. *Nature*, 339, pp. 460–2.

Farman, J. 1987. What hope for the ozone layer now? *New Scientist*, 116, 12 November, pp. 50–4.

Farman, J.C., Gardiner, B.G. and Shanklin, J.D. 1985. Large losses of total ozone in Antarctica reveal seasonal ClO$_x$/NO$_x$ interaction. *Nature*, 315, pp. 207–10.

Ferry, G. 1989. Alzheimer's and aluminium – the guesswork goes on. *New Scientist*, 121, p. 1652.

Fisher, R.V. and Schmincke, H.-U. 1984. *Pyroclastic Rocks.* Berlin: Springer-Verlag.

Flood, M. 1991. *Energy without End.* Friends of the Earth.

Foucault, A. and Stanley, D.J. 1989. Late Quaternary palaeoclimatic oscillations in East Africa recorded by heavy minerals in the Nile Delta. *Nature*, 339, pp. 44–6.

Fowler, D. 1993. A land laid waste. *Surveyor*, 25 February, pp. 19–21.

Freeth, S. 1992. The deadly cloud hanging over cameroon. *New Scientist*, 15 August, pp. 23–7.

Fritts, H.C. 1976. *Tree Rings and Climate.* London: Academic Press.

BIBLIOGRAPHY

Friday, L. and Laskey, R. (eds) 1989. *The Fragile Environment: The Darwin College Lectures.* Cambridge: Cambridge University Press.

Friends of the Earth 1992. *Energy for a Future: Friends of the Earth's Evidence to the Government's Review of Energy Policy.* London: Friends of the Earth.

Frosch, R.A. and Gallopoulos, N.E. 1989. Strategies for manufacturing. *Scientific American*, 261, pp. 144–52.

Fulkerson, W., Judkins, R.R. and Sanghvi, M.K. 1990. Energy from fossil fuels. *Scientific American*, 263, pp. 83–9.

Galloway, J.N. 1990. The intercontinental transport of sulfur and nitrogen. In Knap, A.H. (ed.), *The Long Range Atmospheric Transport of Natural and Contaminant Substances.* Netherlands: Kluwer Academic Publishing, pp. 87–104.

Galloway, J.N. and Rodhe, H. 1991. Regional atmospheric budgets of S and N fluxes: how well can they be quantified? In Last, F.T. and Watling, R. (eds), *Acidic Deposition: Its Nature and Impacts.* Edinburgh: The Royal Society of Edinburgh, pp. 61–80.

Gao, G. 1993. The temperatures and oxygen-isotope composition of early Devonian oceans. *Nature*, 361, pp. 712–14.

Gasse, F., Tehet, R., Durand, A., Gibert, E. and Fontes, J.-C. 1990. The arid–humid transition in the Sahara and Sahel during the last deglaciation. *Nature*, 346, pp. 141–6.

Gates, D.M. 1993. *Climate Change and its Biological Consequences.* Sunderland, Mass.: Sinauer Associates, Inc.

Gavaghan, H. 1989. The problems in policing short-range missiles. *New Scientist*, 3 June, p. 35.

Genthon, C., Barnola, J.M., Raynaud, D., Lorius, C., Jouzel, J., Barkov, N.I., Korotkevich, Y.S. and Kotlyakov, V.M. 1987. Vostok ice core: climatic response to CO_2 and orbital forcing changes over the last climatic cycle. *Nature*, 329, pp. 414–19.

Gerlach, T. 1991. Etna's greenhouse pump. *Nature*, 351, pp. 352–3.

Ghadiri, H. and Payne, D. 1980. A study of soil splash using cine photography. In Boodt, M. de and Gabriels, D. (eds), *Assessment of Erosion.* Chichester: John Wiley & Sons, pp. 185–92.

Gibbons, J.H., Blair, P.D. and Gwin, H.L. 1989. Strategies for energy use. *Scientific American*, September Issue, pp. 105–13.

Gibbons, J.H., Blair, P.D. and Gwin, H.L. 1989. Strategies for energy use. *Scientific American*, 261, pp. 136–43.

Gleick, J. 1987. *Chaos.* London: Heinemann.

Goodman, D. and Redclift, M. 1991. *Refashioning Nature: Food, Ecology and Culture.* London: Routledge.

Gordon, A. and Suzuki, D. 1991. *It's a Matter of Survival.* London: HarperCollins.

Gordon, D., Smart, P.L., Ford, D.C., Andrews, J.N., Atkinson, T.C., Rowe, P.J. and Christopher,

N.S.J. 1989. Dating the late Pleistocene interglacial and interstadial periods in the UK. *Quaternary Research*, 31, pp. 14–26.

Goss, E.G. 1991. The hurricane dilemma in the United States. *Episodes*, 14, pp. 36–45.

Goudie, A.S. 1992. *Environmental Change: Contemporary Problems in Geography.* 3rd edn. Oxford: Clarendon Press.

Goudie, A.S. 1993. *The Human Impact on the Environment.* 4th edn. Oxford: Blackwell.

Gough, D.O. 1981. *Solar Physics*, 74, pp. 21–34.

Gould, S.J. 1991. *Wonderful Life: the Burgess Shale and the Nature of History.* London: Penguin Books.

Gourlay, K.A. 1988. *Poisoners of the Seas.* London: Zed Books.

Gradwohl, J. and Greenberg, R. 1988. *Saving Tropical Forests.* London: Earthscan Publications.

Grainger, A. 1990. *The Threatening Desert: Controlling Desertification.* London: Earthscan Publications.

Greenpeace 1990. *The Greenpeace Report, Global Warming.* Oxford: Oxford University Press.

Gregory, K. and Rowlands, H. 1990. Have global hazards increased? *Geographical Review*, 4, pp. 35–8.

Gribbin, J. 1988. *The Hole in the Sky.* Reading: Corgi Books.

Gribbin, J. 1989. The end of the ice ages? *New Scientist*, 17 June, pp. 48–52.

Gribbin, J. 1991. Climate now. *New Scientist*, Inside Science 44, 16 March, pp. 1–4.

GRIP (Greenland ice-core Project) Members 1993. Climate instability during the last interglacial period recorded in the GRIP ice core. *Nature*, 364, pp. 203–7.

Grove, J.M. 1979. The glacial history of the Holocene. *Progress in Physical Geography*, 3, pp. 1–54.

Grove, J.M. 1988. *The Little Ice Age.* London: Methuen.

Grun, R. and Stringer, C.B. 1991. Electron spin resonance dating and the evolution of modern humans. *Archaeometry*, 33, pp. 153–99.

Guiot, J., Pons, A., Beaulieu, J.L. de and Reille, M. 1989. A 140,000-year continental climate reconstruction from two European pollen records. *Nature*, 338, pp. 309–14.

Hall, D.O., Rosillo-Calle, F. and de Groot, P. 1992. Biomass energy. *Energy Policy*, 20, pp. 62–73.

Hammer, C.U., Clausen, H.B. and Dansgaard, W. 1980. Greenland ice sheet evidence of postglacial volcanism and its climatic impact. *Nature*, 288, pp. 230–5.

Hammer, C.U., Clausen, H.B. and Dansgaard, W. 1981. Past volcanism and climate revealed by Greenland ice cores. *Journal of Volcanology and Geothermal Research*, 11, pp. 3–11.

Hammer, C.U., Clausen, H.B., Friedrich, W.L. and Tauber, H. 1987. The Minoan eruption of Santorini in Greece dated 1645 BC? *Nature*, 328, pp. 517–19.

Hansen, J.E. and Lacis, A.A. 1990. Sun and dust versus greenhouse gases: an assessment of their relative roles in global climate change. *Nature*, 346, pp. 713–19.

Harland, W.B., Armstrong, R.L., Cox, A.V. *et al.* 1989. *A Geological Time Scale*. Cambridge: Cambridge University Press.

Hassard, J. 1992. Arms and the ban. *New Scientist*, 28 November, pp. 38–41.

Hawking, S.W. 1988. *A Brief History of Time: From the Big Bang to Black Holes*. London: Bantam Press.

Heliker, C. 1991. *Volcanic and Seismic Hazards on the Island of Hawaii*. US Department of the Interior/US Geological Survey, Denver. US Government Printing Office.

Henderson-Sellers, A. and Robinson, P.J. 1986. *Contemporary Climatology*. London: Longman Scientific & Technical.

Herndl, G.J., Müller-Niklas, G. and Frick, J. 1993. Major role of ultraviolet-B in controlling bacterioplankton growth in the surface layer of the ocean. *Nature*, 361, pp. 717–19.

Hinrichsen, D. 1990. *Our Common Seas: Coasts in Crisis*. London: Earthscan Publications.

Hoffert, M.I. 1992. Climatic sensitivity, climatic feedbacks and policy implications. In Mintzer, I.M. (ed.), *Confronting Climate Change: Risks, Implications and Responses*. Cambridge: Cambridge University Press, pp. 33–54.

Hofmann, D.J., Deshler, T.L., Aimedieu, P., Matthews, W.A., Johnston, P.V., Kondo, Y., Sheldon, W.R., Byrne, G.J. and Benbrook, J.R. 1989. Stratospheric clouds and ozone depletion in the Arctic during January 1989. *Nature*, 340, pp. 117–21.

Hofmann, D.J., Oltmans, S.J., Harris, J.M., Solomon, S., Deshler, T. and Johnson, B.J. 1992. Observation and possible causes of new ozone depletion in Antarctica in 1991. *Nature*, 359, pp. 283–7.

Holdren, J.P. and Pachauri, R.K. 1992. Energy. In ICSU, *An Agenda of Science for Environment and Development into the Twenty-first Century*. Cambridge: Cambridge University Press.

Holdsworth, G. 1986. Evidence for a link between atmospheric thermonuclear detonations and nitric acid. *Nature*, 324, pp. 551–4.

Hollin, J.T. 1969. Ice-sheet surges and the geological record. *Canadian Journal of Earth Sciences*, 5, pp. 903–10.

Homewood, B. 1993. Will Brazil's cars go on the wagon? *New Scientist*, 9 January, pp. 22–4.

Hornung, M., Stevens, P.A. and Reynolds, B. 1986. The impact of pasture improvement on the soil solution chemistry of some stagnopodzols in mid-Wales. *Soil Use and Management*, 2, pp. 18–26.

House of Commons Energy Committee 1990. *The Cost of Nuclear Power*. Fourth report. 2 volumes. London: HMSO.

House of Commons Energy Committee 1992. *Renewable Energy*. Fourth report. 3 volumes. London: HMSO.

House of Commons Environment Committee 1992. *The Government's Proposals for an Environment Agency*. First report. London: HMSO.

House of Commons Trade and Industry Committee 1993. *British Energy Policy and the Market for Coal*. First report. London: HMSO.

House of Lords Select Committee on the European Communities 1990. *Paying for Pollution: Civil Liability for Damage Caused by Waste*. Twenty-fifth report. London: HMSO.

Housner, G.W. 1987. *Confronting Natural Disasters: An International Decade for Natural Hazard Reduction*. Washington, DC: National Academy Press.

Hovan, S.A., Rea, D.K., Pisias, N.G. and Shackleton, N.J. 1989. A direct link between the China loess and marine records: aeolian flux to the north Pacific. *Nature*, 340, pp. 296–8.

Hsü, K.J. 1983. *The Mediterranean Was a Desert: a Voyage of the Glomar Challenger*. New Jersey: Princeton University Press.

Huhne, C. 1989. Some lessons of the debt crisis: never again? In O'Brien, R. and Datta, T. (eds), *International Economics and Financial Markets: the AMEX Bank Review Prize Essays 1988*. Oxford: Oxford University Press.

Hulme, M. 1989. Is environmental degradation causing drought in the Sahel? An assessment from the recent empirical research. *Geography*, 74, pp. 38–46.

ICSU (International Council of Scientific Unions) 1992. *An Agenda of Science for Environment and Development into the Twenty-first Century*. Cambridge: Cambridge University Press.

Imbrie, J. and Imbrie, K.P. 1979. *Ice Ages: Solving the Mystery*. Harvard, Mass.: Harvard University Press.

Imbrie, J., van Donk, J. and Kipp, N.G. 1973. Paleoclimatic investigation of a late Pleistocene Caribbean deep-sea core: comparison of isotopic and faunal methods. *Quaternary Research*, 3, pp. 10–38.

IPCC (Intergovernmental Panel on Climatic Change) 1990. *Climate Change: The IPCC Scientific Assessment*. Ed. Houghton, J.T., Jenkins, G.J. and Ephraums, J.J. Cambridge: Cambridge University Press.

IPCC (Intergovernmental Panel on Climatic Change) 1992. *Climate Change 1992: The Supplementary Report to the IPCC Scientific Assessment*. Ed. Houghton, J.T., Callander, B.H. and Varney, S.K. Cambridge: Cambridge University Press.

Ives, J.D. and Messerli, B. 1989. *The Himalayan Dilemma*. London: Routledge.

Jacobs, D.K. and Sahagian, D.L. 1993. Climate-induced fluctuations in sea level during non-glacial times. *Nature*, 361, pp. 710–12.

Jacobs, S.S. 1992. Is the Antarctic ice sheet growing? *Nature*, 360, pp. 29–33.

Jelgersma, S. 1966. Sea level changes in the last 10,000

BIBLIOGRAPHY

years. In *International Symposium of World Climate from 8000-0 BC*. Royal Meteorological Society.

Johansson, T.D., Kelly, H., Reddy, A.K.M. and Williams, R.H. (eds) 1993. *Renewable Energy for Fuels and Electricity*. London: Earthscan Publications/ United Nations.

Johnsen, S.J., Clausen, H.B., Dansgaard, W., Fuhrer, K., Gunerstrup, N., Hammer, C.U., Iversen, P., Jouzel, J., Stauffer, B. and Steffensen, J.P. 1992. Irregular glacial interstadials recorded in a new Greenland ice core. *Nature*, 359, pp. 311–13.

Johnson, C., Henshaw, J. and McInnes, G. 1992. Impact of aircraft and surface emissions of nitrogen oxides on tropospheric ozone and global warming. *Nature*, 355, pp. 69–71.

Johnson, D.W., Kilsby, C.G., McKenna, D.S., Saunders, R.W., Jenkins, G.J., Smith, F.B. and Foot, J.S. 1991. Airborne observations of the physical and chemical characteristics of the Kuwait oil smoke plume. *Nature*, 353, pp. 617–21.

Journal of the Geological Society, London vol. 134, part 4, 1986. *Geochemical Aspects of Acid Rain*. Thematic set of papers, pp. 619–720.

Journal of the Geological Society, London vol. 148, part 3, 1991. *Monitoring Active Volcanoes*. Thematic set of papers, pp. 561–93.

Jouzel, J., Lorius, C., Petit, J.R., Genthon, C., Barkov, N.I., Kotlyakov, V.M. and Petrov, V.M. 1987. Vostok ice core: a continuous isotope temperature record over the last climatic cycle (160,000 years). *Nature*, 329, pp. 403–7.

Joyce, C. 1991. Ozone hole linked to sea temperatures. *New Scientist*, p. 22.

Kanamori, H. and Kikuchi, M. 1993. The 1992 Nicaragua earthquake: a slow tsunami earthquake associated with subducted sediments. *Nature*, 361, pp. 714–16.

Kasting, J.F. 1989. *Palaeogeography, Palaeoclimatology, Palaeoecology*, 75, pp. 83–95.

Keeling, C.D., Bacastow, R.B., Carter, A.F., Piper, S.C., Whorf, T.P., Heimann, M., Mook, W.G. and Roeloffzen, H. 1989. A three dimensional model of atmospheric CO_2 transport based on observed winds: I. Analysis of observational data. In D.H. Peterson (ed.) *Aspects of Climate Variability in the Pacific and the Western Americas*. Geophysical Monograph, 55, AGU Washington (USA), pp. 165–236.

Keeling, R.F. and Shertz, S. 1992. Seasonal and interannual variations in atmospheric oxygen and implications for the global carbon cycle. *Nature*, 358, pp. 723–7.

Kelly, P.M., Munro, M.A.R., Hughes, M.K. and Goodess, C.M. 1989. Climate and signature years in west European oaks. *Nature*, 340, pp. 57–60.

Kemp, A.E. and Baldauf, J.G. 1993. Vast Neogene laminated diatom mat deposits from the eastern equatorial Pacific Ocean. *Nature*, 362, pp. 141–4.

Kemp, D.D. 1990. *Global Environmental Issues: a Climatological Approach*. London: Routledge.

Keyfitz, N. 1989. The growing human population. *Scientific American*, 261, pp. 118–26.

Kissinger, H.A. 1982. *Years of Upheaval*. London: Weidenfeld & Nicolson/Michael Joseph.

Knapp, B.J. 1979. *Soil Processes*. London: George Allen & Unwin.

Krogh, T.E., Kamo, S.L. and Bahor, B.F. 1993. Fingerprinting the K/T impact site and determining the time of impact by U-Pb dating of single shocked zircons from distal ejecta. *Earth and Planetary Science Letters*, 119, pp. 425–9.

Kudrass, H.R., Erienkeuser, H., Vollbrecht, R. and Weiss, W. 1991. Global nature of the Younger Dryas cooling event inferred from oxygen isotope data from Sulu Sea cores. *Nature*, 349, pp. 406–9.

Kumar, C., Patel, N. and Bloembergen, N. 1987. Strategic defense and directed-energy weapons. *Scientific American*, 257, pp. 31–7.

Lachenbruch, A.H. and Marshall, B.V. 1986. Changing climate: geothermal evidence from permafrost in the Alaskan Arctic. *Science*, 234, pp. 689–96.

LaMarche, V.C. and Hirschboech, K.K. 1984. Frost rings in trees as records of major volcanic eruptions. *Nature*, 307, pp. 121–6.

Lamb, H.H. 1972. *Climate: Present, Past and Future*, vol. 1. London: Methuen.

Landsberg, J.P., McDonald, B. and Watt, F. 1992. Absence of aluminium in neuritic plague cores in Alzheimer's disease. *Nature*, 360, pp. 65–8.

Langer, J., Rodhe, H., Crutzen, P.J. and Zimmermann, P. 1992. Anthropogenic influence on the distribution of tropospheric sulphate aerosol. *Nature*, 359, pp. 712–16.

Larsen, E., Gulliksen, S., Lauritzen, S.-E., Lie, R., Lovlie, R. and Mangerud 1987. Cave stratigraphy in western Norway: multiple Weichselian glaciations and interstadial vertebrate fauna. *Boreas*, 16, pp. 267–92.

Last, F.T. 1991. Critique. In Last, F.T. and Watling, R. (eds), *Acidic Deposition: Its Nature and Impacts*. Edinburgh: The Royal Society of Edinburgh, pp. 273–324.

Last, F.T. and Watling, R. (eds) 1991. *Acidic Deposition: Its Nature and Impacts*. Edinburgh: The Royal Society of Edinburgh.

Legrand, M., Feniet-Saigne, C., Saltzman, E.S., Germain, C., Barkov, N.I. and Petrov, V.N. 1991. Ice-core record of oceanic emissions of dimethylsulphide during the last climate cycle. *Nature*, 350, pp. 144–6.

Lents, J.M. and Kelly, W.J. 1993. Clearing the air in Los Angeles. *Scientific American*, 269, pp. 18–25.

Leuenberger, M. and Siegenthaler, U. 1992. Ice-age atmospheric concentration of nitrous oxide from an Antarctic ice core. *Nature*, 360, pp. 449–51.

Leuenberger, M., Siegenthaler, U. and Langway, C.C. 1992. Carbon isotope composition of atmospheric CO_2 during the last ice age from an Antarctic ice core. *Nature*, 357, pp. 488–90.

Lorius, C., Jouzel, J., Ritz, C., Merlivat, L., Barkov, N.I., Korotkevich, Y.S. and Kotlyakov, V.M. 1985. A 150,000-year climatic record from Antarctic ice. *Nature*, 310, pp. 591–6.

Lorius, C., Barkov, N.I., Jouzel, J., Korotkevich, Y.S., Kotlyakov, V.M. and Raynaud, D. 1988. Antarctic ice core: CO_2 and change over the last climatic cycle. *EOS*, 68, pp. 681–4.

Lovelock, J.E. 1988. *The Ages of Gaia: A Biography of our Living Earth*. Oxford: Oxford University Press.

Lowe, J.J. and Walker, M.J.C. 1984. *Reconstructing Quaternary Environments*. London: Longman.

MacAyeal, D.R. 1992. Irregular oscillations of the West Antarctic ice sheet. *Nature*, 359, pp. 29–32.

McCormack, J. 1989. *Acid Earth: The Global Threat of Acid Pollution*. London: Earthscan Publications.

McCormick, M.P. and Velga, R.E. 1992. *Journal of Geophysical Research Letters*, 9, pp. 155–8.

McIlveen, R. 1992. *Fundamentals of Weather and Climate*. London: Chapman and Hall.

MacKenzie, D. 1991. Energy answers for North and South. *New Scientist*, 16 February, 48–51.

MacNeill, J. 1989. Strategies for sustainable economic development. *Scientific American*, 261, pp. 154–65.

Maddox, J. 1989. The biggest greenhouse still intact. *Nature*, 338, p. 111.

Maitlis, P. and Rourke, J. 1993. Rich seams for chemicals. *New Scientist*, 23 January, pp. 37–41.

Malthus, T. 1798. *An Essay on the Principle of Population and a Summary View of the Principle of Population*. Ed. Field, A., 1970. Harmondsworth: Pelican.

Manabe, S. and Stouffer, R.J. 1993. Century-scale effects of increased atmospheric CO_2 on the ocean–atmosphere system. *Nature*, 364, pp. 215–18.

Mannion, A.M. 1991. *Global Environmental Change: A Natural and Cultural Environmental History*. Harlow: Longman Scientific & Technical.

Marrow, J.E., Coombs, J. and Lees, E.W. 1987. *An Assessment of Bio-Ethanol as Transport Fuel in the UK*. ETSU-R44, vol. 1. London: HMSO.

Marrow, J.E. and Coombs, J. 1990. *An Assessment of Bio-Ethanol as Transport Fuel in the UK*. ETSU-R55, vol. 2. London: HMSO.

Martyn, C.M., Barker, D.J.P., Osmond, C., Harris, E.C., Edwardson, J.A. and Lacey, R.F. (1989) Geographical relation between Alzheimer's Disease and aluminium in drinking water. *Lancet*, 14 January, 1, 8629, pp. 59–62.

Maurits la Rivire, J.W. 1989. Threats to the world's water. *Scientific American*, 261, pp. 80–94.

Meadows, D.L., Randers, J. and Behrens III, W.W. 1972. *The Limits to Growth: A Report to the Club of Rome's Project on the Predicament of Mankind*. New York: Potomac Associates.

Miller, G.H. and de Vernal, A. 1992. Will greenhouse warming lead to northern hemisphere ice-sheet growth? *Nature*, 355, pp. 244–6.

Miller, J.M. and Ball, T.K. 1969. *Second Progress Report on the Measurement of Radon in Soil Air as a Prospecting Technique*. Institute of Geological Sciences, Metalliferous Minerals and Applied Geochemistry Unit Report 28. Keyworth: British Geological Survey.

Miller, J.M. and Ostle, D. 1973. Radon measurements in uranium prospecting. In *Uranium Exploration Methods*. Vienna: International Atomic Energy Agency.

Milne, R. 1989. North Sea algae threaten British coasts. *New Scientist*, 122 (1663), pp. 37–41.

Mintzer, I.M. (ed.) 1992. *Confronting Climate Change: Risks, Implications and Responses*. Cambridge: Cambridge University Press.

Mohnen, V.A. 1988. The challenge of acid rain. *Scientific American*, August, pp. 14–22.

Montgomery, H., Pessagno, E., Soegaard, K., Smith, C., Munoz, I. and Pessagno, J. 1992. Misconceptions concerning the Cretaceous/Tertiary boundary at the Brazos River, Falls County, Texas. *Earth and Planetary Science Letters*, 109, pp. 593–600.

Mook, W. and Woillard, G. 1982. Carbon-14 dates at Grand Pile. Correlation of land and sea chronologies. *Science*, 215, pp. 159–61.

Moore Lappe, F. and Schurman, R. 1989. *Taking Population Seriously*. London: Earthscan Publications.

Morgan, H. and Simms, D.L. 1988. Setting trigger concentrations for contaminated land. In *Vol. 1 Contaminated Soil '88*, Second International TNO Conference on Contaminated Soil, 11–15 April 1988, Hamburg. Dordrecht: Kluwer Academic.

Morgan, R.P.C. 1986. *Soil Erosion and Conservation*. New York: Longman Scientific & Technical.

Morgan, V.I., Goodwin, I.D., Etheridge, D.M. and Wookey, C.W. 1991. Evidence from Antarctic ice cores for recent increases in snow accumulation. *Nature*, 354, pp. 58–60.

Mortimer, N. 1989. *Evidence to the House of Commons Select Committee on Energy Bearing on the Energy Policy Implications of the Greenhouse Effect and Proof of Evidence*. Friends of the Earth 9, to the Hinckley Point C public inquiry. London: Friends of the Earth.

Mungall, C. and McLaren, D.J. (eds) 1990. *Planet Under Stress: The Challenge of Global Change*. Oxford: Oxford University Press.

Muniz, I.P. 1991. Freshwater acidification: its effects on species and communities of freshwater microbes, plants and animals. In Last, F.T. and Watling, R. (eds), *Acidic Deposition: Its Nature and Impacts*. Edinburgh: The Royal Society of Edinburgh, pp. 227–54.

Murphy, A.P. 1991. Chemical removal of nitrate from water. *Nature*, 350, pp. 223–5.

Mysak, L.A. and Lin, C.A. 1990. The tempering seas. In Mungall, C. and McLaren, D.J. (eds), *Planet*

BIBLIOGRAPHY

Under Stress: The Challenge of Global Change. Oxford: Oxford University Press, pp. 134–48.

NAS (National Academy of Sciences) 1986. *Acid Deposition: Long-Term Trends.* Committee on Monitoring and Assessment of Trends in Acid Deposition.

Nebel, B.J. and Wright, R.T. 1993. *Environmental Science: The Way the World Works.* 4th edn. Englewood Cliffs: Prentice-Hall.

Nelson, B.K., MacLeod, G.K. and Ward, P.D. 1991. Rapid change in strontium isotopic composition of sea water before the Cretaceous/Tertiary boundary. *Nature,* 351, pp. 644–7.

Newhouse, J. 1989. *The Nuclear Age: From Hiroshima to Star Wars.* London: Michael Joseph.

Nilsson, S. and Pitt, D. 1991. *Mountain World in Danger: Climate Change in the Forests and Mountains of Europe.* London: Earthscan Publications.

Ninkovich, D., Shackleton, N.J., Abdel-Monem, A., Obradovich, J.D. and Izett, G. 1978. K-Ar age of the late Pleistocene eruption of Toba, north Sumatra. *Nature,* 276, pp. 574–7.

Obradovich, J., Snee, L.W. and Izett, G.A. 1989. Is there more than one glassy layer in the late Eocene? *Geological Society of America Abstracts with Programs,* 21, p. 134.

OECD 1987. *Pricing of Water Services.* Paris: Organization for Economic Co-operation and Development.

Oeschger, H. and Mintzer, I.M. 1992. Lessons from the ice cores: rapid climate changes during the last 160,000 years. In Mintzer, I.M. (ed.), *Confronting Climate Change: Risks, Implications and Responses.* Cambridge: Cambridge University Press, pp. 55–64.

Officer, C. 1993. Victims of volcanoes. *New Scientist,* 20 February, pp. 34–8.

OFWAT 1990. *Paying for Water – A Time for Decisions.* Birmingham: Office of Water Services.

Oliver, D. 1991. The house that came in from the cold. *New Scientist,* 9 March, pp. 45–9.

Oltmans, S.J. and Levy, H. 1992. Seasonal cycle of surface ozone over the western North Atlantic. *Nature,* 358, pp. 392–4.

O'Neill, B. 1989. Falling out over water. *New Scientist,* 124 (1687), 21 October, pp. 60–4.

Opdyke, N.D., Glass, B., Hays, J.D. and Foster, J. 1966. Paleomagnetic study of Antarctic deep-sea cores. *Science,* 154, pp. 349–57.

Open University 1991. *Case Studies in Oceanography and Marine Affairs.* Oxford: Pergamon Press.

Ottaway, R. 1990. *Less People, Less Pollution: An Answer to Environmental Decline Caused by the World's Population Explosion.* London: Bow Group.

Page, R.A., Boore, D.M., Bucknam, R.C. and Thatcher, W.R. 1992. *Goals, Opportunities, and Priorities for the USGS Earthquake Hazards Reduction Program.* Circular 1079. US Department of the Interior/US Geological Survey, Denver.

Washington: US Government Printing Office.

Pain, S. 1989. Greenhouse warming at nuclear inquiry. *New Scientist,* p. 33.

Pakiser, L.C. 1991. *Earthquakes.* US Department of the Interior/US Geological Survey, Denver. Washington: US Government Printing Office.

Parry, M. 1990. *Climatic Change and World Agriculture.* London: Earthscan Publications.

Patterson, W. 1989. Energy issues another challenge. *New Scientist,* 28 January, pp. 45–50.

Pearce, D., Markandya, A. and Barber, E. 1989. *Blueprint for a Green Economy.* London: Earthscan.

Pearce, F. 1987. *Acid Rain.* Harmondsworth/New York: Penguin Books.

Pearce, F. 1989a. Felled trees deal double blow to global warming. *New Scientist,* 123 (1682), p. 25.

Pearce, F. 1989b. Methane: the hidden greenhouse gas. *New Scientist,* 122 (1663), pp. 37–41.

Pearce, F. 1990. Whatever happened to acid rain? *New Scientist,* 15 September, pp. 57–60.

Pearce, F. 1992. Mirage of the shifting sands. *New Scientist,* 12 December, pp. 38–42.

Pearce, F. 1993. How Britain hides its acid soil. *New Scientist,* 27 February, pp. 29–33.

Peltier, W.R. 1990. Our fragile inheritance. In Mungall, C. and McLaren, D.J. (eds), *Planet Under Stress: The Challenge of Global Change.* Oxford: Oxford University Press, pp. 80–95.

Petit-Maire, N., Fontugne, M. and Rouland, C. 1991. Atmospheric methane ratio and environmental changes in the Sahara and Sahel during the last 130k yrs. *Paleogeography, Paleoclimatology and Paleoecology,* 86, pp. 195–204.

Peto, J. 1990. Radon and the risks of cancer. *Nature,* 345, pp. 389–90.

Pielou, E.C. 1991. *After the Ice Age.* Chicago: The University of Chicago Press.

Pirazzoli, P. 1973. Inondations et niveaux marins à Venise. *Mem. Lab. Geomorph. Ecole Pratique Hautes,* 22, Dinard.

Poag, C.W., Powars, D.S., Poppe, L.J., Mixon, R.B., Edwards, L.C., Folger, D.W. and Bruce, S. 1992. Deep Sea Drilling Project Site 612 bolide event: new evidence of a late Eocene impact-wave deposit and a possible impact site, US east coast. *Geology,* 20, pp. 771–4.

Poore, D. 1989. *No Timber Without Trees: Sustainability in the Tropical Forest.* London: Earthscan Publications.

Porritt, J. and Winner, D. 1988. *The Coming of the Greens.* London: Fontana Paperbacks.

Porter, S.C. 1986. Pattern and forcing of northern hemisphere glacier variation during the last millennium. *Quaternary Research,* 26, pp. 27–48.

Porter *et al.* 1976. Lyell centenary issue. *British Journal of History of Science,* 9, pp. 91–242.

POST (Parliamentary Office of Science and Technology) 1992a. *Clean Coal Technology.* Briefing Note 38.

POST (Parliamentary Office of Science and Technology)

1992b. *The Polluter-Pays Principle and Cost Recovery Charging.*

POST (Parliamentary Office of Science and Technology) 1993a. *Dealing with Drought: Environmental and Technical Aspects of Water Shortages.*

POST (Parliamentary Office of Science and Technology) 1993b. *Biofuels for Transport.* UK Parliamentary Office of Science and Technology, Briefing Note 41.

Prospero, J.M., Savoie, D.L., Saltzman, E.S. and Larsen, R. 1991. Impact of oceanic sources of biogenic sulphur on sulphate aerosol concentrations at Mawson, Antarctica. *Nature*, 350, pp. 221–3.

Purseglove, J. 1991. Liberty, ecology, modernity. *New Scientist*, 28 September, pp. 45–8.

Radford, T. 1990. *The Crisis of Life on Earth: Our Legacy from the Second Millennium.* London: Thorsons Publishing Group.

Ramage, J. 1988. *Energy: A Guidebook.* Oxford: Oxford University Press.

Ramanathan, V. and Collins, W. 1991. Thermodynamic regulation of ocean warming by cirrus clouds deduced from observations of the 1987 El Nino. *Nature*, 351, pp. 27–32.

Rampino, M.R. and Self, S. 1992. Volcanic winter and accelerated glaciation following the Toba super-eruption. *Nature*, 359, pp. 50–2.

Rau, J.L. and Nutalaya, P. 1982. Geomorphology and land subsidence in Bangkok, Thailand. In Craig, R.G. and Crafts, J.L. (eds), *Applied Geomorphology.* London: Mackays & Chatham, pp. 181–201.

Ravven, W. 1991. Asteroid impact emptied Gulf of Mexico. *New Scientist*, p. 14.

Raymo, M.E. and Ruddiman, W.F. 1992. Tectonic forcing of late Cenozoic climate. *Nature*, 359, pp. 117–22.

Redclift, M. 1987. *Sustainable Development: Exploring the Contradictions.* London: Methuen.

Revkin, A. 1990. *The Burning Season: The Murder of Chico Mendes and the Fight for the Amazon Rain Forest.* Collins.

Richards, K. 1989. All gas and garbage. *New Scientist*, 3 June, pp. 38–41.

Ridley, M. 1993. Cleaning up with cheap technology. *New Scientist*, 23 January, pp. 26–7.

Robertson, J. 1990. Alternative futures of cities. In Cadman, D. and Payne, G. (eds), *The Living City: Towards a Sustainable Future.* London: Routledge, pp. 127–35.

Robin, G. 1977. Ice cores and climatic change. *Philosophical Transactions of the Royal Society of London*, B 280, pp. 143–68.

Rodhe, H.E., Cowling, E., Galbally, I., Galloway, J. and Herrera, R. 1988. Acidification and regional air pollution in the tropics. In Rodhe, H.E. and Herrera, R. (eds), *Acidification in Tropical Countries.* Chichester: John Wiley & Sons, pp. 3–39.

Rogers, P., Lydon, P. and Seckler, D. 1989. *Easter Waters Study: Strategies to Manage Flood and Drought in the Ganges-Brahmaputra Basin.* Washington, DC: US Agency for International Development.

Ruckelshaus, W.D. 1989. Towards a sustainable world. *Scientific American*, 261, pp. 166–75.

Ruddiman, W.F. and Kutzbach, J.E. 1991. Plateau uplift and climatic change. *Scientific American*, 264(3), pp. 42–50.

Ruddiman, W.F. and McIntyre 1981. The North Atlantic during the last deglaciation. *Palaeogeography, Palaeoclimatology, Palaeoecology*, 35, pp. 145–214.

Sagan, C. and Turco, R. 1990. *A Path Where No Man Thought: Nuclear Winter and the End of the Arms Race.* London: Century Press.

Sarnthein, M. 1978. Sand deserts during the Glacial Maximum and climatic optimum. *Nature*, 273, pp. 43–6.

Sarre, P. (ed.) 1991. *Environment, Population and Development.* London: Hodder & Stoughton.

Schidlowski, M. 1988. A 3,800-million-year isotopic record of life from carbon in sedimentary rocks. *Nature*, 333, pp. 313–18.

Schlesinger, M.E. and Jiang, X. 1991. Revised projection of future greenhouse warming. *Nature*, 350, pp. 219–21.

Schlesinger, M.E. and Ramankutty, N. 1992. Implications for global warming of intercycle solar irradiance variations. *Nature*, 360, pp. 330–3.

Schweingruber, F.H. 1989. *Tree Rings: Basics and Applications of Dendrochronology.* London: Kluwer Academic Publishing.

Scoging, H. 1993. The assessment of desertification. *Geography*, 339(78, 2), pp. 190–3.

Scurlock, J. and Hall, D. 1991. The Carbon Cycle. *New Scientist*, 2 November, Inside Science 51.

Seers, D. 1977. The new meaning of development. *International Development Review*, 3, pp. 2–7.

Self, S., Pampino, M.A. and Bonbera, J.J. 1981. The possible effects of large nineteenth- and twentieth-century volcanic eruptions on zonal and hemispherical surface temperatures. *Journal of Volcanology and Geothermal Research*, 11, pp. 41–60.

Selby, M.J. 1985. *Earth's Changing Surface.* Oxford: Clarendon Press.

Shackleton, N.J. 1977. Carbon-13 in *Uvigerina*: tropical rainforest history and the equatorial Pacific carbonate dissolution cycles. In Anderson, N.R. and Malahoff, A. (eds), *The Fate of Fossil Fuel CO$_2$ in the Oceans.* New York: Plenum Press, pp. 401–28.

Shackleton, N.J., Hall, M.A., Line, J. and Shuxi, C. 1983. Carbon isotope data in core V19-30 confirm reduced carbon dioxide concentration in the ice age atmosphere. *Nature* 306, pp. 319–22.

Shackleton, N.J. 1987. Oxygen isotopes, ice volume and sea level. *Quaternary Science Reviews*, 6, pp. 183–90.

BIBLIOGRAPHY

Shackleton, N.J. 1990. Estimating atmospheric CO_2. *Nature*, 347, pp. 427–8.

Shackleton, N.J., Berger, A. and Pettier, W.R. 1990. An alternative astronomical calibration of the lower Pleistocene timescale based on ODP site 677. *Transactions of the Royal Society of Edinburgh, Earth Sciences*, 81, pp. 251–61.

Shackleton, N.J. and Opdyke, N.D. 1973. Oxygen-isotope and paleomagnetic stratigraphy of Pacific cores V28-239, late Pliocene to latest Pleistocene. In Cline, R.M. and Hays, J.D. (eds), *Investigation of Late Quaternary Paleogeography and Paleoecology*. Boulder, Col.: Geological Society of America Memoir, 45, pp. 449–64.

Shackleton, N.J. *et al.* 1984. Oxygen isotope calibration of the onset of ice-rafting and history of glaciations in the North Atlantic region. *Nature*, 307, pp. 620–3.

Sharp, A.D., Trudgill, S.T., Cooke, R.U., Price, C.A., Crabtree, R.W., Pickles, A.M. and Smith, D.I. 1982. Weathering of the balustrade on St Paul's Cathedral, London. *Earth Surface Processes and Landforms*, 7, pp. 387–9.

Sharpton, V.L., Dalrymple, G.B., Marin, L.E., Ryder, G., Schuraytz, B.C. and Urrutia-Fucugauchi, J. 1992. New links between the Chicxulub impact structure and the Cretaceous/Tertiary boundary. *Nature*, 359, pp. 819–21.

Sheehan, M.J. 1988. *Arms Control: Theory and Practice*. Oxford: Blackwell.

Shen, G.T., Boyle, E.A. and Lea, D.W. 1987. Cidmium in corals as a tracer of historic upwelling and industrial fallout. *Nature*, 328, pp. 794–6.

Sheppard, C. and Price, A. 1991. Will marine life survive the Gulf War? *New Scientist*, 9 March, pp. 36–40.

Sievering, H., Boatman, J., Gorman, E., Kim, Y., Anderson, L., Ennis, G., Luria, M. and Pandis, S. 1992. Removal of sulphur from the marine boundary layer by ozone oxidation in sea-salt aerosols. *Nature*, 360, pp. 571–3.

Simon, J. 1981. *The Ultimate Resource*. Oxford: Martin Robertson.

Simpson, S. 1990. *The Times Guide to the Environment*. London: Times Books.

Singh, O.N., Borchers, R., Fabian, P., Lal, S. and Subbaraya, B.H. 1988. Measurements of atmospheric BrO_x radicals in the tropical and mid-latitude atmosphere. *Nature*, 334, pp. 593–5.

Slingo, T. 1989. Wetter clouds dampen global greenhouse warming. *Nature*, 341, p. 104.

Small, R.D. 1991. Environmental impact of fires in Kuwait. *Nature*, 350, pp. 11–12.

Smith, K. 1992. *Environmental Hazards*. London: Routledge.

Smith, K. 1993. Riverine flood hazard. *Geography*, 339(78, 2), pp. 182–5.

Smith, P.M. and Warr, K. (eds) 1991. *Global Environmental Issues*. The Open University/Hodder & Stoughton.

Smith, R.A. 1852. *Memoirs and Proceedings of the Manchester Literary and Philosophical Society*, 2, pp. 207–17.

Spencer, R.W. and Christy, J.R. 1990. Precise monitoring of global temperature trends from satellites. *Science*, 247, pp. 1558–62.

Spinks, P. 1990. Plug into the Sun. *New Scientist*, 22 September, pp. 48–51.

Stauffer, B., Lochbronner, E., Oeschger, H. and Schwander, J. 1988. Methane concentration in the glacial atmosphere was only half that of pre-industrial Holocene. *Nature*, 332, pp. 812–14.

Stauffer, P.H., Nishimura, S. and Batchelor, B.C. 1980. Volcanic ash in Malaya from a catastrophic eruption of Toba, Sumatra, 30,000 years ago. In Nishimura, S. (ed.), *Physical Geology of the Indonesian Island Arcs*. Kyoto, pp. 156–64.

Street-Perrott, F.A. and Perrott, R.A. 1990. Abrupt climatic fluctuations in the tropics: the influence of Atlantic Ocean circulation. *Nature*, 343, pp. 607–12.

Stockholm Environment Institute 1993. *Energy without Oil: The Technical and Economic Feasibility of Phasing Out Global Oil Use*.

Swinburne, N. 1993. It came from outer space. *New Scientist*, 20 February, pp. 28–32.

Swisher, C.C. III, Grajales-Nishimura, J.M., Montanari, A., Margolis, S.V., Claeys, P., Alvarez, W., Renne, P., Cedillo-Pordo, E., Maurrasse, F.J.-M.R., Curtis, G.H., Smil, J. and McWilliams, M.O. 1992. Coeval $^{40}Ar/^{39}Ar$ Ages of 65.0 Million Years Ago from Chicxulub Crater Melt Rock and Cretaceous-Tertiary Boundary Teletites. *Science*, 257, pp. 954–8.

Swiss Federal Laboratories for Materials Testing and Research 1992. *Untersuchung des Emissionsverhaltens eines Nutzfahrzeugmotors bei Betrieb mit Rapsolmethylester*.

Thompson, R.D. 1989. Short-term climate change: evidence, cause, environmental consequences and strategies for action. *Progress in Physical Geography*, 13, pp. 315–47.

Thompson, R.D. 1992. The changing atmosphere and its impact on Planet Earth. In Mannion, A.M. and Bowlby, S.R. (eds), *Environmental Issues in the 1990s*. Chichester: John Wiley & Sons, pp. 61–96.

Tivy, J. 1982. *Biogeography*. Harlow: Longman.

Turco, R.P., Toon, O.B., Park, C., Whitten, R.C., Pollack, J.B. and Noerdlinger, P. 1981. Tunguska Meteor Fall of 1908: Effects on Stratospheric Ozone. *Science*, 214, pp. 19–23.

Turco, R.P., Toon, O.B., Ackerman, T.P., Pollack, J.B. and Sagan, C. 1984. The climatic effects of nuclear war. *Scientific American*, 251, pp. 33–43.

Turner, R.H., Nigg, J.M. and Paz, D.H. 1986. *Waiting for Disaster*. Berkeley: University of California Press.

Tyndall, J. 1861. *Philosophical Magazine*, 22, p. 161.

UNEP (United Nations Environmental Programme)

1992. *World Atlas of Thematic Indicators of Desertification*. London: Edward Arnold.

United Nations Population Fund 1991. *The State of the World Population*.

US Congress Office of Technology Assessment 1990. *Replacing Gasoline: Alternative Fuels for Light-Duty Vehicles*. OTA-E-364. Washington, DC: US Government Printing Office.

US Congress Office of Technology Assessment 1991. *US Oil Import Vulnerability: The Technical Replacement Capability*. OTA-E-503. Washington, DC: US Government Printing Office.

US Congress Office of Technology Assessment 1992a. *Managing Industrial Solid Wastes from Manufacturing, Mining, Oil and Gas Production, and Utility Coal Combustion – Background Paper*. OTA-BP-0-82. Washington, DC: US Government Printing Office.

US Congress Office of Technology Assessment 1992b. *Fueling Development: Energy Technologies for Developing Countries*. OTA-E-516. Washington, DC: US Government Printing Office.

Van Geen, A., Luoma, S.N., Fuller, C.C., Anima, R., Clifton, H.E. and Trumbore, S. 1992. Evidence from Cd/Ca ratios in foraminifera for greater upwelling off California 4,000 years ago. *Nature*, 358, pp. 54–6.

Vaughan, D. 1993. Chasing the rogue icebergs. *New Scientist*, 9 January, pp. 24–7.

Veum, T., Jansen, E., Arnold, M., Beyer, I. and Duplessy, J.-C. 1992. Water mass exchange between the North Atlantic and the Norwegian Sea during the past 28,000 years. *Nature*, 356, pp. 783–5.

Vogel, J.S., Cornell, W., Nelson, D.E. and Southon, J.R. 1990. Versuvius/Avellino, one possible source of seventeenth-century BC climatic disturbances. *Nature*, 344, pp. 534–7.

Vogelmann, A.M., Ackerman, T.P. and Turco, D.P. 1992. Enhancement in biologically effective ultraviolet radiation following volcanic eruptions. *Nature*, 359, pp. 47–9.

Volz, A. and Kley, D. 1988. Evaluation of Montsauris series of ozone measurements made in the nineteenth century. *Nature*, 332, pp. 240–2.

Von Gunten, H.R. and Lienert, C. 1993. Decreased metal concentrations in ground water caused by controls of phosphate emissions. *Nature*, 364, pp. 220–2.

Wadleigh, M.A. and Velzer, J. 1992. $^{18}O/^{16}O$ and $^{13}C/^{12}C$ in lower Paleozaic articulate brachiopods. Implications for isotopic composition of seawater. *Geochimica Cosmochimica Acta*, 56, pp. 431–43.

Walker, A.S. 1992. *Deserts: Geology and Resources*. US Department of the Interior/US Geological Survey, Denver. Washington, DC: US Government Printing Office.

Wardley-Smith, J. (ed.) 1976. *The Control of Oil Pollution on the Sea and Inland Waters: The Effect of Oil Spills on the Marine Environment and Methods of Dealing with Them*. London: Graham & Trotman.

Warren Spring Laboratory Report 1992. Armishaw, R., Bardos, R.P., Dunn, R.M., Hill, J.W., Pearl, M., Rampling, T. and Wood, P.A. *Review of Innovative Contaminated Soil Clean-Up Processes*. Warren Spring Laboratory, UK.

Water Services Association 1992. *Waterfacts*. Richmond, UK: Print Management Systems.

Waters, J.W., Froidevaux, L., Read, W.G., Manney, G.L., Elson, L.S., Flower, D.A., Jarnot, R.F. and Harwood, R.S. 1993. Stratospheric ClO and ozone from the microwave limb sounder on the upper atmosphere research satellite. *Nature*, 362, pp. 597–602.

Watkins, N.D., Sparks, R.S.J., Sigurdsson, H., Huang, T.C., Federman, A., Carey, S. and Ninkovich, D. 1978. Volume and extent of the Minoan tephra from Santorini Volcano: new evidence from deep-sea cores. *Nature*, 271, pp. 122–6.

Watson, A. 1991. Carbon dioxide. *New Scientist*, Inside Science 48, 6 July.

Watson, A. 1991. Gaia. *New Scientist*, 6 July, Inside Science, 48, pp. 1–4.

Watson, A.J. and Lovelock, J.E. 1983. Biological homeostasis of the global environment: the parable of the 'daisy' world. *Tellus*, 35b, pp. 282–9.

Westrich, H.R. and Gerlach, T.M. 1992. Magmatic gas source for stratospheric SO_2 cloud from the June 15, 1991 eruption of Mount Pinatubo. *Geology*, 20, pp. 867–70.

Which? 1991. River pollution. *Which?*, August, pp. 458–61.

White, G.F. (ed.) 1974. *Natural Hazards*. Oxford: Oxford University Press.

White, I.D., Mottershead, D.N. and Harrison, S.J. 1984. *Environmental Systems*. London: Allen & Unwin.

Whittow, J. 1980. *Disasters: The Anatomy of Environmental Hazards*. London: Penguin Books.

Wigley, T.M.L. 1981. Climate and paleoclimate: what we can learn about solar luminosity variations. *Solar Physics*, 74, pp. 435–71.

Wigley, T.M.L. and Raper, S.C.B. 1992. Implications for climate and sea level of revised IPCC emissions scenarios. *Nature*, 357, pp. 293–300.

Williams, M.J., Dunkerley, D.L., Deckker, P. de, Kershaw, A.P. and Stokes, T. 1993. *Quaternary Environments*. London: Edward Arnold.

Williams, S.J., Dodd, K. and Gohn, K.K. 1991. *Coasts in Crisis*. US Department of the Interior/US Geological Survey circular 1075. Washington, DC: US Government Printing Office.

Williamson, P. and Gribbin, J. 1991. How plankton change the climate. *New Scientist*, 16 March, pp. 48–52.

Wilson, A.T. 1964. Origin of Ice Ages: an ice shelf theory for Pleistocene glaciation. *Nature*, 201, pp. 147–9.

BIBLIOGRAPHY

Wilson, A.T. 1969. The climatic effects of large-scale surges of ice sheets. *Canadian Journal of Earth Sciences*, 6, p. 911.

Wilson, E.O. 1989. Threats to biodiversity. *Scientific American*, 261, pp. 60–8.

Wolfe, J.A. 1991. Palaeobotanical evidence for a June 'impact winter' at the Cretaceous/Tertiary boundary. *Nature*, 352, pp. 420–3.

Wolfenden, J. and Mansfield, T.A. 1991. Physiological disturbances in plants caused by air pollutants. In Last, F.T. and Watling, R. (eds), *Acidic Deposition: Its Nature and Impacts*. Edinburgh: The Royal Society of Edinburgh, pp. 117–38.

Wood, B. 1992. Origin and evolution of the genus *Homo*. *Nature*, 355, pp. 783–90.

Woods, A.W. and Wohletz, K. 1991. Dimensions and dynamics of co-ignimbrite eruption columns. *Nature*, 350, pp. 225–7.

Woolbach, W. *et al.* 1985. *Science*, 230, pp. 167–70.

World Bank 1979. *Nepal: Development Performance and Prospects. A World Bank Country Study, South Asia Regional Office*. Washington, DC: World Bank.

World Bank 1992. *World Development Report 1992: Development and the Environment*. Oxford: Oxford University Press.

World Commission on Environment and Development (WCED) 1987. *Our Common Future*. Oxford: Oxford University Press.

World Energy Commission 1992. *Energy for Tomorrow's World – the Realities, the Real Options and the Agenda for Achievement*. Draft Summary Global Report.

World Resources Institute 1991. *World Resources 1990–91*. Report produced in collaboration with the United Nations Environment Programme and the United Nations Development Programme. Oxford: Oxford University Press.

World Resources Institute 1992. *World Resources 1992–93*. Report produced in collaboration with the United Nations Environment Programme and the United Nations Development Programme. Oxford: Oxford University Press.

Wright, R.F. and Hauhs, M. 1991. Reversibility of acidification: soils and surface waters. In Last, F.T. and Watling, R. (eds), *Acidic Deposition: Its Nature and Impacts*. Edinburgh: The Royal Society of Edinburgh, pp. 169–91.

Yearley, S. 1991. *The Green Case: A Sociology of Environmental Issues, Arguments and Politics*. London: Routledge.

Yergin, D. 1991. *The Prize: The Quest for Oil, Money, and Power*. London: Simon & Schuster.

Zahnie, K. and Grinspoon, D. 1990. Comet dust as a source of amino acids at the Cretaceous/Tertiary boundary. *Nature*, 348, pp. 157–60.

Zimmerman, P. 1989. A new resource for arms control. *New Scientist*, 23 September, pp. 38–43.

Zwally, H.J., Brenner, A.C., Major, J.A., Bindschadler, R.A. and Marsh, J.G. 1989. Growth of the Greenland ice sheet: measurement. *Science*, 246, pp. 1587–9.

Note: Not all the words in this glossary appear in the text, but they are included here in order to make as complete a listing of useful terms as is possible.

Abiotic – non-living, relating to factors and/or things which are independent/separate from living things.

Accretionary prism – thick wedge-shaped body of sediments formed by both tectonic and sedimentary processes, and associated with the off-scraping of material above a subducting oceanic plate.

Acetogenic bacteria – bacteria which convert sugars into fatty acids.

Acid – chemical substance that releases hydrogen ions when dissolved in water, or an aqueous solution containing an excess of hydrogen ions.

Acid deposition – falling of acids and acid-forming substances from the atmosphere onto the surface of the Earth. Acid rain is a type of acid deposition.

Acid rain – rain and snow with a pH less than 5.6. Acid rain strictly refers to the wet deposition of acids and acid-forming substances.

Acid susceptibility – capacity of a water body to become acidified.

Acid-neutralizing capacity (ANC) – ability of a water body to reduce (or neutralize) the acidity of incoming acid water.

Acidification – increase in acidity (lower pH) in a water body.

Active coke – pellets of treated coal (coke) approximately 5 mm in diameter that are used to catalyse reactions to help remove pollutants in the generation of power in coal fire power stations.

Actinides – fourteen chemical elements in the final period of the periodic table of elements (thorium, protactinium, uranium, neptunium, plutonium, americium, curium, berkelium, californium, einsteinium, fermium, mendelevium, nobelium, lawrencium).

Activated sludge – sludge containing living organisms feeding on the solids to encourage its breakdown, and which are recycled during secondary treatment of sewage.

Acute Radiation Syndrome (ARS) – symptoms resulting from intensive irradiation of the body including nausea, vomiting, abdominal pain, fever, dehydration, loss of hair, infection, haemorrhage, damage to bone marrow, and cancers, notably leukemia and breast cancer.

Acute toxicity – effect of a single high-level exposure to specified chemical/s, for example, due to their accidental release.

Adaption (evolutionary/ecological) – changes in the function and/or structure of a system to produce greater life chances through survivability and reproduction.

Aeolian – pertaining to wind processes, landforms or sediments.

Aerosol – solid or liquid particles suspended or dispersed in a gas.

Albedo – measure of the reflectivity of a body or surface, often used to describe the ability of the Earth's cloud, snow, and vegetation cover to reflect incoming solar radiation.

Algal blooms – proliferation of algae in water bodies as a result of changes in water chemistry and temperature.

Alpha particles – particles emitted from the nucleus of an atom during radioactive decay; an alpha particle has an atomic mass of 4 and an atomic number of 2, and it is equivalent to a helium nucleus.

Alzheimer's Disease – illness that leads to senile dementia, and for which the causes remain poorly understood.

Americium-241 – radioactive isotope of Americium.

GLOSSARY

Amphiboles – group of dark-coloured rock-forming minerals comprising iron-magnesium silicates.

Anaerobic – without free oxygen.

Anaerobic digestion – breakdown of organic matter by organisms in an oxygen-free environment.

Aniline point – the minimum temperature for a complete mixing of aniline and materials such as gasoline, used in some specifications to indicate the aromatic content of oils and to calculate the approximate heat of combustion. Aniline is the $C_6H_5NH_2$- group, for example, benzanilide $C_6H_5NHCOC_6H_5$.

Anions – atoms which by virtue of the imbalance in electrical forces have a net negative charge.

Anoxic – oxygen-deficient environments.

Anthracite – dark hard coal composed of between 92 and 98 per cent carbon.

Anthropogenic – human influenced processes or forms.

Aquifer – body of rock or sediments at depth which is capable of storing groundwater.

Asteroids (asteroid belt) – number of small possibly fragmented planetisimals which orbit around the Sun between the orbits of Mars and Jupiter.

Asthenosphere – layer within the Earth's upper mantle which extends from depths of 5–50 km to at least 300 km and is characterized by a lower mechanical strength and lower resistance to deformation than the region above, the crust.

Atmosphere – gaseous layer surrounding the Earth and bound to it by gravitational attraction.

Atmospheric fluidized-bed combustion – new technique used in coal-fired power stations to reduce noxious gaseous emissions, and involving passing the gases through a bed of coal and limestones which become supported (fluidized) by the upward flow of the gas in a fluidized-bed combustion furnace.

Autotroph – organism that can synthesize the organic substances it requires entirely from inorganic nutrients, by obtaining energy from light and/or various inorganic substances. The main autotrophs are green plants.

Background radiation – radioactivity from non-human sources.

Badlands – intensely dissected landscape produced by natural or human-influenced erosion.

Base (alkali) – any chemical substance which releases hydroxyl ions (OH^-) when dissolved in water, or an aqueous solution containing an excess of hydroxyl ions.

BATNEEC – 'best available techniques not entailing excessive cost.'

Becquerel (Bq) – unit by which radioactivity is measured. 1 Bq = 1 atomic disintegration per second.

Bedload – the sediment load carried along and very near to the bottom of a flowing current (for example, river bed or seafloor) rather than in suspension, and which tends to be the coarser and heavier grain-size fraction.

Benthic – pertaining to bottom-dwelling organisms.

Benzene (C_6H_6) – aromatic hydrocarbon widely used in industry.

Beta activity – release of beta particles (the emission of an electron from the nucleus) during the radioactive decay of an element.

Big Bang – explosion that marked the creation of the Universe, which probably occurred about 15,000–20,000 million years ago.

Biochemical oxygen demand (BOD) – amount of oxygen used ('demanded') in chemical/biological processes during the digestion or oxidation of wastes. The potential environmental impacts of wastes are frequently expressed in terms of their BOD value.

Biodegradable – compound which can be decomposed and/or disintegrated by biological processes. Antonym is 'non-biodegradable'.

Biodiversity – natural variability within the animal and plant kingdoms, the variety of species in any area.

Biofuel – substance produced by organic activity that can be used as an energy source, either in a pure form or refined and blended with conventional fossil fuels, such as petroleum.

Biogas – gas mixture arising from anaerobic digestion of organic matter, and comprising about two-thirds methane, one-third carbon dioxide, and minor amounts of other gases. The methane content in biogas makes it useful as a fuel/energy resource.

Biogenic – pertaining to organic origin.

Biomass – mass of biological matter present per plant or animal, per community, or per unit area. Total dry organic matter or stored energy content of living organisms in a specified area.

Biome – ecosystems linked through similar climatic conditions and vegetation, for example, tropical rainforests, deserts, and high-latitude tundra.

Biomineralization – formation of minerals by living organisms.

Biosphere – layer at the interface of the Earth's crust, ocean, and atmosphere where life is found, that is, the total ecosystem of the Earth.

Biosynthesis – chemical reactions promoted by an organism in order to produce new chemicals.

Biota – general term for any specific or all living organisms and their associated ecosystems.

Bioturbation – burrowing activity of organisms which helps mix up the sediment and soil layers.

Bitumen – black to dark brown solid or semi-solid thermo-plastic material possessing waterproofing and adhesive properties, obtained from processing crude oil. It is a complex combination of higher molecular weight organic compounds containing a relatively high percentage of hydrocarbons having carbon numbers greater than C_{25} with a high carbon to oxygen ratio; there are also trace amounts of metals such as nickel, iron, and vanadium.

Bituminous coal – coal containing 78–90 per cent carbon, which has a moderately high calorific value.

Bivalve - mollusc with a shell composed of two distinct parts (valves), which are generally similar. The two halves of the shell, which house the animal, are joined by a flexible muscle or ligament.

Black smoke – visible smoke comprising particulates, and in an engine it is formed from pyrolysis and incomplete combustion of fuel, typically under high loads.

Black smokers – submarine springs which form chimneys comprising sulphides, typically of zinc, iron, copper, and molybdenum and oxides of manganese which are present along the mid-ocean ridges at depths of about 2.5 km below sea level.

Boreal forests – Northern coniferous forest (including spruce, fir and hemlock) approximately bounded in the north by the 10°C July average isotherm section and transitional to the Tundra which is synonymous with Taiga.

BP – Before the Present Day.

British Thermal Unit (BTU) – amount of heat required to raise the temperature of 1 pound weight (1 lb) of water through 1° Fahrenheit (1°F).

Bromeliad – any member of the pineapple family, typically with fleshy, spiny-leaved epiphytes (plants not rooted in the soil, but growing above ground level, usually on other plants), and notable for their ability to hold water in the cup-shaped centre of the leaf rosette. Bromeliads are particularly common in the canopy of tropical rainforests.

Buffer – chemical that can maintain the pH of a solution by reacting with the excess acid or alkali (base). Limestone is a natural buffer which helps stabilize the pH of groundwater and soil close to neutral.

Buffering chemical reactions – chemical reactions which reduce the likelihood of the solution changing its pH.

Butterfly effect – highly variable knock-on effect (positive feedback) or output produced by a system, as a result of subtle change in the initial inputs.

Caatinga – form of dry thorny woodland found in northeast Brazil.

Caesium-137 – radioactive isotope of caesium.

Calorie – amount of heat energy required to raise 1 gram of water through 1° Celsius (1°C). When used in connection with food, the units are typically kilocalories which is the amount of heat energy required to raise 1 litre of water through 1°C.

Calving – process which involves the fracturing and break-up of ice-sheets or ice-caps, where they enter the sea or a lake, and which leads to the formation of icebergs.

Carbon cycle – transfer of carbon, one of three basic elements (with hydrogen and oxygen) for life, through the biosphere, hydrosphere, atmosphere, and lithosphere. For conversion purposes, 1 million tonnes carbon (MtC) = 3.67 million tonnes of CO_2.

Carbonate compensation depth (CCD) – depth in the oceans where seawater is undersaturated with respect to $CaCO_3$; this leads to the dissolution of carbonate sediment as it falls through the water column, or sits on the seafloor. CCD varies between oceans.

Carbonic acid (H_2CO_3) – acidic rainwater or snow formed by the water combining with atmospheric CO_2.

Carbon dioxide (CO_2) – naturally occurring gas molecule comprising two atoms of oxygen to one atom of carbon. CO_2 is produced during respiration by humans, and is used during photosynthesis by plants.

Carbon monoxide (CO) – gas molecule with one oxygen atom and one carbon atom. CO is highly toxic to humans and many other organisms.

Carbon tax – tax levied on fossil fuels in proportion to the amount of CO_2 produced during combustion.

Carcinogenic – having the potential to cause cancer.

Catalyst – substance which may initiate or speed up a chemical reaction without itself being changed during the reaction. In biological reactions, enzymes act as catalysts.

Catalytic converter – device fitted to motor vehicle exhaust system to reduce the emissions of pollutants.

Catastrophe theory – the hypothesis that important and widespread changes in the physical environment are brought about by major, relatively brief, and sudden events.

Cations – atoms which have lost one or more electrons from their orbiting shells and thus have a net positive charge.

Cation exchange – substitution of one cation for another of a different element in a mineral structure.

Caveat emptor principle – literally, 'buyer beware', in which, for example, responsibility for any environmental harm caused by contaminated land passes from the vendor to purchaser.

Cellulose – complex carbohydrate which forms the basic structural component of plant cell walls and comprises glucose molecules. It cannot be digested by humans.

Centre pivot irrigation – irrigation system utilizing a rotating spray arm up to several hundred metres long and supported by wheels pivoting about a central well from which water is pumped. These systems are commonly used to artificially irrigate desert regions.

Cerrado – form of savannah vegetation comprising mainly grasses and small trees found in Brazil.

Cetane number – indicator of how easily a fuel ignites under compression, that is, the ignition quality. Higher number means easier ignition.

Channelization – straightening of natural stream channels by the construction of artificial banks.

GLOSSARY

Chaos theory – theory explaining phenomena as being a consequence of inherent randomness in a system. There are mathematical models to simulate chaotic systems.

Chemical weathering – see weathering.

Chitinous exoskeleton – external part of an animal, usually an insect or a crustacean, comprising a hard organic substance called chitin.

Chlorinated hydrocarbons – organic compounds containing chlorine.

Chlorofluorocarbons (CFCs) – organic compounds comprising carbon, chlorine, and fluorine atoms, commonly used in industrial processes and manufacturing, and which are very stable in the troposphere, and degraded in the stratosphere by solar radiation which releases the chlorine that may contribute to ozone depletion.

Chlorophyll – green pigment in plant tissues which is essential for photosynthesis.

Chondritic meteorites – iron-rich meteorites.

Chronic toxicity – assessment conducted over lifetime of the test animals to evaluate any late-in-life toxicity caused by exposure to specified chemical/s.

Clay minerals – group of finely crystalline layers silicate minerals.

Claystone – fine grained rock comprising grains less than 0.0039 mm in diameter.

Cloud seeding – artificial addition of condensation nuclei, such as silver iodide crystals, into the atmosphere to help produce rain.

CO_2 emissions – these are commonly measured according to the carbon content, in millions of tonnes of carbon (MtC), where 1 tonne of carbon is equivalent to 3.67 (or 44/12) tonnes of carbon dioxide.

Coastal set-up – term used to describe the meteorological conditions, often associated with storms, along the coast.

Coccolithophorida – microscopic algae which secrete calcareous shell comprising round platelets known as coccoliths. These coccoliths reached their acme in the Cretaceous Period when the deposition of coccoliths led to the formation of extensive and thick Chalk deposits.

Coesite – form (polymorph) of quartz formed under high pressures.

Cogeneration – simultaneous production of both electrical and heat energy, for example, for use in domestic/commercial, industrial, or other purposes.

Coleoptera – genetic name for beetles.

Collagen – group of proteins which have great tensile strength and are present in tendons, ligaments, connective tissues of the skin, dentin and cartilages.

Combustion-dust loading – increased quantities of dust in the atmosphere produced as the by-product of burning fossil fuels.

Continental drift – lateral movement of continental plates around the Earth, a theory superseded by the plate tectonic theory that ascribes this plate movement to sea-floor spreading.

Continental plate – rigid outer layer of the Earth, the lithosphere, which averages about 40 km in thickness and may form relatively stable and long-lived parts of the crust, and constitute the continents. It has an average silica (SiO_2) content of 65 per cent.

Convergent plate boundaries – boundaries between discrete plates, where two plates are colliding. These include the collision between two oceanic plates, two continental plates, and an oceanic and a continental plate.

Cretaceous – period of geological time, *c.* 146 to 65 Ma.

Critical loads – maximum loads of a pollutant which the environment can sustain before damage occurs.

Cryptosporidia – intestinal parasite which causes diarrhoea and vomiting in humans.

Cybernetics – study of regulating and self-regulating mechanisms in nature and technology

Deforestation – conversion of forest land to other uses, for example, pasture, cropland.

Deglaciation – period of time when glaciers start to retreat at the end of a glacial or stadial.

Dendritic cells – cells with branching or tree-like forms.

Demand side management (DSM) – planning, implementation, and monitoring of utility activities, to encourage customers/users to modify their pattern and total consumption of energy use/electricity.

Dendrochronologies – time correlations based on the width of annual growth rings of trees.

Dendroecology – science of the width of annual growth rings of trees in order to interpret specific ecological events that resulted in changes in the tree's ability to photosynthesize and fix carbon.

Desalination plant – industrial plant for purifying seawater and converting it into high-quality drinking water, using distillation and other techniques.

Desertification – spread of desert-like conditions in arid or semi-arid lands, as a result of climatic change or human influence.

Desulfomaculum – organic compound which aids bacterial degradation of oil.

Desulfovibrio – organic compound which aids bacterial degradation of oil.

Deuterium – isotope of hydrogen.

Devensian – last glacial stage in Britain.

Diatoms – microscopic unicellular algae which secrete siliceous walls.

Dichloro-diphenyl-trichloro-ethane (DDT) – poisonous organic compound used as a pesticide.

Dilatancy-diffusion model – theory which is used to explain phenomena and events which occur before and during an earthquake.

Diluvial theory – theory which attributed landforms,

sediments, and fossils to the Noachian deluge.

Dimethyl sulphide – biologically-produced organic chemicals which may act as condensation nuclei.

DNA – deoxyribonucleic acid, an organic compound comprising two polymer strands wrapped round each other in a double helix; DNA carried the genetic information of living organisms.

Divergent oceanic spreading ridges (mid-ocean ridges) – see spreading centres.

Dolomite – rock/mineral comprising carbonate of calcium and magnesium ($CaMg(CO_3)_2$).

Dose-response relationship – relationship between the level of a pollutant and the environmental impact.

Downstream flood – river flood which becomes progressively larger down valley.

Droughts – condition of dryness because of a lack of precipitation.

Dry deposition – the direct transfer of gases and particles to surfaces whether leaves, soil or building materials.

Dust bowl – badland regions where dust is actively being deflated from the ground and being transported.

Dust pneumonia – see pneumoconiosis.

Dutch Elm Disease – widespread fungoid killer of elm trees, first found in the Netherlands and introduced into Europe from Asia during the Second World War.

Dynamic equilibrium – situation which is fluctuating about some apparent average state, where the average state itself is also changing through time.

Earth-surface processes – processes acting on the Earth's surface which includes river, lake, sea, slope, biological and atmospheric processes.

EC – European Community (previously referred to as the EEC or European Economic Community).

Ecology – study of living organisms' habits, modes of life, and relations to their surroundings.

Ecological niche – occupation of space by a community in a particular environmental setting.

Ecosphere – all-encompassing realm which includes the atmosphere, hydrosphere, lithosphere, and biosphere.

Ecosystem – community with interacting organisms of different species, and their relationships with the associated chemical and physical systems.

El Nino Southern Ocean Oscillation (ENSO) event – appearance of unusually warm water off the coasts of Peru, Ecuador, and Chile which causes major shifts in the general circulation of the atmosphere.

Electrical resistivity – ability for a substance to resist the flow of an electrical current through it.

Electromagnetic spectrum – range of electrical and magnetic radiation of varying wavelengths which travels at the same velocity as light.

Endothermic reaction – chemical reaction that involves the absorption of heat energy, which may cool down the surrounding environment.

Entropy – degree of disorder in any physical or chemical system; the greater the entropy, the greater the inherent disorder in a system.

EPA – [US] Environmental Protection Agency in the United States, which is responsible for managing federal efforts aimed at controlling air and water quality, reducing radiation and pesticide hazards, regulating the disposal of hazardous waste, and undertaking/sponsoring environmental research. [UK] Environmental Protection Act.

Epidemiology – study of the pattern of diseases and/or other harmful effects produced by toxic substances in various groups of people, with the purpose of understanding the reasons for certain individuals/groups of people being more susceptible than others to ill health.

Eukaryotic organisms – organisms which require O_2 to biosynthesize.

Eustasic – relating to global changes in sea level.

Eutrophic – pertaining to lakes, ponds or rivers which abound in plant nutrients and are therefore highly productive.

Eutrophication – addition of nutrients to water bodies which increase their productivity.

Evaporites – water-soluble minerals which have been deposited by precipitation from saline water as a result of evaporation, for example, halite (NaCl), gypsum ($CaSO_4.2H_2O$), and anhydrite ($CaSO_4$).

Evapotranspiration – diffusion of water vapour into the atmosphere from vegetated surfaces.

Exajoule – 10^{18} joules (see joule).

FAO – United Nations Food and Agricultural Organization.

Faults (geological) – cracks or fissures in rock produced by Earth (tectonic) movement along which displacement has occurred.

Fission (nuclear) – splitting of larger atoms into two or more lighter elements with the release of energy.

Flash flooding – flood event commonly associated with ephemeral streams in arid and semi-arid environments. The flood is characterized by an almost instantaneous rise in discharge which progresses down-stream as waves, for example, as bores (solitary waves).

Fluidized bed combustion furnaces – see atmospheric fluidized-bed combustion.

Fluvial – pertaining to river processes, landforms, or sediments.

Folds – geological strata or alignments of minerals which have been deformed by compressional forces into bends.

Food chain – transfer of food from one type of organism to another in sequence.

Food web – transfer of food from one type of organism to another within a complex community of organisms.

Foraminifera – single-celled (protozoan) micro-organisms which secrete calcareous skeletons and drift in the seas and oceans.

GLOSSARY

Fossil fuels – energy sources in the form of buried organic matter which generally have undergone chemical and physical changes. Common fossil fuels include petroleum, natural gas (mainly methane), coals, and peats.

Fractal geometry – study of the scales of invariant processes and forms.

Fujita Intensity Scale – scale which describes the damage associated with a tornado in relation to the velocity of the rotating spiral of wind.

Fumaroles – small vents associated with volcanic centres through which liquids erupt.

G7 nations – Group of top seven industrialized nations: Canada, France, Germany, Italy, Japan, UK, USA.

Gaia hypothesis – hypothesis developed by scientist James Lovelock which suggests that the Earth is a self-regulating system, like a living organism, able to maintain its climate, atmosphere, soil, and ocean composition in a stable balance favourable to life.

Galaxy – band of stars to which our Sun belongs and which is seen from the Earth as the Milky Way.

Gasahol – blended alcohol and conventional petroleum products, for example, a blend of 90 per cent gasoline and 10 per cent bio-ethanol.

GATT – General Agreement on Tariffs and Trade. GATT is an international mechanism to control and regulate economic growth through legislative/fiscal policy. The last round, the Uruguay Round, of negotiations lasted from 1988 until early 1994.

General circulation models (GCMs) – simulation of atmospheric circulation involving a system of equations used to describe atmospheric, and ocean-water motion, the heat exchange and fluxes within this system, and the consequences. GCMs usually involve the solution of these equations on a high speed computer or super-computer.

Geochronology – measurement of time intervals or dating of events on a geological time scale.

Geodesy – the study of the shape and size of the Earth by survey and mathematical means.

Geoid – shape of the Earth at mean sea level.

Geomorphologists – scientists who study earth surface processes and landforms.

Geosphere – the Earth.

Geothermal energy – utilization of the Earth's internal heat energy to generate energy.

Geothermal gradient – rate of change of temperature with depth, generally used in relation to the Earth's crust.

Giardia – parasite that lives on the human gut lining and causes dysentery.

Glacial – cold stage during an Ice Age when ice-sheets, glaciers, permafrost, and sea-ice were more extensive.

Glacio-isostatic rebound – uplift by rebound of regions in which the lithosphere was previously depressed by the weight of former glaciers.

Glaciomarine – pertaining to marine environment, landform or sediments, which are/were influenced by glacial processes.

Global warming potentials (GWPs) – effect that a given amount of a trace gas can have on forcing climate compared to the effect of the same amount of CO_2.

Glycerine – hydrocarbon belonging to the alcohol family used in the manufacture of a number of commercial products including cosmetics, soaps, nitroglycerin.

Gondwana – large continent which existed in the southern hemisphere, and which split up about 300 Ma (the late Palaeozoic) to form Africa, Australia, Antarctica, South America, and India.

Greenhouse effect – effect analogous to that which is supposed to operate in a greenhouse, whereby the Earth's surface is maintained at a much higher temperature than the approximate balance conditions with the solar radiation reaching the Earth's surface.

Greenhouse gases – gases which absorb infrared radiation in the atmosphere. These include carbon dioxide, methane, water vapour, nitrous oxide, ozone, and chlorofluorocarbons.

GDP – Gross Domestic Product. The value of output produced in a country.

GNP – Gross National Product. The value of all the final goods and services produced in an economy.

Gypsum – mineral (hydrated calcium sulphate – $CaSO_4.2H_2O$) precipitated from a saline solution.

Half-life – time required for 50 per cent of the atoms of a radioactive isotope to decay to a different element/group of elements, with the associated emission of various sub-atomic particles and the release of energy.

Halite – Mineral rock salt (NaCl) precipitated from a saline solution.

Halocarbons – organic compounds which contain chlorine and bromine, and which are important stratospheric ozone-depleters. Examples of these gases include CFC-11 (CCl_3F) and HCFC-22 ($CHClF_2$).

Halons – organic compounds containing bromine and fluorine, for example, halon 1211 (CF_2BrCl ($CBrClF_2$ – bromodichloromethane)), and halon 2402 ($C_2F_4Br_2$ – dibromo-tetrafluoroethane). These chemicals play a part in stratospheric ozone depletion.

Hard water – water in which certain minerals, particularly calcium carbonate, are dissolved and which tend to precipitate as a 'scum' and may fur-up kettles, water pipes, and other domestic-industrial appliances/machinery.

Heat engine – name given to the mechanism by which tropical cyclones develop and are maintained.

Heat-island effect – relative warmth of a city compared to the surrounding countryside controlled by urban activity.

Heavy metals – metallic elements with high atomic masses such as mercury, lead, arsenic, tin, cadmium, cobalt, selenium, copper, and manganese.

Heinrich layers – layers of ice-rafted sediment which are present in core collected from the north Atlantic.

Hepatitis B – disease that causes inflammation of the liver; Hepatitis B is transmitted by infection, including sexually.

Heterogeneous reactions – reactions of chemicals in different states, for example as between gas and liquid, gas and solid, or solid and liquid.

Holocene – period of time from after the last major glaciation (*c.* 11,000 years BP) to the present.

Hominid – creature of the family Hominidae (primates) of which only one species exists today *Homo sapiens sapiens*.

Hot spot – region of relatively high heat flux on the surface of the Earth, caused by anomalously hot magma rising towards the Earth's surface from the mantle as a plume, causing volcanic-igneous activity at the Earth's surface, uplift, and possibly the splitting apart of the oceanic or continental crust.

Hominoids – upstanding bipedal human-like apes, which are generally considered to be the ancestors of modern human.

Hurricane – name given to tropical cyclones which originate in the Caribbean and mid-Atlantic Ocean.

Hydrocarbon – natural or synthetic chemical compound consisting essentially of carbon and hydrogen, from simple to very complex (long-chain) molecules, for example methane (CH_4) represents an extremely simple hydrocarbon.

Hydroelectricity – or hydropower. Electricity generated by water power.

Hydrological cycle – continuous movement of water (vapour, liquid, and solid) on, in, and above the Earth's surface.

Hydrological flowpath – direction of flow of water beneath the ground.

Hydrology – the study of the movement of water (vapour, liquid, and solid) on, in, and above the Earth's surface.

Hydroperoxyl radical (HO_2^{3-}) – a negatively-charged molecule comprising one atom of hydrogen and two atoms of oxygen.

Hydrosphere – the Earth's water layer which includes liquid, solid, and gaseous phase.

Hydrothermal systems – hot fluids, usually water, which are rich in dissolved gases, nutrients, and metals. These originate from within the lithosphere and emerge at the Earth's surface, commonly on the seafloor, where there are zones of particularly high heat flow, notably associated with volcanic areas.

Hydroxyl radical (OH^-) – negatively charged ion comprising one atom of hydrogen and one atom of oxygen. Hydroxyl ions are very reactive and tend to oxidize many pollutants, in air or water; therefore they are good natural chemical cleansing agents. Hydroxyl ions are the primary oxidizing agents in the atmosphere.

Hypersaline – extremely salty.

Ice Age – period in Earth's history when ice-sheets are extensive and sea-ice and permafrost are widespread in mid- and high latitudes.

Ice shelf – floating sheet of ice attached to the coast which is nourished by snow falling on to its surface and/or by land-based glaciers.

Icehouse effect – conditions which lead to global cooling, the opposite effect to the greenhouse effect.

Ignimbrite – welded to non-welded pyroclastic rock (formed by explosive fragmentation of magma and/or previously solid rock during volcanic eruptions), and comprising mainly pumice and ash.

Impact matrix – multi-dimensional array used to show the effects of policy actions on the environment.

Impact winter – global cooling resulting from the reduction of solar radiation induced by increased atmospheric dust as a result of ejected debris and global fires, for example, as might be caused by a meteorite impact with the Earth.

Infrared radiation – form of electromagnetic radiation with longer wavelengths than visible light.

Inter Tropical Convergence Zone (ITCZ) – zone of nearly continuous atmospheric low pressure with light and variable winds, high humidity, and intermittent heavy rain showers which is present near the equator.

Interglacial – warm stage between glacials in an Ice Age when glaciers retreat, sea-ice and permafrost are of limited extent, and tundra replaced by forest.

Interstadial – warm period during a glacial stage.

Inversion layer – level in the atmosphere which prevents the vertical mixing of air. Such a layer typically shows increasing temperature upwards, the reverse to the normal upward cooling.

Iridium anomaly – high concentrations of the element iridium found in sediments or rocks.

Irrigation – supply of water, usually via channels to agricultural land.

Isostasy – condition of lithostatic equilibrium between sections of the lithosphere with respect to the underlying asthenospheric mantle.

Isotope – different forms of an element with similar chemical properties by virtue of atomic weight, that is, they have the same number of protons in the nucleus but have different number of neutrons.

Joint – fissure in rock along which little or no relative face-parallel displacement has occurred, and which are formed by the relaxation of crustal stresses.

GLOSSARY

Jokulhlaups – catastrophic floods formed by the drainage of a sub-glacial or ice-dammed lake.

Joule (J) – unit of energy expended in order to do work, equivalent to a Newton metre (Nm). 1 Newton = force required to accelerate 1 kg through 1 m per second per second (m s^{-2}). 1 watt of energy (W) is equivalent to 1 Joule per second.

Jurassic – period of geological time, 190–135 Ma.

Kerogen – hydrocarbon occurring in crude oil and formed by the breakdown of organic matter.

Kinetic energy – energy possessed by matter by virtue of its motion, for example, heat, wave energy.

Lacustrine – pertaining to lake environments, processes, or landforms.

Lahars – landslides of volcanic material, usually generated during a volcanic eruption.

Last Glacial Maximum – period of time (*c.* 25,000–16,000 years BP) during the maximum extent of glaciers in the last major glaciation.

Laterization – enrichment of sesquioxides of aluminium and/or iron in a soil which leads to the formation of laterites.

Laurentide Ice Sheet - ice sheet that formed during the last glaciation (known in North America as the 'Wisconsin') and covered vast areas of North America.

Lava – molten rock which originates from the mantle and flows across the Earth's surface.

Law of Gravitation – Newton's Law of Gravitation states that a gravitational force exists between two bodies which is proportional to the product of their masses and inversely proportional to the square of their distance apart.

LDC – 'Less Developed Country', equates with 'Developing Country' or 'Third World'.

Leguminous plants – plants which have bulbous growths (legumes) on their roots which contain symbiotic bacteria which fix nitrogen to form nitrates.

Leptospirosis (Sewerman's Disease) – disease contracted from contact with rat's urine.

Life cycle cost (operating costs) – the cost of goods and/or services over an entire life cycle.

Light year – distance light travels in one year.

Lignite – organic-rich deposit containing about 70 per cent carbon.

Limestone – rock comprising mainly the mineral calcium carbonate ($CaCO_3$), formed either directly by precipitation from solution, or the accumulation of detrital organic or inorganic $CaCO_3$.

Lipid – substance which has similar properties to fats.

Liquefied petroleum gas (LPG) – comprises mainly propane, butane, and isobutane, but may contain C_3 and C_4 unsaturated hydrocarbons. Its main use is as a fuel, but it is also has widespread use as an aerosol propellant, and as a chemical feedstock.

Lithosphere – Earth's crust and upper portion of the mantle, which together constitute a layer of relative mechanical strength compared to the more easily deformable asthenosphere below.

Litter – natural veneer of dead and rotting plant material within an ecosystem such as a forest. Natural litter is characteristically highly biodegradable and recyclable, unlike much human litter.

Little Ice Age – cold period during the seventeenth century when glaciers advanced throughout the world.

Load – in an engine, the resistance that is overcome by the torque delivered (to which it is numerically equivalent).

Loess – silt-size (63–2 microns diameter) sediment deposited from wind.

Longshore sediment drift (littoral drift) – transport of sediment by wave action (sub-)parallel to a coastline.

Lymphocyte cells – white blood cells that are of fundamental importance in the immune system.

Macrophages – stationary cells within living tissue, for example, in the lymph nodes, spleen, bone marrow, alveoli in the lungs, which engulf and destroy bacteria, playing an important role in the immune system.

Magma – fused, molten, rock material found beneath the Earth's crust, and from which volcanic rocks are formed.

Magmatic – processes which involve the formation, movement, emplacement, or crystallization of magma.

Malaria – intermittent and remittent fever transmitted by the bite of a mosquito which conveys the parasite that transmits the disease.

Malthusian views – ideas expressed by Reverend Malthus and others who believe that world population can be kept in check by disease, wars, and natural disasters.

Manganese nodules – spherical precipitations comprising manganese, iron and lower concentrations of nickel, copper, cobalt, and molybdenum present on the ocean floor, associated with areas of high heat flow and black smokers.

Mantle – internal layer of the Earth extending from 11–50 km below the surface to a depth of 2900 km. The uppermost part is essentially rigid (as a very high-viscosity fluid, the lower part of the lithosphere) and the lower part of the asthenosphere is partially molten.

Mass movement – movement of material downslope under the influence of gravity.

MDC – 'More Developed Country', equates with 'Developed Country' or 'Industrialized Country'.

Meltdown – process in a nuclear reactor where the uncontrolled accumulation of heat will increase the temperature of the core until it becomes [goes] critical and literally melts down. If the meltdown is not totally confined within the reactor containment building, core meltdown can release extremely high and dangerous levels of radioactivity thousands of times greater than the nuclear fission bomb that destroyed Hiroshima.

Meteorite – Extra-terrestrial material that may fall to Earth if it travels across the Earth's orbit. Many thousands of small meteorites enter the Earth's atmosphere, but burn up before reaching the ground. Meteorites comprise relatively primitive matter in the Solar System, and therefore provide scientists with an opportunity to study the early history of the Solar System. The principal asteroid belt in our Solar System is between Mars and Jupiter.

Meteorology – the study of the dynamics and composition of the atmosphere.

Methanogenic bacteria – bacteria which covert organic acids into biological gases, notably methane (CH_4).

Messinian salinity crisis – period of time (Messinian) approximately 5 Ma when the waters in the Mediterranean almost dried up which increased the salinity of the waters and led to the precipitation of large salt deposits.

Metamorphism – processes by which the composition, structure, and texture of rocks are altered by the action of heat and pressure.

Methemoglobinaemia – disease responsible for 'blue baby' birth.

Methyl chloroform – organic compound used as a solvent and cleaning fluid, which may act as a greenhouse gas.

Methyl isocyanate – poisonous organic compound which contains cyanide.

Microfossils – extremely small to microscopic remains of past organisms, including foraminifera, diatoms, Coleoptera, and pollen.

Microglial cells – small cells of the central nervous system.

Mid-ocean ridge – submarine, only locally subaerial, linear mountain chain with a central rift valley marking the boundaries between two oceanic plates which are moving apart, and along which basaltic rocks are being formed by the creation of new seafloor.

Milankovitch cyclicity – natural fluctuations in the Earth's orbital parameters and named after the person who first clearly elucidated these, that is, changes in the Earth's precession, obliquity, and eccentricity in orbit around the Sun, and which lead to cyclically varying solar flux, and therefore induce global climate change.

Millisievert (mSv) – a measure of the radiation dose received by an individual, that is, the amount of energy given up by radiation in a particular mass of body tissue through which it passes. The exact relationship between 1 Bq and a mSv is complex and depends, amongst other factors, on the type of radiation (alpha, beta, or gamma) and the sensitivity of different types of tissue.

Monoculture – cultivation of single-species crop.

Mutually Assured Destruction (MAD) – the certain annihilation of the participants in a nuclear war. MAD assumes that a pre-emptive strike by one side will still result in complete destruction.

NASA – US National Aeronautics and Space Administration.

Natural bitumen – similar in physical properties to bitumen (see bitumen), it is naturally occurring and has a different composition to synthetic bitumen.

Negative feedback – return of a fraction of an output from a system to the input of the system which decreases the subsequent output of the system.

Neolithic (New Stone Age) – ancient cultural stage or level of human development characterized by stone tools shaped by polishing or grinding.

Neoplastic disease – medical condition involving the formation of a neoplasm or tumour.

Nephos – name given by the people of Athens for a mixture of toxins which formed in Athens on 1st October 1991.

NIMBY – acronym for 'not in my back yard', used to epitomize the widely held public attitude that undesirable facilities (nuclear power plants, waste disposal utilities, chemical plants etc.) should be sited away from their homes and/or workplaces but rather near somebody else.

Nitrates – chemical compounds containing nitrogen, and which are essential nutrients for life.

Nitrous oxide (N_2O) – nitrogen oxide commonly derived from the use of fertilizers, combustion of fossil fuels, and biomass. Nitrous oxide causes concern because tropospheric N_2O is a greenhouse gas, whereas stratospheric N_2O contributes to ozone depletion.

NOAA – National Oceanic and Atmosphere Administration.

North Atlantic Deep Water (NADW) – cold dense deep ocean current which travels southwards in the North Atlantic.

Nuclear fission – subdivision of a heavy atomic nucleus, for example, uranium or plutonium, into two fragments of roughly equal mass and releasing large amounts of energy.

Nuclear fusion – nuclear reactions between light elements to form heavier ones, whilst releasing large amounts of energy.

Nuclear reprocessing plant – industrial site where radioactive substances, commonly in the form of spent nuclear reactor rods, are refined into either more concentrated material and/or made less harmful.

Nuclear winter – severe deterioration of climate that might take place as a result of multiple nuclear explosion, and may generate great fires and wind; large quantities of smoke and dust may be ejected into the upper atmosphere and remain there for a period of months to years, causing prolonged

GLOSSARY

darkness and reduced incoming solar radiation, resulting in extreme cooling of the Earth's surface, possibly to between –15°C and –25°C.

Nucleus – centre of an atom containing neutrons and protons.

Occult deposition – the deposition by impaction of cloud/fog droplets commonly containing appreciable larger concentrations of major ions than the biggest drops of rain (at the same sites).

OPEC – Organization of Petroleum Exporting Countries.

Orbital parameters – appertaining to the Earth's rotation around the Sun, and including its precession (19,000–23,000 year), obliquity (40,000 year), and eccentricity (100,000 and 400,000 year).

Organization for Economic Cooperation and Development (OECD) – includes most of the world's industrialized market economies, that is, Australia, Austria, Belgium, Canada, Denmark, Finland, France, Germany, Greece, Iceland, Ireland, Italy, Japan, Luxembourg, Netherlands, New Zealand, Norway, Portugal, Spain, Sweden, Switzerland, Turkey, the UK, and the USA.

Organochlorides – organic compounds containing chloride.

Organotin – chemical compound containing tin, used primarily as a fungicide.

Orogeny – tectonic mechanism which creates continental mountain chains.

Oxygenic photosynthesis – biological activity in which O_2 is released to the atmosphere from the splitting up of water molecules.

Ozone (O_3) – form of oxygen, each molecule comprising three atoms of oxygen.

Ozone layer – zone in the atmosphere between 15 and 45 km above the Earth's surface which contains ozone, and which reaches concentrations of 1 in 10^5 parts at 35 km altitude. This layer is also referred to as the ozonesphere.

Ozonesphere – see ozone layer.

Palaeo-oceanography – science of past oceans, their configurations, chemistry and dynamics.

Palaeoclimate – past climate.

Palaeoclimatology – the study of past climates, their rates of change and dynamics.

Palaeoenvironmental – pertaining to past environments.

Palaeolatitude – former latitudinal position of a region which has experienced plate movement because of seafloor spreading.

Palaeontology – the study of past creatures (fossils) which includes actual remains, remains replaced by mineral matter and the trails, tracks, and burrows of past creatures.

Palaeosol – ancient, fossil or relict soil or soil horizon.

Palaeotemperature – past temperature.

Palynologists – scientists who study pollen.

Pangea – large continent (supercontinent) which existed in the northern hemisphere, and which split up during the late Palaeozoic to form Asia, Europe, and North America.

Pathogen – disease-producing organism.

Pedology – the study of soils.

Permafrost – ground which persists below 0°C for two or more years.

Permeability – capacity of a rock or soil to transmit fluid.

Permian – period of geological time from 280–225 Ma.

Petajoule – unit of energy. 1 million tonnes of oil equivalent = 41.87 petajoules.

pH – logarithmic scale which provides a measure of the acidity or alkalinity in a solution. pH7 is neutral, while decreasing values indicate increased acidity, and values greater than 7 represent increasing alkalinity (basicity).

Pheromones – hormones which are produced by organisms to attract the opposite sex.

Phosphates – chemical compounds containing phosphorus, an essential nutrient for life.

Phosphate nodule – concretionary growth, commonly formed early in the burial history of sediments, containing exclusively or mainly phosphate (see phosphates).

Photic zone – surface layer of a lake, sea, or ocean above the maximum depth to which light penetrates.

Photochemical – chemical substance which is sensitive to light, and may involve a change in its chemical composition or energy state when bombarded with solar energy.

Photochemical smog – poor air quality caused by sunlight energy catalysing chemical reactions mainly with nitrogen compounds and hydrocarbons, commonly to produce a reddish-yellow-brown haze. Photochemical smogs commonly form on very warm and sunny days in large urban areas subject to large amounts of tailpipe emissions (exhaust) from motor vehicles.

Photon – a quantum of sunlight energy.

Photosynthesis – biological process in which plants convert CO_2 and H_2O to carbohydrates and release O_2.

Photosynthetic strategy – means by which organisms synthesize chlorophyll using sunlight energy.

Photovoltaic cell – device capable of changing solar energy directly into electricity.

Plankton – drifting or floating organic life, chiefly microscopic, found at various depths in seas, lakes, and rivers.

Plate tectonics – theory which explains the nature of the Earth's surface in terms of continental and oceanic plates. Currently, there are eight major and several minor lithospheric plates which move relative to each other because of convection cells in the Earth's mantle, causing the creation of new oceanic crust at spreading centres, and its destruction at subduction zones.

Pleistocene – interval of geological time which spanned from 1.64 Ma (or 2.4 Ma) to 10,000 years BP and was associated with widespread glaciation.

Plumes (mantle plumes) – see hot spots.

Plutonium – radioactive element used as a fuel in certain nuclear reactors and as an ingredient in nuclear weapons.

Pneumoconiosis – lung disease caused by inhaling fine dust, common in coal miners. Also known as dust pneumonia.

Poliomyelitis – inflammation of the grey matter in the spinal cord which may lead to paralysis.

Pollutant – a substance which causes pollution, typically because of reaching concentrations or levels that pose an environmental problem.

Polychlorinated biphenyls (PCBs) – organic compounds containing chlorine and phenyls, which are used in the manufacture of paints, plastics, adhesives, hydraulic fluids, and electrical components, and which are toxic to humans and other organisms.

Polycyclic aromatic hydrocarbons – organic compounds with carbon ring-chains, and which contribute to atmospheric pollution.

Population density – number of individuals per unit area.

Population explosion – term used to describe an exponential increase in population where conditions favour a very large birth and survival rate.

Porosity – ratio of void volume to bulk volume of rock on soil.

Positive feedback – return of a fraction of an output from a system to the input of the system which strengthens the subsequent output of the system.

Positive point sources – sharp points which have a net positive electrical charge.

Pozzolan material – substance acting like cement, and containing silicates of aluminosilicates which react with lime and water to form stable insoluble compounds; used in disposal of toxic waste.

ppb[v] – parts per billion [by volume]. Unit of measurement typically used to define very low concentrations of chemical elements and compounds.

ppm[v] – parts per million [by volume]. Unit of measurement typically used to define low concentrations of chemical elements and compounds.

Precipitation – all forms of moisture which condense in the atmosphere and are deposited on the Earth's surface.

Precursors – events or phenomena which occur a short time before an earthquake or volcanic eruption.

Primary energy – includes fossil fuels (coal, crude oil, gas), and biomass in a raw state prior to processing into a form suitable for energy consumption.

Proterozoic – period of geological time, 2,500 to 530 Ma.

Punctuated evolution – non-gradual, abrupt development of organisms leading to the generation of new species.

Pyrmnestophyte algae – see coccoliths.

Pyrotoxins – toxic chemicals produced by fires.

Pyroxenes – group of silicate minerals which are rich in iron and magnesium.

Quaternary – geological period from 1.64 Ma to 10,000 yr BP, which was characterized by widespread glaciations.

Radiative forcing – a measure of the ability of greenhouse gases to perturb the heat balance in a simplified model of the Earth-atmosphere system.

Radiation – transmission of electromagnetic energy from a body to its surroundings.

Radioactivity – process of emitting subatomic particles and energy.

Radiocarbon dating – technique used to determine the age of a material in years by measuring the decay of the ^{14}C isotope present within the substance or material.

Radioisotope – isotope of a chemical element that is naturally unstable, and tends to become more stable by the emission of radioactive particles, for example, alpha, beta, and gamma radiation.

Radiometric – prefix to age or the technique used to date a substance in years, by determining the relative proportions of radioactive isotopes and their decay products within that substance.

Radionuclides – atomic nuclei of elements which are capable of breaking down into new isotopes by radioactive decay.

Radium – radioactive chemical element that does not occur naturally on Earth.

Radon – naturally occurring radioactive gas which may build up in houses and other buildings in sufficient concentrations to pose a serious health risk.

Rain gauge – instrument used to measure the quantity of rain falling over a period of time at a particular location.

Raised beach – emerged shoreline represented by stranded beach deposits, marine shells, and wave cut platforms backed by former sea cliffs.

Rape (oilseed rape) – cereal which produces an oilseed which can be used to make rape methyl ester and can be used to power diesel engines.

Rape methyl ester (RME) – ester produced from oilseed rape, which is used for example as motor transport fuel, raw or blended.

Recurrence intervals – time period between successive earthquake events in a particular area.

Redox – reactions involving the loss or gain of electrons.

Retention ponds – artificial ponds which are constructed to collect flood waters to allow the controlled release of water into the main stream.

Retrovirus – virus which has its genetic material in the form of ribonucleic acid.

RNA (ribonucleic acid) – molecule involved in the synthesis of protein.

Roentgen – unit of invisible electromagnetic radiation.

GLOSSARY

Salinization – concentration of salts in the upper layers of a soil due to the drawing of water upwards by the evaporation of near-surface waters.

Sapropel – mud or ooze comprising decomposing organic material, usually present in aqueous environments.

Savanna – grassland region in the tropics or subtropics.

Schistosoma (bilharziasis) – chronic disease found in residents of the tropics, produced by the presence of a flat-worm parasite in the blood and bladder.

Scleractinian corals (Hexacoralla) – group of corals (multicellular organisms which secrete a calcareous skeleton) which first evolved in the middle Triassic and have existed to today; they form an important component of coral reefs.

Seafloor spreading – movement of oceanic crust by plate tectonic processes away from ocean ridge spreading centres.

Sedimentary ironstones – rocks which comprise high concentrations of iron oxides, iron sulphides, and/or iron silicates which were deposited by precipitation and/or deposition of detrital sediments.

Seismic – pertaining to earthquake activity.

Seismic gaps – zones or layers within tectonically active regions which have not experienced large earthquakes during historical time or the recent past.

Seismologists – scientists who study the dynamics of earthquakes, using them to aid the elucidation of the Earth's interior.

Seismometer – instrument used for detecting earthquakes.

Self-purification – natural process in which waste is degraded by microbes in water.

Sensitive clays – fine material (0.0039 mm) which easily deforms and fails under stress.

Sensitivity analyses – analyses of the effects on an appraisal of varying key assumptions and variables.

Seroarchaeology – the study of the history and development of AIDS.

Siderophile elements – (e.g., Ni, Cr, Co, Ag) elements which are soluble in molten iron (Fe).

Silicate minerals – chemical substances which contain silica and oxygen atomic structures: these include the amphiboles, pyroxenes, feldspars, micas, clay minerals, and quartz.

Smart materials – synthetic materials comprising polymers which are able to change their physical properties (e.g., density, shape, tensile strength, colour, etc.) in response to an external stimulus such as the passage of an electric current, temperature, pressure, stress, and strain rate. Smart materials are very much at a research and development stage.

Soft water – water that lacks or has very low concentrations of dissolved salts such as calcium, magnesium, and other ions in solution (commonly as carbonates), which precipitate out to cause the 'furring up' of pipes and appliances.

Solar energy – electromagnetic energy produced by the Sun.

Solar flux – flow of radiation from the Sun to the Earth.

South-East Trade Winds – strong winds which blow from a subtropical high pressure area around 30° south of the equator.

Species – basic unit of classification of living things, which may be defined by morphology, the ability for members of the species to interbreed and by ecological requirements.

Speleothem – deposits, usually of calcite, or less commonly silica, gypsum, or ice, formed by the precipitation from water which has seeped through the ground and which upon contact with the air, generally in a cave, results in the formation of features such as stalactites, stalagmites, and flowstones.

Spheroids – approximately spherical-shaped sand-size substance believed to result from the crystallization at high temperatures of material melted by a meteorite impact and rapidly ejected into the air and water. Spheroids typically may consist of the mineral feldspar.

Spike – term used in Earth Sciences to denote a datum time against which other events may be measured and/or other geological units correlated, for example, 'golden spikes' refer to internationally recognized and correlatable events such as major extinctions and/or radiations of species.

Sporopollenin – highly resistant waxy outer layer of a pollen grain.

Spreading ridges – linear, generally submerged ridges along plate margins which represent zones where the crust is forming and moving away from the ridge.

SSSI – 'Site of Special Scientific Interest', area designated by the UK government's advisers as requiring conservation.

Stadials – cold period during an interglacial stage.

Stishovite – form (polymorph) of quartz (SiO_2) formed under high pressures.

Storm sewers – channels or pipes which drain rainwater into a river, lake, or the sea.

Strain – deformation of a material resulting from an applied force.

Strain gauge – instrument used to measure ground deformation.

Stratigraphy – the study of the order and arrangement of geological strata

Stratosphere – layer of the atmosphere above the troposphere, ranging in altitude from 8–15 km at the lower boundary to about 50 km at the top, accounting for about 10 per cent of the mass of the atmosphere. Between about 15–45 km altitude, the air contains relatively high concentrations of naturally occurring ozone, up to 10 ppm, referred to as the 'ozone layer'. Temperature increases with altitude in the stratosphere, and vertical mixing is relatively slow.

Stratospheric ozone (high-level ozone) – triatomic molecules of oxygen which are present in the atmosphere at heights of between 15 and 45 km.

Stream gauge – instrument used to measure the amount of water passing a particular location.

Stress – force exerted per unit area.

Stromatolites – organic-sedimentary structures produced by sediment trapping, binding, and/or precipitation as a result of the growth and metabolic activity of micro-organisms (principally algae), which live in the seas, marshes or lakes. They still form today, but they reached their acme during the Proterozoic when they formed mounds several hundred metres across and tens of metres high.

Strontium-90 – radioactive isotope of strontium.

Subtropical anticyclones – high pressure atmospheric systems in the subtropics which develop divergent winds.

Sun-spot – area of above-average temperature seen on the Sun's surface, and which develops and decays periodically.

Superfund – popular name for the US Comprehensive Environmental Response, Compensation, and Liability Act (CERCLA) of 1980, which provides a mechanism and funding to help clean up potentially dangerous hazardous waste sites.

Superinsulator – substance which conducts very little heat.

Surficial glacial deposits – veneer of sediments of glacial origin present on the land surface.

Surging glaciers – glaciers which flow at a velocity which is an order of magnitude higher than normal.

Suspended particulate matter – group of air and/or water pollutants comprising solid, or solid and liquid, particles suspended in air or water.

Suspended solids – see suspended particulate matter.

Sustainable development – development of available resources without compromising the ability of future generations to meet their needs.

Symbiotic – mutually beneficial relationship between organisms.

Symbiotic bacteria – bacteria which live in leguminous plants and obtain nutrients from the plants and fix nitrogen to form nitrates which in turn can be used by the plant.

Syngas – mix of H_2 and CO used to synthesize liquid fuels such as pure hydrogen, methanol, and gasoline.

System – sets of interrelated parts, which includes components (elements) and states, and the relationships between the elements and states.

Systematics – methodological study of classification, such as the classification of the diversity of biological systems.

Systematists – scientists who study systematics.

Tectonics – study of the structures of the Earth's lithosphere and the processes involving stress and strain that form them.

Tectonic processes – deformation processes (strain) which is a consequence of continental and oceanic plate movements, and includes volcanicity, earthquakes, mountain building, subsidence, and crustal extension and rotation.

Tektites – spherules of glass generated by a meteorite impact.

Tephra – general term used for all pyroclastic deposits produced by explosive volcanism.

Tertiary – period of geological time which spans from the Cretaceous (65 Ma) to the Quaternary (1.64 Ma).

Terawatt (TW) – unit of energy output or consumption. 1TW = 1 billion kW. 10^{12} watts (joules per second) = 1 TW. In 1990, the total world energy use was 13.16 TW.

Thermocline – zone of water within a lake or the sea which marks a sharp temperature change, typically from an upper warm surface layer to cooler water below.

Thermohaline circulation – movement of ocean water induced by the interaction of waters with different temperature and salinity.

Thermophobic – cold-seeking.

Third World – Nations that are not at an advanced stage of economic development and are not fully industrialized. Third World countries predominate in Africa, Latin America, and in many parts of Asia. Such nations are commonly referred to as less developed countries or LDCs.

Threshold – physical and/or chemical state across which there is a sudden change of conditions brought about by the increase of an input to a critical level.

Till – sediment deposited by glacial ice.

Tornado – violent, rotating storm with winds of up to 100 ms^{-1} circulating round a funnel cloud some 100 m in diameter.

Trace element – chemical elements that occur in relatively low concentrations and/or that are required by organisms in small or trace amounts.

Traction load – material moved by a river along a river bed.

Transpiration – loss of water vapour from the cells of plants.

Treated sludge – sewage sludge which has undergone biological, chemical or heat treatment, long-term storage, or any other appropriate process, so as to significantly reduce its fermentability and the health hazards resulting from its use.

Triassic – period of geological time from 225–190 Ma.

Tributyn (TBT) – type of organotin used as an anti-fouling paint on boats.

Tritium – an isotope of hydrogen (^{3}H), comprising two neutrons and one proton in the nucleus of the atom. Tritium is typically artificially made but is naturally occurring in very small amounts.

Trophic level – feeding stratum in relation to the primary energy source. Organisms, both plant and animal, can be classified into various trophic levels which emphasize their interdependence and hierarchical relationships.

Tropical cyclones – intense cyclonic vortices that form in tropical oceans and have winds which reach 33 ms^{-1}.

Troposphere – lower portion of the atmosphere which extends to heights of 8–15 km above the Earth's surface, accounting for about 90 per cent of the atmospheric mass, and in which most of the weather processes take place causing it to be well mixed. Temperature decreases with altitude.

Tropospheric ozone – triatomic oxygen which forms and is present near ground level to heights of about 12–15 km in the atmosphere.

Tsunami – tidal waves caused by submarine earthquakes or landslides.

Tundra – biologically defined region which is treeless, marshy and usually has permanently frozen subsoil, typical of northernmost Eurasia, the Arctic and parts of the Antarctic.

Typhoid – infectious fever with severe intestinal irritations.

Typhoon – tropical cyclone which originates in the west Pacific Ocean.

Uranium/thorium dating – geochemical technique which compares the ratio of uranium and its daughter element, thorium, to provide an absolute age on a rock or fossil.

Ultraviolet radiation – electromagnetic radiation of short wavelength.

Upwelling – movement of deep cold water towards the sea surface often rich in nutrients and allowing increased productivity in surface waters, for example, of phytoplankton.

Upstream flood – flood whose effects become progressively dissipated down valley.

Urbanization – establishment and/or growth of cities and towns.

Ventilation – term used in oceanography to refer to the circulation of oxygen-rich waters in the world's oceans.

Verification – checking by third parties of the accuracy or truthfulness of experiments or assertions by others, and commonly used in connection with nuclear arms and other international treaties.

Volcano – vent or opening through which magma, ash or volatiles erupt on to the Earth's surface and into the atmosphere.

Volcanic basement – crustal rocks which are volcanic in origin.

Volcanic ash – fragmented rock or crystals which are ejected by volcanic eruptions.

Volcanism – eruption of magma, ash, or volatiles on to or above the Earth's surface.

Waste inventory – document compiled to record the types and amounts of various waste products created by specified industries. Waste inventories may be used to monitor waste production and aid in its control. Such inventories are used both by the producers of waste and [independent] watchdog bodies.

Water cycle – the complete loop system involving water movement from evaporation, precipitation, and run-off through the surface and/or ground.

Water table – surface above an unconfined body of ground water which has totally filled fissures and pores in bedrock or soil.

Watershed – ridge of ground from which surface waters flow in different directions to eventually collect different streams; watersheds delimit the area of catchment for particular rivers.

Watt (W) – unit of power. For electrical energy, units expressed as We; for thermal energy, Wt (1000 watts = 1 kW). 1 unit of electricity contains 1 kWh of energy.

Wavelength – distance between two successive crests or troughs of waveform.

Weathering – decomposition (chemical weathering) and/or disintegration (physical/mechanical weathering) of rock *in situ* by chemical and/or physical processes.

Weichselian – last glacial stage in mainland Europe.

Westerlies – mid-latitude winds which blow from the southwest.

Wet deposition – the incorporation of particles and gases into rain and snow which deposit by gravity – 'acid rain' in the strict scientific sense.

Wetland – ecosystems constantly containing surface water, and which are commonly regularly flooded, for example, marshes and swamps.

White smoke – an aerosol emitted from an engine of totally or partially unburnt fuel, normally emitted only when the engine and ambient air are cold: in worn engines, lubricating oil may provide a source.

Wind farm – array of wind turbines erected to generate electricity.

Wisconsin – last glacial stage in North America.

Xeroscraping – ecosystem containing drought-resistant plants.

Younger Dryas Stadial – global sudden abrupt cold period between about 11,000 and 10,500 years BP during the early part of the present interglacial (Holocene).

Zooplankton – small aquatic animals, most of which are microscopic, comprising mature and larval stages of many animal groups including the Protozoa, Crustacea, and Mollusca.

Index

Note: References to Boxes are in **Bold**; those to plates, tables and figures are in *Italic*. In both cases, there may be other textual references on the same pages.